T0223713

Lecture Notes in Computer Science 1289

Edited by G. Goos, J. Hartmanis and J. van Leeuwen

Advisory Board: W. Brauer   D. Gries   J. Stoer

Springer

Berlin
Heidelberg
New York
Barcelona
Budapest
Hong Kong
London
Milan
Paris
Santa Clara
Singapore
Tokyo

Georg Gottlob   Alexander Leitsch
Daniele Mundici (Eds.)

Computational Logic and Proof Theory

5th Kurt Gödel Colloquim, KGC'97
Vienna, Austria, August 25-29, 1997
Proceedings

 Springer

Series Editors

Gerhard Goos, Karlsruhe University, Germany

Juris Hartmanis, Cornell University, NY, USA

Jan van Leeuwen, Utrecht University, The Netherlands

Volume Editors

Georg Gottlob
Institute for Information Systems, Vienna University of Technology
Paniglgasse 16, A-1040 Vienna, Austria

Alexander Leitsch
Institute for Computer Languages, Vienna University of Technology
Resselgasse 3/1, A-1040 Vienna, Austria

Daniele Mundici
Department of Computer Science, University of Milan
Via Comelico 39-41, I-20135 Milan, Italy

Cataloging-in-Publication data applied for

Die Deutsche Bibliothek - CIP-Einheitsaufnahme

Computational logic and proof theory : proceedings / 5th Kurt
Gödel Colloquium, KGC '97, Vienna, Austria, August 25 - 29, 1997.
Georg Gottlob ... (ed.). - Berlin ; Heidelberg ; New York ; Barcelona
; Budapest ; Hong Kong ; London ; Milan ; Paris ; Santa Clara ;
Singapore ; Tokyo : Springer, 1997
　ISBN 3-540-63385-5

CR Subject Classification (1991): F.4, I.2.3-4, F.2.2

ISSN 0302-9743
ISBN 3-540-63385-5 Springer-Verlag Berlin Heidelberg New York

This work is subject to copyright. All rights are reserved, whether the whole or part of the material is
concerned, specifically the rights of translation, reprinting, re-use of illustrations, recitation, broadcasting,
reproduction on microfilms or in any other way, and storage in data banks. Duplication of this publication
or parts thereof is permitted only under the provisions of the German Copyright Law of September 9, 1965,
in its current version, and permission for use must always be obtained from Springer -Verlag. Violations are
liable for prosecution under the German Copyright Law.

© Springer-Verlag Berlin Heidelberg 1997
Printed in Germany

Typesetting: Camera-ready by author
SPIN 10546422   06/3142 – 5 4 3 2 1 0   Printed on acid-free paper

Preface

KGC'97, the *Fifth Kurt Gödel Colloquium*, held on August 25–29, 1997, at the Vienna University of Technology, Austria, is the fifth in a series of biennial colloquia on logic, theoretical computer science, and philosophy of mathematics, which are organized by the Kurt Gödel Society. The first colloquium took place in Salzburg, Austria (1989), the second one in Kirchberg am Wechsel, Austria (1991), the third in Brno, Czech Republic, (1993) and the fourth in Florence, Italy, (1995). The topic "Computational Logic and Proof Theory" of KGC'97 is the same as that of KGC'93. Again, the aim of this meeting is to bring together researchers working in the fields of computational logic and proof theory. By combining research on provability, analysis of proofs, proof search and complexity, KGC'97 seeks to further tighten the bonds connecting logic and computer science and to bring about a deeper understanding of fundamental concepts.

This volume contains contributions by 27 authors from 11 different countries: 7 invited papers and 20 contributed papers, which were selected from 38 submissions.

Many thanks to the group of referees who, with their competent advice and criticism, were of decisive help in the difficult selection process.

We gratefully acknowledge the financial sponsorship by the Austrian Ministry for Science and Research.

June 1997   G. Gottlob, A. Leitsch, D. Mundici

Program Committee

Matthias Baaz (Vienna University of Technology)
Georg Gottlob (Vienna University of Technology)
Jan Krajíček (Czech Academy of Sciences, Prague)
Alexander Leitsch (chair; Vienna University of Technology)
Daniele Mundici (University of Milan)
David Plaisted (University of North Carolina, Chapel Hill)
Pavel Pudlák (Czech Academy of Sciences, Prague)
Peter Schmitt (University of Karlsruhe)
Andrei Voronkov (University of Uppsala)

Organizing Committee

Mandana Eibegger (Vienna University of Technology)
Christian Fermüller (Vienna University of Technology)
Franziska Gusel (Vienna University of Technology)
Karin Hörwein (chair; Vienna University of Technology)
Georg Moser (Vienna University of Technology)
Katrin Seyr (Vienna University of Technology)
Helmut Veith (co-chair; Vienna University of Technology)
Richard Zach (University of California, Berkeley)

Invited Speakers

Leo Bachmair (Stony Brook, New York)
Wilfried Buchholz (University of Munich)
Samuel R. Buss (Univ. of California, San Diego)
Walter A. Carnielli (University of Campinas)
John A. Robinson (Syracuse University)
Tanel Tammet (University of Göteborg)
Jerzy Tiuryn (University of Warsaw)

Table of Contents

Paramodulation, Superposition, and Simplification[*]

Leo Bachmair

Department of Computer Science
SUNY at Stony Brook
Stony Brook, New York 11794, U.S.A.
leo@cs.sunysb.edu

Abstract

Techniques for equational reasoning are a key component in many automated theorem provers and interactive proof and verification systems. A notable recent success in equational theorem proving has been the solution of an open problem (the "Robbins conjecture") by William McCune with his prover EQP [13].[2]

EQP is one of many equational theorem provers that use *completion* as the main deductive mechanism. Completion derives from the work of Knuth and Bendix [11] and is characterized by the extensive use of rewrite techniques (especially normalization by rewriting) for reasoning about equational theories. More specifically, the Knuth-Bendix procedure attempts to transform a given set of equations into a set of rewrite rules so that any two equivalent terms, and only equivalent terms, have identical normal forms. Not every equational theory can be presented as such a *convergent* rewrite system, but various refinements of the approach have led to the formulation of a refutationally complete method called *ordered completion*; the main contributions can be found in [12, 10, 8, 3, 1, 2].

The deductive inference rule used in completion procedures is *superposition*, which consists of first unifying one side of one equation with a subterm of another, and then applying the two possible equational replacements to this "overlapped" term. In ordered completion the selection of the two terms to be unified is guided by a given term ordering, which imposes certain restrictions on inferences (and thus usually results in a smaller search space, though also potentially longer proofs). The superposition rule is actually a restricted instance of a clausal inference rule, called *paramodulation*, that was proposed by Robinson and Wos [16]. (Informally, certain paramodulation inferences contain a superposition step applied to two equality literals selected from two given clauses.)

Paramodulation is often combined with resolution in clausal theorem provers and provides a refutationally complete inference system for clauses with

---

[*] This research was supported in part by the National Science Foundation under grant CCR-9510072.

[2] The problem consisted of proving that a certain set of equations, originally proposed by Herbert Robbins in the 1930s, forms a basis for the the variety of Boolean algebras; cf. [7].

equality.[3] In its original form, paramodulation was not constrained by any of the restrictions that are considered to be indispensable for the efficiency of completion, but many improvements of paramodulation have been proposed since the inference rule was first introduced. In particular, orderings have been used to control the selection of the literals and subterms in them to be unified; see for instance [15, 9, 20, 19, 17, 4]. The most advanced variant of paramodulation is perhaps *basic paramodulation*, as described in [6, 14], where in addition to ordering restrictions one also prevents selection of subterms that have been obtained solely by instantiation of variables in previous inference steps.[4]

Inference rules naturally form the core of any reasoning system. But the control of the proof search, and hence the theorem proving process, by a judicious use of techniques for simplifying formulas and eliminating (or avoiding) redundant formulas and inferences is often even more important. Typical simplification mechanisms are subsumption (i.e., elimination of subsumed clauses) and normalization by rewriting (of which *demodulation* [18] is essentially a special case). McCune [13], for instance, reports that the (successful) proof search on the Robbins problem required about eight days on a Sparc-5 class UNIX computer: less than 1% of the total search time was spent on deriving equations, while most of the time was spent on simplification. Similar observations pertain not only to equational theorem provers, but to (resolution-based) *saturation* methods in general (in the sense of [5]).

We will discuss (i) the fundamental ideas underlying paramodulation and superposition, (ii) a suitable notion of redundancy, (iii) specific simplification techniques, and (iv) the connection between deduction and simplification.

References

1. L. Bachmair. *Canonical equational proofs*. Birkhäuser, Boston, 1991.
2. L. Bachmair and N. Dershowitz. Equational inference, canonical proofs, and proof orderings. *J. of the Association for Computing Machinery*, 41:236–276, 1994.
3. L. Bachmair, N. Dershowitz, and D. A. Plaisted. Completion without failure. In H. Aït-Kaci and M. Nivat, editors, *Resolution of Equations in Algebraic Structures (Vol. 2: Rewriting Techniques)*, pages 1–30. Boston, Academic Press, 1989.
4. L. Bachmair and H. Ganzinger. On restrictions of ordered paramodulation with simplification. In *Proc. Tenth Int. Conf. on Automated Deduction*, volume 449 of *Lect. Notes in Artificial Intelligence*, pages 427–441. Springer-Verlag, Berlin, 1990.
5. L. Bachmair and H. Ganzinger. Rewrite-based equational theorem proving with selection and simplification. *J. Logic Comput.*, 4:217–247, 1994.
6. L. Bachmair, H. Ganzinger, C. Lynch, and W. Snyder. Basic paramodulation. *Information and Computation*, 121:172–192, 1995.
7. L. Henkin, J.D. Monk, and A. Tarski. *Cylindrical Algebras, Part I*. North-Holland, Amsterdam, 1971.

---

[3] Most automated theorem provers for first-order logic translate formulas into clause format and eliminate quantifiers via Skolemization.

[4] It was the corresponding *basic completion* method that was employed by McCune in the successful proof search on the Robbins problem.

8. J. Hsiang and M. Rusinowitch. On word problems in equational theories. In T. Ottmann, editor, *Proceedings of the Fourteenth EATCS International Conference on Automata, Languages and Programming*, pages 54–71, Karlsruhe, West Germany, July 1987. Vol. 267 of *Lecture Notes in Computer Science*, Springer, Berlin.

9. J. Hsiang and M. Rusinowitch. A new method for establishing refutational completeness in theorem proving. *J. of the Association for Computing Machinery*, 3:133–151, 1991.

10. G. Huet. Confluent reductions: Abstract properties and applications to term rewriting systems. *J. of the Association for Computing Machinery*, 27:797–821, 1980.

11. D. E. Knuth and P. B. Bendix. Simple word problems in universal algebras. In J. Leech, editor, *Computational Problems in Abstract Algebra*, pages 263–297. Pergamon Press, Oxford, U. K., 1970. Reprinted in *Automation of Reasoning 2*, Springer, Berlin, pp. 342–376 (1983).

12. D. Lankford. Canonical inference. Technical Report ATP-32, Dept. of Mathematics and Computer Science, University of Texas, Austin, 1975.

13. W. McCune. Well-behaved search and the Robbins problem. In *Proc. Eighth Int. Conf. on Rewriting Techniques and Applications*, Lect. Notes in Comput. Sci. Springer-Verlag, Berlin, 1997. To appear.

14. R. Nieuwenhuis and A. Rubio. Theorem proving with ordering and equality constrained clauses. *J. Symbolic Computation*, 19:321–351, 1995.

15. G. E. Peterson. A technique for establishing completeness results in theorem proving with equality. *SIAM J. on Computing*, 12:82–100, 1983.

16. G. Robinson and L. Wos. Paramodulation and theorem-proving in first order theories with equality. In B. Meltzer and D. Michie, editors, *Machine Intelligence 4*, pages 135–150. Edinburgh University Press, Edinburgh, Scotland, 1969.

17. M. Rusinowitch. Theorem proving with resolution and superposition: An extension of the Knuth and Bendix procedure as a complete set of inference rules. *J. Symbolic Computation*, 1991. To appear.

18. L. T. Wos, G. A. Robinson, D. F. Carson, and L. Shalla. The concept of demodulation in theorem proving. *J. of the Association for Computing Machinery*, 14:698–709, 1967.

19. H. Zhang. *Reduction, superposition and induction: Automated reasoning in an equational logic*. PhD thesis, Rensselaer Polytechnic Institute, Schenectady, New York, 1988.

20. Hantao Zhang and Deepak Kapur. First-order theorem proving using conditional equations. In E. Lusk and R. Overbeek, editors, *Proceedings of the Ninth International Conference on Automated Deduction*, pages 1–20, Argonne, Illinois, May 1988. Vol. 310 of *Lecture Notes in Computer Science*, Springer, Berlin.

Explaining Gentzen's Consistency Proof within Infinitary Proof Theory

Wilfried Buchholz
Mathematisches Institut
der Ludwig-Maximilians-Universität München
Theresienstr. 39, D-80333 München, Germany
email: buchholz@rz.mathematik.uni-muenchen.de

Introduction

There are two main approaches to ordinal analysis of formal theories: the finitary Gentzen-Takeuti approach on one side, and the use of infinitary derivations initiated by Schütte on the other. Up to now these approaches where thought of as separated and only vaguely related. But in the present paper we will show that actually they are intrisically connected. Using the concept of *notations for infinitary derivations* (introduced in [Bu91]) a precise explanation of Gentzen's reduction steps on derivations in 1st order arithmetic Z (cf. [Ge38]) in terms of (cut-elimination for) infinitary derivations in ω-arithmetic will be given. Even more, Gentzen's reduction steps and ordinal assignment will be *derived* from infinitary proof theory. In a forthcoming paper we will extend the present work to impredicative subsystems of 2nd order arithmetic thereby explaining Takeuti's consistency proof for Π_1^1-CA in terms of the infinitary approach (with $\Omega_{\mu+1}$-rules) from [BS88] (cf. [Bu97]).

Our general idea is that such investigations may perhaps be helpful for the understanding and unification of two of the most advanced achievements in contemporary proof theory, namely the methodically quite different work of T. Arai ([Ar96b], [Ar97a], [Ar97b]) and M. Rathjen ([Ra91], [Ra94], [Ra95]) on the ordinal analysis of very strong subsystems of 2nd order arithmetic and set theory.

Content

In §1 and §2 essential material from [Bu91] is repeated in a somewhat modified form, so that it fits exactly for the present purpose. §1 contains the definition of operators $\mathcal{R}_C$ and $\mathcal{E}$ which make up a cut-elimination procedure for Z^∞ (the infinitary Tait-style sequent calculus for ω-arithmetic) due to Schütte [Sch51], Tait [Ta68] and Mints [Mi75]. In §2 we introduce a finitary Tait-style sequent calculus Z^* for pure number theory Z which differs from the usual version only by a certain additional inference rule (E) $\frac{\Gamma}{\Gamma}$ and the fact that cuts $\frac{\Gamma,C \quad \Gamma,\neg C}{\Gamma}$ are labeled by the symbol R_C (instead of Cut_C). Every Z^*-derivation h with closed endsequent is considered as a notation for a certain Z^∞-derivation h^∞ of the same sequent. In other words, we define a translation $h \mapsto h^\infty$ from Z^* into Z^∞. The definition of h^∞ runs as usual only that cuts and E-inferences are not translated literally but according to the intended meaning of the symbols R_C, E:

$$\left(\frac{\begin{array}{cc} h_0 & h_1 \\ \Gamma, C & \Gamma, \neg C \end{array}}{\Gamma} R_C\right)^{\infty} := \mathcal{R}_C(h_0^{\infty}, h_1^{\infty}) \,, \qquad \left(\frac{\begin{array}{c} h_0 \\ \Gamma \end{array}}{\Gamma} E\right)^{\infty} := \mathcal{E}(h_0^{\infty}).$$

From this interpretation and the properties of $\mathcal{R}_C$ and $\mathcal{E}$ (established in §1) one immediately reads off a definition of ordinals $o(h) < \varepsilon_0$ and $\deg(h) < \omega$ such that $depth(h^{\infty}) \leq o(h)$ and $\sup\{\mathrm{rk}(C)+1 : C$ is cut-formula in $h^{\infty}\} \leq \deg(h)$. Formally the definition of $o(h)$ and $\deg(h)$ proceeds by (primitive) recursion on the build-up of h and does not refer to h^{∞}. Further by looking on the definitions of $\mathcal{R}_C$ and $\mathcal{E}$ (given in §1) we derive (via $h \mapsto h^{\infty}$) a definition which assigns to each $\mathbf{Z}^*$-derivation h a certain inference symbol $\mathrm{tp}(h)$ (corresponding to the last inference of h^{∞}) and, for each $i \in |\mathrm{tp}(h)|$, a new $\mathbf{Z}^*$-derivation $h[i]$ such that $(h[i])^{\infty} = h^{\infty}(i)$, where $\left(h^{\infty}(i)\right)_{i \in |\mathrm{tp}(h)|}$ is the family of immediate subderivations of h^{∞}. The definition of $\mathrm{tp}(h)$ and $h[i]$ also proceeds by recursion on the build-up of h.

In §3 we describe the (Tait-style adaption of) Gentzen's reduction procedure and ordinal assignment (from [Ge38]) in terms of the notions introduced in §2. Let $\mathbf{Z}$ denote the subsystem of $\mathbf{Z}^*$ obtained by omitting the E-rule. So $\mathbf{Z}$ is just ordinary 1st order arithmetic. We consider a (hypothetical) $\mathbf{Z}$-derivation d of the empty sequent. Let d' be the $\mathbf{Z}^*$-derivation which results from d' by filling in E-inferences in such a way that for each node ν of d' (which originates from a node of d) we have $\mathsf{hgt}^*(d',\nu) = \mathrm{H\ddot{o}he}(d',\nu)$, where $\mathsf{hgt}^*(d',\nu)$ is the number of E's below ν, and Höhe is defined as in [Ge38]. Then $o(d')$ is precisely the ordinal $O(d)$ which Gentzen assigns to d, and $d'[0]$ (after deleting all E's) coincides with the result of a Gentzen reduction step applied to d.

Remark. The E-rule is also present in [Ar96a] (under the name "height rule") but there no interpretation of E as cut-elimination operator is given.

§1 Cut-elimination for the infinitary system $\mathbf{Z}^{\infty}$

Preliminaries

We assume a formal language of arithmetic which has predicate symbols for primitive recursive relations, but no function symbols except the constant 0 and the unary function symbol S (successor). *Atomic formulas* are of the form $pt_1...t_n$ where p is an n-ary predicate symbol and $t_1, ..., t_n$ are terms. *Literals* are expressions of the shape A or $\neg A$ where A is an atomic formula. *Formulas* are built up from literals by means of $\wedge, \vee, \forall x, \exists x$. The *negation* $\neg C$ of a formula C is defined via de Morgan's laws. The *rank* $\mathrm{rk}(C)$ of a formula C is defined as usual: $\mathrm{rk}(C) := 0$ if C is a literal, $\mathrm{rk}(A_0 \wedge A_1) := \mathrm{rk}(A_0 \vee A_1) := \max\{\mathrm{rk}(A_0), \mathrm{rk}(A_1)\}+1$, $\mathrm{rk}(\forall x A) := \mathrm{rk}(\exists x A) := \mathrm{rk}(A)+1$. By $\mathrm{FV}(\theta)$ we denote the set of all free variables of the formula or term θ. A formula or term θ is called *closed* iff $\mathrm{FV}(\theta) = \emptyset$. $\theta_x(t)$ (or $\theta(x/t)$) denotes the result of replacing every free occurrence of x in θ by t (renaming bound variables of θ if necessary). The only closed terms are the *numerals* $0, S0, SS0, ...$ We identify numerals and natural numbers. By TRUE_0 we denote the set of all true closed literals. Finite sets of formulas are called *sequents*.

We use the following syntactic variables: s, t for terms, A, B, C, D, F for formulas, Γ, Δ for sequents, α, β, γ for ordinals, i, j, k, l, m, n for natural numbers (and numerals).

As far as sequents are concerned we usually write $A_1, ..., A_n$ for $\{A_1, ..., A_n\}$, and A, Γ, Δ for $\{A\} \cup \Gamma \cup \Delta$, etc.

Proof systems

A *proof system* $\mathfrak{S}$ is given by
- a set of formal expressions called *inference symbols* (syntactic variable $\mathcal{I}$)
- for each inference symbol $\mathcal{I}$ a set $|\mathcal{I}|$, a sequent $\Delta(\mathcal{I})$ and a family of sequents $(\Delta_\iota(\mathcal{I}))_{\iota \in |\mathcal{I}|}$.

NOTATION

By writing $\quad (\mathcal{I}) \quad \dfrac{... \Delta_\iota ... (\iota \in I)}{\Delta}$

we declare $\mathcal{I}$ as an inference symbol with $|\mathcal{I}| = I$, $\Delta(\mathcal{I}) = \Delta$, $\Delta_\iota(\mathcal{I}) = \Delta_\iota$.

If $|\mathcal{I}| = \{0, ..., n-1\}$ we write $\dfrac{\Delta_0 \ \Delta_1 \ ... \ \Delta_{n-1}}{\Delta}$ instead of $\dfrac{... \Delta_\iota ... (\iota \in I)}{\Delta}$.

Up to a few exceptions the sequents $\Delta(\mathcal{I}), \Delta_\iota(\mathcal{I})$ are singletons or empty.

Definition

The figure $\dfrac{... \Gamma_\iota ... (\iota \in I)}{\Gamma} \mathcal{I}$ is called a *(correct) $\mathfrak{S}$-inference* iff

$\mathcal{I} \in \mathfrak{S}$ and $|\mathcal{I}| = I$ and $\Delta(\mathcal{I}) \subseteq \Gamma$ and $\forall \iota \in I (\Gamma_\iota \subseteq \Gamma, \Delta_\iota(\mathcal{I}))$.

The infinitary proof system $\mathbf{Z}^\infty$ (ω-arithmetic)

$(\text{Ax}_A) \quad \dfrac{}{A} \quad$ if $A \in \text{TRUE}_0$.

$(\bigwedge_{A_0 \wedge A_1}) \quad \dfrac{A_0 \qquad A_1}{A_0 \wedge A_1} \qquad\qquad (\bigvee^k_{A_0 \vee A_1}) \quad \dfrac{A_k}{A_0 \vee A_1} \quad (k \in \{0, 1\})$

$(\bigwedge_{\forall x A}) \quad \dfrac{... A_x(i) ... (i \in \mathbb{N})}{\forall x A} \qquad (\bigvee^k_{\exists x A}) \quad \dfrac{A_x(k)}{\exists x A} \quad (k \in \mathbb{N})$

$(\text{Cut}_C) \quad \dfrac{C \qquad \neg C}{\emptyset} \qquad\qquad (\text{Rep}) \quad \dfrac{\emptyset}{\emptyset}$

Note:
To avoid a possible misunderstanding we stress that $|\text{Rep}| = \{0\}$ while $|\text{Ax}_A| = \emptyset$.

Inductive definition of $\mathbf{Z}^\infty$-derivations

If Γ is a sequent, α an ordinal, $\mathcal{I} \in \mathbf{Z}^\infty$, and $(d_i)_{i \in I}$ a family of $\mathbf{Z}^\infty$-*derivations* such that $\dfrac{... \Gamma(d_i) ... (i \in I)}{\Gamma} \mathcal{I}$ is a correct $\mathbf{Z}^\infty$-inference and $\forall i \in I (o(d_i) < \alpha)$

then the tree $d := \left\{ \dfrac{\dots d_i \dots (i \in I)}{I : \Gamma : \alpha} \right.$ is a $\mathbf{Z}^\infty$-*derivation* with

$\Gamma(d) := \Gamma$, $\mathrm{last}(d) := I$, $o(d) := \alpha$, $d(i) := d_i$

and $\deg(d) := \begin{cases} \max\{\mathrm{rk}(C)+1, \deg(d_0), \deg(d_1)\} & \text{if } I = \mathrm{Cut}_C \\ \sup\{\deg(d_i) : i \in I\} & \text{otherwise} \end{cases}$

$\Gamma(d)$ is called the *endsequent* of d, $o(d)$ the *ordinal* of d, $\mathrm{last}(d)$ the *last inference (symbol)* of d, and $d(i)$ the *i-th immediate subderivation* of d.

We use $d, d_0, \dots$ as syntactic variables for $\mathbf{Z}^\infty$-derivations.

Abbreviation $d \vdash^\alpha_m \Gamma \; :\Longleftrightarrow \; \Gamma(d) \subseteq \Gamma \; \& \; \deg(d) \le m \; \& \; o(d) = \alpha$.

Cut-elimination for $\mathbf{Z}^\infty$

Theorem 1 and Definition
Let C be given. We define an operator $\mathcal{R}_C$ such that:

$d_0 \vdash^\alpha_m \Gamma, C \; \& \; d_1 \vdash^\beta_m \Gamma, \neg C \; \& \; \mathrm{rk}(C) \le m \implies \mathcal{R}_C(d_0, d_1) \vdash^{\alpha \# \beta}_m \Gamma$.

Proof by induction on $\alpha \# \beta$:
W.l.o.g. we may assume that $\Gamma = (\Gamma(d_0) \setminus \{C\}) \cup (\Gamma(d_1) \setminus \{\neg C\})$.

Case 1. $C \notin \Delta(I)$ where $I := \mathrm{last}(d_0)$:
Then $\Delta(I) \subseteq \Gamma$, and $d_0(i) \vdash^{\alpha_i}_m \Gamma, C, \Delta_i(I)$ with $\alpha_i < \alpha$, for all $i \in |I|$.
By IH we get $\mathcal{R}_C(d_0(i), d_1) \vdash^{\alpha_i \# \beta}_m \Gamma, \Delta_i(I)$ for all $i \in |I|$.

Hence $\mathcal{R}_C(d_0, d_1) := \left\{ \dfrac{\dots \mathcal{R}_C(d_0(i), d_1) \dots (i \in |I|)}{I : \Gamma : \alpha \# \beta} \right.$ is a derivation as required.

Case 1'. $\neg C \notin \Delta(\mathrm{last}(d_1))$: symmetric to Case 1.
Case 2. $C \in \Delta(\mathrm{last}(d_0))$ and $\neg C \in \Delta(\mathrm{last}(d_1))$:
Then $\mathrm{rk}(C) \ne 0$, since C and $\neg C$ cannot both be true literals.
Case 2.1. $C = \forall x A(x)$: Then $\neg C = \exists x \neg A(x)$, $\mathrm{last}(d_1) = \bigvee^k_{\neg C}$, and
$d_0(i) \vdash^{\alpha_i}_m \Gamma, C, A(i)$ with $\alpha_i < \alpha$, for all $i \in \mathbb{N}$,
$d_1(0) \vdash^{\beta_0}_m \Gamma, C, \neg A(k)$ with $\beta_0 < \beta$.
By IH we get $\mathcal{R}_C(d_0(k), d_1) \vdash^{\alpha_k \# \beta}_m \Gamma, A(k)$ and $\mathcal{R}_C(d_0, d_1(0)) \vdash^{\alpha \# \beta_0}_m \Gamma, \neg A(k)$.
Further $\mathrm{rk}(A(k)) < \mathrm{rk}(C) \le m$.

Hence $\mathcal{R}_C(d_0, d_1) := \left\{ \dfrac{\mathcal{R}_C(d_0(k), d_1) \qquad \mathcal{R}_C(d_0, d_1(0))}{\mathrm{Cut}_{A(k)} : \Gamma : \alpha \# \beta} \right.$.

Case 2.2.–2.4. $C = \exists x A$ or $A_0 \wedge A_1$ or $A_0 \vee A_1$: analogous to Case 2.1.

Theorem 2 and Definition
We define an operator $\mathcal{E}$ such that: $d \vdash^\alpha_{m+1} \Gamma \implies \mathcal{E}(d) \vdash^{\omega^\alpha}_m \Gamma$.

Proof by induction on α:
W.l.o.g. we may assume that $\Gamma = \Gamma(d)$.

Case 1. $\mathrm{last}(d) = \mathrm{Cut}_C$:
Then $\mathrm{rk}(C) \le m$ and $d(0) \vdash^{\alpha_0}_{m+1} \Gamma, C$, $d(1) \vdash^{\alpha_1}_{m+1} \Gamma, \neg C$ with $\alpha_0, \alpha_1 < \alpha$.
By IH we get $\mathcal{E}(d(0)) \vdash^{\omega^{\alpha_0}}_m \Gamma, C$ and $\mathcal{E}(d(1)) \vdash^{\omega^{\alpha_1}}_m \Gamma, \neg C$.

Hence by Theorem 1 $\mathcal{R}_C(\mathcal{E}(\mathbf{d}(0)), \mathcal{E}(\mathbf{d}(1))) \vdash_m^{\omega^{\alpha_0} \# \omega^{\alpha_1}} \Gamma$, and

$$\mathcal{E}(\mathbf{d}) := \left\{ \frac{\mathcal{R}_C(\mathcal{E}(\mathbf{d}(0)), \mathcal{E}(\mathbf{d}(1)))}{\text{Rep} : \Gamma : \omega^\alpha} \right. \quad \text{is a derivation as required.}$$

Case 2. otherwise: $\mathcal{E}(\mathbf{d}) := \left\{ \dfrac{\ldots \mathcal{E}(\mathbf{d}(i)) \ldots (i \in |\mathcal{I}|)}{\mathcal{I} : \Gamma : \omega^\alpha} \right. \quad$ where $\mathcal{I} := \text{last}(\mathbf{d})$.

Remark In the whole paper $\lambda\xi.\omega^\xi$ could be replaced by any ordinal function f such that $\forall \alpha_0, \alpha_1, \alpha(\alpha_0, \alpha_1 < \alpha \Rightarrow f(\alpha_0) \# f(\alpha_1) < f(\alpha))$.

§2 The finitary system Z*

Let Z be the formal system of pure number theory (Peano arithmetic). The mathematical axioms of Z are the scheme of complete induction and finitely many axioms of the shape $\forall \vec{x}(A_0 \vee \ldots \vee A_m)$ where $A_0, \ldots, A_m$ are literals. In our sequent calculus the latter axioms are represented by a (prim. rec.) set $\text{Ax}(Z)$ of sequents such that

(i) $\Delta \in \text{Ax}(Z)$ & $A \in \Delta \Rightarrow A$ is a literal,

(ii) $\Delta \in \text{Ax}(Z) \Rightarrow \Delta_{\vec{x}}(\vec{t}) \in \text{Ax}(Z)$,

(iii) $\Delta \in \text{Ax}(Z)$ & $\text{FV}(\Delta) = \emptyset \Rightarrow \Delta \cap \text{TRUE}_0 \neq \emptyset$.

Definition of the finitary proof system Z*

The inference symbols of **Z*** are

$$(\text{Ax}_\Delta) \; \frac{}{\Delta} \; \text{if } \Delta \in \text{Ax}(Z) \;, \qquad (\bigwedge_{\forall x A}^y) \; \frac{A_x(y)}{\forall x A} \;, \qquad (\bigvee_{\exists x A}^t) \; \frac{A_x(t)}{\exists x A} \;,$$

$$(\text{Ind}_F^{y,t}) \; \frac{\neg F, F_y(Sy)}{\neg F_y(0), F_y(t)} \;, \qquad (\text{R}_C) \; \frac{C \quad \neg C}{\emptyset} \;, \qquad (\text{E}) \; \frac{\emptyset}{\emptyset} \;,$$

and $\bigwedge_{A_0 \wedge A_1}$, $\bigvee_{A_0 \vee A_1}^k$ as in **Z∞**.

Z*-derivations

Z*-derivations are defined in a somewhat different style than **Z∞**-derivations. The difference is that the nodes of a **Z***-derivation h are labeled with inference symbols only, while the endsequent $\Gamma(h)$ and the ordinal $o(h)$ of h will be computed from h by structural recursion. Actually **Z***-derivations will be introduced as *terms* (in prefix notation) built up from inference symbols $\mathcal{I}$ which we consider as n-ary function symbols, where $|\mathcal{I}| = \{0, \ldots, n-1\}$.

Inductive Definition of Z*-quasi-derivations

If $\mathcal{I}$ is an n-ary **Z***-inference symbol and $h_0, \ldots, h_{n-1}$ are **Z***-quasi-derivations then $h := \mathcal{I} h_0 \ldots h_{n-1}$ is a **Z***-quasi-derivation and

$$\Gamma(h) := \Delta(\mathcal{I}) \cup \bigcup_{i<n}(\Gamma(h_i) \setminus \Delta_i(\mathcal{I})) \;,$$

$$o(h) := \begin{cases} o(h_0) \# o(h_1) & \text{if } \mathcal{I} = \text{R}_C \\ o(h_0) \cdot \omega & \text{if } \mathcal{I} = \text{Ind}_F^{y,t} \\ \omega^{o(h_0)} & \text{if } \mathcal{I} = \text{E} \\ (\sup_{i<n} o(h_i)) + 1 & \text{otherwise} \end{cases}$$

$$\deg(h) = \begin{cases} \max\{\mathrm{rk}(C), \deg(h_0), \deg(h_1)\} & \text{if } \mathcal{I} = \mathsf{R}_C \\ \max\{\mathrm{rk}(F), \deg(h_0)\} & \text{if } \mathcal{I} = \mathsf{Ind}_F^{y,t} \\ \deg(h_0) \dot{-} 1 & \text{if } \mathcal{I} = \mathsf{E} \\ \sup_{i<n} \deg(h_i) & \text{otherwise} \end{cases}$$

Remark: The definitions of $o(h)$ and $\deg(h)$ are motivated by the interpretation $h \mapsto h^\infty$ (introduced below) and Theorems 1,2.

Inductive Definition of $\mathbf{Z}^*$-derivations

If $\mathcal{I}$ is an n-ary $\mathbf{Z}^*$-inference symbol and $h_0, ..., h_{n-1}$ are $\mathbf{Z}^*$-derivations then $h := \mathcal{I}h_0...h_{n-1}$ is a $\mathbf{Z}^*$-derivation if the following conditions are satisfied

- $\mathcal{I} = \bigwedge_{\forall x A}^y \Rightarrow y \notin \mathrm{FV}(\Gamma(h))$,
- $\mathcal{I} = \mathsf{Ind}_F^{y,t} \Rightarrow y \notin \mathrm{FV}(\Gamma(h))$,
- $\mathcal{I} = \bigvee_{\exists x A}^t \Rightarrow \mathrm{FV}(t) \subseteq \mathrm{FV}(\Gamma(h))$,
- $\mathcal{I} = \mathsf{R}_C \Rightarrow \mathrm{FV}(C) \subseteq \mathrm{FV}(\Gamma(h))$.

A $\mathbf{Z}^*$-derivation h is called *closed* iff $\mathrm{FV}(\Gamma(h)) = \emptyset$.

Remark: As one easily verifies the last two conditions in the above definition do not restrict the set of provable sequents. They imply the following proposition: If $h = \mathcal{I}h_0...h_{n-1}$ is a closed $\mathbf{Z}^*$-derivation with $\mathcal{I} \neq \bigwedge_{\forall x A}^y, \mathsf{Ind}_F^{y,t}$ then $h_0,..., h_{n-1}$ are closed too. If $h = \bigwedge_{\forall x A}^y h_0$ or $h = \mathsf{Ind}_F^{y,t} h_0$ is closed then $\mathrm{FV}(\Gamma(h_0)) \subseteq \{y\}$.

Definition

Let $\mathbf{Z}$ denote the subsystem of $\mathbf{Z}^*$ which arises by omitting the symbol E. Obviously $\mathbf{Z}$ is nothing else than the Tait-style version of pure number theory Z.

We use d, d_i (h, h_i) as syntactic variables for $\mathbf{Z}(\mathbf{Z}^*)$-derivations.

Definition

In the usual way we define $h(z/i)$, i.e. the result of substituting i for z in h:

$\mathsf{Ax}_\Delta(z/i) := \mathsf{Ax}_{\Delta_z(i)}$,

$(\bigvee_C^t h_0)(z/i) := \bigvee_{C_z(i)}^{t_z(i)} h_0(z/i)$, $(\bigwedge_C h_0 h_1)(z/i) := \bigwedge_{C_z(i)} h_0(z/i) h_1(z/i)$,

$(\bigwedge_C^z h_0)(z/i) := \bigwedge_C^z h_0$, $(\bigwedge_C^y h_0)(z/i) := \bigwedge_{C_z(i)}^y h_0(z/i)$ if $y \neq z$,

$(\mathsf{Ind}_F^{z,t} h_0)(z/i) := \mathsf{Ind}_F^{z,t} h_0$, $(\mathsf{Ind}_F^{y,t} h_0)(z/i) := \mathsf{Ind}_{F_z(i)}^{y,t_z(i)} h_0(z/i)$ if $y \neq z$,

$(\mathsf{R}_C h_0 h_1)(z/i) := \mathsf{R}_{C_z(i)} h_0(z/i) h_1(z/i)$, $(\mathsf{E}h_0)(z/i) := \mathsf{E}h_0(z/i)$.

Proposition If h is a $\mathbf{Z}^*$-derivation then also $h(z/i)$ is a $\mathbf{Z}^*$-derivation and $\Gamma(h(z/i)) \subseteq \Gamma(h)_z(i)$, $\deg(h(z/i)) = \deg(h)$, $o(h(z/i)) = o(h)$.

Interpretation of $\mathbf{Z}^*$ in $\mathbf{Z}^\infty$

For each closed $\mathbf{Z}^*$-derivation h we define its interpretation $h^\infty \in \mathbf{Z}^\infty$ as follows: Let $h = \mathcal{I}h_0...h_{n-1}$, $\Gamma = \Gamma(h)$, $\alpha = o(h)$:

0. $(\mathsf{Ax}_\Gamma)^\infty := \left\{ \dfrac{}{\mathsf{Ax}_A : \Gamma : \alpha} \right.$, where A is the "least" element of $\Gamma \cap \mathrm{TRUE}_0$,

1. $(\bigwedge_{\forall x A}^y h_0)^\infty := \left\{ \dfrac{... h_0(y/i)^\infty ... (i \in \mathbb{N})}{\bigwedge_{\forall x A} : \Gamma : \alpha} \right.$,

2. $(\mathcal{R}_C h_0 h_1)^\infty := \mathcal{R}_C(h_0^\infty, h_1^\infty)$,

3. $(\mathcal{E} h_0)^\infty := \mathcal{E}(h_0^\infty)$,

4. $(\mathsf{Ind}_F^{y,n} h_0)^\infty := \left\{ \dfrac{\mathbf{e}_n}{\mathsf{Rep}:\Gamma:\alpha} \right.$ with

   $\mathbf{e}_1 := h_0(y/0)^\infty$, $\mathbf{e}_{i+1} := \mathcal{R}_{F(i)}(\mathbf{e}_i, h_0(y/i)^\infty)$ for $i > 0$, and $\mathbf{e}_0$ is the canonical $\mathbf{Z}^\infty$-derivation with $\Gamma(\mathbf{e}_0) = \{\neg F(0), F(0)\}$, $\deg(\mathbf{e}_0) = 0$, $o(\mathbf{e}_0) = 2\mathrm{rk}(F)$.

5. Otherwise: $(\mathcal{I} h_0...h_{n-1})^\infty := \left\{ \dfrac{h_0^\infty \ldots h_{n-1}^\infty}{\mathcal{I}:\Gamma:\alpha} \right.$

Remark With the help of Theorems 1,2 one easily verifies that h^∞ is a $\mathbf{Z}^\infty$-derivation with $h^\infty \vdash_{\deg(h)}^{o(h)} \Gamma(h)$.

Definition of $\mathrm{tp}(h)$ and $h[i]$ for closed $\mathbf{Z}^*$-derivations h and $i \in |\mathrm{tp}(h)|$

By (prim.) recursion on the build-up of h we define an inference symbol $\mathrm{tp}(h) \in \mathbf{Z}^\infty$ and closed $\mathbf{Z}^*$-derivation(s) $h[i]$ in such a way that $\mathrm{tp}(h) = \mathrm{last}(h^\infty)$ and $(h[i])^\infty = h^\infty(i)$. The definition clauses for $h = \mathcal{R}_C h_0 h_1$ and $h = \mathcal{E} h_0$ can be read off from the corresponding clauses in the definitions of $\mathcal{R}_C$ and $\mathcal{E}$.

1.1. $h = \mathsf{Ax}_\Delta$: $\mathrm{tp}(\mathsf{Ax}_\Delta) := \mathsf{Ax}_A$ where A is the "least" element of $\Delta \cap \mathrm{TRUE}_0$.

1.2. $h = \bigwedge_C h_0 h_1$: $\mathrm{tp}(h) := \bigwedge_C$, $h[i] := h_i$.

1.3. $h = \bigwedge_C^y h_0$: $\mathrm{tp}(h) := \bigwedge_C$, $h[i] := h_0(y/i)$.

1.4. $h = \bigvee_C^k h_0$: $\mathrm{tp}(h) := \bigvee_C^k$, $h[0] := h_0$.

2. $h = \mathsf{Ind}_F^{y,n} h_0$: $\mathrm{tp}(h) := \mathsf{Rep}$, $h[0] := \mathbf{e}_n$ with

   $\mathbf{e}_1 := h_0(y/0)$, $\mathbf{e}_{i+1} := \mathcal{R}_{F(i)} \mathbf{e}_i h_0(y/i)$ for $i > 0$, and $\mathbf{e}_0$ is the canonical $\mathbf{Z}$-derivation with $\Gamma(\mathbf{e}_0) = \{\neg F(0), F(0)\}$, $\deg(\mathbf{e}_0) = 0$, $o(\mathbf{e}_0) = 1+2\mathrm{rk}(F)$.

3. $h = \mathcal{E} h_0$:

3.1. $\mathrm{tp}(h_0) = \mathsf{Cut}_C$: $\mathrm{tp}(h) := \mathsf{Rep}$, $h[0] := \mathcal{R}_C \mathcal{E} h_0[0] \mathcal{E} h_0[1]$.

3.2. otherwise: $\mathrm{tp}(h) := \mathrm{tp}(h_0)$, $h[i] := \mathcal{E} h_0[i]$.

4. $h = \mathcal{R}_C h_0 h_1$:

4.1. $C \notin \Delta(\mathrm{tp}(h_0))$: $\mathrm{tp}(h) := \mathrm{tp}(h_0)$, $h[i] := \mathcal{R}_C h_0[i] h_1$.

4.2. $\neg C \notin \Delta(\mathrm{tp}(h_1))$: $\mathrm{tp}(h) := \mathrm{tp}(h_1)$, $h[i] := \mathcal{R}_C h_0 h_1[i]$.

4.3. $C \in \Delta(\mathrm{tp}(h_0))$ and $\neg C \in \Delta(\mathrm{tp}(h_1))$:

   Then $\mathrm{rk}(C) \neq 0$, since C and $\neg C$ cannot both be true literals.

4.3.1. $C = \forall x A$: Then $\mathrm{tp}(h_1) = \bigvee_{\neg C}^k$ for some $k \in \mathbb{N}$.

   $\mathrm{tp}(h) := \mathsf{Cut}_{A_x(k)}$, $h[0] := \mathcal{R}_C h_0[k] h_1$, $h[1] := \mathcal{R}_C h_0 h_1[0]$.

4.3.2. $C = \exists x A$ or $A_0 \wedge A_1$ or $A_0 \vee A_1$: analogous to 4.3.1.

Theorem 3

For each closed $\mathbf{Z}^*$-derivation h the following holds:

a) $\dfrac{\ldots \Gamma(h[i]) \ldots (i \in |\mathrm{tp}(h)|)}{\Gamma(h)} \mathrm{tp}(h)$ is a correct $\mathbf{Z}^\infty$-inference,

b) $\mathrm{tp}(h) = \mathsf{Cut}_C \Rightarrow \mathrm{rk}(C) < \deg(h)$,

c) $\deg(h[i]) \leq \deg(h)$ for all $i \in |\mathrm{tp}(h)|$,

d) $o(h[i]) < o(h)$ for all $i \in |\mathrm{tp}(h)|$.

Proof by straightforward induction on the build-up of h:

We only consider two cases.

Abbreviation: $h \vdash_m^\alpha \Gamma :\Leftrightarrow \Gamma(h) \subseteq \Gamma \;\&\; \deg(h) \leq m \;\&\; o(h) = \alpha$.

1. $h = R_C h_0 h_1$ with $C = \forall x A$, $\mathrm{tp}(h_0) = \bigwedge_C$, $\mathrm{tp}(h_1) = \bigvee_{\neg C}^k$, $\mathrm{tp}(h) = \mathsf{Cut}_{A(k)}$:

Let $\Gamma := \Gamma(h)$, $\alpha := o(h_0)$, $\beta := o(h_1)$, and $m := \deg(h)$.

Then $h_0 \vdash_m^\alpha \Gamma, C$ and $h_1 \vdash_m^\beta \Gamma, \neg C$ and $\mathrm{rk}(A(k)) < \mathrm{rk}(C) \leq \deg(h)$.

By IH we obtain $h_0[k] \vdash_m^{\alpha_k} \Gamma, C, A(k)$ with $\alpha_k < \alpha$,

and $h_1[0] \vdash_m^{\beta_0} \Gamma, \neg C, \neg A(k)$ with $\beta_0 < \beta$.

Hence $h[0] = R_C h_0[k] h_1 \vdash_m^{\alpha_k \# \beta} \Gamma, A(k)$ and $h[1] = R_C h_0 h_1[0] \vdash_m^{\alpha \# \beta_0} \Gamma, \neg A(k)$

with $\alpha_k \# \beta$, $\alpha \# \beta_0 < \alpha \# \beta = o(h)$.

2. $h = E h_0$ with $\mathrm{tp}(h_0) = \mathsf{Cut}_C$: Then $\mathrm{tp}(h) = \mathsf{Rep}$ and $h[0] = R_C E h_0[0] E h_0[1]$.

Let $\Gamma := \Gamma(h_0) = \Gamma(h)$, $\alpha := o(h_0)$ and $m := \deg(h_0) \dot{-} 1 = \deg(h)$.

By IH we have $\mathrm{rk}(C) < \deg(h_0) \leq m+1$ and $h_0[0] \vdash_{m+1}^{\alpha_0} \Gamma, C$, $h_0[1] \vdash_{m+1}^{\alpha_1} \Gamma, \neg C$

with $\alpha_0, \alpha_1 < \alpha$. Hence $E h_0[0] \vdash_m^{\omega^{\alpha_0}} \Gamma, C$ and $E h_0[1] \vdash_m^{\omega^{\alpha_1}} \Gamma, \neg C$.

From this together with $\mathrm{rk}(C) \leq m$ we get $h[0] = R_C E h_0[0] E h_0[1] \vdash_m^{\omega^{\alpha_1} \# \omega^{\alpha_1}} \Gamma$

and $\omega^{\alpha_0} \# \omega^{\alpha_1} < \omega^\alpha = o(h)$.

Corollary

Let $\mathbf{Z}_\perp^*$ be the set of all $\mathbf{Z}^*$-derivations h with $\Gamma(h) = \emptyset \;\&\; \deg(h) = 0$.

a) $h \in \mathbf{Z}_\perp^* \;\Rightarrow\; h[0] \in \mathbf{Z}_\perp^* \;\&\; o(h[0]) < o(h)$,

b) There is no $\mathbf{Z}$-derivation d with $\Gamma(d) = \emptyset$.

Proof:

a) $h \in \mathbf{Z}_\perp^* \overset{\text{Th.3}}{\Rightarrow} h \in \mathbf{Z}_\perp^* \;\&\; \mathrm{tp}(d) = \mathsf{Rep} \overset{\text{Th.3}}{\Rightarrow} h[0] \in \mathbf{Z}_\perp^* \;\&\; o(h[0]) < o(h)$.

b) By transfinite induction up to ε_0 from a) we get $\mathbf{Z}_\perp^* = \emptyset$. Now assume that d is a $\mathbf{Z}$-derivation with $\Gamma(d) = \emptyset$. Let $m := \deg(d)$. Then $E^m d = E...Ed \in \mathbf{Z}_\perp^*$. *Contradiction.*

Conclusion

In this section we have proved the consistency of $\mathbf{Z}$ in a Gentzen style manner (i.e., by defining reduction steps on finite derivations in such a way that the assigned ordinals decrease), but we have not yet achieved a literal reconstruction of Gentzen's original consistency proof in [Ge38]. This is contained in §3.

§3 Connection to Gentzen's consistency proof

Notation:

If d is a $\mathbf{Z}$-derivation and ν a node (position) in d then :

(i) $d|_\nu$ denotes the subderivation of d determined by ν. (Especially $d|_{()} = d$.)

(ii) $hgt(d, \nu)$ is Gentzen's height (Höhe) of ν in d.

(iii) $O(d, \nu)$ is the ordinal which Gentzen assigns to ν in d.

(The definition of $hgt(d, \nu)$ and $O(d, \nu)$ can be found in the proof of Lemma 1.)

Definition
For each $\mathbf{Z}^*$-derivation h let $\phi(h)$ denote the $\mathbf{Z}$-derivation which results from h by deleting all E's.

Definition of a $\mathbf{Z}^*$-derivation $\psi_n(d)$ for each $\mathbf{Z}$-derivation d
1. $\psi_n(\mathsf{R}_C d_0 d_1) := \mathsf{E}^{l-n}\mathsf{R}_C\psi_l(d_0)\psi_l(d_1)$, where $l := \max\{n, \mathrm{rk}(C)\}$,
2. $\psi_n(\mathsf{Ind}_F^{y,t} d_0) := \mathsf{E}^{l-n}\mathsf{Ind}_F^{y,t}\psi_l(d_0)$, where $l := \max\{n, \mathrm{rk}(F)\}$,
3. Otherwise: $\psi_n(\mathcal{I}d_0 \ldots d_{m-1}) := \mathcal{I}\psi_n(d_0)\ldots\psi_n(d_{m-1})$.

Proposition
$\Gamma(\psi_n(d)) = \Gamma(d)$, $\deg(\psi_n(d)) \le n$, $\phi(\psi_n(d)) = d$.

Remark
As we will see below (cf. Lemma 5) $\psi_n(d)$ has the following minimality property:
$\forall h(\deg(h) \le n \ \& \ \phi(h) = d \Rightarrow \mathrm{o}(\psi_n(d)) \le \mathrm{o}(h))$.

The rest of this section is occupied with the proof of the following Theorem.

Theorem 4
For each $\mathbf{Z}$-derivation d we have
a) $\mathrm{o}(\psi_0(d)) = O(d, \langle\rangle)$.
b) If $\Gamma(d) = \emptyset$ then $\mathrm{red}(d) := \phi(\psi_0(d)[0])$ results from d by a Gentzen reduction step, and $O(\mathrm{red}(d), \langle\rangle) < O(d, \langle\rangle)$.

Lemma 1
If $n = hgt(d, \nu)$ then $\mathrm{o}(\psi_n(d|_\nu)) = O(d, \nu)$.

Proof by induction on $d|_\nu$:
1. $d|_\nu = \mathsf{R}_C d_0 d_1$: Then $d_i = d|_{\nu*\langle i\rangle}$ and $hgt(d, \nu*\langle i\rangle) = l := \max\{n, \mathrm{rk}(C)\}$. Hence by IH $\mathrm{o}(\psi_l(d_i)) = O(d, \nu*\langle i\rangle)$, and thus $\mathrm{o}(\psi_n(d|_\nu)) = \omega_{l-n}(\mathrm{o}(\psi_l(d_0))\#\mathrm{o}(\psi_l(d_1))) = \omega_{l-n}(O(d, \nu*\langle 0\rangle)\#O(d, \nu*\langle 1\rangle)) = O(d, \nu)$.
2. $d|_\nu = \mathsf{Ind}_F^{y,t}d_0$: Then $d_0 = d|_{\nu*\langle 0\rangle}$ and $hgt(d, \nu*\langle 0\rangle) = l := \max\{n, \mathrm{rk}(F)\}$. Hence $\mathrm{o}(\psi_n(d|_\nu)) = \omega_{l-n}(\mathrm{o}(\psi_l(d_0)) \cdot \omega) \overset{\mathrm{IH}}{=} \omega_{l-n}(O(d, \nu*\langle 0\rangle) \cdot \omega) = O(d, \nu)$.
3. $d|_\nu = \mathcal{I}d_0 \ldots d_{k-1}$ otherwise: Then $d_i = d|_{\nu*\langle i\rangle}$ and $hgt(d, \nu*\langle i\rangle) = n$. Hence by IH $\mathrm{o}(\psi_n(d_i)) = O(d, \nu*\langle i\rangle)$ and thus $\mathrm{o}(\psi_n(d|_\nu)) = (\sup_{i<k} \mathrm{o}(\psi_n(d_i))) + 1 = (\sup_{i<k} O(d, \nu*\langle i\rangle)) + 1 = O(d, \nu)$.

From Lemma 1 we get $\mathrm{o}(\psi_0(d|_{\langle\rangle})) = O(d, \langle\rangle)$, and thus Theorem 4a is proved.

Abbreviation: $\mathsf{E}^m h := \underbrace{\mathsf{E}\ldots\mathsf{E}}_{m} h$.

Definition (Nominal forms for derivations)
1. $*$ is a nominal form. $\mathrm{Cut}(*) := \emptyset$, $\mathrm{hgt}^*(*) := 0$.
2. If a is a nominal form, $m \in \mathbb{N}$, and h a $\mathbf{Z}^*$-derivation then $\mathsf{E}^m\mathsf{R}_C ah$ and $\mathsf{E}^m\mathsf{R}_C ha$ are nominal forms.
$\mathrm{Cut}(\mathsf{E}^m\mathsf{R}_C ah) := \mathrm{Cut}(a) \cup \{C\}$, $\mathrm{Cut}(\mathsf{E}^m\mathsf{R}_C ha) := \mathrm{Cut}(a) \cup \{\neg C\}$,
$\mathrm{hgt}^*(\mathsf{E}^m\mathsf{R}_C ah) := \mathrm{hgt}^*(\mathsf{E}^m\mathsf{R}_C ha) := m + \mathrm{hgt}^*(a)$.

We use a, b, c as syntactic variables for nominal forms.

Definition
$\text{hgt}(a) := \sup\{\text{rk}(C) : C \in \text{Cut}(a)\}$,
$a\{q\} :=$ the result of substituting q for $*$ in a (q a nominal form or $\mathbf{Z}^*$-derivation).

Lemma 2
$\psi_n(d) = a\{h'\} \Rightarrow n + \text{hgt}^*(a) = \max\{n, \text{hgt}(a)\}$.

Proof by induction on a:
1. $a = *$: $n + \text{hgt}^*(a) = n = \max\{n, \text{hgt}(a)\}$.
2. $a = E^m R_C a_0 h_1$:
Then $d = R_C d_0 d_1$ and $\psi_n(d) = E^{l-n} R_C \psi_l(d_0)\psi_l(d_1)$ with $l := \max\{\text{rk}(C), n\}$.
This yields $E^m R_C a_0\{h'\}h_1 = a\{h'\} = E^{l-n} R_C \psi_l(d_0)\psi_l(d_1)$ and then $m = l - n$
and $a_0\{h'\} = \psi_l(d_0)$. Hence $n + \text{hgt}^*(a) = l + \text{hgt}^*(a_0) \overset{\text{IH}}{=} \max\{l, \text{hgt}(a_0)\} = \max\{n, \text{rk}(C), \text{hgt}(a_0)\} = \max\{n, \text{hgt}(a)\}$.

Corollary
$\psi_0(d) = a\{b\{h'\}\} \Rightarrow \text{hgt}(a\{b\}) = \text{hgt}(a) + \text{hgt}^*(b)$.

Proof: $\text{hgt}(a\{b\}) \overset{\text{L.2}}{=} \text{hgt}^*(a\{b\}) = \text{hgt}^*(a) + \text{hgt}^*(b) \overset{\text{L.2}}{=} \text{hgt}(a) + \text{hgt}^*(b)$

Definition
A $\mathbf{Z}^*$-derivation h is called *regular* iff for every subterm Eh_0 of h we have $\text{last}(h_0) \in \{E, R_C, \text{Ind}_F^{y,t}\}$. – Obviously each $\psi_n(d)$ is regular.
$$C[k] := \begin{cases} A_x(k) & \text{if } C = QxA \text{ with } Q \in \{\forall, \exists\} \\ A_k & \text{if } C = A_0 \circ A_1 \text{ with } \circ \in \{\wedge, \vee\} \text{ and } k \in \{0, 1\} \end{cases}$$

Lemma 3
Let h be a closed $\mathbf{Z}^*$-derivation.
a) If $\text{tp}(h) = \text{Rep}$ then there are a, h' such that $h = a\{h'\}$, $h[0] = a\{h'[0]\}$ and
   either $h' = E^m \text{Ind}_F^{y,t} h''$ or $h' = E^{m+1} h''$ & $\text{tp}(h'') = \text{Cut}_B$.
b) If $\text{tp}(h) = \text{Cut}_B$ then there are b, C, h_0, h_1 such that $\text{hgt}^*(b) = 0$,
   $h = b\{R_C h_0 h_1\}$, $h[i] = b\{(R_C h_0 h_1)[i]\}$, and either
   (1) $\text{tp}(h_0) = \bigwedge_C$ & $\text{tp}(h_1) = \bigvee_{\neg C}^k$ & $B = C[k]$ or
   (2) $\text{tp}(h_0) = \bigvee_C^k$ & $\text{tp}(h_1) = \bigwedge_{\neg C}$ & $B = C[k]$.
c) If h is regular and $\text{tp}(h) = \bigwedge_C$ or $\bigvee_C^k$ then there are $c, h_0[, h_1]$ such that
   $C \notin \text{Cut}(c)$ and
   $[h = c\{\bigwedge_C^y h_0\}$ & $h[i] = c\{h_0(y/i)\}]$ or $[h = c\{\bigwedge_C h_0 h_1\}$ & $h[i] = c\{h_i\}]$ or
   $[h = c\{\bigvee_C^k h_0\}$ & $h[0] = c\{h_0\}]$.

Proof:
a) By definition of $\text{tp}(h)$ one of the following cases holds:
1. $h = E^m \text{Ind}_F^{y,t} \tilde{h}$: Then $a := *$, $h' := h$.
2. $h = E^n \tilde{h}$ with $\text{last}(\tilde{h}) \neq E, \text{Ind}$:

2.1. $\text{tp}(\tilde{h}) = \text{Cut}_B$ & $n > 0$: Then $a := *$, $h' := h$.

2.2. $\text{tp}(\tilde{h}) = \text{Rep}$: Then $\tilde{h} = R_C h_0 h_1$ and (w.l.o.g) $\text{tp}(h_0) = \text{Rep}$. By IH $h_0 = a_0\{h'\}$ with $h_0[0] = a_0\{h'[0]\}$ and $h' = E^m \text{Ind}_F^{y,t} h''$ or $h' = E^{m+1} h''$ & $\text{tp}(h'') = \text{Cut}_B$. Now for $a := E^n R_C a_0 h_1$ we have $h = a\{h'\}$ and $h[0] = E^n R_C h_0[0] h_1 = E^n R_C a_0\{h'[0]\} h_1 = a\{h'[0]\}$.

b) Assume that $\text{tp}(h) = \text{Cut}_B$. Then one of the following cases holds:

1. $h = R_C h_0 h_1$ and $[(\text{tp}(h_0) = \bigwedge_C$ & $\text{tp}(h_1) = \bigvee_{\neg C}^k$ & $B = C[k])$ or $(\text{tp}(h_0) = \bigvee_C^k$ & $\text{tp}(h_1) = \bigwedge_{\neg C}$ & $B = C[k])]$: The claim holds for $b := *$.

2. $h = R_D h_0' h_1'$ and (w.l.o.g.) $\text{tp}(h_0') = \text{tp}(h)$ & $h[i] = R_D h_0'[i] h_1'$:
By IH there are b_0, C, h_0, h_1 such that $\text{hgt}^*(b_0) = 0$, $h_0' = b_0\{R_C h_0 h_1\}$, $h_0'[i] = b_0\{(R_C h_0 h_1)[i]\}$ and one of the subcases (1),(2) holds. Let $b := R_D b_0 h_1'$. Then $h = b\{R_C h_0 h_1\}$, $h[i] = R_D h_0'[i] h_1' = R_D b_0\{(R_C h_0 h_1)[i]\} h_1' = b\{(R_C h_0 h_1)[i]\}$, and $\text{hgt}^*(b) = \text{hgt}^*(b_0) = 0$.

c) Assume that h is regular, and $\text{tp}(h) = \bigwedge_C$ with $C = \forall x A$.
Then one of the following cases holds:

1. $h = \bigwedge_{\forall x A}^y h_0$: Then the claim holds for $c := *$.

2. $h = E^m R_D h_0' h_1'$ with (w.l.o.g.) $\text{tp}(h) = \text{tp}(h_0')$ and $D \neq C$: By IH $h_0' = c_0\{\bigwedge_C^y h_0\}$ and $h_0'[i] = c_0\{h_0(y/i)\}$ with $C \notin \text{Cut}(c_0)$. Let $c := E^m R_D c_0 h_1'$. Then $h = c\{\bigwedge_{\forall x A}^y h_0\}$, $h[i] = E^m R_D h_0'[i] h_1' = E^m R_D c_0\{h_0(y/i)\} h_1' = c\{h_0(y/i)\}$ and $C \notin \{D\} \cup \text{Cut}(c_0) = \text{Cut}(c)$.

Theorem 5

Assume that $\Gamma(d) = \emptyset$ and let $h := \psi_0(d)$.
Then $\text{tp}(h) = \text{Rep}$ and one of the following two cases holds:

(I) $h = a\{E^m \text{Ind}_F^{y,t} h_0\}$, $h[0] = a\{E^m (\text{Ind}_F^{y,t} h_0)[0]\}$,

(II) $h = a\{E^{m+1} b\{R_C h_0 h_1\}\}$, $h[0] = a\{E^m R_{C[k]} E b\{R_C h_0^- h_1\} E b\{R_C h_0 h_1^-\}\}$
and either

    (1) $\text{tp}(h_0) = \bigwedge_C$ & $\text{tp}(h_1) = \bigvee_{\neg C}^k$ & $h_0^- = h_0[k]$ & $h_1^- = h_1[0]$ or

    (2) $\text{tp}(h_0) = \bigvee_C^k$ & $\text{tp}(h_1) = \bigwedge_{\neg C}$ & $h_0^- = h_0[0]$ & $h_1^- = h_1[k]$.

    Moreover $\text{hgt}^*(b) = 0$ and $\text{hgt}(a) + m + 1 = \text{rk}(\tilde{C}) = \max\{\text{hgt}(b), \text{rk}(C)\}$

    with $\tilde{C} := \begin{cases} C & \text{if } b = * \\ D & \text{if } \text{last}(b) = R_D \end{cases}$.

Proof:
We have $\Gamma(h) = \emptyset$ & $\deg(h) = 0$ and therefore (by Theorem 3) $\text{tp}(h) = \text{Rep}$.
Further h is regular. Now let us assume that (I) does not hold.
Then according to L.3a) $h = a\{E^{m+1} h''\}$ with $\text{tp}(h'') = \text{Cut}_B$ and
$h[0] = a\{(E^{m+1} h'')[0]\} = a\{E^m (E h'')[0]\} = a\{E^m R_B E h''[0] E h''[1]\}$.
By L.3b) we get $h'' = b\{R_C h_0 h_1\}$, $h''[i] = b\{(R_C h_0 h_1)[i]\}$ with $\text{hgt}^*(b) = 0$, and
— in subcase (1) — $\text{tp}(h_0) = \bigwedge_C$ & $\text{tp}(h_1) = \bigvee_{\neg C}^k$ & $B = C[k]$.
Putting things together yields

$h[0] = \mathfrak{a}\{E^m R_B E h''[0] E h''[1]\} = \mathfrak{a}\{E^m R_{C[k]} E \mathfrak{b}\{R_C h_0[k] h_1\} E \mathfrak{b}\{R_C h_0 h_1[0]\}\}$.

It remains to prove that $\mathrm{hgt}(\mathfrak{a}) + m + 1 = \mathrm{rk}(\tilde{C}) = \max\{\mathrm{hgt}(\mathfrak{b}), \mathrm{rk}(C)\}$.

Let $\mathfrak{b}' := \mathfrak{b}\{R_C * h_1\}$. Then $\mathrm{last}(\mathfrak{b}') = R_{\tilde{C}}$ and $\max\{\mathrm{hgt}(\mathfrak{a}), \mathrm{hgt}(\mathfrak{b}')\} =$

$\mathrm{hgt}(\mathfrak{a}\{E^{m+1}\mathfrak{b}'\}) \overset{\mathrm{Cor. L. 2}}{=} \mathrm{hgt}(\mathfrak{a}) + \mathrm{hgt}^*(E^{m+1}\mathfrak{b}') = \mathrm{hgt}(\mathfrak{a}) + m + 1$.

Hence $\mathrm{hgt}(\mathfrak{a}) + m + 1 = \mathrm{hgt}(\mathfrak{b}') = \max\{\mathrm{hgt}(\mathfrak{b}), \mathrm{rk}(C)\}$.

Similarly we obtain $\mathrm{rk}(\tilde{C}) = \mathrm{hgt}(\mathfrak{a}) + m + 1$.

Remark

With the above Theorem at hand the reader may now go through the relevant parts of [Ge38] and convince him/herself that indeed $\mathrm{red}(d) := \phi(\psi_0(d)[0])$ results from d by a reduction step in the sense of [Ge38]. To facilitate this task let us take a closer look at case (II)(1) with $C = \forall x A$. In doing so we use the following abbreviation: $d \approx h :\Leftrightarrow d = \phi(h)$. Then by combining Lemma 3c with Theorem 5 and writing derivations as trees we obtain the following presentation of d and $\mathrm{red}(d)$ which (apart from weakenings, contractions and permutations) is exactly as in [Ge38] (pp. 34,35):

In traditional notation with sequents displayed this is:

The relation $\mathrm{hgt}(\mathfrak{a}) < \mathrm{rk}(\tilde{C}) = \max\{\mathrm{hgt}(\mathfrak{b}), \mathrm{rk}(C)\}$ (proved above) implies that our "Höhenlinie" coincides with Gentzen's.

Now the last part of Theorem 4, i.e. the relation $O(\mathrm{red}(d), \langle\rangle) < O(d, \langle\rangle)$, immediately follows from [Ge38]. But we think it may be useful to include an independent proof here.

Lemma 4 $n \leq k \Rightarrow o(\psi_n(d)) \leq \omega_{k-n} o(\psi_k(d))$.

Proof:
Abbreviation: $o_n(d) := o(\psi_n(d))$.
1. $d = R_C d_0 d_1$ and $l := \max\{n, \mathrm{rk}(C)\}$:

1.1. $l \leq k$: $o_n(d) = \omega_{l-n}(o_l(d_0) \# o_l(d_1)) \overset{\mathrm{IH}}{\leq} \omega_{l-n}(\omega_{k-l} o_k(d_0) \# \omega_{k-l} o_k(d_1)) \leq$
$\omega_{l-n}(\omega_{k-l}(o_k(d_0) \# o_k(d_1))) = \omega_{k-n} o_k(d)$.
1.2. $n \leq k < l$:
$o_n(d) = \omega_{l-n}(o_l(d_0) \# o_l(d_1)) = \omega_{k-n} \omega_{l-k}(o_l(d_0) \# o_l(d_1)) = \omega_{k-n} o_k(d)$.
2. $d = \mathrm{Ind}_F^{y,t}$: analogous to 1.

3. $d = \mathcal{I} d_0 ... d_{m-1}$ otherwise: $o_n(d) = (\sup_{i<m} o_n(d_i)) + 1 \overset{\mathrm{IH}}{\leq}$
$(\sup_{i<m} \omega_{k-n} o_k(d_i)) + 1 \leq \omega_{k-n}((\sup_{i<m} o_k(d_i)) + 1) = \omega_{k-n}(o_k(d))$.

Lemma 5
$\deg(h) \leq n \Rightarrow o(\psi_n \phi(h)) \leq o(h)$.

Proof:

1. $h = Eh_0$ with $\deg(h_0) \leq n+1$: $o(\psi_n \phi(h)) = o(\psi_n \phi(h_0)) \overset{\mathrm{L.4}}{\leq}$
$\omega_1 o(\psi_{n+1} \phi(h_0)) \overset{\mathrm{IH}}{\leq} \omega_1 o(h_0) = o(h)$.
2. $h = R_C h_0 h_1$ with $\max\{\mathrm{rk}(C), \deg(h_0), \deg(h_1)\} \leq n$:
$o(\psi_n \phi(h)) = o(\psi_n R_C \phi(h_0) \phi(h_1)) \overset{\mathrm{rk}(\underline{C}) \leq n}{=} o(R_C \psi_n \phi(h_0) \psi_n \phi(h_1)) =$
$o(\psi_n \phi(h_0)) \# o(\psi_n \phi(h_1)) \overset{\mathrm{IH}}{\leq} o(h_0) \# o(h_1) = o(h)$.
3. $h = \mathrm{Ind}_F^{y,t} h_0$: analogous to 2.
4. $h = \mathcal{I} h_0 ... h_{m-1}$ otherwise: immediately by IH.

Proof of $O(\mathrm{red}(d), \langle\rangle) < O(d, \langle\rangle)$: Let $h := \psi_0(d)[0]$.
From $\deg(\psi_0(d)) = 0$ it follows by Theorem 3 that $\deg(h) = 0$. Hence

$O(\mathrm{red}(d), \langle\rangle) \overset{\mathrm{L.1}}{=} o(\psi_0 \mathrm{red}(d)) = o(\psi_0 \phi(h)) \overset{\mathrm{L.5}}{\leq} o(h) \overset{\mathrm{Th.3}}{<} o(\psi_0(d)) \overset{\mathrm{L.1}}{=} O(d, \langle\rangle)$.

References

[Ar96a] Arai, T.: Consistency Proof via Pointwise Induction. Preprint (1996)

[Ar96b] Arai, T.: Proof Theory for Theories of Ordinals I: Reflecting Ordinals. Preprint (1996)

[Ar97a] Arai, T.: Proof Theory for Theories of Ordinals II: Σ_1-stability. Preprint (1997)

[Ar97b] Arai, T.: Proof Theory for Theories of Ordinals III: Π_1-collection. Preprint (1997)

[BS88] Buchholz, W. and Schütte, K.: Proof Theory of Impredicative Subsystems of Analysis. Studies in Proof Theory, Monographs 2. Napoli: Bibliopolis 1988

[Bu91] Buchholz, W.: Notation systems for infinitary derivations. Arch. Math. Logic 30, pp. 277-296 (1991)

[Bu97] Buchholz, W.: Explaining the Gentzen-Takeuti reduction steps. Preprint (1997)

[Ge38] Gentzen, G.: Neue Fassung des Widerspruchsfreiheitsbeweises für die reine Zahlentheorie. Forschungen zur Logik und zur Grundlegung der exakten Wissenschaften. Neue Folge 4, pp. 19-44 (1938)

[Mi75] Mints, G.: Finite Investigations of Transfinite Derivations. In: Mints,G., Selected Papers in Proof Theory. Studies in Proof Theory, Monographs 3. Napoli: Bibliopolis 1992. Russian original: Zapiski nauchnykh seminarov, LOMI 49 (1975) pp. 67-122

[Ra91] Rathjen, M.: Proof Theoretic Analysis of KPM. Arch. Math. Logic 30 (1991) pp. 377-403

[Ra94] Rathjen, M.: Proof Theory of Reflection. APAL 68 (1994) pp. 181-224

[Ra95] Rathjen, M.: Recent Advances in Ordinal Analysis: Π_2^1-CA and related systems. The Bulletin of Symbolic Logic 1/4 (1995) pp. 468-485

[Sch51] Schütte, K.: Beweistheoretische Erfassung der unendlichen Induktion in der Zahlentheorie. Math. Ann. 122, pp. 369-389 (1951)

[Ta68] Tait, W.W.: Normal Derivability in Classical Logic. In: Barwise, J. (ed.) The syntax and semantics of infinitary languages. (Lect. Notes Math., vol.72, pp. 204-236) Berlin Heidelberg New York: Springer 1968

Alogtime Algorithms for Tree Isomorphism, Comparison, and Canonization

Samuel R. Buss*

Departments of Mathematics & Computer Science,
Univ. of California, San Diego,
La Jolla, CA 92093-0112

Abstract. The tree isomorphism problem is the problem of determining whether two trees are isomorphic. The tree canonization problem is the problem of producing a canonical tree isomorphic to a given tree. The tree comparison problem is the problem of determining whether one tree is less than a second tree in a natural ordering on trees. We present alternating logarithmic time algorithms for the tree isomorphism problem, the tree canonization problem and the tree comparison problem. As a consequence, there is a recursive enumeration of the alternating log time tree problems.

1 Introduction

A tree is a finite, connected, acyclic graph with a distinguished root node. An *isomorphism* between two trees T_1 and T_2 is a bijection between the nodes of T_1 and the nodes of T_2 which preserves the edges and which maps the root of T_1 to the root of T_2. Two trees are *isomorphic* if there is an isomorphism between them.

An implicit component of our definition of "isomorphism" is that the children of a node in a tree are *unordered*. Of course, whenever a tree is drawn, or especially, is represented in on a computer, there is an ordering specified by the representation of the tree. For instance, we define a string representation of trees below, and any string representation of a tree orders the subtrees and nodes of T. This means of course that a given tree may have many different representations. The question of whether two different string representations are representations of the same tree is called the *tree isomorphism problem*. One way to solve the isomorphism is define canonical representations for trees in such way that every (unordered) tree has exactly one canonical representations. The problem of converting an arbitrary string representation of a tree into a canonical representation is called the *tree canonization* problem.

In addition to the tree isomorphism and tree canonization problems, we will consider the tree comparison problem. Below we define a linear ordering, $\prec$, on trees: the *tree comparison* problem is the problem of, given string representations of two trees, to determine whether the first tree is less than ($\prec$) the second.

* Supported in part by NSF grants DMS-9503247 and DMS-9205181.

The main results of this paper give alternating logtime (Alogtime) algorithms for the tree isomorphism, tree comparison and tree canonization problems. There have been several prior related results. First, Aho-Hopcroft-Ullman[1] gave a linear time algorithm for tree isomorphism, based on comparing two trees in a bottom-up fashion. Obviously, linear time is the best possible sequential run time for tree isomorphism, but it is possible to consider refined algorithms in smaller complexity classes, where one restricts the circuit depth, the parallel time, or the Turing machine space usage. Recall that the complexity class NC is the class of predicates solvable by polynomial size, polylog depth circuits; as is well known, uniform NC algorithms are exactly the algorithms that can be solved on PRAM's in polylogarithmic time with polynomially many processors. Ruzzo [13] mentions an NC-algorithm for solving the tree isomorphism problem for trees of logarithmic degree. Miller and Reif [12] later gave an NC-algorithm for solving the tree isomorphism and tree canonization problems for trees of arbitrary degree and depth, based on tree-contraction methods. Finally, in the best prior result, Lindell [8] gave deterministic logarithmic space algorithms for the tree isomorphism, tree comparison and tree canonization problems. It was the logarithmic space algorithms that motivated the work of this paper to give alternating logarithmic time algorithms for these algorithms.

One application of the Alogtime algorithms presented in this paper is to a question that arises from finite model theory. One would like to consider algorithms that compute *intrinsic* properties of trees, which do not depend on the particular representation of the trees. Let C be a natural complexity class containing Alogtime (e.g., C may be logspace, nondeterministic logspace, NC^k or Alogtime itself). With our alternating logtime algorithm for tree canonization, one can immediately give a recursive enumeration of all intrinsic tree predicates which are computable in C. Namely, one enumerates the C-algorithms which first create a canonical representation of their input tree and then operate only on the canonical representation. It is clear that every algorithm of this type computes a property of trees which is independent of the representation of the tree, and conversely, every C-algorithm which computes a property of trees independent of their representations is equivalent to an algorithm in this enumeration.

In the next section we give technical definitions regarding trees, including the definition of the linear ordering of trees and the string representation of trees. Section 3 is devoted to the alternating logarithmic time algorithm for tree isomorphism. Then, in sections 4 and 5, we present the algorithms for tree comparison and tree canonization.

It is easy to check that our results apply also to labeled trees, with only relatively minor changes to the definitions and algorithms.

2 Technical Definitions for Trees

The *parent* of a node x in a tree T is the unique node adjacent to x which is closer to the root. The *children* of a node x are the nodes of which x is the parent. A node without any children is a *leaf*. A node x is an *ancestor* of a node y if the shortest path from y to the root contains x; in this case we also say y is a *descendent* of x. If x is a node of T, there is a *subtree* of T rooted at x, namely the node x plus all of its descendents. (Some authors use the terminology "maximal subtree", but for this paper, a subtree is always maximal.) An *immediate subtree* of T is a subtree whose root is a child of T's root node. If S is a proper subtree of T, then the *parent tree* of S is the subtree of T of which S is an immediate subtree.

A second, more formal, inductive definition of tree isomorphism or tree equality, $\equiv$, is given next. The *size* of T, denoted $|T|$, equals the number of leaf nodes in T.

Definition *(Tree equality). Let S and T be trees. We define $S \equiv T$ by induction on the number of nodes in S and T by defining that $S \equiv T$ holds if and only if*

(a) $|S| = |T| = 1$ *or*
(b) *S and T both have the same number, m, of immediate subtrees, and there is some ordering $S_1, \ldots, S_m$ of the immediate subtrees of S and some ordering $T_1, \ldots, T_m$ of the immediate subtrees of T such that $S_i \equiv T_i$ for all $1 \le i \le m$.*

It is easy to check that $S \equiv T$ if and only if there is an isomorphism of S and T. There are many possible ways to define a linear ordering, $\prec$, on trees. The one we give here seems to us to be the most elegant, however, the results of this paper would hold under many other choices for the linear ordering, for instance the linear ordering of [8].

Definition *(Linear ordering of trees). Let S and T be trees. We define $S \preccurlyeq T$ and $S \prec T$ simultaneously by induction on the size of S and T. $S \preccurlyeq T$ holds if and only either $S \prec T$ or $S \equiv T$. $S \prec T$ holds if and only if either $|S| < |T|$ holds or the following conditions hold:*

(a) $|S| = |T|$, *and*
(b) *Let $S_1, \ldots, S_m$ be the immediate subtrees of S ordered so that $S_1 \succcurlyeq S_2 \succcurlyeq S_3 \succcurlyeq \cdots \succcurlyeq S_m$, and let $T_1, \ldots, T_n$ be the immediate subtrees of T, similarly ordered with $T_i \succcurlyeq T_{i+1}$ for all i. Then*
(b.i) *For some $i \le \min\{m, n\}$, $S_i \prec T_i$ and for all $1 \le j < i$, $S_j \equiv T_j$, or*
(b.ii) *$m < n$ and $T_i \equiv S_i$ for all $1 \le i \le m$.*

Following [12], we shall represent trees by strings over the two symbol alphabet containing open and close parentheses. The tree with a single node is denoted by the string "()". If T is a tree with more than one node, if $S_1, \ldots, S_m$ are the immediate subtrees of T (in any order), and if $\alpha_1, \ldots, \alpha_m$ are strings which represent the immediate subtrees, then

$$\text{``}(\alpha_1, \ldots, \alpha_m)\text{''}$$

is a string representing T. Note that different orderings of the subtrees in T can give rise to different string representations for T. Thus trees have more than one string representation, except for highly symmetric trees in which each node has all its children isomorphic.

Definition *The* Tree Isomorphism Problem *is the problem of determining whether two input string representations represent isomorphic trees. The* Tree Comparison Problem *is the problem of, given string representations of two trees* T_1 *and* T_2, *determining whether* $T_1 \prec T_2$. *The* Tree Canonization Problem *is the problem of, given an input string representation of a tree* T, *finding a (usually different) string representation,* $canon_T$, *so that the output* $canon_T$ *is the same for all input string representations of* T.

The definition of the tree canonization problem allows considerable flexibility in how $canon_T$ is defined. However, for this paper, we use the following, natural, definition for $canon_T$.

Definition *If* $|T| = 1$, *then* $canon_T$ *is "()", the only string representation of* T. *If* $|T| > 1$, *then let* $S_1, \ldots, S_m$ *be the immediate subtrees of* T *ordered so that* $S_1 \succcurlyeq S_2 \succcurlyeq \cdots \succcurlyeq S_m$. *Then* $canon_T$ *is equal to*

$$\text{``}(canon_{S_1}\, canon_{S_2} \cdots canon_{S_m})\text{''}.$$

A canonical *string representation is a string which is of the form* $canon_T$ *for some tree* T.

It is worth discussing the importance of using string representations of trees. An alternative to string representations would have been a pointer representation, where the nodes of T are assigned integer values and where each node has a pointer to its parent node and a linked list of pointers to its children. The problem with the pointer representation is that we are working with the very weak complexity class of Alogtime, and Alogtime algorithms are incapable of parsing trees specified by pointers (unless Alogtime is equal to logspace). For instance, evaluating the predicate "node x is a descendent of node y" from the pointer representation of a tree is believed to be outside the computational power of Alogtime; in fact, the problem DTC, deterministic transitive closure, can be reduced to the descendent predicate and DTC is known to be complete for logspace [7].

Therefore, we use the more explicit string representation for trees. It is known that Alogtime algorithms are capable of parsing parenthesis languages. In particular, using counting of parentheses, an Alogtime algorithm can determine the depth of a node in a tree, can determine the i-th child of any node in a tree, can compute the ancestor/descendent predicates, etc. In addition, Alogtime algorithms are capable of converting between prefix and infix notations. We will assume that the reader is familiar with these capabilities of Alogtime, and also is familiar with both the circuit characterization and the game characterization of Alogtime. For more information on these aspects of alternating logtime, the reader can consult the introductory portions of [3, 5, 4].

There are logspace algorithms for converting pointer representations of trees into string representations of tree, and vice-versa, so for logspace and for more powerful complexity classes, the use of string representations is equivalent to the use of pointer representations.

Definition *When T is a tree represented by a string, the leaves are ordered by their occurrence in the string; the leaves are numbered from left to right with consecutive integers in the range 0 to $|T| - 1$.*

Let S be a subtree of T. The domain, $dom(S)$, is the interval $[i, j]$ where S contains the leaves numbered i through j. We let $mindom(S) = i$, the minimum leaf number in S.

There are Alogtime algorithms which, given the value of $|S|$ and the value $mindom(S)$, can find the depth of the root of S, the last leaf node in S, the parent tree of S, the i-th subtree of S, etc.

For the next section, we will need an Alogtime algorithm which, given a value ℓ and given a leaf x in a tree representation, finds the unique maximum-size subtree S such that x is in the domain of S and $logsize(S) = \ell$ (if any such subtree exists). This can be readily done in Alogtime by guessing the first and last leaves of S, and then checking that S is a maximal size subtree with logsize ℓ by universally branching over all other possible subtrees containing leaf x.

3 The Alogtime Algorithm for Tree Isomorphism

We let $\log n$ denote the logarithm base 2 of n rounded down to the nearest integer. The logsize, $logsize(T)$, of a tree T is defined to equal $\log |T|$.

Definition *Let S be a subtree of a tree T. Let $T = T_0, T_1, \ldots, T_k = S$ be the (unique) sequence of subtrees of T such that each T_{i+1} is an immediate subtree of T_i. The size-signature of S in T is defined to equal the sequence $\langle |T_0|, |T_1|, \ldots, |T_k| \rangle$.*

If S' is a subtree of a (possibly different) tree T', then S and S' are similar provided that both:

(1) *They have the same size-signature (as subtrees of their respective trees), and*
(2) *They are isomorphic, i.e., $S \equiv S'$.*

It is obvious that the notion of size-signature is invariant under isomorphism. That is to say, if trees T_1 and T_2 are isomorphic under the mapping f, and if S_1 is a subtree of T_1 and $S_2 = f(S_1)$ is its isomorphic copy in T_2, then S_1 and S_2 have the same size-signatures in T_1 and T_2, respectively.

By the usual parsing and counting techniques, there is an Alogtime procedure which, given a string representation of a tree T and given a subtree S of T, produces the size signature of S in T.

Definition *Let $T_1 \not\equiv T_2$ be trees. Let S be a subtree of T_1. We say that S distinguishes T_1 from T_2 provided that S is a proper subtree of T and:*

(1) *The logsize of S is strictly less than the logsize of the parent tree of S, and*

(2) *The number of subtrees of T_1 which are similar to S is not equal to the number of subtrees of T_2 which are similar to S.*

Lemma 1. *Suppose $|T_1| = |T_2|$ and $T_1 \not\equiv T_2$. Then there is a subtree S of T_1 which distinguishes T_1 from T_2.*

Proof. Since the trees are not isomorphic, there must be some immediate subtree S_1 of T_1 such that the number of immediate subtrees of T_1 which are isomorphic to S_1 is not equal to the number of immediate subtrees of T_2 which are isomorphic to S_1.

Now if the logsize of S_1 is less than the logsize of T_1, then S_1 is a subtree distinguishing T_1 from T_2. In particular, suppose that there is another immediate subtree S_1' of T_1 such that $|S_1'| \geq |S_1|$: then $|T_1| \geq 2 \cdot |S_1|$ which implies that the logsize of S_1 is strictly less than the logsize of T_1. Similarly, if there is more than one immediate subtree of T_2 with size greater than or equal to the size of S_1, then $|T_1| = |T_2| \geq 2 \cdot |S_1|$ and again the logsize of S_1 is less than the logsize of T_1.

Therefore, either S_1 distinguishes T_1 from T_2 or the following conditions hold: (1) S_1 is the unique maximum-sized proper subtree of T_1 and all other immediate subtrees of T_1 have size less than the size of S_1, and (2) T_2 has at most one subtree with size equal to the size of S_1. If T_2 has no subtree with size equal to the size of S_1, then any leaf of S_1 distinguishes T_1 from T_2, since no leaf of T_2 can have the same size signature. Finally, suppose T_2 has a subtree S_2 which has size equal to the size of S_1. By the choice of S_1, $S_1 \not\equiv S_2$. Thus, by the induction hypothesis, there is a subtree R of S_1 which distinguishes S_1 from S_2. Since S_1 and S_2 are the unique immediate subtrees of T_1 and T_2 of size $|S_1|$, the subtree R also distinguishes T_1 from T_2. □

We shall shortly present a precise definition of uniform logarithmic depth circuits for solving the tree isomorphism problem, which will suffice to show that the tree isomorphism problem is in Alogtime. However, as an intuitive aid, we first sketch the general idea of the alternating logarithmic time algorithm for tree isomorphism. As mentioned above, one can view an Alogtime algorithm as a game between players: the first player is asserting that the two trees are non-isomorphic, while the second player is asserting that the two trees *are* isomorphic. The input to the game consists of two string representations of two trees T_1 and T_2, and we denote this instance of the game $G[T_1, T_2]$. The game begins with the first player identifying a subtree S_1 that distinguishes T_1 from T_2. Then the two players play a log time game to count the number of subtrees of T_1 and of T_2 which are similar to S_1. If these numbers are equal, the second player wins and otherwise the first player wins. The inputs to the counting game consists of assertions of the form "S is similar to S_1" where S ranges over all subtrees of T_1 and T_2 (or at least over all subtrees which are the same size of S_1). Determining the truth of these assertions involves (a) comparing the size-signatures of S and S_1 and (b) checking whether $S \equiv S_1$. Part (a) is easily done in Alogtime, and part (b) involves a recursive call to the tree-isomorphism algorithm, $G[S, S_1]$.

Of course, care must be taken to show that the entire game $G[T_1, T_2]$, including the recursive calls to the game, uses only $O(\log n)$ rounds, where n is the maximum size of T_1 and T_2. For this, it is important to remember that the number of rounds in the play of game is defined to equal the number of bits of information exchanged by the two players (i.e., if one of the player sends ℓ bits, that counts as ℓ rounds). The fact that only $O(\log n)$ rounds are needed will depend crucially on two facts: firstly, that the logsize of S_1 is strictly less than the logsize of T_1 so the recursive calls are nested at most to the depth $\log n$, and secondly, that only $O(logsize(T_1) - logsize(S_1))$ many rounds of the game elapse between the beginning of the game $G[T_1, T_2]$ and the recursive call to $G[S, S_1]$.

The first step to establishing the last two facts is that we need a succinct way to specify the subtree S_1 of T_1. Our specification scheme will depend heavily on the fact that S_1 has logsize strictly less than the logsize of its parent tree. The first player can therefore uniquely specify S_1 as a subtree of T_1 by specifying (a) the logsize of S_1 and (b) the minimum number w such that $w \cdot 2^{logsize(S_1)} \le mindom(S_1)$. In particular, this means that the numbers of the leaves of S_1 are a subinterval of

$$[w \cdot 2^{logsize(S_1)}, (w + 3) \cdot 2^{logsize(S_1)} - 2]$$

Actually, since the game invokes itself recursively, we need more than just a succinct representation of S_1, we need to be able to succinctly represent a nested series of subtrees, each of successively smaller logsize. We will do this by using *trinary* strings over the alphabet $\{0, 1, 2\}$ to represent the substrings, according to the following construction.

Definition *We define now the trinary strings that define subtrees that arise during the play of the game above.*

Let $S_0 = T, S_1, S_2, \ldots, S_k$ be a sequence of subtrees of T, with each S_{i+1} a proper subtree of S_i (not necessarily an immediate subtree) and each S_i having logsize strictly less than its parent tree. Let $n = logsize(T)$ and let $i_m = logsize(S_m)$. The sequence $\langle S_0, \ldots, S_k \rangle$ has succinct representation defined as follows:

(1) *If $k = 0$, the empty string, ϵ, represents $\langle S_0 \rangle = \langle T \rangle$.*

(2) *If $k = 1$, the trinary string $2^{n-i_1} w_1$ where $w_1 \in \{0, 1\}^{n-i_1+2}$ is the binary representation of the largest number (both the number and its binary representation are denoted w_1) such that $w_1 \cdot 2^{i_1} \le mindom(S_1)$.*

(3) *For $k = 2$, the trinary string $2^{n-i_1} w_1 2^{i_1-i_2} w_2$ where w_1 satisfies the conditions of (2) and where $w_2 \in \{0, 1\}^{i_1-i_2+2}$ is the binary representation of the largest number such that $w_1 2^{i_1} + w_2 2^{i_2} \le mindom(S_2)$.*

(4) *More generally, for any value of $\ell \ge 0$, the trinary string representing the sequence of trees $\langle S_0, \ldots, S_\ell \rangle$ is the string $\sigma_1 \cdots \sigma_\ell$, where each σ_p is a*

string $2^{i_{p-1}-i_p}w_p$ where $w_p \in \{0,1\}^{i_{p-1}-i_p+2}$ is the binary representation of the largest value, also denoted w_p, such that

$$m_p = \sum_{j=1}^{p} w_j 2^{logsize(S_j)} \leq mindom(S_p).$$

It is easy to verify, by induction on p, that $dom(S_p)$ is contained in the interval $[m_p, m_p + 3 \cdot 2^{logsize(S_p)} - 2]$. Therefore the value of w_p will satisfy

$$w_p \cdot 2^{logsize(S_p)} \leq 3 \cdot 2^{logsize(S_{p-1})},$$

from which it is immediate that $i_{p-1} - i_p + 2$ bits are sufficient for the binary representation of w_p.

When w represents a sequence $\langle S_0, \ldots, S_k \rangle$, we sometimes abuse notation and say that w represents the subtree S_k.

It is easy to check, but very important to note, that the number of symbols in the trinary string representing the sequence $\langle S_0, \ldots, S_k \rangle$ is $O(\log n - logsize(S_k))$. For the rest of this section, we let $\mu > 0$ be a constant such that the trinary strings w representing subtrees S_k of T_1 and T_2 have length $|w| \leq \mu(\log n - logsize(S_k))$.

In addition, it is important that there is an Alogtime algorithm which, given a string representation of a tree T and given a trinary string α, determines whether the string properly represents a subtree of T and, if so, which subtree if represents. This is not however, so difficult: By counting the total number, ℓ, of 2's in α, one obtains the logsize, $n - \ell$, of the represented subformula S_k (if it exists). Then decompose the string α into substrings of 2's and binary substrings. For each binary string w, if it is preceded by total of j 2's, then view this as the integer $w \cdot 2^{n-j}$; adding up these integers gives the multiple of $2^{n-\ell}$ less than or equal to $mindom(S_k)$. From this, straightforward Alogtime parsing techniques give the formula S_k.

Theorem 2. *The tree isomorphism problem is in alternating logarithmic time.*

Proof. It will suffice to construct logarithmic depth, Alogtime-uniform, bounded fan-in circuits for the tree isomorphism problem. We present a construction of the circuit; but will omit the straightforward proof that the construction can be made Alogtime uniform. The circuit is essentially a direct implementation of the game described informally above.

Fix a tree size n. We may assume that the inputs T_1 and T_2 both have the same size $|T_1| = |T_2| = n$, since otherwise they are not isomorphic. We describe the construction of a circuit C_n which recognizes whether the trees are isomorphic. We will use names such as "$C_{1,2}[u, v]$" for gates in C_n where u and v are trinary strings of length $\leq \mu \log n$ which potentially name subtrees of T_1 or T_2. The circuit C_n will contain the following gates:

$C_{1,a}^+[u,v]$: Here $a \in \{1,2\}$ and u and v are equal length strings which are intended to represent subtrees of T_1 and T_a. This gate $C_{1,a}[u,v]$ is a two input AND gate. The first input to the AND gate is a logarithmic depth circuit which outputs *True* provided u represents a subtree R_1 of T_1 and v represents a subtree R_2 of T_a and provided that R_1 and R_2 have identical size signatures. The second input to the AND gate is the circuit $C_{1,a}[u,v]$.

$C_{1,a}[u,v]$: Here u, v and a are as above. This gate is relevant only if u and v represent subtrees R_1 and R_2 of T_1 and T_a (respectively) with R_1 and R_2 having the same size signature. The gate $C_{1,a}[u,v]$ is intended to compute whether R_1 is isomorphic to R_2. The circuit consists of the conjunction of gates $D_{1,a}^+[uw,v]$ for all trinary strings w such that w is in $2^i\{0,1\}^{i+2}$ for some $i \geq 1$ and such that $|uw| \leq \mu \log n$. The conjunction is implemented as a tree of two-input AND gates with $D_{1,a}^+[uw,v]$ being an input at depth $|w|+2$.

$D_{1,a}^+[uw,v]$: Here u, v, w and a are as above. The gate $D_{1,a}^+[uw,v]$ outputs *True* provided either (i) it is not the case that uw represents a subtree of T_1 which has logsize strictly less than the logsize of its parent tree, OR (ii) the gate $D_{1,a}[uw,v]$ computes *True*.

$D_{1,a}[uw,v]$: Here u, v, w and a are as above. The gate $D_{1,a}[uw,v]$ outputs *True* provided that the number of trinary strings x in $2^i\{0,1\}^{i+2}$ with $|x| = |w|$ for which $C_{1,a}^+[uw,vx] =$ *True* is equal to the number for which $C_{1,1}^+[uw,ux] =$ *True*. There are $3^{|w|}$ many trinary strings $x \in \{0,1,2\}^{|w|}$; therefore, using the Alogtime circuits for counting, the gate $D_{1,a}[uw,v]$ can be computed by a circuit of depth $O(|w|)$ which has as inputs the gates $C_{1,a}^+[uw,vx]$ and $C_{1,1}^+[uw,ux]$.

The description of the circuit C_n is now complete, since the output of C_n is just the gate $C_{1,2}^+[\epsilon, \epsilon]$.

There are a couple things which must be verified to see that C_n is a correct, logarithmic depth circuit. The fact that C_n has logarithmic depth is immediate from the construction. Indeed any gate $X[u,v]$ with X one of $C_{1,a}^+$, $C_{1,a}$, $D_{1,a}^+$ or $D_{1,a}$ will be at depth $\Omega(|u|+|v|) = O(\log n)$ in the circuit. It is important for the depth analysis, that the conjunction for the subcircuit $C_{1,a}[u,v]$ be correctly balanced so that the inputs $D_{1,a}^+[uw,v]$ occur at depth $|w|+2$ in the conjunction. This balancing of AND's is illustrated in Figure 1, where we show the top few levels of the left half of a conjunction of inputs D_w, showing where the three inputs D_0, D_{00} and D_{01} occur.

To verify that the circuit correctly computes tree isomorphism, one proceeds by induction on $(\mu \log n) - |u|$ that $C_{1,a}^+[u,v] =$ *True* if and only if u and v specify isomorphic subtrees. Suppose u has maximum length for which there is a v, $|u| = |v|$, such that $C_{1,a}^+[u,v]$ incorrectly computes the isomorphism predicate for the two subtrees R_1 and R_2 represented by u and v. It is immediate from the definitions that if R_1 and R_2 are isomorphic, then $C_{1,a}^+[u,v]$ must be true. So suppose R_1 and R_2 are not isomorphic. By Lemma 1, there is a subtree S of R_1 which distinguishes R_1 from R_2. Let uw be the string which represents S: $D_{1,a}^+[uw,v]$ will be false and therefore $C_{1,a}[u,v]$ is false.

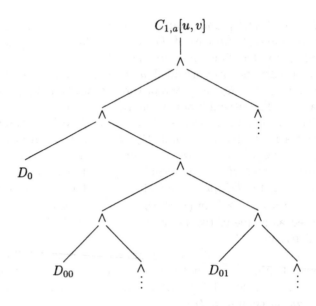

Fig. 1. This figure illustrates the balancing of the conjunction computing $C_{1,a}[u,v]$. The D_w's are intended to be replaced by inputs $D^+_{1,a}[uw,v]$, although these are relevant only for $|w| \geq 2$; and the above circuit is intended only to illustrate the idea of the construction of the conjunction.

□

4 An Alogtime Algorithm for Tree Comparison

In the next two sections, we give Alogtime algorithms for the tree comparison and tree canonization problems. We'll start with the tree comparison problem, and then see that a solution to the tree comparison problem can be used to give a solution to the tree canonization problem.

First, we need some technical definitions about trees.

Definition *Let T be a tree. If S_1 and S_2 are subtrees of T, then S_2 is the maxsize subtree of S_1 if and only if S_2 is an immediate subtree of S_1 and every other subtree of S_1 has size strictly smaller than the size of S_2.*

A subtree S of T is called a unimax *subtree of T if and only if there is a sequence of subtrees $S_0 = T, S_1, \ldots, S_k = S$ such that each S_{i+1} is the maxsize subtree of S_i.*

Note there is at most one unimax subtree S_k at a given depth k in a tree T. Let T_1 and T_2 be two trees which we are trying to compare and suppose $|T_1| = |T_2|$.

Definition *Suppose S is a unimax subtree of T_1 and let R_1 be the parent tree of S. We say S unimax $\succ$-distinguishes T_1 and T_2 if and only if there is a unimax subtree R_2 of T_2 with the same size signature as R_1 such that every immediate subtree of R_2 has size strictly less than the size of S.*

Lemma 3. *If S unimax $\succ$-distinguishes T_1 and T_2, then $T_1 \succ T_2$.*

Proof. This is immediate from the definition of $\succ$, the definition of unimax subtree, and induction on the depth of S. □

Lemma 4. *The problem of determining whether there is a subtree S which unimax $\succ$-distinguishes T_1 and T_2 is in alternating logarithmic time.*

Proof. This is very simple. The Alogtime algorithm first guesses the identity of the subtree S and then verifies the syntactic conditions of unimax $\succ$-distinguishing using the usual Alogtime counting and parsing techniques. □

When we can't unimax $\succ$-distinguish two trees, then we must instead "weakly $\succ$-distinguish" them.

Definition *If S and R are trees, then the multiplicity of S in R is equal to the number of immediate subtrees of R which are isomorphic to S.*

Since tree isomorphism is in Alogtime, and by the usual Alogtime parsing techniques, there are Alogtime algorithms for computing and comparing the multiplicities of subtrees.

Definition *We say S weakly $\succ$-distinguishes T_1 and T_2 if and only if the following conditions hold:*

(1) *The parent tree R_1 of S is a unimax subtree of T_1, but S is not a unimax subtree.*
(2) *There is a (necessarily unique) unimax subtree R_2 of T_2 with the same size signature as R_1.*
(3) *For each subtree S_2 of R_2 such that $|S_2| > |S|$, the multiplicity of S_2 in R_1 is greater than or equal to its multiplicity in R_2.[3]*
(4) *For each subtree S_2 of R_2 such that $|S_2| = |S|$ and $S_2 \succ S$, the multiplicity of S_2 in R_1 is greater than or equal to its multiplicity in R_2.*
(5) *The multiplicity of S in R_1 is strictly greater than the multiplicity of S in R_2.*

Lemma 5. *If S weakly $\succ$-distinguishes T_1 and T_2, then $T_1 \succ T_2$.*

Proof. This is immediate from the definition of $\succ$.

Lemma 6. *If $|T_1| = |T_2|$ and $T_1 \succ T_2$, then there is a subtree S of T_1 which either unimax or weakly $\succ$-distinguishes T_1 from T_2.*

---

[3] We could use "equal to" in place of "greater than or equal to" in (3) and (4), but our choice will save some implementation details in the circuit constructed below.

Proof. The proof is by induction on $|T_1|$. Order the immediate subtrees of T_1 as $R_1 \succcurlyeq R_2 \succcurlyeq \cdots \succcurlyeq R_m$ and the immediate subtrees of T_2 as $S_1 \succcurlyeq S_2 \succcurlyeq \cdots \succcurlyeq S_n$. Find the unique i such that $R_j \equiv S_j$ for all $1 \leq j < i$ and such that either $i = n + 1 \leq m$ or $R_i \succ S_i$. If $i > 1$, then $S = R_i$ is not a unimax subtree of T_1 and obviously weakly $\succ$-distinguishes T_1 and T_2. If $i = 1$ and $|R_1| > |S_1|$, then $S = R_1$ either unimax or weakly $\succ$-distinguishes T_1 and T_2 depending on whether $logsize(R_1) = logsize(T_1)$. Finally, if $i = 1$ and $|R_1| = |S_1|$, then the induction hypothesis gives a subtree S of R_1 which unimax or weakly $\succ$-distinguishes R_1 and S_1. This S also unimax or weakly $\succ$-distinguishes T_1 and T_2. □

We now give an informal definition of the Alogtime algorithm for tree comparison by describing a two player game that runs for logarithmically many rounds: the first player is asserting that $T_1 \succ T_2$, whereas the second player is asserting that $T_1 \preccurlyeq T_2$. The game, denoted $G'[T_1, T_2]$ will invoke itself recursively on subtrees of T_1 and T_2. Since we already have an Alogtime algorithm for tree isomorphism, we are allowed to invoke that algorithm during the play of the game for tree comparison. The game $G'[T_1, T_2]$ starts with the first player specifying a subtree S of T_1 and asserting either that S unimax $\succ$-distinguishes T_1 and T_2 or weakly $\succ$-distinguishes T_1 and T_2. In the first case, the two players verify whether S actually does unimax $\succ$-distinguish T_1 and T_2, and then the first player wins iff it does so. In the second case, the second player will choose one of the conditions (1)-(5) of the definition of "S weakly $\succ$-distinguishes T_1 and T_2" that must be verified. Conditions (1)-(3) and (5) are easily checked with an Alogtime algorithm — in this case, the first player wins iff the one of these conditions picked by the second player is determined to be valid. If the second player picks condition (4), then the second player also specifies a subtree S_2 of R_2 as a candidate subtree for showing that condition (4) fails. Then the first player chooses either to check either that S_2 does not satisfy $|S_2| = |S|$ or has multiplicity in R_1 greater than or equal to its multiplicity in R_2 or that $S \succ S_2$ (anyone of these checks would be sufficient to show that S_2 does not refute condition (4)). If any of these checks succeed the first player wins. The check that $S \succ S_2$ involves a recursive invocation of the game $G'[S, S_2]$.

Lemma 6 shows the correctness of the above game for tree comparison. It remains to see that the game only takes $O(\log |T_1|)$ rounds. A careful analysis of the number of rounds is best done by examining the circuit for tree comparison which is given below in the next proof; however, there are certainly reasons to hope that the above game can be formalized so as to use only logarithmically many bits of communication between the players. First, we have already developed succinct representations for subformulas. Second, whenever S weakly $\succ$-distinguishes two trees, it must be the case that $logsize(S)$ is strictly less than the logsize of the parent tree of S. This latter fact is because the parent tree of S is unimax, whereas S is not, and therefore the parent tree of S is at least twice the size of S.

Theorem 7. *The tree comparison problem is in alternating logarithmic time.*

Proof. It will suffice to construct logarithmic depth, Alogtime uniform, bounded fanin circuits for the tree comparison problem. We present a construction of the circuit, but omit the straightforward proof that the construction can be made Alogtime uniform. The circuit is essentially a direct implementation of the game G' above.

Fix a tree size n. We may assume that $|T_1| = |T_2|$ since otherwise the tree comparison problem is trivial. We describe the construction of a circuit C'_n which performs tree comparison. Some of the gates in the circuit will be given names such as $E[u, v]$ where u and v are trinary strings of length $\leq \mu \log n$ which potentially name subtrees of T_1 and T_2. C'_n contains gates:

$E[u, v]$: Here u and v are equal length strings which represent subtrees U_1 and U_2 of T_1 and T_2, respectively. This gate is relevant to the circuit computation only when u and v represent subtrees U_1 and U_2 which have the same size signature.

The gate $E[u, v]$ is intended to compute whether $U_1 \succ U_2$. The topmost gate of $E[u, v]$ is a disjunction. The first input to the disjunction is a logarithmic depth circuit that checks whether there is a subtree of U_1 which unimax $\succ$-distinguishes U_1 and U_2. This circuit exists by Lemma 4. The second input to the disjunction is a disjunction (actually a tree of two-input disjunctions) of the gates $F^+[uw, v]$ for all trinary strings w such that such that $w \in 2^i\{0, 1\}^{i+2}$ for some $i \geq 1$ and such that $|uw| \leq \mu \log n$. The disjunction is implemented as a tree of two-input OR gates with $F[uw, v]$ being an input at depth $|w| + 2$.

$F^+[uw, v]$: Here u, v, and w are as above. The gate $F^+[uw, v]$ is a conjunction of (i) the circuit $F[uw, v]$ and of (ii) a logarithmic depth circuit which outputs *True* iff the following conditions hold:

the string uw represents a non-unimax subtree S of T_1, which has a unimax parent tree R_1; there is a unimax subtree R_2 of T_2 with the same size signature as R_1; the multiplicity of S in R_1 is greater than the multiplicity of S in R_2; and the maxsize subformula of R_2 (if it exists) has multiplicity in R_1 greater than or equal to its multiplicity in R_2.

$F[uw, v]$: Here u, v and w are as above. The gate $F[uw, v]$ is the conjunction of the gates $H[uw, vw^*]$ for all $w^* \in 2^i\{0, 1\}^{i+2}$ with $i \geq 1$ and $|w^*| \leq |w|$. The gate $H[uw, vw^*]$ is at depth $|w^*| + 2$ in the conjunction (it is actually a tree of two-input conjunctions).

$H[uw, vw^*]$: Here u, v, w and w^* are as above. This gate is the disjunction of (i) the circuit $E[uw, vw^*]$ provided $|w| = |w^*|$, or otherwise this input can be set to *False*. and (ii) a logarithmic depth circuit with outputs *True* iff at least one of the following conditions hold:

The string vw^* is not a valid representation of subtree S_2 in T_2; or $S_2 \equiv S$; or $|S_2| < |S|$; or the multiplicity of S_2 in R_1 is greater than or equal to its multiplicity in R_2.

The description of the circuit C'_n is now complete, since the output of C_n is just the gate $E[\epsilon, \epsilon]$.

There are a couple things which must be verified to see that C'_n is a correct, logarithmic depth circuit. The fact that C'_n has logarithmic depth is immediate from the construction. Indeed any gate $X[u, v]$ with X one of E, F^+, F, or H will be at depth $\Omega(|u| + |v|) = O(\log n)$. For this, it is important that the disjunctions for $E[u, v]$ and the conjunctions for $F[uw, v]$ be correctly balanced with inputs at the correct depths.

The correctness of the circuit C'_n is proved by induction on the lengths of the strings u, v. The gate $E[u, v]$ is intended to compute *True* iff u and v represent subtrees U_1 and U_2 of T_1 and T_2 (resp.) with $U_1 \succ U_2$. The gate $F^+[uv, w]$ is intended to output *True* iff uv represents a subtree S of U_1 which weakly $\succ$-distinguishes U_1 and U_2. The gate $H[uw, vw^*]$ outputs *True* iff it is not the case that vw^* represents a subtree S_2 which violates one of the conditions (3) or (4) of the definition of "S weakly $\succ$-distinguishes U_1 and U_2". In view of Lemma 6, it is straightforward to verify that the circuit correctly performs tree comparison. One point that deserves further justification is the manner in which w^* is chosen in $F^+[uw, v]$: the string w^* is chosen so that $H[uw, vw^*]$ can decide whether vw^* represents a subtree S_2 which violates condition (4) of the definition of "S weakly $\succ$-distinguishes U_1 and U_2". The reason we can restrict w^* to have $|w^*| \leq |w|$ is that condition (4) only considers formulas S_2 which have size greater than or equal to S; this means that $logsize(S_2) \geq logsize(S)$ and therefore, by the definition of trinary representation, $|w^*| \leq |w|$. In addition, if S_2 is not a maxsize subformula of R_2, then S_2 has logsize strictly less than the logsize of R_2 and therefore S_2 does have a trinary string representation. The verification of condition (3) for S_2 the maxsize subtree of R_2 was already taken care by the subcircuit (ii) described under the heading $F^+[uw, v]$. $\square$

5 An Alogtime Algorithm for Tree Canonization

Given the Alogtime algorithms for tree isomorphism and tree contraction, it will be easy to prove the existence of an Alogtime algorithm for tree canonization. Recall that the definition of $canon_T$, the canonization of T, was given in section 2.

Theorem 8. *The tree canonization problem is in alternating logarithmic time.*

Proof. It will suffice to show that there is an Alogtime algorithm which determines the orders of the leaves of an input tree in the canonization of T. As usual, the input tree T is specified by its string representation, i.e., a string of parentheses. Consider two leaves, x and y, with leaf numbers i and j, with $i < j$. Find the node z in the tree which is the least common ancestor of x and y. Let S_x be the immediate subtree of z containing x and S_y be the immediate subtree of z containing y. If $S_x \succ S_y$, then let x precede y in a reordering of the tree; otherwise let y precede x. Applying this construction to all leaf nodes of T, we get a consistent reordering of the leaves of T which induces a canonization of T.

Operations such as finding least common ancestors can be done in Alogtime, so the Alogtime algorithm for tree comparison gives an Alogtime algorithm for tree canonization. $\Box$

6 Conclusions

Our main results show that tree isomorphism, tree comparison and tree canonization are in alternating logarithmic time; improving on the logarithmic space algorithms of [8].

There are number of other problems known to be in Alogtime, including the Boolean Formula Value Problem [3, 5, 4], and the word problem for S_5 [2]. These problems are also known to be complete for Alogtime under deterministic log time reductions. It is still open whether the tree isomorphism, comparison and canonization problems are also complete for Alogtime.

A problem with a similar name to the tree isomorphism problem is the "subtree isomorphism problem". This is the problem of, given two trees T_1 and T_2, determining whether there is injection from T_1 to T_2 which preserves the edge relation. In other words, whether T_1 is isomorphic to a subgraph of T_2 induced by some subset of the nodes of T_2. A polynomial-time deterministic algorithm for the subtree isomorphism problem was first found by [11]; and [6, 9] gave random NC (RNC) algorithms for subtree isomorphism. [6] also proved that bipartite matching is reducible to subtree isomorphism; this means that subtree isomorphism cannot be in logspace or Alogtime unless bipartite matching is. Thus it appears that subtree isomorphism is substantially more difficult than tree isomorphism.

We conclude by mentioning one of our favorite open problems about alternating logtime: is the 2-sided Dyck language recognizable in alternating logarithmic time. Equivalently, is the word problem for free groups with two generators decidable in alternating logarithmic time? This language was shown in [10] to be in logspace.

Acknowledgement. We thank Steven Lindell for helpful discussions during the formulation of the algorithms in this paper.

References

1. A. V. AHO, J. E. HOPCROFT, AND J. D. ULLMAN, *The Design and Analysis of Computer Algorithms*, Addison-Wesley, 1974.
2. D. A. M. BARRINGTON, *Bounded-width polynomial-size branching programs recognize exactly those languages in NC^1*, J. Comput. System Sci., 38 (1989), pp. 150–164.
3. S. R. BUSS, *The Boolean formula value problem is in ALOGTIME*, in Proceedings of the 19-th Annual ACM Symposium on Theory of Computing, May 1987, pp. 123–131.

4. ——, *Algorithms for Boolean formula evaluation and for tree contraction*, in Arithmetic, Proof Theory and Computational Complexity, P. Clote and J. Krajíček, eds., Oxford University Press, 1993, pp. 96–115.

5. S. R. BUSS, S. A. COOK, A. GUPTA, AND V. RAMACHANDRAN, *An optimal parallel algorithm for formula evaluation*, SIAM J. Comput., 21 (1992), pp. 755–780.

6. P. B. GIBBONS, R. M. KARP, G. L. MILLER, AND D. SOROKER, *Subtree isomorphism in in random NC*, Discrete Applied Mathematics, 29 (1990), pp. 35–62.

7. N. IMMERMAN, *Languages that capture complexity classes*, SIAM Journal on Computing, 16 (1987), pp. 760–778.

8. S. LINDELL, *A logspace algorithm for tree canonization*, in Proceedings of the 24th Annual ACM Symposium on Theory of Computing, 1992, pp. 400–404.

9. A. LINGAS AND M. KARPINSKI, *Subtree isomorphism is NC reducible to bipartite perfect matching*, Information Processing Letters, 30 (1989), pp. 27–32.

10. R. J. LIPTON AND Y. ZALCSTEIN, *Word problems solvable in logspace*, J. Assoc. Comput. Mach., 24 (1977), pp. 522–526.

11. D. W. MATULA, *Subtree isomorphism in $O(n^{5/2})$*, Annals of Discrete Mathematics, 2 (1978), pp. 91–106.

12. G. L. MILLER AND J. H. REIF, *Parallel tree contraction part 2: Further applications*, SIAM Journal on Computing, 20 (1991), pp. 1128–1147.

13. W. L. RUZZO, *On uniform circuit complexity*, J. Comput. System Sci., 22 (1981), pp. 365–383.

Ultrafilter Logic and Generic Reasoning

W. A. Carnielli*   P. A. S. Veloso**

Abstract

We examine a new logical system, capturing the intuition of 'most' by means of generalised quantifiers over ultrafilters, with the aim of providing a basis for generic reasoning. This monotonic ultrafilter logic is a conservative extension of classical first-order logic, with which it shares several properties, including a simple sound and complete deductive system. For reasoning about generic objects, we introduce 'generic' individuals as those possessing the properties that most individuals have. We examine some properties of these 'generic' individuals and internalise them as generic constants, which produces conservative extensions where one can correctly reason about generic objects as intended. A many-sorted version of our ultrafilter logic is also introduced and employed to handle correctly distinct notions of 'large' subsets. Examples similar to ones in the literature illustrate the presentation. We also comment on some perspectives for further work: interesting connections with fuzzy logic, inductive reasoning and empirical reasoning suggest the possibility of other applications for our logic.

Contents

*   Dept. of Philsophy-IFCH and Center for Logic, Epistemology and the History of Science,

State University of Campinas. Work supported by FAPESP and CNPq research grants.

**   COPPE and Institute of Mathematics, Federal University of Rio de Janeiro (on leave from

Dept. of Informatics, PUC- Rio de Janeiro), Brazil.

1. Introduction

This paper examines a logical system which captures the intuition of 'most' by means of generalised quantifiers over ultrafilters. The primary motivation is to obtain a monotonic approach to generic reasoning. The resulting ultrafilter logic is a conservative extension of classical logic.

The nonmonotonic approach via default rules takes a *negative view* of defaults as 'absence of information to the contrary'. In contrast, our alternative approach favours a *positive interpretation*, capturing the intuitive sense of 'most' or 'generally' (a property holds for 'most' objects when the set of exceptions is 'small') in a precise manner via ultrafilters.

We briefly review the motivations for generic reasoning, and an earlier analysis of the nonmonotonic approach via default rules (developed by the present authors in collaboration with A. M. Sette in [Sette et al. 1997]), which suggests the possibility of facing directly the underlying quantificational problem by means of a generalised quantifier. We set up, both semantically and axiomatically, a new logical system based on this idea. This system, which will be shown to be sound and complete, proves to be a conservative extension of classical first-order logic. We introduce the idea of 'generic' individuals, examine some of its properties, and show how one can reason correctly about them within our formalism. We also indicate the desirability of having several notions of 'large' and how this can be simply formulated in a many-sorted version of our ultrafilter logic. The presentation is illustrated with some examples similar to ones often presented in the literature. We also comment on some perspectives and directions for further work, specifically some interesting connections with fuzzy logic and inductive and empirical reasoning, which suggest the possibility of other applications for our ultrafilter logic.

2. Approaches to Generic Reasoning

We now examine the nonmonotonic approach to generic reasoning, emphasising Reiter's default logic (referring to [Sette et al. 1997] for a more detailed analysis), and propose an alternative view.

2.1 Nonmonotonic Defaults and Generic Reasoning

A logic for default reasoning was explicitly introduced by Reiter [1980]. As he emphasizes, a good deal of what we know about the world is 'almost true' with a few exceptions. Such facts usually assume the form most P's are Q's. "For example, most birds fly, except for penguins, ostriches,...etc. Given a particular bird, we will conclude that it flies unless we happen to know that it satisfies one of these exceptions. How is the fact that most birds fly to be represented?" According to Reiter [1980], what we have is a question of quantification, namely how to represent and understand a generalised quantifier of the kind 'almost always'. Reiter convincingly argues that traditional first-order logic is not sufficient for handling this problem and then proceeds to offer his interpretation to the question of 'reasoning by default'.

A natural first-order representation explicitly lists the exceptions to fly: $\forall x[(Bird(x)\wedge\neg Penguin(x)\wedge\neg Ostrich(x)...)\rightarrow Fly(x)]$. But from this one cannot conclude that a 'general' bird can fly. An attempt to prove Fly(Tweety) (when all we know about Tweety is that he is a bird) is blocked. "What is required", continues

Reiter, "is somehow to allow Tweety to fly *by default* ... 'if x is a bird, then in absence of any information to the contrary, infer that x can fly'. The problem then is to interpret the phrase 'in absence of any information to the contrary'." He proposes this solution: just interpret '*in absence of any information to the contrary* that x can fly' as '*it is consistent to assume* that x can fly'. Thus, "if x is a bird and it is consistent to assume that x can fly, then infer that x can fly", can be expressed by a default rule, which can be read as 'if x is a bird and it is consistent to believe that x can fly, then one may believe that x can fly'.

Reiter [1980] formalises his theory by adding to first-order logic certain expressions called default rules. He then defines extensions of a (closed) default theory as fixed points of an extension operator. As he himself points out, not all default theories are closed under extensions, and it is difficult to know which default theories have a (consistent) extension. Also, a default theory may have several extensions, and in this case it is not clear how the resulting (nonmonotonic) consequence operator should be defined: should one consider σ as a consequence when (sceptically) σ is in *all* extensions, or when (credulously) σ is in only *some* extensions, or perhaps in some 'intermediate' case?

Attempts to found generic reasoning on nonmonotonic logic have been extensively criticised [Besnard 1989]. Also, nonmonotonic logic itself has also been the target of severe criticisms [Israel 1980]. The controversial character of this issue is apparent [Besnard et al. 1991]. Despite such criticisms, the default approach to generic reasoning based on nonmonotonic inference has gained influence. Certain points, however, seem to deserve further analysis.

Reiter's analysis of generic reasoning appears to involve a series of simplifications. The situation -that a great part of our knowledge of the world is 'almost always' true - gives rise to the original problem: given a 'generic' bird (an object about which all that we know is just that it is a bird) we wish to conclude that it flies. This problem is then simplified in a number of ways (as we discuss in [Sette et al. 1997]) until it can be formulated in terms of default rules.

The original problem and some of the considerations about the nature of knowledge and generic reasoning have their own interest. Nevertheless, one gets the impression that the simplifications involve taking somewhat arbitrary decisions. The final step of simplifications is only weakly connected, if at all, to the original formulation. Moreover, default rules appear to involve extra-logical notions. One may interpret 'it is consistent to assume β' as '¬β cannot be deduced' [Besnard [1989]. But, the application of such a rule based on non-deducibility becomes non-effective.

In the sequel we motivate and outline an alternative approach to generic reasoning, based on making precise the notion of 'most'.

2.2 Ultrafilters for Default Reasoning and Generic Statements

Returning to the initial question, the nonmonotonic approach via default rules originated by Reiter bases generic reasoning on the idea of 'in the absence of any information to the contrary', and in this sense it favours a *negative interpretation*. Our approach goes in the opposite direction: we propose a *positive interpretation* of generic reasoning. (Our approach bears some similarity to the treatment of default reasoning proposed by Schlechta [1995] and Ben-David and Ben-Aliyahu [1994].)

The idea is as follows: for a *generic individual* p we wish to be able to deduce Q(p) from 'Q(x) holds for most x (in universe P)'. Such a generic p can be regarded as a kind of representative of the individuals (in P). An example is 'everyone is innocent until proof to the contrary'. From 'almost' everyone is innocent, we wish to conclude that a 'generic' person is innocent, without commitment to the innocence of a specific person.

In formalising this kind of reasoning - as an extension of classical first-order logic - we wish to capture the intuition of 'most' present in the original formulation. We need, of course, a clear account of 'almost all'.

In the example of flying birds, let B be the universe of birds and consider the set $F=\{b\in B: F(b)\}$ of birds that fly. The account we propose for 'almost all birds fly' is 'F is almost as large as B'. By considering the universe B to be 'large', we can reduce 'F is almost as large as B' to simply 'F is as large'. The complement $F^c=\{b\in B: \neg F(b)\}$ (non-flying birds) may be regarded as the set of exceptions. The basic intuition is that $F\subseteq S$ is 'almost as large as' B exactly when the set F^c of exceptions is 'small' with respect to B.

Some - reasonable - criteria for subsets of a universe S to be considered 'small' with respect to S, or simply small, are: subsets of small sets are small (in particular, $\varnothing$ is small), and the union of small sets is small.

We have, dually, reasonable criteria for subsets of a universe S to be considered 'almost as large as' S, or simply large. For subsets $X, Y \subseteq S$:

($\supseteq$) if X is large and $Y\supseteq X$, then Y is large as well (in particular, S is large);

($\cap$) if X and Y are both large, then so is their intersection $X\cap Y$ large.

Finally, in order to have a decisive criterion, we also require:

(^{c}) either X is large or X^c is large.

These intuitive ideas lead naturally to the concept of (ultra)filter (see, e. g. [Chang and Keisler 1969]). There are many distinct (ultra)filters over a given universe, each one providing a distinct notion of 'large'.

A *filter* over a universe S is a family $\mathcal{F}\subseteq \wp(S)$ of subsets of S that is closed under intersections and supersets. A filter $\mathcal{F}$ is *principal* if $\mathcal{F}$ is generated by some element $X\in \wp(S)$ ($\mathcal{F}=\{Y\subseteq S:X\subseteq Y\}$) and *non-principal* if not principal. An *ultrafilter* $\mathcal{U}$ is a filter which is maximal with respect to inclusion, i. e. $\mathcal{U} = \mathcal{F}$ for any filter $\mathcal{U}\subseteq \mathcal{F}$. A *proper* ultrafilter is one with $\mathcal{U}\neq \wp(S)$.

Ultrafilters thus generalise the idea that either a subset is 'large' or its complement is 'large', as shown in the following well-known equivalences for a filter $\mathcal{F}$ over a universe S: $\mathcal{F}$ is an ultrafilter iff, for every $X\subseteq S$, either $X\in \mathcal{F}$ or $X^c\in \mathcal{F}$ iff $\mathcal{F}$ is maximal (with respect to inclusion).

A collection $\mathcal{F} \subseteq \wp(S)$ is said to have *fip* (short for *finite intersection property*) when $X_1 \cap .. \cap X_n \neq \varnothing$ for any finite family $\{X_1,...,X_n\} \subseteq \mathcal{F}$. A collection $\mathcal{F}$ $\subseteq \wp(S)$ has fip iff it can be extended to a proper ultrafilter over S.

Ultrafilters are also connected to a notion of *measure* in that what is outside an ultrafilter has 'measure zero'. A basic intuition underlying our approach is what lies outside an ultrafilter is small, independently of the size of the universe, which is the main point in applying ultrafilters.

To illustrate the application of such ideas to generic reasoning, consider a universe B of birds and suppose that we have certain knowledge about birds, including the information that "birds 'generally' fly", which we formulate as $\nabla xF(x)$. Then we would like to be able to conclude that a 'generic' bird flies. We shall see that, by introducing an appropriate new constant b for a 'generic' bird, we shall be able to derive F(b), as desired. We may also have some extra information, including exceptions, such as that there exist penguins, which are non-flying birds. If we know that birds 'generally' fly, we shall still be able to conclude that a 'generic' bird b does fly. The point here is that, since $\mathcal{U}$ is a proper ultrafilter over B, if the set $F=\{b \in B : F(b)\}$ of flying birds belongs to ultrafilter $\mathcal{U}$, its complement $F^c=\{b \in B : \neg F(b)\}$ is small, but not necessarily empty, and possible exceptions may very well exist.

The new logical operator ∇ can be seen as a generalised quantifier (see, e. g. Barwise and Feferman [1985]). The logic obtained by adding ∇ to classical first-order logic will provide a tool for generic reasoning. In the next section we formalise this logic semantically and present a sound and complete deductive calculus for it. We then show how to use this logic for reasoning about 'generic' individuals. In some cases, the behaviour of the operator ∇ as a generalised-quantifier may introduce some undesired side effects because of some unwanted inheritances. We later show how a sorted version of our approach can be employed to take care of distinct notions of 'large', thereby circumventing these undesirable inheritances.

3. Ultrafilter Logic and Generic Statements

Our ultrafilter logic extends first-order logic by means of a generalised quantifier ∇, whose behaviour can be seen to be intermediate between $\forall$ and $\exists$.

3.1 Syntax, Semantics and Axiomatics of ∇

Consider a type λ with symbols for predicates, functions and constants. Let $L_{\omega\omega}(\lambda)$ be the usual first-order language[1] of type λ, closed under the propositional connectives, as well as under the quantifiers $\forall$ and $\exists$. We use $L_{\omega\omega}(\lambda)^{\nabla}$ for the extension of $L_{\omega\omega}(\lambda)$ obtained by adding the operator ∇.

The formulae of $L_{\omega\omega}(\lambda)^{\nabla}$ are built by the usual formation rules and the following new formation rule (enabling iterated applications of ∇):

---

[1] We use $\equiv$ for equality.

(∇) for each variable x, if φ is a formula in $L_{\omega\omega}(\lambda)^{\nabla}$ then so is $\nabla x\varphi$.

The next example illustrates the expressive power of the language $L_{\omega\omega}(\lambda)^{\nabla}$.

Example 1: (Expressive power of ∇)

A. Consider a binary predicate L (with L(x,y) standing for x loves y).

1. "Almost everybody loves somebody": $\nabla x \exists y L(x,y)$.

2. "Everybody loves almost everybody": $\forall x \nabla y L(x,y)$.

3. "Generally people love each other": $\nabla x \nabla y L(x,y)$.

B. Consider a type λ having binary predicates S, D and O (with S(x,y), D(z,y) and O(x,z) standing, respectively, for x supports y, z defeats y and x opposes z), as well as a constant b (standing for Brazil).

1. "Most supporters of Brazil oppose most teams that defeat Brazil":

$$\nabla x[S(x,b)\rightarrow\nabla z(D(z,b)\rightarrow O(x,z))], \text{ or } \nabla x\nabla z[S(x,b)\wedge D(z,b)\rightarrow O(x,z)].$$

2. "Generally one opposes defeaters of supported teams"

$$\nabla x\nabla y\nabla z[S(x,y)\wedge D(z,y)\rightarrow O(x,z)].$$

The semantic interpretation for the formulae in $L_{\omega\omega}(\lambda)^{\nabla}$ is defined by extending the usual first-order interpretation to ultrafilters. To provide a semantical version $\mathcal{L}_{\omega\omega}(\lambda)^{\nabla}$ of ultrafilter logic, we enrich first-order structures with ultrafilters. An *ultrafilter structure* $M^{\nabla}=(M, \mathcal{U}^M)$ for $L_{\omega\omega}(\lambda)^{\nabla}$ consists of a first-order structure M for $L_{\omega\omega}(\lambda)$ together with a proper ultrafilter $\mathcal{U}^M$ over the universe M of M.

We now define *satisfaction* of a formula φ in a structure M^{∇} under an assignment s to variables ($M^{\nabla}|= \varphi [s]$) by extending the usual definition:

- for formulae not involving ∇, satisfaction is as usual, i. e. for a formula φ of $L_{\omega\omega}(\lambda)$: $(M, \mathcal{U}^M) |= \varphi[s]$ iff $M |= \varphi[s]$;

- for a formula $\varphi(\underline{x})$ of the form $\nabla y\theta(\underline{x},y)$: we define $M^{\nabla}|= \nabla y\, \theta(\underline{x},y)[\underline{a}]$ iff the set $\{b\in M: M^{\nabla} |= \theta(\underline{x},y)[\underline{a},b]\}$ is in the ultrafilter $\mathcal{U}^M$.

In particular, for a sentence $\nabla y\theta(y)$: $M^{\nabla}|= \nabla y\theta(y)$ iff $M^{\nabla}[\theta(y)]\in \mathcal{U}^M$, where $M^{\nabla}[\theta(y)]=\{b\in M:M^{\nabla}|= \theta(y)[b]\}$ is the set defined by $\theta(y)$ in M^{∇}.

Other usual semantic notions, such as model ($M^{\nabla}|= \sigma$), logical consequence ($\Gamma |= \sigma$), theory, validity, etc. can be appropriately adapted. In fact, satisfaction does not depend on the entire ultrafilter; but on an ultrafilter of the Boolean algebra of subsets definable from points.

We set up a deductive system for our logic by considering a sound and complete axiomatisation $A_{\omega\omega}(\lambda)$ for classical first-order logic, with the usual logic rules Modus Ponens and Generalisation[2] The axioms of $A_{\omega\omega}(\lambda)^\nabla$ are those of $A_{\omega\omega}(\lambda)$ together with the following *ultrafilter axioms*[3] (where ψ and θ are formulae of $L_{\omega\omega}(\lambda)^\nabla$):

($\nabla\exists$) $\nabla x\psi\to\exists x\psi$	[large sets are nonempty],
($\nabla\wedge$) $(\nabla x\psi\wedge\nabla x\theta)\to\nabla x(\psi\wedge\theta)$	[intersections of large sets are large],
($\nabla\neg$) $\nabla x\psi\vee\nabla x\neg\psi$	[a set or its complement is large].

One sees immediatly that $A_{\omega\omega}(\lambda)^\nabla$ is consistent, by using a translation from $L_{\omega\omega}(\lambda)^\nabla$ into the propositional calculus which deletes all quantifiers (including ∇).

The following formulae are provable in $A_{\omega\omega}(\lambda)^\nabla$: $\forall x\varphi\to\nabla x\varphi$, $\neg\nabla x\varphi\leftrightarrow\nabla x\neg\varphi$, and $(\nabla x\psi\wedge\nabla x\theta)\leftrightarrow\nabla x(\psi\wedge\theta)$. Thus, we obtain prenex forms: every formula of $L_{\omega\omega}(\lambda)^\nabla$is provably equivalent to one consisting of a prefix of quantifiers ($\forall$, $\exists$ and ∇) followed by a quantifier-free matrix.

Adaptations of the usual proofs establish some properties of $A_{\omega\omega}(\lambda)^\nabla$.

• Consider a set $\Gamma\cup\{\sigma,\tau\}$ of sentences and a formula φ in $L_{\omega\omega}(\lambda)^\nabla$.

1. (Deduction Theorem) If $\Gamma\cup\{\sigma\}\vdash\varphi$ then $\Gamma\vdash \sigma\to\varphi$.
2. Set Γ is consistent iff every finite subset of Γ is consistent.
3. For a maximal consistent Γ: $\Gamma\vdash\sigma$ iff $\Gamma\vdash\neg\sigma$,and $\Gamma\vdash\sigma\vee\tau$ iff $\Gamma\vdash\sigma$ or $\Gamma\vdash\tau$.
4. A consistent Γ can be extended to a maximal consistent theory.
5. $\Gamma\cup\{\sigma\}$ is consistent iff $\Gamma\vdash\neg\sigma$.

The next example illustrates an application of the quantifier ∇.

Example 2: (Smoking in restaurants and in churches)

You are a smoker in a country whose habits and laws are not familiar to you. If you are in a restaurant and do not see any sign to the contrary, you may try a cigarette.

But if you are in a church in this country, you probably would not smoke. How can the reasoning involved in this story be formalised ?

Consider unary predicates S, R, C, and F over places (with S(x), R(x), C(x), and F(x), respectively, representing "I am allowed to smoke at place x", "x is a restaurant", "x is a church", and "place x has a no-smoking sign".

---

[2] One can also replace the Generalisation rule by universal generalisations of the axioms.

[3] A previous formulation also had axiom $\forall x(\psi\to\theta)\to(\nabla x\psi\to\nabla x\theta)$, which can be derived from the others.

The information about churches may be formulated as follows:

$\nabla x[C(x) \rightarrowtail \neg S(x)]$ [generally I am not allowed to smoke in churches].

The information concerning restaurants may be formulated as follows:

$\nabla x[\neg F(x) \rightarrowtail S(x)]$ [generally I may smoke unless it is prohibited],

$\nabla x[R(x) \rightarrowtail \neg F(x)]$ [generally restaurants do not prohibit smoking];

whence $\nabla x[R(x) \rightarrow S(x)]$ [generally I may smoke in restaurants].

The usage of ultrafilters here models the available knowledge, so that the restaurants where I know I may smoke is a large set (with respect to my knowledge), whereas the collection of the churches where I know I may smoke is small. This fine distinction would not be captured by the traditional default theory, where, in the absence of information, one would not know whether or not one may smoke

3.2 The Logic of ∇: Soundness and Completeness

We now establish the soundness and completeness of the deductive system $A_{\omega\omega}(\lambda)^\nabla$ with respect to ultrafilter structures.

Soundness of $A_{\omega\omega}(\lambda)^\nabla$ with respect to ultrafilter structures is clear, since the axioms are valid in such structures, and the rules preserve validity. Completeness is usually harder, but we can adapt Henkin's well-known proof for classical predicate calculus, by providing an adequate ultrafilter.

We proceed to outline how this can be done. Given a consistent set Γ, extend it to a maximal consistent set Δ, with witnesses in a set C of new constants. The canonical structure H has universe $H=C/\approx$ where $c\approx d$ iff $\Delta \vdash c \equiv d$. Henkin's proof establishes by induction on formulae of $L_{\omega\omega}(\lambda)$ that

$$H \models \varphi[c_1/\approx,\ldots,c_m/\approx] \quad \text{iff} \quad \Delta \vdash \varphi(c_1,\ldots,c_m).$$

In our case, we need an extra inductive step to deal with the new quantifier ∇. This can be handled as follows.

We define, for each formula $\varphi(\underline{v})$ with set $\underline{v} = \{v_1,\ldots,v_m\}$ of free variables, $\varphi(\underline{v})^\Delta := \{<c_1/\approx,\ldots,c_m/\approx> \in H^m : \Delta \vdash \varphi(c_1,\ldots,c_m)\}$, and, considering formulae with a single free variable, $\mathcal{R}^\Delta := \{\theta(x)^\Delta \subseteq H : \Delta \vdash \nabla x\theta(x)\}$. In view of our axioms, $\mathcal{R}^\Delta \subseteq \mathcal{P}(H)$ has fip, so $\mathcal{R}^\Delta$ can be extended to a proper ultrafilter $\mathcal{U}^H \subseteq \mathcal{P}(H)$. We use $\mathcal{U}^H$ to expand H to an ultrafilter structure $H^\nabla = (H, \mathcal{U}^H)$ for $L_{\omega\omega}(\lambda)^\nabla$.

We now show by induction on formulae of $L_{\omega\omega}(\lambda)^\nabla$ that $H^\nabla \models \varphi[\underline{h}]$ iff $\underline{h} \in \varphi^\Delta$.

The inductive steps for the connectives and quantifiers $\forall$ and $\exists$ are as in the classical proof. The inductive step $H^\nabla \models \nabla y\varphi[\underline{h}]$ iff $\underline{h} \in (\nabla y\varphi)^\Delta$ for the new quantifier ∇ now follows from $\{g/\approx \in H : H^\nabla \models \varphi(\underline{h},y)[g/\approx]\} \in \mathcal{U}^H$ iff $(\varphi(\underline{h},y))^\Delta \in \mathcal{R}^\Delta$.

We thus have a Löwenheim-Skolem Theorem for our logical system.

- A consistent set of sentences of $A_{\omega\omega}(\lambda)^{\nabla}$ has an ultrafilter model with cardinality at most that of its language. □

Hence, we have the desired result for ultrafilter logic.

- $A_{\omega\omega}(\lambda)^{\nabla}$ is sound and complete with respect to $\mathcal{L}_{\omega\omega}(\lambda)^{\nabla}$: □

3.3 Properties of Ultrafilter Logic

Completeness transfers the finitary character of derivability $\vdash$ to compactness of semantical consequence $\vDash$. Thus, our ultrafilter logic is a proper extension of classical first-order logic with both compactness and Löwenheim-Skolem properties, a feature which confers on $\mathcal{L}_{\omega\omega}(\lambda)^{\nabla}$ an independent interest. The apparent contradiction to Lindström's theorem (see, e. g. Barwise and Feferman [1985]) is explained because we are using a non-standard notion - due to the ultrafilters - of model.

It can be seen that the notion of consequence within $\mathcal{L}_{\omega\omega}(\lambda)^{\nabla}$ has the usual properties of consequence operator [Tarski 1930], in particular, it is monotonic. Since a first-order structure can be expanded by a proper ultrafilter over its universe, we have the pleasing fact that our ultrafilter logic $\mathcal{L}_{\omega\omega}(\lambda)^{\nabla}$ is a conservative extension of classical first-order logic $\mathcal{L}_{\omega\omega}(\lambda)$.

- The logic $\mathcal{L}_{\omega\omega}(\lambda)^{\nabla}$ is a conservative extension of first-order logic $\mathcal{L}_{\omega\omega}(\lambda)$, i. e. for every set $\Sigma \cup \{\tau\}$ of sentences of $L_{\omega\omega}(\lambda)$: $\Sigma \vdash \tau$ iff $\Sigma \cup A_{\omega\omega}(\lambda)^{\nabla} \vdash \tau$. □

Since any nonempty set is included in some proper ultrafilter, the 'almost all' consequences of a pure first-order theory coincide with the universal ones.

- Given a set Σ of sentences of $L_{\omega\omega}(\lambda)$, for every formula $\varphi(x)$ of $L_{\omega\omega}(\lambda)$: $\Sigma \vdash \nabla x\varphi(x)$ iff $\Sigma \vdash \forall x\varphi(x)$. □

This result corroborates the feeling that 'almost all' requires positive information, otherwise it reduces to 'all'. This observation will become clearer in the context of generic reasoning, to which we now turn.

4. Generic Reasoning

We now wish to argue that our logic supports generic reasoning, as in the familiar Tweety example: from the facts that birds 'generally' fly and that Tweety is a 'generic' bird we can conclude that Tweety does fly. Our approach involves two steps: formulating 'generally' as ∇x and regarding 'generic' as 'prototypical'. The former - formulating "birds 'generally' fly" as $\nabla xF(x)$ - looks quite natural, in view of our interpretation of ∇x as 'holding almost universally'. The latter - regarding 'generic' as 'prototypical' - may require some explanation. How should one picture a 'prototypical' bird: with or without wings, with or without beak? We propose to interpret a 'prototypical' bird as "a bird that exhibits the properties that most birds exhibit". If we formulate "birds 'generally' fly" as $\nabla xF(x)$, and "Tweety is a 'generic' bird" as "Tweety has the properties that most birds have", in particular $\nabla x F(x) \rightarrow F(Tweety)$, then we conclude F(Tweety), i. e. "Tweety does fly".

It remains to give a precise formulation for this idea of a 'generic' object as one "possessing the properties that most objects possess". We proceed to explain how this can be done. Our approach can be understood as a symbolic form of *generic reasoning*, in that the quantifier ∇ can be used to capture precisely the meaning of 'generic' or 'prototypical' elements.

4.1 Reasoning about Generic Individuals

Given a type λ, consider an ultrafilter structure $M^\nabla = (M, \mathcal{U}^M)$ for $L_{\omega\omega}(\lambda)^\nabla$.

Given a formula $\psi(x)$ of $L_{\omega\omega}(\lambda)^\nabla$, consider the subset $M^\nabla[\psi(x)] \subseteq M$ defined by it. The idea of an element $p \in M$ being 'prototypical' for $\psi(x)$ can be expressed by $p \in M^\nabla[\psi(x)]$ whenever $M^\nabla[\psi(x)] \in \mathcal{U}^M$. Recalling that $\neg\nabla x\psi(x)$ and $\nabla x \neg\psi(x)$ are equivalent sentences, we may as well consider a more definite version.: a *generic element for* formula $\psi(x)$ is an element $g \in M$ such that $M^\nabla \models \nabla x\psi(x)$ iff $M^\nabla \models \psi(x)[g]$. We call element $g \in M$ *generic for* a set Ψ of formulae of $L_{\omega\omega}(\lambda)^\nabla$ (with single free variable x) iff g is generic for each formula $\psi(x) \in \Psi$. By a *generic* element we mean one that is generic for the set of all formulae of $L_{\omega\omega}(\lambda)^\nabla$ (with single free variable x).

Generic elements are indiscernible among themselves: they cannot be separated by formulae. In a broad intuitive sense, generic elements are somewhat reminiscent of Hilbert's ideal elements, or even of Platonic forms. So, it is not surprising that some ultrafilter structures fail to have generic elements. For instance, in the natural numbers with zero and successor and a non-principal ultrafilter, say one containing the cofinite subsets, any generic element turns out to be non-standard. This can be seen in view of the following connections.

• Consider an ultrafilter structure $M^\nabla = (M, \mathcal{U}^M)$.

1. If some generic element of M^∇ is definable then the ultrafilter $\mathcal{U}^M$ is principal and generated by a definable subset.

2. The following assertions are equivalent:

  a) the ultrafilter $\mathcal{U}^M$ is generated by a definable subset;

  b) the set of generic elements is nonempty and definable;

  c) there exists a formula $v(x)$ with $M^\nabla[v(x)] \neq \emptyset$, such that for every formula $\psi(x)$:
  $$M^\nabla \models \nabla x\psi(x) \text{ iff } M^\nabla \models \forall x[v(x) \rightarrow \psi(x)];$$

Indeed, if $\delta(x)$ defines generic g, then $[\delta(x)] = \{g\}$ generates an ultrafilter $\mathcal{U}^M$. If $\mathcal{U}^M$ is generated by $M^\nabla[v(x)]$, then $M^\nabla[v(x)] \neq \emptyset$ and $v(x)$ defines the generic elements. In this case, $M^\nabla[\psi(x)] \in \mathcal{U}^M$ iff $M^\nabla \models \forall x[v(x) \rightarrow \psi(x)]$, for every formula $\psi(x)$. The latter yields $\mathcal{U}^M$ to be generated by $M^\nabla[v(x)]$. $\square$

This result explains why this case is somewhat uninteresting: the operator ∇ reduces to universal quantifier relativised to the set of generic elements. But interestingly enough, such generic elements have much better behaviour in theories, where they can be regarded as generic witnesses.

First, in view of fip, finite sets of formulae have generic elements.

• A *finite* set Φ of formulae of $L_{\omega\omega}(\lambda)^\nabla$ has a generic element in M^∇.

Partition Φ into two sets of formulae Φ_+ and Φ_- depending on whether the subset represented or its complement is in $\mathcal{U}^M$. By fip, we have some $g \in M$ such that $M^\nabla |= \varphi(x)[g]$ for each $\varphi(x) \in \Phi_+$ and $M^\nabla |= \neg\varphi(x)[g]$ for each $\varphi(x) \in \Phi$.

We now internalise the previous ideas in an extension by a new constant.

Given a type λ and a new constant c *not* in λ, consider the extension $\lambda[c]$ of type λ obtained by adding the new constant c. The *genericity axiom* $\gamma(c/\psi(x))$ of c *for* formula $\psi(x)$ of $L_{\omega\omega}(\lambda)^\nabla$ is the *sentence* $\nabla x\varphi(x) \leftrightarrow \varphi(c)$ of $L_{\omega\omega}(\lambda[c])^\nabla$. The *genericity condition on c for* set Ψ of formulae of $L_{\omega\omega}(\lambda)^\nabla$ (with single free variablle x) is the set $\gamma(c/\Psi):=\{\gamma(c/\psi(x)): \psi(x) \in \Psi\}$. In particular, when Ψ is the set of formulae of $L_{\omega\omega}(\lambda)^\nabla$ with free variable x, we use $\gamma(c)$ for the *genericity condition on c*. These conditions extend conservatively theories in $\mathcal{L}_{\omega\omega}(\lambda)^\nabla$ to $L_{\omega\omega}(\lambda[c])^\nabla$:

• Consider sets Σ, of sentences, and Ψ, of formulae of with single free variable x in $L_{\omega\omega}(\lambda)^\nabla$. Then, $\Sigma[c/\Psi]=\Sigma \cup \gamma(c/\Psi)$ is a conservative extension of Σ (i.e., for a sentence τ of $L_{\omega\omega}(\lambda)$, $\Sigma \vdash \tau$ iff $\Sigma[c/\Psi] \vdash \tau$), such that for every formula $\psi(x)$ of Ψ: $\Sigma[c/\Psi] \vdash \psi(c)$ iff $\Sigma \vdash \nabla x\psi(x)$. □

This result establishes the correctness of reasoning with generic constants. In particular, the extension $\Sigma[c]$ of Σ by generic constant c is conservative and $\Sigma \vdash \nabla x\varphi(x)$ iff $\Sigma[c] \vdash \varphi(c)$ for every formula $\varphi(x)$ of $L_{\omega\omega}(\lambda)^\nabla$.

An example similar to flying birds and Tweety is as follows.

Example 3: (White swans)

From "Most swans are white", one concludes "a generic swan is white".In this case the type λ has unary predicate W and Σ has $\nabla xW(x)$.Considering a new constant s (for generic swan), the generic extension $\Sigma[s]$ has the genericity axiom $\nabla xW(x) \leftrightarrow W(s)$. Hence $\Sigma[s] \vdash W(s)$.

Notice that, if b is a non-white swan, in $\Sigma[s] \cup \{\neg W(b)\}$ one has both $\neg W(b)$ and $W(s)$, so $\neg b \equiv s$ (this non-white swan b is not a generic swan).

This example illustrates the monotonic nature of our logic: we do not have to retract conclusions in view of new axioms. Given that "Most swans are white", we conclude that "a generic swan is white", a conclusion which we may hold even if further evidence indicates the existence of some non-white swans.

4.2 Reasoning with several Generic Constants

The next example illustrates the usage of several generic constants.

Example 4: (Generic supporters and teams)

Assuming that "people generally oppose defeaters of each team they suppor", one would conclude that "a generic supporter of the Brazilian team opposes a generic team that defeats the Brazilian team".

Here type λ is as in example 1.

Theory Σ has axiom $\nabla x \forall y \nabla z[S(x,y) \wedge D(z,y) \rightarrow O(x,z)]$.

1. Consider a new constant p (for a generic person) and extension $\Sigma[p]$:

$\Sigma[p] \circ \forall y \nabla z[S(p,y) \wedge D(z,y) \rightarrow O(p,z)]$, so $\Sigma[p] \circ \nabla z[S(p,b) \wedge D(z,b) \rightarrow O(p,z)]$.

2. Consider new constant t (for a generic team) and extension $\Sigma[p][t]$:

$\Sigma[p][t] \circ [S(p,b) \wedge D(t,b) \rightarrow O(p,t)]$.

The kind of reasoning in the previous example involves several generic constants, which can be introduced by iterating our construction.

Given a type λ and a (denumerable) list $\underline{c} = <c_1,\dots,c_n,\dots>$ of new constants not in λ, consider the iterated extensions $\lambda_n = \lambda_{n-1}[c_n]$ of λ by new constants $c_1,\dots,c_n$ (with $\lambda_0 = \lambda$ for the case n=0), and set $\lambda[\underline{c}] := \cup_{n>0} \lambda_n$. As before, given sets Ψ_m of formulae of $L_{\omega\omega}(\lambda_m)^\nabla$, we have the genericity condition on c_{m+1} for Ψ_m

$\gamma(c_{m+1}/\Psi_m) = \{\nabla x \psi(x) \leftrightarrow \psi(c_{m+1}) : \psi(x) \in \Psi_m\} \subseteq L_{\omega\omega}(\lambda_{m+1})^\nabla$. In particular, we now use $\gamma(c_{m+1}/\lambda_m)$ for the genericity condition on c_{m+1}. The next result establishes the conservativeness of iterated extensions.

• Given a type λ and a (denumerable) list $\underline{c} = <c_1,\dots,c_n,\dots>$ of new constants not in λ, form iterated extensions $\lambda = \lambda_0 \subseteq \dots \subseteq \lambda_n = \lambda[c_1,\dots,c_n] \subseteq \dots \subseteq \lambda[\underline{c}] = \cup_{n>0} \lambda_n$. Given a set Σ of sentences of $L_{\omega\omega}(\lambda)^\nabla$ and a list $\underline{\Psi} = <\Psi_0,\dots,\Psi_m,\dots>$ of sets of formulae with $\Psi_m \subseteq L_{\omega\omega}(\lambda_m)^\nabla$ (with single free variable x), for each m≥0, we have as conservative extensions of Σ:

a) $\Sigma \cup \gamma(c_1/\Psi_0,) \cup \dots \cup \gamma(c_1/\Psi_0,\dots,c_{m+1}/\Psi_m) \subseteq L_{\omega\omega}(\lambda_{m+1})^\nabla$, for each m≥0;

b) the union $\Sigma \cup \cup_{m\geq 0} \gamma(c_{m+1}/\Psi_m) \subseteq L_{\omega\omega}(\lambda[\underline{c}])^\nabla$. □

We thus have correct reasoning with several generic constants. For each n>0, the conservative extension $\Sigma[c_1/\lambda_0,\dots,c_n/\lambda_{n-1}] = \Sigma \cup \gamma(c_1/\lambda_0) \cup \dots \cup \gamma(c_n/\lambda_{n-1})$ of Σ by n generic constants $c_1,\dots,c_n$ is such that $\Sigma \circ \nabla x_1 \dots \nabla x_n x \varphi(x_1,\dots,x_n)$ iff $\Sigma[c_1,\dots,c_n] \circ \varphi(c_1,\dots,c_n)$ for every formula $\varphi(x_1,\dots,x_n)$ of $L_{\omega\omega}(\lambda)^\nabla$.

The next example illustrates reasoning with relativised generic constants.

Example 5: (Eagles and penguins)

Assuming that "eagles generally fly" and "penguins generally do not fly", one concludes that "generic eagles fly" and "generic penguins do not fly".

Consider a type λ with unary predicates E, P and F (standing, respectively, for 'is an eagle', 'is a penguin' and 'flies'). Assume that Σ includes the following pieces of information:

1. $\nabla x[E(x) \rightarrow F(x)]$ {generally eagles fly},

2. $\nabla x[P(x) \rightarrow \neg F(x)]$ {generally penguins do not fly}.

Considering new constants and genericity axioms $\nabla x[E(x) \rightarrow F(x)] \leftrightarrow [E(e) \rightarrow F(e)]$ and $\nabla x[P(x) \rightarrow \neg F(x)] \leftrightarrow [P(p) \rightarrow \neg F(p)]$, the consequences of $\Sigma[e,p]$ include:

(e) $[E(e) \rightarrow F(e)]$ {if e is a generic eagle then e does fly},

(p) $[P(p) \rightarrow \neg F(p)]$ {if p is a generic penguin then p does not fly}.

A variation of this example may serve to illustrate why we do not have multiple extensions. A theory asserting both "generally birds fly", as $\nabla xF(x)$, and "generally birds do not fly", as $\nabla x \neg F(x)$, would be inconsistent.

The next example illustrates some aspects of the transitive behaviour of ∇.

Example 6: (Birds, wings, beaks and penguins)

Consider a universe of animals.

A. From the facts $\nabla x[B(x) \rightarrow W(x)]$ {generally birds have wings} and $\nabla x[W(x) \rightarrow F(x)]$ {generally winged animals fly}, we conclude $\nabla x[B(x) \rightarrow F(x)]$ {generally birds fly}, which is reasonable.

B. From the facts $\forall x[B(x) \rightarrow K(x)]$ {all birds have beaks} and $\nabla x[K(x) \rightarrow F(x)]$ {generally animals with beaks fly}, we conclude $\nabla x[B(x) \rightarrow F(x)]$ {generally birds fly}, which is reasonable

Now, contrast the above situations with the following case.

C. From the facts $\forall x[P(x) \rightarrow B(x)]$ {all penguins are birds} and $\nabla x[B(x) \rightarrow F(x)]$ {generally birds fly}, we conclude $\nabla x[P(x) \rightarrow F(x)]$ {generally penguins fly}, which is *not* expected.

The third conclusion in this example is strange, and it arises because 'generally' is inherited by subsets (as (ultra)filters are closed under supersets). This problem is connected to the so-called 'Confirmation Paradox' in Philosophy of Science. Each flying eagle is considered as evidence in favour of "eagles fly", whereas a non-flying non-eagle is not, even though "eagles are fliers" and "non-fliers are non-eagles" are logically equivalent.

5. Sorted Quantifiers

As mentioned, our new generalised quantifier ∇ may exhibit somewhat unexpected behaviour in some cases. We shall now examine these undesirable side effects and propose a way to overcome this difficulty. Let us reconsider our preceding example. We see that situations C and B have the same form (and B is a special case of A). So, how can one regard one conclusion - in B - reasonable and another one -

in C - unexpected? This seems to be due to a misledind reading of the formulae involved.

Indeed, both conclusions have the form $\nabla x[P(x) \to Q(x)]$ and the reading "generally P's are Q's" is quite appropriate. But we must bear in mind that what this asserts is "for most *animals* x, if $P(x)$ then $Q(x)$", i. e., given the meaning of material implication, the set $\underline{P}^c \cup Q$ is a large set of *animals*.

Our distinct attitudes towards the conclusions of B and C appears to stem from misunderstandings, namely reading $\nabla x[B(x) \to F(x)]$ as "most_*birds* fly" (rather than "most *animals* that are birds fly") and $\nabla x[P(x) \to F(x)]$ as "most *penguins* fly" (rather than "most *animals* that are penguins fly").

One can consistently hold that "all penguins are birds", "most birds fly", and "most penguins do not fly" (or even "no penguin flies"). One believes that the set of penguins, being a small set of birds, does constitute a sizeable set of exceptions to the belief that "most birds fly".

This suggests considering distinct notions of 'large'. Then, we could interpret "most birds fly" as 'flying birds form a large set of *birds*' ($B \cap F$ almost as large as $\underline{B}$), whereas most penguins fly" would mean 'flying penguins form a large set of *penguins*' ($P \cap F$ almost as large as $\underline{P}$).

5.1 Distinct Notions of 'Large' Subsets

We now take a closer look at the proposal of employing distinct notions of 'large' subsets. We shall use variations of the preceding example to introduce this idea and examine some of its features. The next example illustrates how undesired conclusions can be blocked.

Example 7: (Birds and penguins with distinct 'large' sets)

Consider the following assertions:

1. "All penguins are birds": $P \subseteq B$
2. "Most birds fly": $B \cap F \subseteq B$ is 'almost as large as' B
3. "Most penguins fly": $P \cap F \subseteq P$ is 'almost as large as' P

In the presence of the first assertion, neither 2 entails 3 (since we may even have $P \cap F = \varnothing$), nor 3 entails 2 (since $P \subseteq B$ may very well be a 'small' set of *birds*), if the different notions of large subsets are not connected.

This example illustrates the idea that we may have independent notions of 'large' subsets. If the set of penguins is not a 'large' set of *birds* ($P \subseteq B$ not 'almost as large as' B), then a set $Y \subseteq P$ may be a 'large' set of *penguins* without being a 'large' set of *birds*. It is this independence that blocks the undesired conclusion: for $T \subseteq S$, given ultrafilters $\mathcal{U}^T$, over T, and $\mathcal{U}^S$, over S, if $T \notin \mathcal{U}^S$ then, for every $Y \subseteq T$, we have $Y \notin \mathcal{U}^S$ (even for those Y in $\mathcal{U}^T$).

The next example illustrates how desirable conclusions can be derived.

Example 8: (Birds, wings and beaks with distinct 'large' sets)

A. Assume that we have the following pieces of information:

(0) "Most birds have wings": $B \cap W \subseteq B$ is 'almost as large as' B.

Consider also the following assertions:

(1) "Most winged birds fly": B∩W∩F⊆B∩W is 'almost as large as' B∩W;

(2) "Most birds fly": B∩F⊆B is 'almost as large as' B.

We probably expect that, in the presence of assertion (0), we can conclude (1) from (2). What about the converse?

B. Consider the following pieces of information:

0. "All birds have beaks": B⊆K

Consider also the following assertions:

1. "Most birds with beaks fly": B∩K∩F⊆B∩K is 'almost as large as' B∩K

2. "Most birds fly": B∩F⊆B is 'almost as large as' B

What connection is there among these assertions? We probably expect that we can conclude the last assertion from the preceding ones.

Now, the situation in case B is, as expected, a special case of that in case A (since B⊆K entails B∩K=B 'almost as large as' B). So, let us consider the former.

Let us resort to the interpretation with small sets of exceptions. We have:

(0) the set of non-winged birds is a small set of *birds*;

(1) the set of non-flying winged birds is a small set of *winged birds*;

(2) the set of non-flying birds is a small set of *birds*.

We now have distinct notions of 'small'/'large' subsets. Is a small set of *winged birds* also a small set of *birds*? Is a set of winged birds that happens to be a small set of *birds* also a small set of *winged birds*?

What we need here is a kind of coherence between these distinct notions of 'small'/'large' subsets that enables us to transfer properties..Consider a universe S and sub-universe T⊆S with notions of 'small'/'large' subsets. Assume that T⊆S is 'almost as large as' S. Then, it appears intuitively plausible that the 'small' subsets of T are the subsets of T that are 'small' subsets of S. We state a *coherence principle* for T⊆S as follows: if T is a 'large' subset of S, then for any set X⊆T: X is a 'large' subset of T iff X is a 'large' subset of S. In terms of ultrafilters, this means: for T⊆S with T∈ $\mathcal{U}$ ^{S}, $\mathcal{U}^T$={X⊆T:X∈ $\mathcal{U}^S$}.

Summing up the preceding discussion, consider T⊆S and an ultrafilter $\mathcal{U}^S$ over S. In case T∉ $\mathcal{U}^S$, $\mathcal{P}$(T)∩ $\mathcal{U}^S$=∅, and we have independent notions of 'large' subsets. In case T∈ $\mathcal{U}^S$, the relativised family {X⊆T:X∈ $\mathcal{U}^S$} is an ultrafilter over T, so we may take $\mathcal{U}^T$={X⊆T:X∈ $\mathcal{U}^S$} and enforce coherence.

5.2 A Many-sorted Approach

A many-sorted approach seems to be appropriate for this intuition of distinct notions of 'large'. The basic ideas are as follows.

We consider many-sorted types where the extra-logical symbols come classified according to the sorts of their arguments and results. Quantifiers are relativised to sorts, as expressed in the formation rules:

- for each variable x over sort s, if φ is a formula in $L_{\omega\omega}(\lambda)^\nabla$ then so are (∀x:s)φ, (∃x:s)φ and (∇x:s)φ.

Recall that a many-sorted structure assigns to each sort a nonempty set as its universe. A *many-sorted ultrafilter structure* M^∇ for $L_{\omega\omega}(\lambda)^\nabla$ assigns to each sort s of S-sorted type λ a nonempty set $M^\nabla[s]$ (its universe) and a proper ultrafilter $\mathcal{U}^\nabla[s]$ over set $M^\nabla[s]$ (giving the 'large' subsets of $M^\nabla[s]$).

Accordingly , the *ultrafilter axioms* for $A_{\omega\omega}(\lambda)^\nabla$ become sorted:

$(\nabla\exists{:}s)$ $(\nabla x{:}s)\psi\rightarrow(\exists x{:}s)\psi$

$(\nabla\wedge s)$ $((\nabla x{:}s)\psi\wedge(\nabla x{:}s)\theta)\rightarrow(\nabla x{:}s)(\psi\wedge\theta)$

$(\nabla\neg s)$ $(\nabla x{:}s)\psi\vee(\nabla x{:}s)\neg\psi$

To express that sort t is a subsort of sort s, we can resort to the following idea [Meré and Veloso 1992, 1995]: we use a unary function i from t to s together with an axiom asserting its injectivity.

The injection i from t to s, establishes a bijection between each subset $X\subseteq t$ and its image $i(X)\subseteq s$. This suggests stating our previous coherence principle for t a subsort of s as: if $i(t)$ is a 'large' subset of s, then for any set $X\subseteq t$: X is a 'large' subset of t iff $i(X)$ is a 'large' subset of s.

This principle can be formulated by the *coherence axiom schema*:

$(\nabla i{:}t\subseteq s/\varphi)$ $(\nabla y{:}s)(\exists x{:}t)y\equiv i(x)\rightarrow[(\nabla y{:}s)\varphi(y)\leftrightarrow(\nabla x{:}t)\varphi(i(x))]$.

Many-sorted logic reduces to unsorted logic with relativisation predicates. So, we can adapt our (unsorted) soundness and completeness to the sorted version of $A_{\omega\omega}(\lambda)^\nabla$. Similarly, generic elements carry over to this case.

Let us now examine our preceding example in this sorted formulation.

Example 9: (Sorted birds, wings and penguins)

Consider three sorts: b (for birds), w (for winged birds) and p (for penguins) and a unary predicate F (for flies) over sort b.

a. Consider the formulation, with $j{:}\underline{w}\rightarrow\underline{b}$.

0. $(\forall x, x'{:}\underline{w})[j(x)\equiv j(x')\rightarrow x\equiv x']$ {all winged birds are birds},

1. $(\nabla z{:}\underline{b})(\exists x{:}\underline{w})z\equiv j(x)$ {most birds are winged}.

Then, from the instance $(\nabla z{:}\underline{b})(\exists x{:}\underline{w})z\equiv j(x)\rightarrow[(\nabla z{:}\underline{b})F(z)\leftrightarrow(\nabla x{:}\underline{w})F(j(x))]$ of coherence axiom schema, we can conclude the equivalence between $(\nabla z{:}\underline{b})F(z)$ {most birds fly} and $(\nabla x{:}\underline{w})F(j(x))$ {most winged birds fly}.

We thus see that, since winged birds form a 'large' set of birds, "generally fly" is inherited both downwards and upwards.

b. Consider the formulation, with $k{:}\underline{p}\rightarrow\underline{b}$.

0. $(\forall y, y'{:}\underline{p})[k(y)\equiv k(y')\rightarrow y\equiv y']$ {all penguins are birds},

1. $(\nabla z{:}\underline{b})F(z)$ {most birds fly}.

We have the instance $(\nabla k{:}\underline{p}\subseteq\underline{b}/F(z))$ of the coherence axiom schema. So, if we had $(\nabla z{:}\underline{b})(\exists y{:}\underline{p})z\equiv k(y)$ {most birds are penguins}, we would be able to conclude $(\nabla y{:}\underline{p})F(k(y))$ {most penguins fly}; but otherwise this conclusion is not forced upon us. In fact, if we also know that $(\nabla x:\underline{p})\neg F(k(y))$ {most penguins do not fly}, then $(\nabla k:\underline{p}{\subseteq}\underline{b}/F(z))$ yields $(\nabla z:\underline{b})\neg(\exists y:\underline{p})z{\equiv}k(y)$ {most birds are not penguins}.

We now consider sorted formulations of the so-called 'Nixon example'.

Example 10: (Sorted Quakers and Republicans)

Assume that "Most Quakers are pacifist" while "Most Republicans are not pacifist". What can one conclude about a Republican Quaker?

Assume that the type λ has sorts $\underline{h}$ (for humans), $\underline{q}$ (for Quakers) and $\underline{r}$ (for Republicans) and a unary predicate P over sort h (for "is pacifist"). Assume also that "Some Quakers are Republicans", i. e. $(\exists y:\underline{q})(\exists z:\underline{r})i(y){\equiv}j(z)$.

We can then introduce a new sort $\underline{m}$ (for Republican Quakers), together with injective functions $k:\underline{m}{\rightarrow}\underline{q}$ and $l:\underline{m}{\rightarrow}\underline{r}$, as well as the intersection axiom $(\forall y:\underline{q})(\forall z:\underline{r})[i(y){\equiv}j(z){\leftrightarrow}(\exists w:\underline{m})(y{\equiv}k(w)\wedge z{\equiv}l(w))]$. Consider also a new generic constant d of sort $\underline{m}$ (a 'generic witness').

In the absence of information about relative sizes, we cannot conclude anything about the pacifist attitude to pacifism of such a Republican Quaker.

Assume that "Most Quakers are Republicans", i. e. $(\nabla y:\underline{q})(\exists z:\underline{r})i(y){\equiv}j(z)$. Then, the intersection axiom yields $(\nabla y:\underline{q})(\exists w:\underline{m})y{\equiv}k(w)$, whence the coherence principle for $\underline{m}$ a subsort of $\underline{q}$ gives $(\nabla w:\underline{m})P(i(k(w)))$ {Most Republican Quakers are pacifist}, whence $P(i(k(d)))$.

On the other hand, if we assume $(\nabla z:\underline{r})(\exists y:\underline{q})j(z){\equiv}i(y)$ {Most Republicans are Quakers}, we can conclude similarly $(\nabla w:\underline{m})\neg P(j(l(w)))$, whence $\neg P(j(l(d)))$ {a generic Republican Quaker is not pacifist}.

6. Other Applications and Extensions

We have already seen that our ultrafilter logic supports generic reasoning, as illustrated by several examples. We now briefly examine a few other applications, exploiting the expressive power of ∇, and some possible extensions.

A first possible application is to the realm of imprecise reasoning, in the spirit of fuzzy logic. Consider a fuzzy concept, such as 'very tall'. One may consider a definition as follows: a 'very tall' person is a person that is taller than 'almost' everybody (else). This indicates a manner of extracting a fuzzy predicate P from a binary relation R: defining P(x) as $\nabla y\ P(x,y)$. Dually, very_short(x) would be defined as ∇y taller(y,x). This may provide qualitative foundations for (versions of) fuzzy logic.

Another possible application would be to the area of inductive reasoning, as in empirical experiments and tests. This arises from the observation that, whereas laws of pure mathematics may be of the form "All P's are Q's", one can argue that laws of natural sciences are really assertions of the - more cautious - form "Most P's are Q's", or at least should be regarded in this manner. Here, the expressive power of ∇ may be helpful.

Consider such an empirical law $\nabla x\theta(x)$. Let us assume that $\theta(x)$ is quantifier-free - a reasonable assumption in the context of experiments. Then, $\theta(x)$ can be put into disjunctive normal form. Thus $\theta(x)$ is equivalent to a disjunction

$\delta_1(x)\vee...\vee\delta_m(x)$, so $\nabla x\theta(x)$ will be established iff some disjunct $\nabla\delta_k(x)$ is established. Such a disjunct $\delta_k(x)$ is in turn equivalent to a conjunction $\kappa_1(x)\wedge...\wedge\kappa_n(x)$. Thus, establishing $\nabla x\theta(x)$ is reduced to establishing $\nabla\kappa_1(x),...,\nabla\kappa_n(x)$, which are independent tasks, each one involving a literal (an atomic formula, perhaps negated).

Now, consider an inductive jump: having established $\varphi(x)$ for a (small) set of objects $(\forall x[\epsilon(x)\rightarrow\varphi(x)])$, one wishes $\nabla x\varphi(x)$. This would follow from $\nabla x\epsilon(x)$, but involving large experimental evidence. The case of program testing may be illustrative: one tests the behaviour of a program for a (small) set of data and then argues that the program will exhibit this behaviour in general. Here, the rationale is that the set of test data is 'representative' in the sense that it covers the possible execution paths.

This idea suggests that, using some appropriate strategy, one may be able to conclude $\nabla x\varphi(x)$ from $\forall x[\epsilon(x)\rightarrow\varphi(x)]$.

The application outlined above suggests two other interesting avenues. First, one basic idea (as above) is weakening some universal quantifiers to ∇. For instance, one might consider the concept of 'almost equivalence', obtained by weakening symmetry and transitivity and replacing reflexivity by $\nabla y\exists x\ R(x,y)$. This would give rise to an 'almost partition' as a set of blocks 'almost' covering the universe where intersecting blocks would have 'almost' the same elements. A similar weakening of some mathematical concepts, such as topological space, might be of interest. (Notice that we are not proposing a programme; one can expect that only some such weakenings - typically with qualitative flavour - will be of interest.)

The second avenue comes from the idea that we generally wish not merely a small set of experiments, but really a finite one. This suggests considering an extension of our logic to deal with finiteness. The idea is to consider another quantifier ϕ, the intended interpretation of $\phi x\varphi(x)$ being "$\varphi(x)$ holds only for finitely many elements". The semantic interpretation involves an ultrafilter structure:

$(M, \mathcal{U}^M)$ extended by the ideal I^M of the finite subsets of M. As axioms we may take the following formulae:

$(\phi\nabla)$	$\phi x\psi\rightarrow\neg\nabla x\psi$	[finite sets are not large],
$(\phi\rightarrow)$	$\forall x(\psi\rightarrow\theta)\rightarrow(\phi x\theta\rightarrow\phi x\psi)$	[subsets of finite sets are finite],
$(\phi\vee)$	$(\phi x\psi\wedge\phi x\theta)\rightarrow\phi x(\psi\vee\theta)$	[unions of finite sets are finite],
(ϕ_0)	$\neg\exists x\psi\rightarrow\phi x\psi$	[the empty set is finite],
(ϕ_n)	$\exists x_1..\exists x_n\forall x(\psi\rightarrow x\equiv x_1\vee...\vee x\equiv x_n)\rightarrow\phi x\psi$	[bounded sets are finite]..

This extension will be of more interest for infinite structures.

7. Conclusions

We have examined a logical system, capturing the intuition of 'most' by means of generalised quantifiers over ultrafilters, as a basis for generic reasoning. This monotonic logical system is a conservative extension of classical first-order logic, with which it shares several properties. The nonmonotonic approach via default rules takes a *negative view* of defaults as 'absence of information to the contrary'. In contrast, our alternative approach favours a *positive interpretation*, capturing the intuitive sense of 'most' or 'generally' (a property holds for 'most' objects when the set of exceptions is 'small') in a precise manner via ultrafilters.

This view of defaults as 'large' subsets leads to an extension of classical first-order logic by the addition of a generalised quantifier for 'almost universally'. Its semantics is given by adding an ultrafilter to a usual first-order structure. By formulating the characteristic properties of ultrafilters, we obtain an axiomatics, which is shown to be sound and complete with respect to this semantics. As a result, this ultrafilter logic turns out to be a conservative extension of classical first-order logic.

For reasoning about generic objects, we introduce the idea of 'generic' individuals as those possessing the properties that most individuals have. The addition of generic constants produces a conservative extension, where one can correctly reason about generic objects as intended.

The interpretation of 'most' as a generalised quantifier captures the idea of 'large' subsets of a given universe. Unfortunately, it fails to capture adequately the idea of 'large' subsets of sub-universes, which may lead to undesired inheritances due to misinterpretations. A many-sorted version of this ultrafilter logic is then introduced and employed to handle correctly distinct notions of 'large' subsets and inheritances.

We have also commented on some perspectives for further work. Some interesting connections with fuzzy logic and inductive and empirical reasoning suggest the possibility of other applications for our logic. Ultrafilter logic shares several properties, such as compactness and Löwenheim-Skolem, with classical first-order logic, and appears to merit further investigation. For instance, if it has interpolation, then it can be used for stepwise development of specifications [Veloso 1996].

References

Barwise, J. and Feferman, S. [1985] - *Model-Theoretic Logics*. Springer-Verlag, Berlin.

Ben-David, S. and Ben-Aliyahu, R. [1994] - A modal logic for subjective default reasoning. In *Proc. 9th IEEE Symp. on Logic in Computer Science (LICS94)*, Paris (p. 477-486).

Besnard, P. [1989] - *An Introduction to Default Logic*. Springer-Verlag, Berlin.

Besnard, P., Brewka, G., Froidevaux, C., Grégoire, E. and Siegel, P. [1991] - Nonmonotonicity. *J. Applied Non-Classical Logics*, vol. 1 (n° 2), p. 267-310.

Chang, C. C. and Keisler, H. J. [1973] - *Model Theory*. North-Holland, Amsterdam.

Hempel, C. [1965] - *Aspects of Scientific Explanation and Other Essays in the Philosophy of Science*. Free Press, New York.

Keisler, H. J. [1970] - Logic with the quantifier "there exist uncountably many". *Annals of Math. Logic*, vol. 1, p. 1-93.

Meré, M. C. and Veloso, P. A. S. [1992] - On extensions by sorts. PUC-Rio, Dept. Informática, Res. Rept. MCC 38/92, Rio de Janeiro, December 1992.

Meré, M. C. and Veloso, P. A. S. [1995] - Definition-like extensions by sorts. *Bull. IGPL*, vol. **3** (n° 4), p. 579-595.

Reiter, R [1980] - A logic for default reasoning. *J. Artificial Intelligence*, vol. **13** (n°1), p. 81-132.

Schlechta, K. [1995] - Defaults as generalised quantifiers. *J. Logic and Computation*, vol. **5** (n° 4), p. 473-494.

Sette, A. M., Carnielli, W. A. and Veloso, P. A. S. [1997] - An alternative view of default reasoning and its logic. Submitted for publication.

Tarski, A. [1930] - Fundamentale Begriffe der Methodologie der deduktiven Wissenschaften. *Monatshefte für Mathematik und Physik*, vol. **37**, p. 361-404 {English translation in Woodger, J. H. (ed.) *Logic, Semantics and Metamathematics*; Oxford, 1956, p. 60-109}.

Veloso, P. A. S. [1996] - On pushout consistency, modularity and interpolation for logical specifications. *Inform. Process. Letters*, vol. **60** (n° 2), p. 59-66.

Informal Rigor and Mathematical Understanding

J.A. Robinson

Syracuse University
Syracuse, New York
robinson@equinox.shaysnet.com

1 Introduction

Even the proponents of the "formalist" view (that it is derivations within formal systems which are the most rigorous mathematical proofs to which we can aspire) will sometimes concede that an unformalized proof can be rigorous. However, they believe that this simply means that the unformalized proof could (in principle, if not in practice) be completely formalized. Formalizing it presumably involves writing it out explicitly in the notation of (say) the first order predicate calculus as a logically correct formal derivation from (say) the axioms of set theory.

This explanation of the rigorousness of rigorous unformalized proofs amounts to saying that informal proofs really are, so to speak, no more than sketches or outlines of formal proofs. But on closer examination this view seems unsatisfactory, and is rejected by most mathematicians.

In actual mathematical work, formal proofs are rarely if ever used. Moreover, the unformalized proofs which are the common currency of real mathematics are judged to be rigorous (or not) directly, on the basis of criteria which are intuitive and semantic – not simply based on syntactic form alone. Although construction of a corresponding formal proof is rarely in practice undertaken, one sometimes attempts it anyway, if only for the sake of the exercise, or perhaps for the sake of submitting it to a computer proof-checking system. Formalization of a given informal proof then often turns out to be surprisingly difficult. The translation from informal to formal is by no means merely a matter of routine. In most cases it requires considerable ingenuity, and has the feel of a fresh and separate mathematical problem in itself. In some cases the formalization is so elusive as to seem to be impossible.

Nor is it always obvious that a given formalization is correct. There is no formal criterion for judging the correctness of a formalization – indeed, how could there be? The judgement is necessarily intuitive, and it is not at all clear why the intuition should be granted the final word here when it is deemed untrustworthy as an arbiter of the validity of the proof itself.

There is an analogous problem concerning the analysis and formal definition of concepts. How, for example, do we know that a given intuitive concept has been adequately captured (formalized) by a proposed formal definition? Again, the judgement must be based on the intuitive understanding one already has (prior to the definition) of the concept.

For example, consider the usual formal definition of the property of "infinite" as applied to sets. The definition is this: a set is infinite if and only if there is a one-to-one mapping of the set onto one of its proper subsets.

To illustrate this definition, we already know intuitively that the set J of positive integers $1, 2, 3, \ldots$, is infinite. The definition works for J because (for example) there is indeed a one-to-one mapping of J onto $J - 1 = 2, 3, 4, \ldots$, namely, the successor function. One might say that such a mapping witnesses the infinity of the set.

That is clever and neat: one should have no hesitation in accepting the definition as stating a fact about all infinite sets. But is this what we actually mean by an infinite set? Does this definition capture fully, or indeed at all, our intuitive understanding of the very notion? This is quite doubtful. The formally defined notion is rarely used in practice. It is our rather strongly intuitive informal or primitive notion which in practice is used.

For example, on the basis of the informal notion, the following fact is completely obvious: if an infinite set is partitioned into two disjoint subsets, then at least one of those subsets must also be an infinite set. However, when one translates this result according to the formal definition, it is not at all obvious and indeed is very difficult to prove. I have tried to prove it, but have not yet succeeded. It may of course be possible to prove it (for one more ingenious than I) but my point is that the formal definition (for me, and I suspect for most people) destroys the obviousness of elementary facts involving the informal notion.

Consider briefly the problem one faces in looking for such a proof. We noted that successor function witnesses the infinity of J. But if we partition J into the disjoint sets $A = 1, 3, 5, \ldots$, and $B = 2, 4, 6. \ldots$, we find that the successor function sends A into B and B into a proper subset of A. So although both A and B are (intuitively, and obviously) infinite, the successor function witnesses the infinity of neither A nor B. This merely shows that it is not in general true that if f witnesses the infinity of S and if $S = A + B$, then f witnesses the infinity of A or of B. The formal translation of the "obvious" fact says that if there is a f which witnesses the infinity of S, and if $S = A + B$, then there exists a function g which witnesses the infinity of A or that of B. We have just seen that we cannot in general take g to be f. What, then, in general, is g? What has happened to the obviousness of the obvious fact?

This phenomenon (catastrophic loss of obviousness) is reminiscent of the notorious complexity and difficulty of the formal Jordan Curve Theorem compared with the obviousness of the informal version. It is intuitively obvious that a closed curve separates the plane into two regions, the inside and the outside, and that a line joining a point inside the curve to a point outside the curve will cross the curve. However, once the intuitive notions are replaced by formal definitions, the obviousness disappears and the theorem becomes a difficult one to prove.

The outstanding pragmatic, epistemic, or cognitive (call them what you will) discrepancies between an informal proof and its formal counterpart (if it has one) are the loss of obviousness and the lack of explanatory power and intuitive ideas in formal proofs.

I believe that these discrepancies account for why working mathematicians prefer to use their unformalized intuitions in constructing proofs – because it is only by so doing can they achieve their main objective, that of mathematical understanding. Saunders Mac Lane expresses this preference as follows:

> ...there are good reasons why mathematicians do not usually present their proofs in fully formal style. It is because proofs are not only a mean to certainty, but also a means to undertanding. Behind each substantial formal proof there lies an idea, or perhaps several ideas. The idea, initially perhaps tenuous, explains why the result holds. ...Proofs serve both to convince and to explain - and they should be so presented. ([2], p.377)

One might summarize this double function of real proofs by a whimsical conceptual equation in the style of Wirth and Kowalski:

$$\text{PROOF} = \text{GUARANTEE} + \text{EXPLANATION}.$$

Formalizing a proof (assuming that it can be done) may well assure us, by a sort of symbolic computation, that its conclusion must be true if its premises are. Thus formalizing a proof may help to fulfil its role as a *guarantee*. But it is dreadful to contemplate the effort, let alone the enormous and cognitively useless complexity of the result, of formalizing and checking a big deep proof such as Professor Wiles' recent proof of Fermat's Last Theorem. It would of course be an excellent and welcome thing to be able to know that the Wiles proof is correct and has no logical gaps.

Does anyone believe that constructing a formal translation of his proof would accomplish that?

The problem is fundamentally that formalizing a proof has nothing whatsoever to do with its cognitive role as an *explanation* - indeed, it typically destroys all traces of the explanatory power of the informal proof.

This is borne out by the fact that much mathematical work consists of seeking to understand and appreciate theorems which have already been proved, and to organize the ideas so as to prove them in new and better ways. As Mac Lane puts it:

> ...there is a continual search to get better proofs of known theorems - not just shorter proofs, but ones which do more to reveal *why* the theorem is true. ([2], p 432)

The working mathematical brain – Brouwer's "thinking subject" – is primarily a proof-understanding cognitive system. It is this thinking subject that logicians ought to be seeking to understand better. We will make progress in this direction only if we try learn more about the second half of the Mac Lane equation: how proofs work as explanations. How might we do this?

Wittgenstein suggested that much of the force of a good proof is often found in the side comments and "stage business" that accompany it:

Pay attention to the patter by which we convince someone of the truth of a mathematical proposition. It tells us something about the function of this conviction. I mean the patter which awakens intuition. ([3], p. 121)

One way to proceed, then, would be to perform what might be called introspective cognitive experiments, using actual proofs together with the accompanying patter. To perform such an experiment, one would go through an intuitive, rigorous proof which has a high degree of explanatory power, and by introspection try to observe how it makes its effect on the mathematical thinking subject, i.e., on one's own knowledge and understanding. Put more simply, we try to watch the proof doing its cognitive work.

Of course, one wants to share these experiments with others through communication. Brouwer seems to have believed that this is not possible, that mathematical thought and experience are essentially private and uncommunicable. I do not agree with him on this.

Let us now carry out, as best we can, two little introspective cognitive experiments.

2   First experiment

This is a simple illustration of the power of our mental system of elementary "wired in" spatial perceptions in two dimensions. They are the means by which the experience of becoming rigorously convinced of a nontrivial and indeed surprising geometrical fact actually takes place. In the following account I reproduce as faithfully as I can the nature of my own actual experience which I introspectively noted on first studying the proof of this theorem as given by Kac and Ulam.

Theorem 1. *It is possible to decompose a square into finitely many subsquares, all of different sizes. ([1], p. 20)*

Ulam and Kac could hardly have chosen a more convincing way to prove this theorem than to exhibit such one such decomposition explicitly. Even though there may be more than one (they do not discuss this possibility), only one is needed in order to show that it can be done.

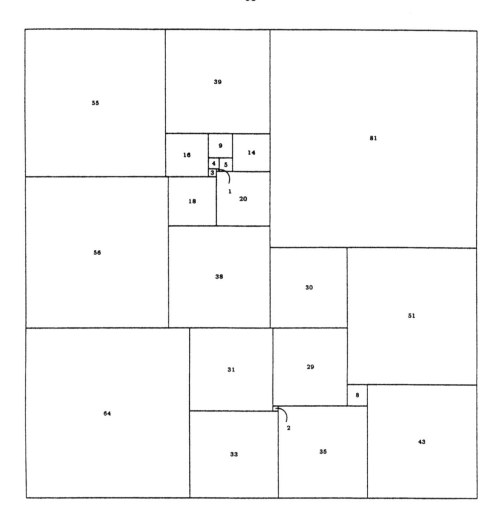

The accompanying diagram is a representation of such a decomposition. Is this part of the patter? Probably Wittgenstein would consider as patter the gestures and pointing that the expositor uses to draw attention to various details in the diagram. The diagram shows twenty three differently-sized squares positioned relative to each other in a certain layout, and gives their relative sizes. The size of each square is considered to be the common length of its four sides, expressed as an integral multiple of the size of the smallest square.

So, in outline, the proof is: "Here, look at this diagram".

We are supposed to study the details shown by the diagram, and thereby arrive at the conviction that it really does represent a decomposition with the required properties.

Running one's eye over the the diagram gradually (but by no means instantaneously) convinces one that it does indeed represent a decomposition of the required kind. But what is it that one's eye sees, in the course of this process? It is not entirely straightforward to put this into words, because at least some of the information that the eye picks up is what it is accustomed to detecting without our supplying any accompanying verbal description. The two-dimensional pattern of linked adjacencies between the various squares, comprising their spatial configuration, is easy to observe but awkward to describe.

It is interesting to speculate how necessary a visual diagram is in conveying the essence of this decomposition of the square. For the visually impaired person, a tactile diagram could provide essentially the same aid to cognition. Spatial relationships are certainly as capable of being perceived by touch and kinaesthesia as by vision, if not quite as rapidly and efficiently. But could one appreciate and understand this decomposition from a bare description alone, with no sensory aids at all? This would be more in keeping with the way in which traditional ideas about formalization would call for the proof to be given. It seems very doubtful that conveying the same information as the diagram conveys could be done in anything like as direct and assimilable a fashion.

In the second experiment, described below, we try to digest a proof by Ulam and Kac dealing with three-dimensional decompositions without the benefit of a diagram.

The geometrical or topological knowledge of the eye, virtually wired in by years of spatial observations and by their correlations with tactile and kinaesthetic perceptions, can deal with a spatial structure either directly in real time, so to speak, or else off-line, in the imagination.

In the present example what the eye sees, loosely speaking, is that a straight horizontal line drawn across the big square, avoiding coincidence with any of the horizontal edges of the subsquares, would pass successively through what might be called a "horizontal spanning chain of subsquares", a chain which bridges the width of the big square, connecting its left edge to its right edge.

By mentally running through all the different such spanning chains, one sees that adding together the lengths of the successive subsquares in a chain always yields the name total length, that of the width of the big square. Eventually one sees, by persisting in the mental survey, that one has exhausted the possibilities – that there are in fact only twelve different such horizontal spanning chains of subsquares.

The realization that these twelve exhaust the various possible spanning chains is the outcome of something like the following mental process. One mentally slides the horizontal line down the square, from top to bottom, noting the changes in the spanning chains through which it passes, as they occur. These changes are, so to speak, discontinuities which occur only at the points where the sliding line coincides with a horizontal edge of one or more of the subsquares. Thus the line initially passes through the top three subsquares, the ones which the diagram shows as having sizes 55, 39, 81.

As the line descends further, it encounters the bottom edge of the size 39 subsquare, whereupon the chain of squares through which the line passes changes to the chain of subsquares of sizes $55, 16, 9, 14, 81$. Continuing downward in this manner the line runs through all the changes and reaches the final (bottommost) chain, namely the chain of subsquares whose sizes are $64, 33, 35, 43$.

The entire sliding-line survey reveals the following complete list of twelve different ways of seeing the width (175) of the big square as the sum of the widths of the subsquares in a horizontal spanning chain. It is interesting to note that the mental work has also produced the conviction that these are all the horizontal spanning chains there are:

$$
\begin{aligned}
175 &= 55 + 39 + 81 \\
&= 55 + 16 + 9 + 14 + 81 \\
&= 55 + 16 + 4 + 5 + 14 + 81 \\
&= 55 + 16 + 3 + 1 + 5 + 14 + 81 \\
&= 55 + 16 + 3 + 20 + 81 \\
&= 56 + 18 + 20 + 81 \\
&= 56 + 38 + 81 \\
&= 56 + 38 + 30 + 51 \\
&= 64 + 31 + 29 + 51 \\
&= 64 + 31 + 29 + 8 + 43 \\
&= 64 + 31 + 2 + 35 + 43 \\
&= 64 + 33 + 35 + 43
\end{aligned}
$$

A precisely analogous mental survey using an imaginary sliding vertical line reveals that there are exactly thirteen different vertical spanning chains, yielding the following complete list of ways of seeing the height of the big square as a sum of the heights of the subsquares in a vertical spanning chain. Again the work done produces the conviction that these are all the vertical spanning chains there are:

$$
\begin{aligned}
175 &= 55 + 56 + 64 \\
&= 39 + 16 + 56 + 64 \\
&= 39 + 16 + 18 + 38 + 64 \\
&= 39 + 16 + 18 + 38 + 31 + 33 \\
&= 39 + 9 + 4 + 3 + 18 + 38 + 31 + 33 \\
&= 39 + 9 + 4 + 1 + 20 + 38 + 31 + 33 \\
&= 39 + 9 + 5 + 20 + 38 + 31 + 33 \\
&= 39 + 14 + 20 + 38 + 31 + 33
\end{aligned}
$$

$$= 81 + 30 + 31 + 33$$
$$= 81 + 30 + 29 + 2 + 33$$
$$= 81 + 30 + 29 + 35$$
$$= 81 + 51 + 8 + 35$$
$$= 81 + 51 + 43.$$

The information uncovered by this mental scrutiny of the pattern of horizontal and vertical spanning sequences of subsquares, in recognizing the topological connectivity map of adjacencies among the subsquares, and in carrying out the addition operations on the various sets of subsquare sizes, is substantial. Yet amassing this information is, at the conscious level at least, cognitively primitive.

The brain is obviously very busy on the task over a fairly long period of time. There is a nonnegligible amount of mental and visual work involved in scanning, connecting, detecting adjacencies and intersections, keeping running totals of partial chain lengths, and in general running through all the many relevant details. The total job can be done rapidly, but not instantaneously. The mental arithmetic required in checking the sums of sizes of subsquares in each spanning chain, is simple and intuitive but not effortless. One simply cannot perform all elementary mental arithmetic tasks at a glance.

However, it is not necessary for mental operations to be instantaneous, or even very fast, in order to be cognitively primitive. They need only be automatic and basic. They can even be assisted by augmenting one's limited mental storage capacity by the use of external scratch-pad memory.

It is, for example, immediate that the product of 8473926374847283 and 11928374651928347 is 101080168572035398661733583631201 even though it takes considerable time to carry out the multiplication work. The immediacy is epistemological. The verification (or "proof") of the corresponding equation

$$8473926374847283 \times 11928374651928347 = 101080168572035398661733583631201$$

is as epistemologically primitive as that of the identity

$$101080168572035398661733583631201 = 101080168572035398661733583631201,$$

which is simply an instance of the axiom "$x = x$". The work needed to check that it is an instance of "$x = x$" is nonnegligible, but that does not alter the situation. The comparison of two strings, and the multiplication of two integer numerals coded as strings of decimal digits, are mental operations that have been "wired into" our brains by education and cultural inheritance. That they take extended time to perform is no reason to disavow them as part of the basic (bottom) level of our cognitive machinery.

What we should notice about the square decomposition example is how inappropriate a formalization would be of the content of our cognitively primitive appraisal of its "visible" properties. I maintain that not only would it be very difficult to state formally what it is we see in checking out the decomposition diagram, but that it would badly distort –indeed, that it would destroy– the cognitive effect of displaying the diagram itself for an intuitive, unformalized scrutiny.

3 Second experiment

It is highly illuminating to consider next, in a similar introspective fashion, the three-dimensional analog of the theorem in our first experiment. Instead of considering the decomposition of a square into finitely many subsquares of different sizes, we consider the decomposition of a cube into finitely many subcubes of different sizes. By the size of a cube we shall here mean the common length of its twelve edges. This time, the theorem to be proved is negative.

Theorem 2. *It is impossible to decompose a cube into finitely many subcubes all of different sizes. ([1], pp. 20 -21)*

The rigorous proof of an impossibility theorem would seem to call for a different approach from that which worked so straightforwardly in the case of the positive possibility result we examined in the first experiment. We of course cannot in this case proceed by displaying a diagram representing a decomposition. We must somehow be brought to see that (and more importantly, to see why) there is, and can be, no such decomposition.

The time-honored proof technique appropriate here is to assume the negation of the theorem and deduce a contradiction from it.

Suppose: It is possible to decompose a cube into finitely many subcubes all of different sizes.

A "proof by contradiction" does not have to be formal. We can deploy much the same intuitive perceptual and cognitive machinery as was done in the positive example. As we shall see, in this case we must summon up our mental ability to imagine two- and three-dimensional structures without the use of diagrams, and to recognize their properties by scrutinizing the images in the mind's eye.

Picture, then, the inside of the cube which is to be filled with finitely many smaller subcubes, all of different sizes. It is like being inside a cubic room. Fix attention on one of the faces, and think of it as the bottom face, or floor. Think of the four faces sharing an edge with the floor, as the four walls.

Among the subcubes which are going to go on the floor (let us call them "1-subcubes", suggesting the first, or bottom, layer of subcubes) there is a smallest one, and we fix our attention on it. What can we say about the place where it will go on the floor?

Well: it cannot go in a bottom corner of the cube, where the floor forms a solid right angle with two walls. Why not? Well, if it were in such a corner, two of its four vertical faces would be flat against two walls, leaving the other two vertical faces to be covered by two other 1-subcubes. But both of these other 1-subcubes would be (necessarily) larger than the smallest 1-subcube in the corner. Each of them will need to protrude into the room beyond the inner vertical edge of the smallest 1-subcube and so will get in each other's way.

Nor can the smallest 1-subcube go on the floor against a wall but away from the corner. For then one of the four vertical faces of the smallest 1-subcube will be flat against the wall, leaving three to be covered by three (necessarily larger) 1-subcubes. But these larger 1-subcubes also would get in each other's way.

Both of these little space-filling impossibilities are reasonably easy to imagine. If one mentally tries to fit the larger cubes around the smallest one, it is intuitively quite obvious why this cannot be done.

The only possible place, therefore, where the smallest 1-subcube can go on the floor of the cubic room is one that is somewhere out in the middle, away from the walls, so that all four of vertical faces are covered by four other (larger) 1-subcubes. These four neighbors can cover the vertical faces of the smallest 1-subcube by protruding beyond it to the north, east, south and west, as it were.

So we can now picture the smallest 1-subcube and its relationship to its neighbors. It sits at the bottom of a square hole formed by the four vertical faces of the four larger 1-subcubes which are its immediate neighbors to the north, east, south and west. In your mental picture you will observe that the vertical faces of each of the four surrounding 1-subcubes all rise above the top face of the smallest 1-subcube, and thus together form the walls of a sort of "well" whose flat square bottom is the top face of the smallest 1-subcube.

This flat-bottomed well, being part of the interior space of the big cubic room, will itself be filled with subcubes. But we can see that this reproduces almost exactly, except for the scale, the situation we were picturing when we considered the floor of the big cube itself. Just as the 1-subcubes cover the floor, so the 2-subcubes cover the square top face of the smallest 1-subcube.

Just as we fixed our attention on the smallest 1-subcube, we now fix our attention on the smallest 2-subcube. Precisely the same reasoning leads us to see that the smallest 2-subcube must be placed out in the middle somewhere of the floor of the well, that is to say the top face of the smallest 1-subcube.

But we can as before see that the top face of the smallest 2-subcube is the square bottom of a well. Covering its top face is a set of 3-subcubes, among which we fix our attention on the smallest 3-subcube, ..., and so on, and so on.

We can repeat this reasoning indefinitely. This fact (which we recognize in the intuitive meta-judgement that an action once performed can be performed again) means that any decomposition of the cube into subcubes of different sizes will contain an infinite vertical chain of successively smaller subcubes, namely the chain whose ith member is the smallest i-subcube, for $i = 1, 2, 3, \ldots,$.

Since we supposed that there are only finitely many subcubes in the decomposition, this is impossible. We therefore conclude that the supposition is false. So we see that there is no decomposition of a cube into finitely many unequal subcubes, and what is more we see why.

Note that the preceding argument does not establish that there is a decomposition of a cube into infinitely many unequal subcubes. It shows only that if a cube were to be decomposed into subcubes all of different sizes, there would be infinitely many of them. It remains possible (as far as this argument is concerned) that there is no decomposition at all of a cube into subcubes all of different sizes.

4 Conclusion

These two introspective cognitive experiments have shown how the rational imagination can be a satisfactory mental arena for the conduct of rigorous mathematical thinking. The diagrams and other representative imagery used in such thinking need not be external and public, but are capable of being privately constructed, viewed, and analyzed "in the mind's eye".

The conclusion I draw from this and other similar introspective cognitive experiments with actual proof scenarios is that such examples of rigorous but unformalized mathematical thinking provide a rich source of material for the development of a more authentic model of mathematical reasoning and communicating results than the more limited model offered by the rigid formal systems of traditional mathematical logic.

It seems clear that in any such realistic model there will have to be a central role for the thinking subject. That is to say, we must explicitly include features in the model which capture the way the intuition and imagination deal directly (without the mediation of formal definitions) with spatial and temporal relationships, with order relations, with comparison and combination of quantities, and with elementary manipulations of symbols.

Understanding the rules and conventions involved in the engineering and operation of formal systems of deduction and computation is itself an example of intuitive and informal mathematics, just as Hilbert long ago stressed in explaining the ideas behind his metamathematics.

In the end, we are left with the question: how does the mathematical mind actually function? To answer, we must watch real minds in action, rather than guessing how idealized minds would work if they existed.

References

1. Kac, M. and Ulam, S. M. Mathematics and Logic. Dover, 1992.
2. Mac Lane, S. Mathematics: Form and Function. Springer-Verlag, 1986.
3. Wittgenstein, L. Remarks on the Foundations of Mathematics. Blackwell, 1956.

Resolution, Inverse Method and the Sequent Calculus

Tanel Tammet

Department of Computer Sciences,
University of Göteborg and Chalmers University of Technology,
41296 Göteborg, Sweden
email: tammet@cs.chalmers.se

Abstract. We discuss the general scheme of building resolution calculi (also called the inverse method) originating from S. Maslov and G. Mints. A survey of resolution calculi for various nonclassical logic is presented, along with several common properties these calculi possess.

1   Introduction

The generic resolution method (also called the "inverse method") originally developed by S. Maslov and G. Mints (see [6, 7]) is a forward-chaining proof search method. Search starts with the set of axioms and produces new sequents from the already derived ones by applying the sequent calculus rules in the "downwards" direction, until the formula we want to prove is eventually derived.

Differently from leaves of a backward-chaining tableau search tree, the sequents derived during the forward-chaining search are independent of each other. Substituting into a variable x in a sequent Γ does not change the value of any variable in any other sequent. The resolution method is generally characterised as a *local method.*

In order to distinguish the generic resolution method (inverse method) from the specific well-known resolution method for classical logic introduced by J. Robinson in [11], we refer to the latter as *Robinson's resolution.*

Definition 1. A *resolution method* for proving a sequent S in some sequent calculus enjoying a subformula property is a forward-chaining proof search method for this calculus with the additional restriction: any derived sequent must contain only subformulas of S.

The resolution method is obviously complete for any sequent calculus enjoying a subformula property, like classical or intuitionistic logic.

In the following we will consider resolution calculi for several logics along with suitable search strategies strategies and general properties of the calculi. Detailed analysis of resolution for intuitionistic logic can be found in [9, 14], for linear logic in [8, 13], for modal logic in [15, 10]. Descriptions of implemented provers can be found in [13, 14, 15].

Different methods for constructing resolution calculi for intuitionistic and multiple-valued logics are presented, correspondingly, in [4] in [1].

1.1 Resolution vs The Tableau Method

By the "tableau method" we mean backward-chaining proof search where the sequent calculus rules are applied bottom-up (observe that premisses are written above the line and the conclusion below the line). Search starts with the formula to be proved and branches backwards using the sequent calculus rules in a bottom-up manner. It is assumed that the quantifier rules are not applied as is – metavariables are used instead (see [12]).

The common feature of tableau methods is that due to the use of metavariables the choices done in the search tree have a global effect: a substitution σ instantiating a (meta)variable x in one branch also changes all the other branches where x occurs. Therefore we have to use backtracking through the proof tree when searching for the proof.

For linear logic similar global effects do occur already on the propositional level. Tableau methods are generally characterised as *global methods*.

1.2 Terminology

For the basic terminology of resolution (term, atom, literal, clause, substitution, most general unifier (denoted *mgu*)) see e.g. [3, 2].

Let us fix terminology concerning Gentzen-type systems (sequent calculus). In each inference rule $\frac{P_1 \; P_2 \; \cdots \; P_n}{C}$ the sequent C written under the line is the *conclusion*, and the sequents $P_1, \ldots, P_n$ over the line are *premises*. The formula constructed by the rule and shown explicitly in the conclusion is the *main formula*, the components of the main formula shown explicitly in the premises are *side* formulas, and the remaining formulas are *parametric* formulas.

We use the cut-free Gentzen-type system (sequent calculus) as the basic formalism for representing the rules of logic. The objects derivable in logic are *sequents*.

2 Building a resolution calculus

2.1 A Suitable Sequent Calculus

Recall the structural rules *weakening* and *contraction*:

$$\frac{\Gamma \vdash \Delta}{A, \Gamma \vdash \Delta}\text{weakening} \qquad \frac{A, A, \Gamma \vdash \Delta}{A, \Gamma \vdash \Delta}\text{contraction}$$

When using forward reasoning for proof search it is important to minimize the number of weakening applications and maximize the number of contraction applications. It is also important to limit the number of axioms. In particular, we do not want to use an axiom schema which can produce axioms of unrestricted size:

$$A, \Gamma \vdash A, \Delta$$

As an example we bring the intuitionistic sequent calculus GJ' from [9] which avoids explicit applications of structural rules. Part of the following material considering the intuitionistic logic is adapted from [14].

We present a modification GJm of GJ':

Logical axioms. $A \vdash A$ for any atom A.

Inference rules.

$$\frac{A, \Gamma \vdash D}{(A \& B, \Gamma) \vdash D} \& \vdash \qquad \frac{\Gamma \vdash A \quad \Sigma \vdash B}{(\Gamma, \Sigma) \vdash A \& B} \vdash \& \qquad \frac{A, \Gamma \vdash}{\Gamma \vdash A \Rightarrow B} \vdash \Rightarrow'$$

$$\frac{B, \Gamma \vdash D}{(A \& B, \Gamma) \vdash D} \& \vdash \qquad \frac{\Gamma \vdash A \quad B, \Sigma \vdash D}{(A \Rightarrow B, \Gamma, \Sigma) \vdash D} \Rightarrow \vdash \qquad \frac{A, \Gamma \vdash B}{\Gamma \vdash A \Rightarrow B} \vdash \Rightarrow$$

$$\frac{\Gamma \vdash B}{\Gamma \vdash A \Rightarrow B} \vdash \Rightarrow'' \qquad \frac{A, \Gamma \vdash}{\Gamma \vdash \neg A} \vdash \neg \qquad \frac{\Gamma \vdash A}{\Gamma \vdash A \lor B} \vdash \lor$$

$$\frac{\Gamma \vdash A}{(\neg A, \Gamma) \vdash} \neg \vdash \qquad \frac{A, \Gamma \vdash D \quad B, \Sigma \vdash D}{(A \lor B, \Gamma, \Sigma) \vdash D} \lor \vdash \qquad \frac{\Gamma \vdash B}{\Gamma \vdash A \lor B} \vdash \lor$$

$$\frac{A[t], \Gamma \vdash D}{(\forall y A[y], \Gamma) \vdash D} \forall \vdash$$

$$\frac{\Gamma \vdash A[t]}{\Gamma \vdash \exists x A} \vdash \exists \qquad \frac{\Gamma \vdash A[y]}{\Gamma \vdash \forall x A} \vdash \forall \ (*)$$

$$\frac{A[y], \Gamma \vdash D}{(\exists x A[x], \Gamma) \vdash D} \exists \vdash \ (*)$$

where $(*)$ denotes the eigenvariable condition: y does not have any free occurrences in Γ or D. D stands for a single atom or no atom at all. Exchange rule is implicit. If Γ and Σ are lists of formulas, then (Γ, Σ) is the result of concatenating them and contracting repetitions modulo names of bound variables. The following theorem is proved easily:

Theorem 2. *A closed sequent $\vdash F'$ is derivable in GJm iff the sequent $\vdash F$ is derivable in GJ'. F' is obtained from F by replacing subformulas $A \Rightarrow \bot$ with $\neg A$ and renaming bound variables. F' may not contain $\bot$.*

2.2 Labelling

One of the main ideas of the general resolution framework for logics with a subformula property (eg. classical, intuitionistic and linear logics) in [7], is to label subformulas of any investigated formula F with new atomic formulas in order to reduce the depth of a formula. Since labelling atomic formulas cannot reduce the depth of the formula, atomic formulas are usually not labelled.

A formula $(A \star (B \circ C))$ where $\star$ and $\circ$ are arbitrary binary connectives, can be labelled as $(A \star (B \circ C)_{L_1})_{L_2}$ and the derivability of the labelled formula can generally be encoded as

$$(B \circ C) \Rightarrow L_1, L_1 \Rightarrow (B \circ C), (A \star L_1) \Rightarrow L_2, L_2 \Rightarrow (A \star L_1) \vdash L_2$$

in the two-sided sequent calculus, provided that $\Rightarrow$ is an implication. It is possible to keep only one of the defining implications $(\ldots \Rightarrow L)$ and $(L \Rightarrow \ldots)$ for each label L.

In case of predicate calculus it is useful to label subformulas S with the atoms formed by taking a new predicate (say, P) and giving it the set of free variables $x_1, \ldots, x_n$ in S as arguments: $P(x_1, \ldots, x_n)$.

Example 1. Consider the formula F_1:

$$(\forall x P(x, b)) \,\&\, (\forall x \forall y (P(x, y) \Rightarrow B(x, y))) \Rightarrow \forall x \exists y (B(x, y) \vee P(y, x))$$

We label all the nonatomic subformulas of F_1. The following is a labelled form of F_1, with the labels attached to leading connectives of subformulas for better readability:

$$(\forall_{L_1} x P(x, b)) \,\&_{L_7}\, (\forall_{L_3} x, y(P(x, y) \Rightarrow_{L_2(x,y)} B(x, y)))$$
$$\Rightarrow_{L_8}$$
$$\forall_{L_6} x \exists_{L_5(x)} y (B(x, y) \vee_{L_4(x,y)} P(y, x))$$

2.3 Instantiating Derivation Rules

We are now going to present the second main idea of the generic resolution method proposed in [6] and developed in [7], [9]: starting the search with maximally general axioms and building unification into derivation rules. Unification is also the essential idea behind Robinson's resolution.

Let F be a formula we are trying to prove. We are interested in finding an efficient implementation for the resolution restriction of sequent calculus: any derived sequent must contain only subformulas of F. Using standard sequent rules and throwing away all the derived sequents which do not satisfy the restriction will be wasteful. Instead, we will create a new instantiated sequent calculus for F. Since each label in the labelled form of F corresponds to a subformula, and this subformula has a certain leading connective, we can create a set of instances of ordinary sequent rules in which allow us to derive the labels (not subformulas themselves) directly. Every sequent derived in this instantiated calculus satisfies the resolution restriction.

Each occurrence of a label has a mapping to the set of instances of the rules in sequent calculus corresponding to the leading connective of the labelled subformula.

Rule Derivation Algorithm RR Consider a subformula S of F in the abstract form: $C(A_1, \ldots, A_n)$ where C is the leading connective and $A_1, \ldots, A_n$ are argument formulas. By $label(X)$ we denote the label of an arbitrary subformula X. Instances of the sequent rules corresponding to $label(S)$ are built from the sequent rules $R_1, \ldots, R_m$ for the connective C in the following way. For each rule R_i replace the main formula of the rule by $label(S)$. Replace the modified conclusion $\Gamma \vdash B$ of the rule by $(\Gamma \vdash B)\sigma$ and add the side condition: σ is obtained by unifying all the side formulas in the modified rule with the corresponding labels from the set $label(A_1), \ldots, label(A_n)$. The eigenvariable condition (y does not occur freely in the conclusion G) is translated as: $y\sigma$ is a variable and $y\sigma$ does not occur in the substitution instance $G'\sigma$ of the modified conclusion G'.

The following *polarity optimisation* is obvious: remove these rules which introduce $label(S)$ with a polarity different from the polarity of S in F.

Axiom Derivation Algorithm RA The set of possible axioms for the formula F is obtained by taking one axiom of the most general form $P(x_1, \ldots, x_n) \vdash P(x_1, \ldots, x_n)$ for each such predicate symbol P in F which has both a negative and a positive occurrence in F. Completeness is preserved if we use *axiom instantiation*: form the set of axioms by taking a set of instances instead: for every positive occurrence of an atom A and every negative occurrence of an atom B form the axiom $(B \vdash A)\sigma$ where $\sigma = mgu(A, B)$.

Proof Search Algorithm RD Proof search is carried out by applying the derivation rules obtained by the algorithm RR to both the axioms obtained by RA and the sequents derived earlier in the search. The proof is found if the sequent $\vdash label(F)$ has been derived, where $label(F)$ is the label of the whole formula F. Before applying any derivation rule, all the variables in the premisses are renamed, so that the sets of variables in the premisses and the rule do not overlap. After the rule has been successfully applied, all the *factors* and factors of factors of the conclusion will be derived using the factorization rule, which is essentially the contraction rule of the sequent calculus combined with necessary substitution. For the intuitionistic logic we have the following factorization rule:

$$\frac{X, X', \Gamma \vdash Y}{(X, \Gamma \vdash Y)\sigma} \quad \sigma = mgu(X, X')$$

Finally, all the repetitions in the left side of the derived sequent are deleted.

Example 2. We continue the example 1. Recall the formula F_1 and its labelling. The following is a set of labelled instances formed from the sequent rules in GJm. Recall implicit exchange and contraction in GJm: the left side of a sequent is a set of atoms.

$$\forall_{L_1} \vdash: \frac{X, \Gamma \vdash Y}{(L_1, \Gamma \vdash Y)\sigma} \; \sigma = mgu(X, P(x, b))$$

$$\Rightarrow_{L_2} \vdash: \frac{\Gamma \vdash X \quad Y, \Delta \vdash Z}{(\Gamma, \Delta, L_2(x, y) \vdash Z)\sigma} \; \mu = mgu(X, P(x, y)), \quad \sigma = mgu(Y\mu, B(x, y)\mu)$$

$$\forall_{L_3} \vdash: \frac{X, \Gamma \vdash Y}{(L_3, \Gamma \vdash Y)\sigma} \; \sigma = mgu(X, L_2(x, y))$$

$$\vdash \forall_{L_4}: \frac{\Gamma \vdash Y}{(\Gamma \vdash L_4(x, y))\sigma} \; \sigma = mgu(Y, B(x, y)) \text{ or } \sigma = mgu(Y, P(y, x))$$

$$\vdash \exists_{L_5}: \frac{\Gamma \vdash Y}{(\Gamma \vdash L_5(x))\sigma} \; \sigma = mgu(Y, L_4(x, y))$$

$$\vdash \forall_{L_6}: \frac{\Gamma \vdash Y}{(\Gamma \vdash L_6)\sigma} \; \sigma = mgu(Y, L_5(x)), VAR(x\sigma), NOOCC(x\sigma, (\Gamma \vdash L_6)\sigma)$$

$$\&_{L_7} \vdash: \frac{X, \Gamma \vdash Y}{L_7, \Gamma \vdash Y} \; X \equiv L_1 \text{ or } X \equiv L_3$$

$$\vdash \Rightarrow''_{L_8}: \frac{\Gamma \vdash Y}{\Gamma \vdash L_8} \; Y \equiv L_6$$

$$\vdash \Rightarrow_{L_8}: \frac{X, \Gamma \vdash Y}{\Gamma \vdash L_8} \; X \equiv L_7, \quad Y \equiv L_6$$

The instance of the rule $\vdash \forall$ corresponding to L_6 contains the eigenvariable condition: $x\sigma$ must be a variable and it must not occur in $(\Gamma \vdash L_6)\sigma$. The form of the rules $\&_{L_7} \vdash$ and $\vdash \Rightarrow_{L_8}$ has been simplified, since one of the literals to be resolved upon is a predicate symbol with no arguments and thus the substitution, if it exists, will be always empty.

The set of possible axioms for the labelled F_1 contains two sequents:

$$P(x, y) \vdash P(x, y) \qquad B(x, y) \vdash B(x, y)$$

Recall that all the variables in each premiss sequent must be different from variables both in the other premiss sequents and the rule. Variable renaming is used implicitly.

The label-form derivation of F_1:

$$\frac{\dfrac{\dfrac{\dfrac{\dfrac{\dfrac{\dfrac{\dfrac{\dfrac{\dfrac{P(x, y) \vdash P(x, y) \quad B(x', y') \vdash B(x', y')}{P(x, y), L_2(x, y) \vdash B(x, y)} \Rightarrow_{L_2} \vdash}{P(x, y), L_3 \vdash B(x, y)} \forall_{L_3} \vdash}{P(x, y), L_7 \vdash B(x, y)} \&_{L_7} \vdash}{L_1, L_7 \vdash B(x, b)} \forall_{L_1} \vdash}{L_7 \vdash B(x, b)} \&_{L_7} \vdash}{L_7 \vdash L_4(x, b)} \vdash \forall_{L_4}}{L_7 \vdash L_5(x)} \vdash \exists_{L_5}}{L_7 \vdash L_6} \vdash \forall_{L_6}}{\vdash L_8} \vdash \Rightarrow_{L_8}$$

Consider, for a change, an attempt to derive F_1 using the right-side argument $P(y, x)$ of the disjuction $B(x, y) \vee P(y, x)$:

$$\frac{\dfrac{\dfrac{\dfrac{P(x, y) \vdash P(x, y)}{P(x, y) \vdash L_4(y, x)} \vdash \vee_{L_4}}{P(x, y) \vdash L_5(y)} \vdash \exists_{L_5}}{P(x, y) \vdash L_6} \vdash \forall_{L_6}??? :}$$

The last step of the derivation is not allowed due to the eigenvariable condition in the rule $\vdash \forall_{L_6}$: $x\sigma = y$, and y occurs in the conclusion $P(x, y) \vdash L_6$. An attempt to derive F_1 using $\forall_{L_1} \vdash$ higher in the derivation also fails due to the eigenvariable condition: $x\sigma = b$, and b is not a variable.

$$\frac{\dfrac{\dfrac{\dfrac{\dfrac{P(x, y) \vdash P(x, y)}{L_1 \vdash P(x, b)} \forall_{L_1} \vdash}{L_1 \vdash L_4(b, x)} \vdash \vee_{L_4}}{L_1 \vdash L_5(b)} \vdash \exists_{L_5}}{L_1 \vdash L_6} \vdash \forall_{L_6}??? : \neg VAR(b)}$$

2.4 Clause Notation

It is convenient to formalize the axioms and the system of instances of GJm rules produced by the algorithms RA and RR using the *clause notation* familiar from Robinson's resolution. The calculus RJm is obtained from RJp given in [9] by the following inessential modifications: (1) the redundant clauses $\Delta \Rightarrow p$ and $\Delta \Rightarrow q_1$ are removed from the rules $\vee \vdash$ and $\exists \vdash$, respectively, (2) two rules for the negation connective $\neg$ are added, (3) the rule *Res* is split into a number of different cases: $\vdash (\vee, \exists, \Rightarrow'')$, $(\&, \forall) \vdash$ and $\vdash \&$. Such a modification was considered in a system RIp in [7]. (4) The notational difference: we write $\neg L_1, \ldots, \neg L_m, R$ instead of $L_1, \ldots, L_m \Rightarrow R$.

Derivable objects of RJm are *clauses* which represent sequents. Negative literals (if any) in the clause represent atoms on the left of the sign $\vdash$, positive literal (if any) represents the atom on the right side of $\vdash$ in the corresponding sequent.

The left premise in all of the following resolution rules except factorization is a *rule clause* obtained by translating the corresponding rule in the instantiation of GJm.

Other premises are derived clauses. The rule clauses are analogous to *nucleons* and the derived clauses are analogous to *electrons* of the hyperresolution strategy of ordinary classical resolution, see [2]. All the literals to the left of | in a rule clause have to be resolved upon. The literal to the right of | will go to the conclusion. The rule $\vdash \Rightarrow$ is different from the usual rules of classical resolution: two literals to the left of | are resolved upon with two literals in a *single* non-rule premiss clause. The rules $\vdash \forall$ and $\exists \vdash$ are also nonstandard, since they contain the eigenvariable condition. $\neg\Gamma$ denotes a set of negative literals. $mgu(p_1, p_1'; \ldots; p_m, p_m')$ denotes the result of unifying the terms $f(p_1, \ldots, p_m)$

and $f(p'_1, \ldots, p'_m)$. R denotes either a positive literal or no literal at all. L denotes the label introduced by the rule. $VAR(t)$ denotes that t is a variable, $NOOCC(v, t)$ denotes that v does not occur in t.

$$\vdash \Rightarrow : \frac{p, \neg q | L \quad \neg \Gamma, \neg p', q'}{(\neg \Gamma, L)\sigma} \quad \sigma = mgu(p, p'; q, q')$$

$$\vdash (\Rightarrow', \neg) : \frac{p | L \quad \neg \Gamma, \neg p'}{(\neg \Gamma, L)\sigma} \quad \sigma = mgu(p, p')$$

$$\neg \vdash : \frac{\neg p | \neg L \quad \neg \Gamma, p'}{(\neg \Gamma, \neg L)\sigma} \quad \sigma = mgu(p, p')$$

$$\Rightarrow \vdash : \frac{p, \neg q | \neg L \quad \neg \Gamma, q' \quad \neg \Sigma, \neg p', R}{(\neg \Gamma, \neg \Sigma, \neg L, R)\sigma} \quad \sigma = mgu(p, p'; q, q')$$

$$\vee \vdash : \frac{p, q | \neg L \quad \neg \Gamma, \neg p', R \quad \neg \Sigma, \neg q', R'}{(\neg \Gamma, \neg \Sigma, \neg L, R)\sigma} \quad \sigma = mgu(p, p'; q, q'; R, R')$$

$$\vdash \forall_x : \frac{\neg p | L \quad \neg \Gamma, p'}{(\neg \Gamma, L)\sigma} \quad \sigma = mgu(p, p'), \quad VAR(x\sigma), NOOCC(x\sigma, (\neg \Gamma, L)\sigma)$$

$$\exists_x \vdash : \frac{p | \neg L \quad \neg \Gamma, \neg p', R}{(\neg \Gamma, \neg L, R)\sigma} \quad \sigma = mgu(p, p'), VAR(x\sigma), \quad NOOCC(x\sigma, (\neg \Gamma, \neg L, R)\sigma)$$

$$\vdash (\vee, \exists, \Rightarrow'') : \frac{\neg p | L \quad \neg \Gamma, p'}{(\neg \Gamma, L)\sigma} \quad \sigma = mgu(p, p')$$

$$\vdash \& : \frac{\neg p, \neg q | L \quad \neg \Gamma, p' \quad \neg \Sigma, q'}{(\neg \Gamma, \neg \Sigma, L)\sigma} \quad \sigma = mgu(p, p'; q, q')$$

$$(\&, \forall) \vdash : \frac{p | \neg L \quad \neg \Gamma, \neg p', R}{(\neg \Gamma, \neg L, R)\sigma} \quad \sigma = mgu(p, p')$$

$$Fact \vdash : \frac{\neg \Gamma, \neg p, \neg p', R}{(\neg \Gamma, \neg p, R)\sigma} \quad \sigma = mgu(p, p')$$

Example 3. We continue the example 2. Recall the formula F_1, the labelled form of F_1 and the corresponding instantiation of GJm. 'conc' will denote the conclusion of applying the rule clause with the substitution σ. The following is the set of instantiated rules from GJm in the clause form:

axiom clauses: $\neg P(x, y), P(x, y)$ and $\neg B(x, y), B(x, y)$.
rule clauses:

$$\forall_{L_1} \vdash : P(x, b) | \neg L_1$$
$$\Rightarrow_{L_2} \vdash : \neg P(x, y), B(x, y) | \neg L_2(x, y)$$
$$\forall_{L_3} \vdash : L_2(x, y) | \neg L_3$$
$$\vdash \vee_{L_4} : \neg B(x, y) | L_4(x, y) \quad \text{and} \quad \neg P(y, x) | L_4(x, y)$$
$$\vdash \exists_{L_5} : \neg L_4(x, y) | L_5(x)$$
$$\vdash \forall_{x, L_6} : \neg L_5(x) | L_6 \quad VAR(x\sigma), NOOCC(x\sigma, \text{conc})$$
$$\&_{L_7} \vdash : L_1 | \neg L_7 \quad \text{and} \quad L_3 | \neg L_7$$
$$\vdash \Rightarrow''_{L_8} : \neg L_6 | L_8$$
$$\vdash \Rightarrow_{L_8} : L_7, \neg L_6 | L_8$$

Now consider the derivations presented in the example 2. Each of these is trivially translated to the clause-notation derivation using the clause rules: replace the sequent notation $\Gamma \vdash R$ with the clause notation $\neg\Gamma, R$.

3 Linear logic

Linear logic, introduced by J.-Y.Girard, is a refinement of classical logic providing means for controlling the allocation of "resources". It has aroused considerable interest both from proof theorists and computer scientists.

The first step in deriving linear logic from classical logic is to drop two structural rules, *contraction* and *weakening*, from the standard set of Gentzen-type rules. This gives a system where each resource must be used *exactly once*.

To restore the power of intuitionistic and classical logic, two modal operators ? and ! are introduced. $!A$ is equivalent to $(?A^\perp)^\perp$ and $?A$ is equivalent to $(!A^\perp)^\perp$. $?A$ allows unlimited use of the resource A and $!A$ allows unlimited consumption of A. Accordingly, logical constants **True** and **False** are split into four: $\top$, 1, $\perp$ and 0.

The following is a set of sequent calculus rules for linear logic as given by J.-Y. Girard in [5]:

Logical axioms:

$$\vdash a, a^\perp$$

(it is enough to restrict a to be a propositional atom)

Multiplicative rules:

$$\vdash 1 \quad \text{(1-axiom)}$$

$$\frac{\vdash A, \Gamma \quad \vdash B, \Delta}{\vdash A \otimes B, \Gamma, \Delta} \otimes$$

$$\frac{\vdash \Gamma}{\vdash \perp, \Gamma} \perp$$

$$\frac{\vdash A, B, \Gamma}{\vdash A \otimes B, \Gamma} \otimes$$

Additive rules:

$$\vdash \top, \Gamma \quad \text{(axiom, } \Gamma \text{ arbitrary)}$$

$$\frac{\vdash A, \Gamma \quad \vdash B, \Gamma}{\vdash A \& B, \Gamma} \&$$

$$\frac{\vdash A, \Gamma}{\vdash A \oplus B, \Gamma} 1\oplus \quad \frac{\vdash B, \Gamma}{\vdash A \oplus B, \Gamma} 2\oplus$$

Exponential (modal) rules:

$$\frac{\vdash A, \Gamma}{\vdash ?A, \Gamma} \text{ (dereliction)}$$

$$\frac{\vdash \Gamma}{\vdash ?A, \Gamma} \text{ (weakening)} \qquad \frac{\vdash A, ?\Gamma}{\vdash !A, ?\Gamma} \text{ (!-storage)}$$

$$\frac{\vdash ?A, ?A, \Gamma}{\vdash ?A, \Gamma} \text{ (contraction)}$$

($?\Gamma$ denotes a sequent where every element is prefixed by ?)

Ordinary exchange rule is implicit: the order of elements in a sequent can be freely changed.

Notice that there is no rule for the constant 0.

3.1 Resolution for linear logic

The first version of resolution for linear logic was presented by G.Mints in [8]. We present a modified version along with special strategies from [13].

The first step, labelling of subformulas, is standard.

We can see that there is a one-to-one correspondence between the rule clauses for connectives $\otimes-$, $\oplus-$, $\&-$, $\wp-$, ? and ! for some formula F and the sets of admissible $\otimes-$, $\oplus-$, $\&-$, $\wp-$, dereliction and !-store inferences in the resolution proof search for F.

Indeed, other kinds of $\otimes-$, $\oplus-$, $\&-$, $\wp-$, dereliction and !-store rule applications are prohibited in the resolution method, since they do not introduce the subformulas of F. For example, if the set of rule clauses for a formula F contains a rule clause $(L_1 \wp B^\perp) = L_2$, we know that the only way to introduce L_2 in the resolution derivation of F is to apply a $\wp$-rule to some sequent containing both L_1 and $B^\perp$.

Example 4. Take $F = ((A \otimes B) \wp B^\perp) \wp A^\perp)$. The labelling gives a following set of rule clauses:

1) $(A \otimes B) = L_1$
2) $(L_1 \wp B^\perp) = L_2$
3) $(L_2 \wp A^\perp) = L_3$

There are two axiom clauses for F:

4) $A, A^\perp$
5) $B, B^\perp$

The resolution method gives a following derivation from clauses 1-5:

6) from 1, 4, 5: $L_1, A^\perp, B^\perp$
7) from 2, 6: $L_2, A^\perp$
8) from 3, 7: L_3

corresponding to the sequent calculus derivation:

$$\frac{\dfrac{\vdash A, A^{\perp} \quad \vdash B, B^{\perp}}{\vdash (A \otimes B), A^{\perp}, B^{\perp}}}{\dfrac{\vdash ((A \otimes B) \otimes B^{\perp}), A^{\perp}}{\vdash (((A \otimes B) \otimes B^{\perp}) \otimes A^{\perp})}}$$

3.2 Problems with ?-prefixed subformulae and the constant ⊤

Definition 3. The set of ?-*labels* for some formula F is the set of labels for all the subformulas of F with a top connective ?.

The set of !-*labels* for some formula F is the set of labels for all the subformulas of F with a top connective !.

Obviously the contraction rule is applicable only to the ?-labels, weakening and dereliction can introduce only ?-labels and the !-store rule can introduce only !-labels.

The most unpleasant rules for the unrestricted resolution formulation are weakening and the axiom schema containing ⊤:

$$\frac{\vdash \Gamma}{\vdash ?A, \Gamma} \quad \text{(weakening)}$$

and

$$\vdash \top, \Gamma \quad \text{(axiom, } \Gamma \text{ arbitrary)}.$$

Applications of weakening introduce arbitrary ?-labels. Thus, for every clause Γ not containing ?-labels it is possible to derive by weakening $2^n - 1$ clauses of the form $?\Delta_i, \Gamma$, where n is the number of ?-labels for the formula and $?\Delta_i$ is a subclause containing some subset of all ?-labels of the formula.

The fact that linear logic contains a ⊤-axiom schema entails that whenever ⊤ is contained in some input formula F, the number of different sequents produced by this schema during the unrestricted proof search for some formula F is generally unbounded.

In principle, the weakening rule and the axiom schema with ⊤ could be handled explicitly, like all the other rules. However, in order to avoid the above-mentioned exponential explosion by weakening and the nonterminating process of generating new ⊤-containing axioms we shall drop the explicit weakening and ⊤-schema and handle those rules implicitly.

We will first introduce following rules containing implicit weakening applications, called implicit-weakening rules:

Additional multiplicative rules:

$$\frac{\vdash \Gamma \quad \vdash B, \Delta}{\vdash A_? \otimes B, \Gamma, \Delta} \otimes_{?l}
\qquad\qquad
\frac{\vdash B, \Gamma}{\vdash A_? \mathbin{\otimes} B, \Gamma} \otimes_{?l}$$

$$\frac{\vdash A, \Gamma \quad \vdash \Delta}{\vdash A \otimes B_?, \Gamma, \Delta} \otimes_{?r}
\qquad\qquad
\frac{\vdash A, \Gamma}{\vdash A \mathbin{\otimes} B_?, \Gamma} \otimes_{?r}$$

$$\frac{\vdash \Gamma \quad \vdash \Delta}{\vdash A_? \otimes B_?, \Gamma, \Delta} \otimes_{?lr}
\qquad\qquad
\frac{\vdash \Gamma}{\vdash A_? \mathbin{\otimes} B_?, \Gamma} \otimes_{?lr}$$

Additional additive rules:

$$\frac{\vdash \Gamma_l \quad \vdash B, \Gamma_r}{\vdash A_? \mathbin{\&} B, \Gamma_{lr}} \&_{?l}$$

$$\frac{\vdash A, \Gamma_l \quad \vdash \Gamma_r}{\vdash A \mathbin{\&} B_?, \Gamma_{lr}} \&_{?r}
\qquad
\frac{\vdash \Gamma}{\vdash A_? \oplus B, \Gamma} 1\oplus_?
\qquad
\frac{\vdash \Gamma}{\vdash A \oplus B_?, \Gamma} 2\oplus_?$$

$$\frac{\vdash \Gamma_l \quad \vdash \Gamma_r}{\vdash A_? \mathbin{\&} B_?, \Gamma_{lr}} \&_{?lr}$$

Additional exponential (modal) rules:

$$\frac{\vdash ?\Gamma}{\vdash !A_?, ?\Gamma} \ (!\text{-storage}_?)$$

where:

- $?\Gamma$ denotes a sequent containing only ?-labels
- $A_?$ and $B_?$ denote ?-labels and a constant $\perp$
- $\Gamma_l = \Delta, ?\Sigma, ?\Theta_l$ and $\Gamma_r = \Delta, ?\Sigma, ?\Theta_r$ where $?\Theta_l \bigcap ?\Theta_r = \emptyset$ and $\Gamma_{lr} = \Delta, ?\Sigma, ?\Theta_l, ?\Theta_r$
- A and B denote arbitrary formulas (including ?-labels and $\perp$)

In [13] several restrictions are introduced to the implicit-weakening rules and the correctness and completeness of the system with implicit-weakening rules is proved.

Also, the explicit T-axiom rule is dropped and analogously to weakening is used implicitly. In order that this were possible, a new type of clauses - *weak-enable clauses* - is introduced along with a modified resolution calculus. We will skip the details here.

4 Classical Logic

The generic resolution method can be applied to the classical logic in the same way as previously applied to intuitionistic and linear logic. Indeed, several methods of *nonclausal resolution* presented in literature use labelling of subformulas and are similar to the generic resolution.

However, since formulas of classical logic can be Skolemized and converted to various normal forms, it is also possible to first convert a formula to a certain normal form and only then apply the generic resolution method.

One of the earliest versions of generic resolution (inverse method) proposed by S. Maslov for classical logic has the following form.

As a first step, a formula F is converted to the clause form C (conjunctive normal form of the negated Skolemized formula) exactly as done for Robinson's resolution.

Since we use the clause form, there is no need to use labels of subformulas. Hence labelling is skipped.

Derivable objects are clauses, ie disjunctions of literals. F is proved iff an empty clause is derived from C. The following axioms and rules are used in the derivation.

- **Axioms:** Initial unit clauses plus tautological clauses $A(x_1, \ldots, x_n) \lor \neg A(x_1, \ldots, x_n)$ for any n-ary predicate symbol A occurring i F.
  Alternatively, a set of such axioms instantiated by all substitutions obtained by unifiying all pairs of positive and negative occurrences of A in F.
- **Rules:** Let C be the initial set of nonunit clauses: $\{D_1, D_2, \ldots, D_m\}$. Each initial nonunit clause $D_i = (A_1 \lor \ldots \lor A_m)$ gives the following rule:

$$\frac{A_1 \lor \ldots \lor A_m \quad B_1 \lor \Gamma_1, \quad \ldots \quad , B_m \lor \Gamma_m}{(\Gamma_1 \lor \ldots \lor \Gamma_m)\sigma}$$

where Γ_i is a (sub)clause, each B_i is unifiable with a negated A_i and the substitution $\sigma = mgu(\neg A_1, B_1; \ldots; \neg A_m, B_m)$.
In addition, factorization is used:

$$\frac{B_1 \lor B_2 \lor \Gamma}{(B_1 \lor \Gamma)\sigma} \sigma = mgu(B_1, B_2)$$

It is easy to see that the derivation rule is very similar to the *hyperresolution* derivation rule. The only difference is that in our case not only the negative literals, but *all* the literals of a "hyperresolution nucleus" have to be cut off. This is always possible due to the presence of tautologous clauses.

Example 5. Let Δ be a clause set

$$\{\neg A \lor B \lor C, \ A \lor B, \ \neg B, \ \neg C\}$$

The set of axioms for Δ is:

1) $A \lor \neg A$
2) $B \lor \neg B$
3) $C \lor \neg C$
4) $\neg B$
5) $\neg C$

The following is the derivation of the empty clause from Δ:

6) initial clause $A \vee B$ with (1) and (4): A.

7) initial clause $\neg A \vee B \vee C$ with (6) and (4) and (5): $\bot$ (empty clause).

Consider the corresponding sequent calculus derivation. Δ stands for the input clause set, treated here as a part of a sequent.

$$\cfrac{\cfrac{\cfrac{\neg A, A, \Delta \vdash \bot \quad B, \Delta \vdash \bot}{\neg A, A \vee B, \Delta \vdash \bot}(6)}{\neg A, \Delta \vdash \bot}(contraction) \quad B, \Delta \vdash \bot \quad C, \Delta \vdash \bot}{\cfrac{\neg A \vee B \vee C, \Delta \vdash \bot}{\Delta \vdash \bot}(contraction)}(7)$$

Let us build a new tree from the last derivation by making contraction implicit and removing Δ from each sequent (ie considering Δ to be implicity present in each sequent).

$$\cfrac{\cfrac{\neg A, A, \vdash \bot \quad B \vdash \bot}{\neg A \vdash \bot}(6) \quad B \vdash \bot \quad C \vdash \bot}{\vdash \bot}(7)$$

The resulting tree corresponds exactly to the resolution derivation constructed previously – with derived clauses corresponding to *negated* left sides of sequents.

Indeed, this correspondence between the resolution and sequent calculus derivations can be easily seen to be one-to-one. Thus we can prove the completeness of the considered resolution calculus for the propositional case in a pure proof-theoretic manner, without any semantical considerations like semantic trees. For the predicate case the lifting lemma must be additionally employed.

It is also easy to see that each step of the considered resolution method can be constructed as a sequence of Robinson's resolution steps. In this way we can prove completeness of Robinson's resolution without any semantical considerations.

5 Strategies of Resolution

The general principle of modifying the calculus is the same for the resolution as for the tableau calculus: diminishing the branching factor at search nodes. However, in the resolution case the search is carried on in the top-down manner and thus we want to diminish branching in the "down" direction.

A number of modifications and lemmas are shared with the tableau method. For example, invertibility of rules can be employed by both. It is frequently (though not always) the case that a lemma justifying a strategy which is natural for one search procedure turns out to be usable also for the other procedure.

The main device of proving the completeness of a developed resolution method M incorporating some restriction strategies is to show that there is a restricted form R for sequent calculus derivations such that R is complete: all formulas derivable in the original calculus have a derivation with the form R and all the possible derivations satisfying R are reachable by M.

5.1 Subsumption

In general, a clause Γ is said to subsume a clause Δ if Δ is derivable from Γ by a chain of weakenings and substitions into variables.

Subsumption strategy: every derived clause Γ which is subsumed by some existing clause Δ is immediately removed. In case a newly derived clause Δ subsumes some existing clauses $\Gamma_1, \ldots, \Gamma_n$, then all the latter are immediately removed from search space.

Unrestricted subsumption can be used for calculi which contain an unrestricted weakening rule (either implicit or explicit), like classical and intuitionistic logics.

Definition 4 (unrestricted subsumption). The clause Γ *subsumes* the clause Δ iff $\Gamma \subseteq \Delta\sigma$ for some substitution σ.

However, in case weakening is prohibited or restricted, as is the case for linear logic, then subsumption has to be restricted in the same way.

Definition 5 (subsumption for linear logic as presented above).
Let $\Gamma = ?\Gamma_1, \Gamma_2$ and $\Delta = ?\Delta_1, \Delta_2$, where $?\Gamma_1$ and $?\Delta_1$ are sets containing all the ?-labels in Γ and Δ, accordingly.

The clause Γ subsumes a clause Δ iff $?\Gamma_1 \subseteq ?\Delta_1$ and $\Gamma_2 = \Delta_2$, where $=$ is an ordinary equality relation for multisets.

Notice that with an explicit weakening instead of an implicit weakening we should require $?\Gamma_1 = ?\Delta_1$ instead of $?\Gamma_1 \subseteq ?\Delta_1$.

The following lemma is an old result, see [9]:

Lemma 6. *The subsumption strategy preserves completeness of resolution.*

5.2 Inversion Strategy

A rule is called *invertible* iff its inversion is true: each premiss is derivable from the conclusion. There exist sequent calculi for classical logic where all the rules are invertible. This is impossible for intuitionistic logic, however.

Invertibility for intuitionistic logic

Definition 7. A label is called *invertible* iff it corresponds to one of the following intuitionistic rules: $\vdash \neg, \vdash \&, \& \vdash, \vee \vdash, \vdash \Rightarrow, \vdash \forall, \exists \vdash$ and $\forall \vdash$. The *inversion strategy* for RJm:

- introduce an arbitrary complete order $\succ_i$ for all the invertible labels,
- prohibit to use a rule clause introducing a label L for the derivation of a clause $(\Gamma, R, L)\sigma$ such that either R is a an invertible label and L is not an invertible label or $R \succ_i L$.

The following lemma is proved in [14]:

Lemma 8. *The inversion strategy preserves completeness for RJm.*

Invertibility for linear logic The linear &- and ⅋-rules are invertible.

The *inversion strategy* for linear resolution:

Whenever some clause Γ contains a label V of a formula with a top connective & or ⅋, prohibit any application of ⊗- and ⊕-rules to any other element of Γ except V.

Whenever some clause Γ contains a label V of a formula with a top connective ⅋, prohibit any application of &-rule to any other element of Γ except V.

5.3 Reduction Strategy

For several kinds of derived clauses we can immediately see that it is possible to allow only a single rule to be applied to the clause and not to consider any other rule applications. A general scheme of reduction strategies for the resolution method is proposed in [15]. The reduction strategy for linear logic developed independently in [13] was of crucial importance for efficiency of the linear resolution prover in [13] and the intuitionistic prover in [14].

Definition 9. We say that a clause Γ is *reducible* by a reduction strategy iff only one clause Δ can be derived from Γ, according to this strategy. Any such derivation figure (reduction) $\Gamma \overset{*}{\rightsquigarrow} \Delta$ consists of one or several ordinary single-premiss inference steps, called reduction steps, $\Gamma \rightsquigarrow \Gamma_1 \rightsquigarrow \ldots \rightsquigarrow \Gamma_n \rightsquigarrow \Delta$ where Δ is the single clause derivable from Γ and all the intermediate clauses $\Gamma, \Gamma_1, \ldots \Gamma_n$ are discarded.

Definition 10. We say that some clause Γ is *fully reduced* to a clause Δ iff immediately after the derivation of Γ a chain of n reduction steps is applied to Γ, producing a reduction $\Gamma \overset{*}{\rightsquigarrow} \Delta$ where Δ is not reducible any more. The *reduction strategy* stands in converting every derived clause to a fully reduced form. A rule clause has *unique premisses* iff all its premisses are either labels or such predicates which have only a single occurrence with the given polarity.

As an example we bring a completeness theorem for the intuitionistic logic:

Theorem 11. *Completeness is preserved if the following rules with unique premisses are, whenever applicable, used as reduction rules for clauses:* ⊢ ∨, & ⊢, ∀ ⊢, ⊢ ∃, ¬ ⊢, ∃ ⊢. *The rule* ⊢ ∀ *with unique premisses can be used as a reduction rule in case the formula to be proved does not have any negative occurrences of* ∨.

Proof. By induction over the derivation tree: reduction preserves applicability of all the other rules.

We note that the reduction strategy is not fully compatible with the inversion strategy.

5.4 Nesting Strategy

We present the nesting strategy for intuitionistic logic. Analogous strategies exist for other logics as well.

Definition 12 (nesting for intuitionistic logic). A clause Γ, R, L is *nested* iff R is a label of subformula F_L in F such that F_L does not occur in the sope of negation or left side of implication and L is a label of a subformula F_L of F_R. *Nesting strategy*: all the nested clauses are immediately discarded.

Theorem 13. *The nesting strategy preserves completeness.*

The idea of a proof we omit here due to the lack of space is to show that the nested clauses cannot occur in the derivation of F.

6 Decidability

The resolution method and its special strategies have been used for showing decidability of a number of subclasses of various logics. The early paper [6] used resolution for showing decidability of Maslov's Class and Class K for classical logic. Special ordering strategies of Robinson's resolution are used as decision strategies in [3].

We will present a class *Near–Monadic* of predicate logic formulas which is decidable for all such logics for which we can build a resolution method such that full subsumption preserves completeness. For example, Near–Monadic formulas are decidable in intuitionistic and modal logics, but not decidable in linear logic.

Definition 14. By *Near–Monadic* we denote the class of formulas without function symbols such that no occurrence of a subformula contains more than one free variable.

The Near–Monadic class is similar to the Monadic class (the function-free class where all the predicates have an arity 1) and its extension, the Essentially Monadic class (the function-free class where all the atoms contain no more than one variable). It is known that although the Essentially Monadic class is decidable for classical logic, even the Monadic class is undecidable for intuitionistic logic. However, the Near–Monadic class does not contain the whole Monadic class.

For intuitionistic logic the following generalised version of Near–Monadic class is decidable: *positive occurrences* of a subformula are allowed to contain an arbitrary number of variables.

Theorem 15. *The resolution method incorporating subsumption strategy is a decision procedure for the* Near–Monadic *class.*

Definition 16. The *splitting* of a clause R is a set $R_1, \ldots, R_n$ of subsets (called blocks) of R, such that: (1) each literal in R occurs in a single block, (2) no two blocks R_i and R_j ($i \neq j$) share variables, (3) no splitting of R contains more than n blocks.

In the following lemma we assume clause language without function symbols: atoms are built of predicate symbols, variables and constants.

Lemma 17. *Assume that we have a finite set of predicate symbols P and a finite set of constant symbols C. Consider the set A of all atoms built using only variables and elements of P and C. Let $A^{01} \subset A$ such that no member of A^{01} contains more than one variable. Let $A^{2+} = A - A^{01}$. Let S be a set of clauses built from the elements of A^{01} and A^{2+} so that no clause in S contains more than a single element of A^{2+}, S contains all the factorizations of clauses in S and and no two clauses in S subsume each other. Then S cannot be infinite.*

Proof. Let a clause R be built from the elements of A^{01} and A^{2+}, so that it contains no more than one element of A^{2+}. Build a splitting $R_1, \ldots, R_n$ of R. No two blocks R_i and R_j ($i \neq j$) share variables. Each block R_i contains either (0) no variables, (1) a single variable or (2) more than one variable. Due to the construction of R, the splitting $R_1, \ldots, R_n$ can contain no more than one block of type (2). In case a block R_i of type (1) can be obtained from another block R_j of type (1) by renaming the variable occurring in R_j ($R_i = R_j\{x/y\}$, where x is the variable occurring in R_i and y is the variable occurring in R_j), then the factor $R\{x/y\}$ of R subsumes R.

In order to show that clauses in S have a limited length, we have to show that among the splittings of clauses in S there is only a finite number of possible blocks modulo renaming of variables. This follows from the limitations on the construction of clauses in S: only elements of A^{01} and A^{2+} may be used, no clause may contain more than one element of A^{2+}.

We are now in the position to prove the theorem 15.

Proof. Consider a near–monadic formula F. All the atoms N labelling subformulas in F contain no more than one variable. Since F is function–free, it is impossible for resolution to derive an instance $N\sigma$ (including input tautologies) such that $N\sigma$ contains more than one variable. Thus all the derivable clauses satisfy conditions of the lemma 17. Thus the number of derivable clauses is finite.

For example, consider the following four formulas from [12]. Each of these is provable classically, but not intuitionistically. Since they fall into the Near–Monadic class, a resolution theorem prover exhausts the search space and stops.

$$\forall x(p(x) \lor q(c)) \Rightarrow (q(c) \lor \forall x p(x)) \quad (a(c) \Rightarrow \exists x b(x)) \Rightarrow \exists x(a(c) \Rightarrow b(x))$$
$$\neg \forall x a(x) \Rightarrow \exists x \neg a(x) \quad ((\forall x a(x)) \Rightarrow b(c)) \Rightarrow \exists x(a(x) \Rightarrow b(c))$$

References

1. M. Baaz, C. Fermüller. Resolution-Based Theorem Proving for Many-Valued Logics. *Journal of Symbolic Logic*, **19**, 353–391 (1995).
2. C.L. Chang, R.C.T. Lee. Symbolic Logic and Mechanical Theorem Proving. Academic Press, (1973).

3. C. Fermüller, A. Leitsch, T. Tammet, N. Zamov. Resolution methods for decision problems. LNCS 679, Springer Verlag, (1993).
4. M. Fitting. Resolution for Intuitionistic Logic. Paper presented at ISMIS '87, Charlotte, NC. 1987.
5. J.-Y. Girard. Linear Logic. *Theoretical Computer Science* **50**, 1-102 (1987).
6. S.Ju. Maslov. An inverse method of establishing deducibility in the classical predicate calculus. Dokl. Akad. Nauk. SSSR 159 (1964) 17-20=Soviet Math. Dokl. 5 (1964) 1420, MR 30 #3005.
7. G. Mints. Gentzen-type Systems and Resolution Rules. Part I. Propositional Logic. In *COLOG-88*, pages 198-231, LNCS 417, Springer-Verlag, 1990.
8. G. Mints. Resolution Calculus for The First Order Linear Logic. *Journal of Logic, Language and Information,* **2**, 58-93 (1993).
9. G. Mints. Resolution Strategies for the Intuitionistic Logic. In *Constraint Programming*, NATO ASI Series F, v. 131, pp. 289–311, Springer Verlag, (1994).
10. G. Mints, V. Orevkov, T. Tammet. Transfer of Sequent Calculus Strategies to Resolution. In *Proof Theory of Modal Logic*, Studies in Pure and Applied Logic 2, Kluwer Academic Publishers, 1996.
11. J.A. Robinson. A Machine-oriented Logic Based on the Resolution Principle. *Journal of the ACM* **12**, 23-41 (1965).
12. N. Shankar. Proof Search in the Intuitionistic Sequent Calculus. In *CADE-11*, pages 522-536, LNCS 607, Springer Verlag, (1992).
13. T. Tammet. Proof Strategies in Linear Logic. *Journal of Automated Reasoning* **12**(3), 273–304 (1994).
14. T. Tammet. Resolution Theorem Prover for Intuitionistic Logic. In *CADE-13*, pages 2-16, LNCS 1104, Springer Verlag, (1996).
15. A. Voronkov. Theorem proving in non-standard logics based on the inverse method. In *CADE-11*, pages 648-662, LNCS 607, Springer Verlag, (1992).

Subtyping over a Lattice
(Abstract)

Jerzy Tiuryn[*]
Institute of Informatics, Warsaw University
Banacha 2, 02-097 Warsaw, POLAND
(tiuryn@mimuw.edu.pl)

Abstract

This talk, in the first part, will overview the main advances in the area of subtyping for the simply typed lambda calculus. In the second part of the talk we will propose a new system of notations for types, which we call *alternating direct acyclic graphs*, and show that for a system of sybtype inequalities over a lattice, if it has a solution then there is a solution whose alternating dag is of polynomial size in the size of the original system. There are examples showing that the well known dag representation of types is not good enough for this purpose, already for the two-element lattice.

The concept of subtyping still attracts a lot of attention in research. Despite being a hot topic of research many fundamental questions remain open. In the first half of the talk we will survey the results on subtyping for simply typed lambda calculus, with main emphasis on the mathematical problem of *satisfiability of subtype inequalities*. The second part will be devoted to solving this problem over a finite lattice. It should be noticed that the results of this paper carry over without any difficulty to the more general case of a disjoint union of lattices.

The name "subtype inequality" appears for the first time in [Tiu92]. However, this concept implicitly exists in some earlier papers. Let $\langle C, \leq \rangle$ be a poset. This poset will be kept fixed throughout the talk. The elements of C can be thought as *atomic types*, and for $c, d \in C$, the relationship $c \leq d$ expresses the property of c being a *subtype* of d. Let X be a countable set of variables and let $\mathcal{T}_C[X]$ denote the the least set of expressions satisfying:

- $C \cup X \subseteq \mathcal{T}_C[X]$;

[*]Partly supported by NSF grant CCR-9417382 and by Polish KBN Grant 8 T11C 034 10.

- if $\sigma, \tau \in \mathcal{T}_C[X]$, then $(\sigma \to \tau) \in \mathcal{T}_C[X]$.

The elements of $\mathcal{T}_C[X]$ are called *type schemes*. Let $\mathcal{T}_C \subseteq \mathcal{T}_C[X]$ be the set of type schemes which do not contain variables. The elements of $\mathcal{T}_C$ are called *types*.

The partial order on $\langle C, \leq \rangle$ is extended to $\mathcal{T}_C$. This is the least partial order which contains the relation $\leq$ on C and satisfies:

$$(\sigma_1 \to \sigma_2) \leq (\tau_1 \to \tau_2) \qquad \text{iff} \qquad \tau_1 \leq \sigma_1 \text{ and } \sigma_2 \leq \tau_2.$$

Let us observe that if two types are comparable under the extended partial order, then they are of the same *shape*, where the shape of a type $\sigma \in \mathcal{T}_C$ is a type obtained from σ by replacing every occurence of every constant from C by the same constant, say $*$. Thus shapes are types over a one-element set of atomic types, i.e. shapes are elements of $\mathcal{T}_*$. For every $sh \in \mathcal{T}_*$, if $\mathcal{T}_C^{sh}$ denotes the set of all types $\sigma \in \mathcal{T}_C$ whose shape is sh, then $\mathcal{T}_C^{sh}$ is order-isomorphic to a direct product of a certain number of copies of the poset $\langle C, \leq \rangle$ and the dual to $\langle C, \leq \rangle$. For example, $\mathcal{T}_C^*$ is isomorphic to $C = \langle C, \leq \rangle$, and $\mathcal{T}_C^{(*\to*)}$ is isomorphic to $C^{op} \times C$, where C^{op} is a poset dual to C. In particular let us notice that if C is a lattice, then every $\mathcal{T}_C^{sh}$ is a lattice too.

An instance of the problem of satisfiability of sybtype inequalities (SSI) over $\langle C, \leq \rangle$ is a finite set $\Sigma \subseteq \mathcal{T}_C[X] \times \mathcal{T}_C[X]$. Σ is said to be *subtype satisfiable* if there is a substitution $S : X \to \mathcal{T}_C$, such that for every $(\sigma, \tau) \in \Sigma$,

$$\mathcal{T}_C \models S(\sigma) \leq S(\tau).$$

Let's notice that in the above definition $\langle C, \leq \rangle$ is a parameter and the complexity of the problem of satisfiability of sybtype inequalities may crucially depend on this poset.

It is known (see [HM95]) that for every $\langle C, \leq \rangle$, the problem of satisfiability for subtype inequalities over $\langle C, \leq \rangle$ is polynomial time equivalent to the problem of type reconstruction for the simply typed lambda calculus with subtyping and atomic types $\langle C, \leq \rangle$. Thus our prime interest in studying complexity of the SSI problem.

A recent result of A. Frey (see [Fre97]) shows that for every poset $\langle C, \leq \rangle$ the problem of SSI can be solved in PSPACE. A naive algorithm yields NEXPTIME upper bound. The PSPACE upper bound cannot be improved in general since it was shown in [Tiu92] that there exists finite posets $\langle C, \leq \rangle$ (all "crowns" have this property) such that the problem of SSI over $\langle C, \leq \rangle$ is PSPACE-hard.

On the other hand one may expect a better upper bound for some particular posets. A degenerate example is when the poset $\langle C, \leq \rangle$ is discrete (i.e., no different elements are comparable). In this case the SSI problem is just the unification problem since $\leq$ is in this case the equality relation. The unification problem is known to be PTIME complete (see [DKM84]). This result was substantially generalized in [Tiu92]. It was shown there that if $\langle C, \leq \rangle$ is a disjoint union of lattices, then there is a polynomial time algorithm which solves the SSI problem.[1]

The point of departure of the second half of this talk is the observation that the algorithm of [Tiu92], which works in polynomial time, doesn't construct a solution — it merely checks some conditions which are shown to be equivalent to the existence of a solution. It is well known that already for the unification problem it can happen that every solution (i.e. a unifier) can be of size exponential in the size of the original system. However, the trick used for unification is that the size of the solution viewed as an *acyclic directed graph* (dag) is in general linear in the size of the original system. As we will show in this talk the situation dramatically changes for the case of SSI problem, already for the two-element lattice $\{0, 1\}$ with $0 \leq 1$. We will expose a system Σ which has a unique solution S, and for one of the variables x of Σ, the size of the dag of $S(x)$ is exponential in the size of Σ.

In this situation a natural and important for potential practical applications is the question of whether there is a suitable system of notation for types, in which solutions can be represented more succinctly. Let's explain what we mean by a system of notation. A *system of notation* is a subset $\Delta \subseteq A^*$ (where A is a certain finite alphabet) together with a function (called the *meaning function*)

$$V : \Delta \rightarrow \mathcal{T}_C.$$

For example, the class of dags which we mention above can be presented using this formalism as follows. Elements of Δ_{dag} are finite directed acyclic graphs[2] such that every node is reachable from one node, called the root, and the nodes are labelled by $\{\rightarrow\} \cup C$. Every node labelled by $\rightarrow$ has out-degree 2 and the two edges leaving this node are labelled: one with 0 and another with 1. Every node labelled by an element of C has out-degree 0. For every dag $D \in \Delta_{dag}$, the type $V_{dag}(D)$ is obtained from D by the obvious operation of unfolding (i.e. an edge of D labelled 0 (resp. 1) leads in $V_{dag}(D)$ to the left (resp. right) child. As our example shows this system of notation is not succint enough to represent solutions of subtype inequalities.

We conclude this abstract with the technical notion of an *alternating directed acyclic graph*. First let us introduce an auxiliary notion of an *alternating pre-dag*. It is a dag such that every node is reachable from one node, called the root, and the nodes are labelled by $\{\rightarrow, \sqcup, \sqcap\} \cup C$. The nodes labelled by $\{\rightarrow\} \cup C$ satisfy the same conditions as in the above definition of a dag. Every node labelled

---

[1] This is indeed a generalization since a discrete poset is just a disjoint union of one-element lattices.

[2] We use the standard technique of encoding graphs in binary, so we can view dags as strings over $\{0, 1\}$.

by $\sqcup$ or by $\sqcap$ has positive out-degree,[3] and the edges leaving such a node are not labelled. Now we are in a position to define the class of alternating dags. The definition is by induction on the number of nodes in the alternating pre-dag D.

If D has only one node, then its node has to be labelled by an element of C, say c. Such a dag is alternating and we set

$$V_{adag}(D) = c.$$

Now suppose that D has more than one node and let v be the root of D. If v is labelled by $\rightarrow$, then let v_0, v_1 be the nodes of D such that there is an edge labelled i $(i = 0, 1)$, which leads from v to v_i. It is possible that $v_0 = v_1$. Let D_i be the sub-dag of D consisting of all nodes reachable from v_i. Then D is an alternating dag iff D_0 and D_1 are alternating dags and if this is the case, then we set

$$V_{adag}(D) = V_{adag}(D_0) \rightarrow V_{adag}(D_1).$$

If v is labelled by $\sqcup$, then let $v_1, \ldots, v_n$ be all nodes of D such that there is an edge from v to v_i. Let D_i be the sub-dag of D consisting of all nodes reachable from v_i. Then D is an alternating dag iff $D_1, \ldots D_n$ are alternating dags and shape of $V_{adag}(D_i)$ is the same as shape of $V_{adag}(D_j)$, for $1 \le i, j \le n$. Moreover, if $D_1, \ldots D_n$ are alternating dags and sh is the common shape of every $V_{adag}(D_i)$, then we set

$$V_{adag}(D) = lub\{V_{adag}(D_i) \mid i = 1, \ldots, n\}$$

where the above least upper bound is taken in $\mathcal{T}_C^{sh}$.

If v is labelled by $\sqcap$, then we proceed as above, except that the lub in the definition of $V_{adag}(D)$ is replaced by glb, i.e., the greatest lower bound.

Let Δ_{adag} denote the set of alternating dags. The basic properties of the above system of notation are listed below:

1. Δ_{adag} is in PTIME.

2. For every subtype satisfiable system Σ, there is a finite function $S : var(\Sigma) \rightarrow \Delta_{adag}$, such that

    (a) for every $x \in var(\Sigma)$, the word $S(x)$ can be constructed in time polynomial in $|\Sigma|$ (hence the length of $S(x)$ is polynomial in $|\Sigma|$);

    (b) $V_{adag}S : var(\Sigma) \rightarrow \mathcal{T}_C$ is a solution of Σ.

3. The relation $\{(w, D, a) \in \{0, 1\}^* \times \Delta_{adag} \times (C \cup \{\rightarrow\}) \mid w$ leads in $V_{adag}(D)$ to a node labelled $a\}$ is computable in polynomial time.

---

[3]We could allow also out-degree 0, but then it would be little more difficult to generalize this approach to the case of a disjoint union of lattices.

4. For every $D_1, D_2 \in \Delta_{adag}$ there is $D \in \Delta_{adag}$, computable in polynomial time from D_1 and D_2 such that

$$V_{adag}(D) = V_{adag}(D_1) \rightarrow V_{adag}(D_2).$$

Unfortunately, the relation $\{(D_1, D_2) \in \Delta_{adag} \times \Delta_{adag} \mid V_{adag}(D_1) \le V_{adag}(D_2)\}$ is *not* computable in polynomial time, unless $P = co - NP$. It can be shown that this relation is $co - NP$−complete.

The above approach is easily extendable to the case when $\langle C, \le \rangle$ is a disjoint union of lattices. The only change one has to make is the notion of a shape. As in [Tiu92], if C is a disjoint union of lattices $C_1 \cup C_2 \cup \ldots \cup C_n$, then shapes are types in $\mathcal{T}_{\{*_1,\ldots,*_n\}}$, where $*_i$ represents the class of constants from the lattice C_i. We still have that for every shape $sh \in \mathcal{T}_{\{*_1,\ldots,*_n\}}$, the set $\mathcal{T}_C^{sh}$ forms a lattice and we can repeat the above definition of an alternating dag. It would be interesting to see whether this approach can be extended to classes of posets different than disjoint union of lattices. For example in [Ben93] it is shown that the PTIME algorithm of [Tiu92] can be extended to a bigger class of posets called *Helly posets*. It would be interesting to see wheteher Helly posets admit a succint system of notation. On the other hand, it follows from the PSPACE lower bound of [Tiu92] that crowns do not admit a succint system of notation which satisfies the above properties (1–4).

References

[Ben93]   M. Benke. Efficient type reconstruction in the presence of inheritance. In Eds. A. M. Borzyszkowski, S. Sokolowski, editor, *MFCS'93: Mathematical Foundations of Computer Science, Proc. 18th Intern. Symp.*, volume 711 of *Lecture Notes in Computer Science*, pages 272–280. Springer Verlag, 1993.

[DKM84] C. Dwork, P. Kanellakis, and J.C. Mitchell. On the sequential nature of unification. *J. Logic Programming*, 1:35–50, 1984.

[Fre97]   A. Frey. Solving subtype inequalities in polynomial space. available from http://www.ensmp.fr/ frey/Notes/frey-02-v1.2.ps.gz, 1997.

[HM95]   M. Hoang and J.C. Mitchell. Lower bounds on type inference with subtypes. In *Conf. Rec. 22nd ACM Symposium on Principles of Programming Languages*, pages 176–185, 1995.

[Tiu92]   J. Tiuryn. Subtype inequalities. In *Proc. IEEE Symp. on Logic in Computer Science*, pages 308–315, 1992.

A New Method for Bounding the Complexity of Modal Logics

David Basin[1], Seán Matthews[2], Luca Viganò[2]

[1] Institut für Informatik, Universität Freiburg
Am Flughafen 17, D-79110 Freiburg, Germany
basin@informatik.uni-freiburg.de
[2] Max-Planck-Institut für Informatik
Im Stadtwald, D-66123 Saarbrücken, Germany
sean@mpi-sb.mpg.de, luca@mpi-sb.mpg.de

Abstract. We present a new proof-theoretic approach to bounding the complexity of the decision problem for propositional modal logics. We formalize logics in a uniform way as sequent systems and then restrict the structural rules for particular systems. This, combined with an analysis of the accessibility relation of the corresponding Kripke structures, yields decision procedures with bounded space requirements. As examples we give $O(n \log n)$ space procedures for the modal logics K and T.

1 Introduction

We present a new proof-theoretic approach to bounding the complexity of the decision problem for propositional modal logics. We formalize logics in a uniform way as cut-free labelled sequent systems and then restrict the structural rules for particular systems. This, combined with an analysis of the accessibility relation of the corresponding Kripke structures, yields decision procedures with space requirements that are easily bounded. As examples we give $O(n \log n)$ space decision procedures for the modal logics K and T.

This paper is based on previous work where we give a *labelled deduction* framework, based on Kripke semantics, in which we are able to formalize a large class of propositional modal logics (essentially those with accessibility relations axiomatizable using Horn-clauses; e.g. K, T, S4, S5, etc.) as labelled natural deduction systems [2]. There we show that modal logics in this class can be decomposed into two separated parts: a fixed base logic, and a relational theory, which we extend to generate particular logics in the class.

Now we use this framework to develop complexity results. We begin by showing a bijection between our labelled natural deduction systems and cut-free labelled sequent systems with a subformula property. This provides us with a bound on the number of different formulas that can appear in a proof. However it does not bound the number of times a formula can appear in a sequent. To do this we prove that we can bound applications of contraction. In general this is impossible (in the same way that it is sometimes necessary to contract universally quantified formulas in first-order logic [5], or implication in intuitionistic propositional logic [7]). However we show that bounds do exist for particular modal logics. Furthermore, our systems provide for restricted reasoning about the accessibility relation that we exploit to factor complexity questions about the accessibility relation out of our analysis. Taken together these results allow us to bound the depth of proofs and the size of sequents arising in them, and thus, by 'resource aware' programming, to provide PSPACE upper bounds.

The space bounds we arrive at are not new; they were recently discovered, using different techniques, by Hudelmaier [8]. What is new is the use of a labelled sequent calculus to provide a framework for complexity bounds combined with the analysis of contraction for particular logics in this general setting. We view this as a first step towards a general framework for both applying proof theory to the analysis of the decision problem for families of modal, and other similarly formalized non-classical, logics, see [1,14], and for implementing decision procedures for these logics.

The remainder of this paper is organized as follows. In §2 we review the essentials of our labelled natural deduction systems and show that they are intertranslatable with the cut-free labelled sequent systems we use here. In §3 we show how contraction can be bounded in our sequent systems for the modal logics K and T. In §4 we show that these results can be combined with bounds on relational reasoning to bound the space needed in the search for a proof. In §5 we discuss related work, and in §6 draw conclusions. Due to lack of space, proofs have been omitted or considerably shortened; see [14] for details.

2 Natural deduction and sequent systems

In this paper we use cut-free labelled sequent systems for modal logics. To show their soundness and completeness with respect to standard Kripke semantics, we use sound and complete natural deduction systems from earlier work [2] and show that our sequent systems are intertranslatable with them. In this way it follows that cut is admissible; i.e., rather than eliminating cut directly, we exploit results we already have for natural deduction systems.

2.1 Labelled natural deduction systems

Let W be a denumerable set of variables, called *labels*, and R a binary relation over W. We factor our presentations into two parts, a fixed *base* logic and a varying *relational theory*, corresponding to the two sorts of syntactic entities we consider: labelled well-formed formulas (*lwffs*), $x{:}A$, which pair a label x and a modal formula A, and relational well-formed formulas (*rwffs*), $x\,R\,y$. The lwff $x{:}A$ expresses that the formula A holds at world x, and the rwff $x\,R\,y$ expresses that (x,y) is in the accessibility relation R. The following rules define the base logic for the (functionally complete) $\bot/{\to}/\Box$ fragment of our natural deduction systems.[1]

---

[1] Other connectives can be defined in the usual manner, e.g. $\neg A$ as $A \to \bot$ and $\Diamond A$ as $\neg\Box\neg A$, and their corresponding rules derived.

$$\frac{y{:}\bot}{x{:}A}\ \bot E \qquad \frac{x{:}B}{x{:}A \to B}\ \to I \qquad \frac{x{:}A \to B \quad x{:}A}{x{:}B}\ \to E \qquad \frac{y{:}A}{x{:}\Box A}\ \Box I \qquad \frac{x{:}\Box A \quad x\,Ry}{y{:}A}\ \Box E$$

$$[x{:}A \to \bot] \qquad [x{:}A] \qquad\qquad\qquad\qquad\qquad [x\,Ry]$$

($\Box I$ has the side condition that y is different from x and does not occur in any assumption on which $y{:}A$ depends other than $x\,Ry$.) We then generate natural deduction systems for modal logics by extending the base logic with a relational theory, i.e., by adding rules for the accessibility relation R. For instance, if we take the base logic alone we get the system K_{ND} formalizing the modal logic K, from which we can produce the system T_{ND} or $K4_{ND}$ by adding respectively a rule for reflexivity or transitivity,

$$\frac{}{x\,Rx} \qquad \text{or} \qquad \frac{x\,Ry \quad y\,Rz}{x\,Rx}$$

This framework is based on a deliberate compromise where we restrict ourselves to modal logics with accessibility relation axiomatizable in terms of Horn-clauses. In return we have that the formalization of any logic in this class is both sound and complete with respect to the corresponding Kripke semantics and has proof-normalization properties. More specifically, in [2] we prove the following: (i) for logics where the theory of the binary accessibility relation characterizing the Kripke structure is Horn-clause axiomatizable (this is a very large class including almost all the Geach hierarchy, in particular the logics K, T, K4, S4, S5, etc.) the natural deduction systems are sound and complete with respect to the corresponding Kripke semantics, and (ii) proofs can be transformed into an equivalent normal form similar to that defined by Prawitz [12], suitably modified for labelled and relational formulas.

2.2 Labelled sequent systems

Let Γ and Δ (possibly annotated) vary over finite multisets of lwffs and rwffs respectively. The cut-free labelled sequent system K consists of the following axioms and rules, where $\Box R$ has the side condition that y does not occur in $\Gamma, \Delta \vdash \Gamma', x{:}\Box A$.

$$\frac{}{x{:}A \vdash x{:}A}\ AXl \qquad \frac{}{x\,Ry \vdash x\,Ry}\ AXr \qquad \frac{}{y{:}\bot \vdash x{:}A}\ \bot L$$

$$\frac{\Gamma, \Delta \vdash \Gamma', x{:}A \quad x{:}B, \Gamma, \Delta \vdash \Gamma'}{x{:}A \to B, \Gamma, \Delta \vdash \Gamma'}\ \to L \qquad \frac{x{:}A, \Gamma, \Delta \vdash \Gamma', x{:}B}{\Gamma, \Delta \vdash \Gamma', x{:}A \to B}\ \to R$$

$$\frac{\Delta \vdash x\,Ry \quad y{:}A, \Gamma, \Delta \vdash \Gamma'}{x{:}\Box A, \Gamma, \Delta \vdash \Gamma'}\ \Box L \qquad \frac{\Gamma, \Delta, x\,Ry \vdash \Gamma', y{:}A}{\Gamma, \Delta \vdash \Gamma', x{:}\Box A}\ \Box R$$

$$\frac{\Gamma, \Delta \vdash \Gamma'}{\Gamma, \Delta, x\,Ry \vdash \Gamma'}\ WrL \qquad \frac{\Gamma, \Delta \vdash \Gamma'}{x{:}A, \Gamma, \Delta \vdash \Gamma'}\ WlL \qquad \frac{\Gamma, \Delta \vdash \Gamma'}{\Gamma, \Delta \vdash \Gamma', x{:}A}\ WlR$$

$$\frac{\Delta, x\,Ry, x\,Ry \vdash u\,Rv}{\Delta, x\,Ry \vdash u\,Rv}\;CrL \qquad \frac{x{:}A, x{:}A, \Gamma, \Delta \vdash \Gamma'}{x{:}A, \Gamma, \Delta \vdash \Gamma'}\;ClL \qquad \frac{\Gamma, \Delta \vdash \Gamma', x{:}A, x{:}A}{\Gamma, \Delta \vdash \Gamma', x{:}A}\;ClR$$

Analogously to the natural deduction systems, we can then, e.g., obtain the cut-free labelled sequent system T or K4 by extending the relational theory with reflexivity or transitivity, formalized by the rules

$$\frac{}{\Delta \vdash x\,Rx}\;refl \qquad \text{or} \qquad \frac{\Delta \vdash x\,Ry \quad \Delta \vdash y\,Rz}{\Delta \vdash x\,Rz}\;trans$$

Normalizing natural deduction systems and sequent systems without cut are closely related [4,12,13]. If we write $\Gamma, \Delta \vdash_{ND} x{:}A$ to indicate a derivation of $x{:}A$ from assumptions Γ, Δ in our natural deduction systems, then we have:

Theorem 1. $\Delta \vdash_{ND} x\,Ry$ iff $\Delta \vdash x\,Ry$; and $\Gamma, y_1{:}\neg B_1, \ldots, y_n{:}\neg B_n, \Delta \vdash_{ND} x{:}A$ iff $\Gamma, \Delta \vdash x{:}A, y_1{:}B_1, \ldots, y_n{:}B_n$.

Proof. (Sketch) We adapt and extend the proof given by Prawitz [12, App. A], to which the reader is referred for details.

Left to right: By induction on the length of normal natural deduction derivations. We consider an application of $\Box I$ as an illustrative case: suppose that from $\Gamma, y_1{:}\neg B_1, \ldots, y_n{:}\neg B_n, \Delta, x\,Ry \vdash_{ND} y{:}A$ we infer $\Gamma, y_1{:}\neg B_1, \ldots, y_n{:}\neg B_n, \Delta \vdash_{ND} x{:}\Box A$. By the induction hypothesis we have $\Gamma, \Delta, x\,Ry, \vdash y{:}A, y_1{:}B_1, \ldots, y_n{:}B_n$, and thus $\Gamma, \Delta \vdash x{:}\Box A, y_1{:}B_1, \ldots, y_n{:}B_n$ by $\Box R$ (the side condition for $\Box R$ holds by the side condition for $\Box I$).

Right to left: By induction on the length of the proof of the sequent. We consider an application of $\Box R$ as an illustrative case. By the induction hypothesis, given $\Gamma, \Delta, x\,Ry \vdash y{:}A, y_1{:}B_1, \ldots, y_n{:}B_n$ we have $\Gamma, y_1{:}\neg B_1, \ldots, y_n{:}\neg B_n, \Delta, x\,R\,y \vdash_{ND} y{:}A$, and we derive $\Gamma, y_1{:}\neg B_1, \ldots, y_n{:}\neg B_n, \Delta \vdash_{ND} x{:}\Box A$ by $\Box I$ (the side condition for $\Box I$ holds by the side condition for $\Box R$).

From Theorem 1 and previous results in [2], it follows that our sequent systems are sound and complete with respect to the corresponding Kripke semantics.

3 Contraction

In this section we show that for particular logics in our framework applications of the contraction rules can be eliminated (in K) or bounded (in T).

We begin by introducing some terminology. We call the formulas that pass through the application of a rule unchanged (they appear in the premises and in the conclusion) the *parametric formulas* of the rule; the formula contracted or introduced in the conclusion is the *principal formula* of the rule, and the formulas from which the principal formula derives are the *active formulas* of the rule. We call *lwff-rules* the rules that have an lwff as principal formula, namely *WlL*, *WlR*, *ClL*, *ClR*, $\rightarrow L$, $\rightarrow R$, $\Box L$ and $\Box R$. Suppose that a sequent S results from sequents $S_1, \ldots, S_n$ by first applying the lwff-rule ρ_1 and then applying the lwff-rule ρ_2, where (i) each of the premises of ρ_2 results from an application of

ρ_1, (ii) each application of ρ_1 introduces or contracts the same lwff, and (iii) the principal formula of ρ_1 is parametric in the application of ρ_2. We then say that ρ_1 is *permutable* with respect to ρ_2 (ρ_1 *permutes* with respect to ρ_2), if the original inference can be replaced by one in which the sequent S is derived from $S_1, \ldots, S_n$ by applying first ρ_2 and then ρ_1. By inspection of the rules, we have that:

Lemma 2. *Every lwff-rule is permutable with respect to any other lwff-rule, with the exception of $\Box L$, which is permutable with respect to every lwff-rule other than $\Box R$.*

For example:

$$\dfrac{\dfrac{u{:}A, \Gamma, \Delta, x\,Ry \vdash \Gamma', u{:}B, y{:}C}{\dfrac{\Gamma, \Delta, x\,Ry \vdash \Gamma', u{:}A \to B, y{:}C}{\Gamma, \Delta \vdash \Gamma', u{:}A \to B, x{:}\Box C}\ \Box R}\ {\to}R}{} \quad \text{permutes to} \quad \dfrac{\dfrac{u{:}A, \Gamma, \Delta, x\,Ry \vdash \Gamma', u{:}B, y{:}C}{\dfrac{u{:}A, \Gamma, \Delta \vdash \Gamma', u{:}B, x{:}\Box C}{\Gamma, \Delta \vdash \Gamma', u{:}A \to B, x{:}\Box C}\ {\to}R}\ \Box R}{}$$

The reason that $\Box L$ is, in general, not permutable with respect to $\Box R$ is that $\Box L$ may have the same active rwff $x\,Ry$ as $\Box R$; examples of this situation are given below, e.g. in the proof-fragment (1).

Before showing that we can eliminate ClL and ClR in K, we observe that:

Lemma 3. *The rule CrL is eliminable in K and T.*

This follows because contracted rwffs can be introduced in the antecedent of a sequent only by applications of the rule WrL; the principal rwff of *refl* is introduced in the succedent of the sequent. Therefore, we just need to delete both the CrL and the corresponding WrL. For example, we transform

$$\dfrac{\dfrac{\dfrac{\Pi_1}{\Delta_1 \vdash u\,Rv}}{\Delta_1, x\,Ry \vdash u\,Rv}\ WrL}{\dfrac{\dfrac{\Pi_2}{\Delta_2, x\,Ry, x\,Ry \vdash u\,Rv}}{\Delta_2, x\,Ry \vdash u\,Rv}\ CrL}{} \qquad \text{to} \qquad \dfrac{\dfrac{\Pi_1}{\Delta_1 \vdash u\,Rv}}{\dfrac{\Pi_2^\dagger}{\Delta_2, x\,Ry \vdash u\,Rv}}{}$$

where, here and in the following, we use '$\dagger$' to denote that in the derivation $\Pi_i^\dagger$ we apply the same sequence of rules applied in the derivation Π_i; in other words, Π_i and $\Pi_i^\dagger$ differ only in their parametric formulas.

We define the *grade* of an lwff $x{:}A$ to be the number of connectives in A. We define the *rank* of a contraction of $x{:}A$ to be the largest number of steps immediately preceding the conclusion of the contraction and containing at least one of the two contracted instances of $x{:}A$. Since, by the permutability of the rules, we can always transform a proof so that the contraction of $x{:}A$ immediately follows the step introducing the second instance of $x{:}A$, the rank measures how many steps stand between the introduction of the first and that of the second instance of $x{:}A$. Thus, the minimum possible rank is 2.

3.1 Contraction elimination in K

Theorem 4. *The rules ClL and ClR are eliminable in K, i.e., every theorem provable in K has a proof in which there are no applications of ClL and ClR.*

Proof. (Sketch) We adapt and extend the proof for classical propositional logic given by Zeman [16], and, given a proof Π of $\vdash d{:}D$ in K, proceed by three nested inductions. The first induction is on the number of contractions in Π. Suppose that Π contains $i + 1$ contractions, and pick a 'highest' contraction in Π, i.e. consider a subproof of Π that ends with a contraction of $x{:}A$ and such that the proof above the contraction is contraction-free. We show how to eliminate this contraction to obtain a proof Π' of $\vdash d{:}D$ that contains i contractions; we proceed first by induction on the grade of $x{:}A$ and then by induction on the rank of the contraction. The base case, $grade(x{:}A) = 0$ is straightforward; for example, if one instance of $x{:}A$ is introduced by weakening, then we conclude by deleting this weakening and the contraction. There is one step for each rule introducing $x{:}A$, and we consider an illustrative case; the other cases follow similarly. Suppose that A is $\Box B$, both instances of $x{:}\Box B$ are introduced by applications of $\Box L$, and the contraction is an application of ClL with rank $j + 1$. Then the only interesting subcase is when the rwff $x\,R\,z$, active in the uppermost application of $\Box L$, is active also in an application of $\Box R$ before the lowest application of $\Box L$, i.e.

$$\frac{\dfrac{\dfrac{\dfrac{\overline{x\,R\,z \vdash x\,R\,z}\ AXr}{\vdots\ WrL}}{\Delta_1, x\,R\,z \vdash x\,R\,z}\quad \Pi_1 \atop z{:}B, \Gamma_1, \Delta_1, x\,R\,z \vdash \Gamma_1'}{\dfrac{x{:}\Box B, \Gamma_1, \Delta_1, x\,R\,z \vdash \Gamma_1'}{\Pi_2}}\ \Box L \qquad \dfrac{\dfrac{\dfrac{\overline{x\,R\,y \vdash x\,R\,y}\ AXr}{\vdots\ WrL}}{\Delta, x\,R\,y \vdash x\,R\,y}\quad \dfrac{x{:}\Box B, \Gamma_2, \Delta_2, x\,R\,z \vdash z{:}C, \Gamma_2'}{\dfrac{x{:}\Box B, \Gamma_2, \Delta_2 \vdash x{:}\Box C, \Gamma_2'}{\Pi_3}}\ \Box R \quad y{:}B, x{:}\Box B, \Gamma, \Delta, x\,R\,y \vdash \Gamma'}{}}{}$$

$$\frac{x{:}\Box B, x{:}\Box B, \Gamma, \Delta, x\,R\,y \vdash \Gamma'}{\dfrac{x{:}\Box B, \Gamma, \Delta, x\,R\,y \vdash \Gamma'}{\begin{array}{c}\Pi_0\\ \vdash d{:}D\end{array}}\ ClL} \ \Box L \qquad (1)$$

(Vertical dots labelled with *WrL* denote a sequence of applications of *WrL*.)

As explained for Lemma 2, we cannot, in general, reduce the rank of ClL by permuting the uppermost $\Box L$ with respect to the rule following it, since, if Π_2 is empty, the uppermost $\Box L$ is not permutable with respect to the $\Box R$ with active rwff $x\,R\,z$. We can however permute this $\Box R$ with respect to the lowest $\Box L$. To show this, we reason as follows. Since we are proving $\vdash d{:}D$, Π_0 must contain an application of $\Box R$ with active rwff $x\,R\,y$, and for this to be possible y must be an arbitrary world accessible from x. In particular, $y \neq x$. Thus, since $x\,R\,y$ and since there are no relational rules, $x{:}\Box C$ cannot follow from $y{:}B$, i.e. $x{:}\Box C$ and $y{:}B$ are independent. Then $x{:}\Box C$ must follow from some formula in Γ or Γ', and Π_3 can be divided into two separate subproofs, Π_4 and Π_5. Π_4 introduces $y{:}B$

and $x\,Ry$, and $x{:}\Box C$ is active in Π_5. (If $y{:}B$ and $x\,Ry$ are contained in Γ_2 and Δ_2, then Π_4 is empty and $\Pi_5 = \Pi_3$.) Given this separation, we can transform (1) to:

$$
\cfrac{
\cfrac{
\cfrac{
\cfrac{\overline{x\,Rz \vdash x\,Rz}\ AXr}{\vdots\ WrL}
}{\Delta_1, x\,Rz \vdash x\,Rz \qquad
\cfrac{\cfrac{z{:}B, \Gamma_1, \Delta_1, x\,Rz \vdash \Gamma_1'}{\ }\ \Pi_1}{\ }
}\ \Box L
}{x{:}\Box B, \Gamma_1, \Delta_1, x\,Rz \vdash \Gamma_1'}
}{\ }
$$

$$
\frac{\overline{x\,Rz \vdash x\,Rz}\ AXr}{\vdots\ WrL}
$$

$$
\begin{array}{c}
\dfrac{\overline{x\,Rz \vdash x\,Rz}\ AXr}{\vdots\ WrL} \\[2pt]
\Delta_1, x\,Rz \vdash x\,Rz \qquad z{:}B, \Gamma_1, \Delta_1, x\,Rz \vdash \Gamma_1' \\
\hline
x{:}\Box B, \Gamma_1, \Delta_1, x\,Rz \vdash \Gamma_1' \\
\Pi_2 \\
x{:}\Box B, \Gamma_2, \Delta_2, x\,Rz \vdash z{:}C, \Gamma_2' \\
\Pi_4^{\dagger} \\
\dfrac{\overline{x\,Ry \vdash x\,Ry}\ AXr}{\vdots\ WrL} \quad y{:}B, x{:}\Box B, \Gamma_3, \Delta_3, x\,Ry, x\,Rz \vdash z{:}C, \Gamma_3' \\
\Delta_3, x\,Ry, x\,Rz \vdash x\,Ry \qquad x{:}\Box B, x{:}\Box B, \Gamma_3, \Delta_3, x\,Ry, x\,Rz \vdash z{:}C, \Gamma_3' \\
\hline
x{:}\Box B, x{:}\Box B, \Gamma_3, \Delta_3, x\,Ry \vdash \Gamma_3' \\
\Pi_5^{\dagger} \\
x{:}\Box B, x{:}\Box B, \Gamma, \Delta, x\,Ry \vdash \Gamma' \\
\hline
x{:}\Box B, \Gamma, \Delta, x\,Ry \vdash \Gamma' \\
\Pi_0 \\
\vdash d{:}D
\end{array}
\tag{2}
$$

(During the transformation it might be necessary to rename some labels to avoid clashes.) The rank of $C\!lL$ is now less than $j + 1$ and we can then conclude by applying the induction hypothesis. Reasoning similarly, we can eliminate all applications of $C\!lL$ and $C\!lR$, and thus conclude the proof of the theorem.

Theorem 4 provides the basis for showing that K is decidable (see §4). Furthermore, the proof of Theorem 4 is extensible: when we add relational rules to K, we only need to consider the new cases that are generated by these rules. In particular, we must only investigate the eliminability of $C\!lL$ when applied with principal formula of the form $x{:}\Box B$; by Lemma 3 and Theorem 4 it immediately follows that in T we can eliminate $C\!lR$, as well as all applications of $C\!lL$ with principal formula different from $x{:}\Box B$. This generalization holds also for other logics extending K, as shown in [14]. However, it turns out that, even for a simple logic like T, applications of $C\!lL$ with principal formula $x{:}\Box B$ are not eliminable; for example, the T-theorem $x{:}\neg\Box\neg(A \to \Box A)$ cannot be proved in our system without one application of $C\!lL$ with principal formula $x{:}\Box\neg(A \to \Box A)$.

3.2 Bounded contraction in T

Although applications of $C\!lL$ with principal formula $x{:}\Box B$ are needed in T, if we can bound their use then we need not give up decidability. To this end, we show that when proving a theorem in T, we need not contract each lwff $x{:}\Box B$ more than once in every branch. This amounts to restricting $C\!lL$ with two side conditions: that the contracted lwff has the form $x{:}\Box B$, and that $x{:}\Box B$ can be contracted only once in a branch. We enforce the second condition by marking every $x{:}\Box B$ with a bit, a 1 or a 0, depending on whether a contraction has already been applied or not. We call this index the *contraction index* of the lwff, and we introduce the convention that when, in a forwards proof, we introduce a new instance of $x{:}\Box B$ in the antecedent of a sequent, we set its contraction index to 0; symmetrically, when in a backwards proof we first obtain $x{:}\Box B$ in the antecedent of a sequent, we set its contraction index to 1. Thus we restrict ClL to the rule

$$\frac{x{:}(\Box B)^0, x{:}(\Box B)^0, \Gamma, \Delta \vdash \Gamma'}{x{:}(\Box B)^1, \Gamma, \Delta \vdash \Gamma'} \; ClL_T,$$

which explicitly requires that both contracted lwffs have contraction index 0. The following theorem shows that these restrictions preserve completeness.

Theorem 5. *Every theorem provable in* T *has a proof in which there are no contractions, except for applications of* ClL *with principal formula of the form* $x{:}\Box B$. *However,* ClL *need not be applied more than once with the same principal formula* $x{:}\Box B$ *in every branch, and we can restrict* ClL *to be* ClL_T.

Proof. (Sketch) We extend Theorem 4 by considering the additional cases that arise in T when ClL is applied with principal formula $x{:}\Box B$. In particular, to show that ClL can be restricted to ClL_T, we show that if ClL is applied with principal formula $x{:}\Box B$ more than once in a branch $\mathcal{B}$ of a proof of $\vdash d{:}D$, then $\vdash d{:}D$ has a proof in which there is at most one application of ClL with principal formula $x{:}\Box B$ in $\mathcal{B}$. By the permutability of the rules, we can assume that if ClL is applied n times with principal formula $x{:}\Box B$ in $\mathcal{B}$, then these n contractions are performed as an uninterrupted sequence of steps immediately following the rule that introduces the n-th instance of $x{:}\Box B$. Then for each branch $\mathcal{B}$ we proceed as follows. If lwffs of different grade are contracted in $\mathcal{B}$, then we pick a 'lowest' sequence of n applications of ClL with contracted lwff of greatest grade. If $n = 1$, we move on to the next sequence of contractions. If $n > 1$, we consider the first two applications in the sequence, and we either eliminate the uppermost one, or we transform it into an application of ClL with principal formula of smaller grade. By iterating this transformation, we obtain the desired proof.

As an illustrative case, let both instances of $x{:}\Box B$ be introduced by an application of $\Box L$ with active rwff $x\,R\,x$, and consider the following branch where Π_1 and Π_2 do not contain applications of ClL with principal formula $x{:}\Box B$ or an lwff of grade greater than the grade of $x{:}\Box B$.

$$\cfrac{\cfrac{\cfrac{\overline{\Delta \vdash x\,R\,x}\;\textit{refl} \quad \cfrac{\cfrac{\overline{\Delta_1 \vdash x\,R\,x}\;\textit{refl} \quad \cfrac{\Pi_1}{x{:}B, x{:}\Box B, \Gamma_1, \Delta_1 \vdash \Gamma_1'}}{x{:}\Box B, x{:}\Box B, \Gamma_1, \Delta_1 \vdash \Gamma_1'}\;\Box L \quad \cfrac{\Pi_2}{x{:}B, x{:}\Box B, x{:}\Box B, \Gamma, \Delta \vdash \Gamma'}}{x{:}\Box B, x{:}\Box B, x{:}\Box B, \Gamma, \Delta, \vdash \Gamma'}\;\Box L}{x{:}\Box B, x{:}\Box B, \Gamma, \Delta \vdash \Gamma'}\;ClL}{x{:}\Box B, \Gamma, \Delta \vdash \Gamma'}\;ClL}{\genfrac{}{}{0pt}{}{\Pi_0}{\vdash d{:}D}} \qquad (3)$$

Since applications of $\Box L$ with active rwff introduced by *refl* are permutable with respect to every rule (even with respect to $\Box R$), we can transform (3) to:

$$
\cfrac{
\cfrac{\Delta \vdash x R x}{}\;refl
\quad
\cfrac{
\cfrac{
\cfrac{\Delta \vdash x R x \; refl \quad
\cfrac{
\Pi_1 \\
x{:}B, x{:}\Box B, \Gamma_1, \Delta_1 \vdash \Gamma_1' \\
\Pi_2^\dagger \\
x{:}B, x{:}B, x{:}\Box B, \Gamma, \Delta \vdash \Gamma'
}{x{:}B, x{:}\Box B, x{:}\Box B, \Gamma, \Delta \vdash \Gamma'}\;\Box L
}{x{:}B, x{:}\Box B, \Gamma, \Delta \vdash \Gamma'}\;C\!lL
}{x{:}\Box B, x{:}\Box B, \Gamma, \Delta \vdash \Gamma'}\;\Box L
}{x{:}\Box B, \Gamma, \Delta \vdash \Gamma'}\;C\!lL
$$
$$
\Pi_0 \\
\vdash d{:}D
$$

$$(4)$$

Then, we transform (4) by replacing the uppermost applications of $\Box L$ and $C\!lL$ with principal formula $x{:}\Box B$ with an application of $C\!lL$ with principal formula $x{:}B$, which has smaller grade, i.e.

$$
\cfrac{
\cfrac{\Delta \vdash x R x}{}\;refl
\quad
\cfrac{
\cfrac{
\Pi_1 \\
x{:}B, x{:}\Box B, \Gamma_1, \Delta_1 \vdash \Gamma_1' \\
\Pi_2^\dagger \\
x{:}B, x{:}B, x{:}\Box B, \Gamma, \Delta \vdash \Gamma'
}{x{:}B, x{:}\Box B, \Gamma, \Delta \vdash \Gamma'}\;C\!lL
}{x{:}\Box B, x{:}\Box B, \Gamma, \Delta \vdash \Gamma'}\;\Box L
}{x{:}\Box B, \Gamma, \Delta \vdash \Gamma'}\;C\!lL
$$
$$
\Pi_0 \\
\vdash d{:}D
$$

4 Complexity of proof search

In this section we show the decidability of our sequent systems by combining bounds on contraction with bounds on the number of worlds and relational formulas that can be generated in proofs. In particular, we show that contraction elimination for K and bounded contraction elimination for T give rise to a proof procedure for these logics that uses $O(n \log n)$ space. Hence, these results, combined with the soundness and completeness of our systems with respect to the standard Kripke semantics for the modal logics K and T, tell us that the provability (validity) problems for these logics are decidable in Polynomial Space.

For clarity, we first consider labelled propositional logic, and then extensions to K and T.

4.1 Propositional logic

Labelled propositional logic, PL, consists of the axioms and rules AXl, $\bot L$, $\rightarrow L$, $\rightarrow R$, WlL and WlR, operating over sequents that contain only lwffs (i.e., Δ is always empty). The rules can be applied backwards to build a proof 'top-down', starting with the end-sequent and working towards the axioms (leaves) of the proof. To show that provability is in Polynomial Space, we employ the following definitions and facts.

Let the degree of an lwff $x{:}A$ simply be the size of A, $|A|$ (the number of symbols in its string representation), and the degree of a sequent the sum of the degrees of its lwffs. We say that $y{:}B$ is a subformula of $x{:}A$ if B is a subformula of A according to the standard definition in [12].

Fact 6. *In each rule of* PL, *each premise has smaller degree then the conclusion.*

Fact 7. *The subformula property holds:* PL *proofs of a sequent S generate only sequents that contain subformulas of S.*

Fact 6 tells us that every proof of a sequent S is bounded in length by the degree of S, which is in turn bounded by its size $n = |S|$. Fact 7 tells us that we can represent any generated sequent in $O(n^2)$ space, since there are at most $O(n)$ subformulas of size $O(n)$. We can reduce this to $O(n \log n)$ space, since we can represent a subformula by an index into the end-sequent S, which requires only $\log n$ bits.[2]

To test provability in Polynomial Space, we must first overcome problems related to two kinds of branching in the search space for proofs.

Conjunctive branching: applying rules with multiple premises builds a branching tree, where all branches must be proven.

Disjunctive branching: more than one rule may be applicable and a given rule may be applicable in different ways.

Conjunctive branching, caused by rules like $\rightarrow L$, leads to proofs that are exponential in size. Disjunctive branching arises when more than one rule may be applied to a sequent or when a single rule can be applied in different ways (e.g., a weakening).[3] This is reflected by a branching point in the search space for proofs rather than in proofs themselves.

To bound space requirements, we adapt and extend standard techniques [7,8]. In particular, rather than storing entire proofs, we store a sequent and a stack that maintains information sufficient to reconstruct branching points. Each stack entry is a triple: the name of the rule applied, the principal formula of the sequent, and an index. The index records sufficient information such that on return to the branching point another branch (if one remains) can be generated. For example, for $\rightarrow L$, the index is a bit indicating the first or second premise; for weakening, the index is a pointer indicating which premise is weakened.

A proof begins with the end-sequent S and the empty stack. Each rule application (we assume rules are ordered and we apply them in order) generates

---

[2] Formulae are labelled but since there are no modality rules, $\Box L$ and $\Box R$, the labels play no essential role and do not change during proofs.

[3] If we are concerned with average time complexity, as opposed to space complexity, we can exploit the fact that all PL rules, other than weakening, are invertible and permute with each other. Thus we can apply rules in some fixed order (without backtracking) and apply weakening only before axioms.

a new sequent and appropriately extends the stack. If the generated sequent is an axiom and the stack contains no conjunctive branching points that still need to be explored, then S is provable. Otherwise we pop entries off the stack until we find a conjunctive branching point that must be further explored and then generate the next branch (first incrementing the index on the stack to record this). Alternatively, if we arrive at a sequent that is not an axiom and no rule applies, then we pop stack entries and continue at the first available disjunctive branching point with the choices that remain (e.g., apply a different weakening, determined by the index, or apply the next available rule). If no such branching point remains, then S is not provable.

This procedure terminates since the stack depth is $O(n)$ and the branching is bounded (by n, the number of rules, and the maximum number of premises). Furthermore, since each entry requires $O(\log n)$ space (we can store the principal formula as an index into the end-sequent S), the algorithm requires $O(n \log n)$ space. Since a formula A of propositional logic is provable iff the sequent $\vdash x{:}A$ is provable in PL, we have the following lemma:

Lemma 8. *Provability for propositional logic is decidable in $O(n \log n)$ space.*

4.2 Extension to K

The sequent system K extends PL with rules for $\Box$, as well as AXr and the weakening rule for rwffs, WrL. This proof system also admits an $O(n \log n)$ space provability procedure, but the details are a bit different.

Fact 6 no longer applies since WrL, although it eliminates an rwff, does not reduce the number of connectives. So we define instead the degree of an lwff $x{:}A$ to be twice the size of A, $2 \times |A|$, and the degree of a sequent S to be the sum of the degrees of the lwffs in S plus the number of rwffs in S. Now we can extend Fact 6 for the rules of K, and since the degree function is linear, any proof will have depth $O(n)$, where $n = |S|$.

Fact 7 holds for K, although subformulas may be labelled differently. Unlike for PL we now must take labels seriously, since applications of $\Box R$ introduce new labels (and rwffs) and applications of $\Box L$ label subformulas with different (currently in existence) labels. Note that we can only generate a new label and a new rwff for each application of $\Box R$. This is bounded above in any branch by the number of $\Box$ operators appearing in the end-sequent S, i.e., $O(n)$. Hence the number of lwffs and rwffs in the generated sequent is also $O(n)$, since it can contain $O(n)$ lwffs and $O(n)$ rwffs. Moreover, since there are only $O(n)$ labels, we can represent a label using $O(\log n)$ bits. Thus, each lwff and rwff can be represented using $O(\log n)$ bits and any generated sequent can be represented using $O(n \log n)$ space.

The proof procedure for K is essentially the same as for PL. The only interesting difference is that the rule for $\Box L$ introduces disjunctive branching: if the goal is $x{:}\Box A$, any label y currently in existence is a possible candidate for producing $y{:}A$ in the generated sequent. We use the index to record which label we choose so that on backtracking we can generate the next possible candidate.

Note that we can determine what the possible candidates are by looking into the stack and examining previous applications of $\Box R$ and WrL. Note also that $\Box L$ is not invertible, and its application introduces a choice point and other rules can be tried on backtracking. From the soundness and completeness of K and the fact that the entire search space is navigated, if need be, on backtracking, this proof procedure is complete. Termination is analogous to PL and the space requirement is the same. Thus we have:

Theorem 9. *Provability for the modal logic* K *is decidable in* $O(n \log n)$ *space.*

The essence of this result is that contraction elimination, the subformula property, and a linear number of rwffs give us linear depth proofs as for propositional logic. Moreover, given our bound on the number of worlds, the space complexity to represent sequents remains $O(n \log n)$. Hence we get an identical overall upperbound.

Note that K is possibly harder then PL given that provability for the former is PSPACE complete whereas the latter is co-NP complete. One intuition for this is that the $\Box L$ rule introduces a necessary disjunctive branching point in the search space; all such branching points for PL are eliminable by analysis of permutabilities, as indicated above.

4.3 Extension to T

The proof system for T differs from K in two respects:

1. Rwffs used in applications of $\Box L$ arise not only from applications of $\Box R$ but also from applications of *refl*.
2. We need possibly a single instance of contraction on the left for an lwff $x{:}\Box A$ in every branch (Theorem 5).

Since the number of worlds is still bounded to be linear in n, the size of the end-sequent S, the first difference means that we can have up to n additional rwffs in generated sequents, namely $x\, R\, x$ for each of the n possible labels x. This does not change the $O(n \log n)$ space bound required to store sequents.

As described in §3 we can introduce controlled contraction by associating a bit with each lwff $x{:}\Box A$ indicating if it has been contracted or not (this also does not change the above space complexity for storing sequents), and thus restrict ClL to only contract lwffs $x{:}\Box A$ that have not yet been contracted. To accommodate possibly longer proofs, we modify the measure for K such that the degree of an lwff $x{:}A$ is four times the size of A plus 1, $(4 \times |A|) + 1$, when it has not been contracted and twice the size of A, $2 \times |A|$, when it has. Then every rule, including ClL_T, reduces this measure; this means that the length of branches of proofs is still $O(n)$. Thus it follows that:

Theorem 10. *Provability for the modal logic* T *is decidable in* $O(n \log n)$ *space.*

5 Related work

Modal logics are typically shown to be decidable semantically, by showing that they possess the finite model property. However, for many logics, including K and T, there are classes of satisfiable formulas in which every satisfying structure contains exponentially many worlds [6]. Thus it is necessary to analyze the complexity of particular proof procedures to get more refined complexity results.

Ladner [11], building upon the decision procedures of Kripke [10], showed that K and T have respective space complexity of n^2 and n^3. Improved bounds (identical to ours) have been recently given by Hudelmaier [8]. He starts with standard sequent calculi (related to those found in [3,16]) in which cut is admissible and contraction is built into the rules for the modalities. He then uses proof transformations to obtain equivalent systems with new modality rules, which are measure decreasing. Although we end up with the same complexity results, our systems and, in particular, the analysis behind them, are completely different from his; for example, unlike Hudelmaier, we do not need to introduce new modalities.

Labelling has often been used as a way to introduce additional information that can be exploited in the presentation of modal logics and in the analysis of their decidability. For instance, Fitting [3] generalizes the procedures described in [9] to give systematic decision procedures for a wide variety of prefixed modal tableaux systems, including K and T. Although our labelled sequent systems share characteristics with Fitting's systems (the main difference is that in Fitting's systems the different properties of the accessibility relation are expressed procedurally as side conditions on the application of the same set of rules, while we use relational rules to extend the fixed base logic), the analysis of decidability relies on different mechanisms. We show that contraction can be eliminated in K and bounded to a single application in T. Fitting avoids explicit repetitions (contractions) of formulas of the form $x{:}\Box A$ in the antecedent of a sequent (i.e., in his notation, formulas of the form $x \top \Box A$ in a tableau) by having a $\Box L$ rule that does not delete $x{:}\Box A$ but prevents its further use until a new y such that $x\,R\,y$ is introduced in the tableau; termination of the decision procedure is then argued by exploiting König's Lemma. Aside from this difference, this kind of procedure leads to implementations based on loop-checking, as opposed to measure-decreasing rule applications (as in Hudelmaier's and our work). Finally, this analysis, although it establishes decidability, does not result in good complexity bounds.

Other work on proof-theoretic decision procedures for modal logic, related to that of Fitting, is that of Wallen [15], who uses matrix-characterizations to investigate a range of standard modal logics, where duplication is achieved by increasing an index associated to the formulas. However, while he stresses the importance of 'computationally sensitive' characterizations of logics, he does not explicitly address the complexity bounds associated with the systems he investigates. Also, the central point of his work concerns identifying pairs of complementary terms that can be used as the basis of a search procedure, rather than analyzing the structural rules.

6 Conclusion and further work

We have presented a new proof-theoretic approach to bounding the complexity of the decision problem for propositional modal logics. As examples, we have given $O(n \log n)$ space decision procedures for the modal logics K and T, by combining restrictions on the structural rules of labelled sequent systems with an analysis of the accessibility relation of the corresponding Kripke structures. This work is a first step towards a general framework for both applying proof theory to the analysis of complexity of families of modal, and other similarly formalized non-classical, logics, see [1,14], and for implementing decision procedures for these logics. Current work includes extending this approach to other modal logics (extensions to K4 and S4 are given in [14]) and to substructural, e.g. relevance, logics, as well as exploring implementation issues.

References

1. D. Basin, S. Matthews, and L. Viganò. Implementing modal and relevance logics in a logical framework. In L. Carlucci Aiello, J. Doyle, and S. Shapiro, editors, *Proceedings of KR'96*, pages 386–397. Morgan Kaufmann, San Francisco, 1996.
2. D. Basin, S. Matthews, and L. Viganò. Labelled propositional modal logics: theory and practice. *Journal of Logic and Computation*, 1997. To appear.
3. M. Fitting. *Proof methods for modal and intuitionistic logics*. Kluwer, Dordrecht, 1983.
4. G. Gentzen. Investigations into logical deduction. In M. Szabo, editor, *The collected papers of Gerhard Gentzen*, pages 68–131. North Holland, Amsterdam, 1969.
5. J.-Y. Girard, Y. Lafont, and P. Taylor. *Proofs and types*. Cambridge University Press, Cambridge, 1989.
6. J. Y. Halpern and Y. O. Moses. A guide to the completeness and complexity for modal logics of knowledge and belief. *Artificial Intelligence*, 54(3):319–379, 1992.
7. J. Hudelmaier. A $O(n \log n)$-space decision procedure for intuitionistic propositional logic. *Journal of Logic and Computation*, 3(1):63–75, 1993.
8. J. Hudelmaier. Improved decision procedures for the modal logics K, T and S4. In H. Kleine Büning, editor, *Proceedings of CSL'95*, LNCS 1092, pages 320–334. Springer, Berlin, 1996.
9. G. Hughes and M. Cresswell. *An introduction to modal logic*. Routledge, London, 1972.
10. S. Kripke. Semantical analysis of modal logic I. *Zeitschrift für mathematische Logik und Grundlagen der Mathematik*, 9:67–96, 1963.
11. R. E. Ladner. The computational complexity of provability in systems of modal propositional logics. *SIAM Journal of Computing*, 6(3):46–65, 1977.
12. D. Prawitz. *Natural deduction*. Almqvist and Wiksell, Stockholm, 1965.
13. A. S. Troelstra and H. Schwichtenberg. *Basic proof theory*. Cambridge University Press, Cambridge, 1996.
14. L. Viganò. *A framework for non-classical logics*. PhD thesis, Universität des Saarlandes, Saarbrücken, Germany, 1997. Forthcoming.
15. L. Wallen. *Automated deduction in non-classical logics*. MIT Press, Cambridge, Massachusetts, 1990.
16. J. J. Zeman. *Modal logic: the Lewis-modal systems*. Oxford University Press, Oxford, 1973.

Parameter free induction and reflection

Lev D. Beklemishev*

Steklov Mathematical Institute
Gubkina 8, 117966 Moscow, Russia
e-mail: lev@bekl.mian.su

Abstract. We give a precise characterization of parameter free Σ_n and Π_n induction schemata, $I\Sigma_n^-$ and $I\Pi_n^-$, in terms of reflection principles. This allows us to show that $I\Pi_{n+1}^-$ is conservative over $I\Sigma_n^-$ w.r.t. boolean combinations of Σ_{n+1} sentences, for $n \geq 1$. In particular, we give a positive answer to a question by R. Kaye, whether the provably recursive functions of $I\Pi_2^-$ are exactly the primitive recursive ones.

1 Introduction

In this paper we shall deal with arithmetical theories containing Kalmar elementary arithmetic EA or, equivalently, $I\Delta_0 + Exp$. Σ_n and Π_n formulas are prenex formulas obtained from the bounded ones by n alternating blocks of similar quantifiers, starting from '∃' and '∀', respectively. $\mathcal{B}(\Sigma_n)$ denotes the class of boolean combinations of Σ_n formulas. Σ_n^{st} and Π_n^{st} denote the classes of Σ_n and Π_n sentences.

Parameter free induction schemata have been introduced and investigated by Kaye, Paris, and Dimitracopoulos [11], Adamowicz and Bigorajska [1], Ratajczyk [13], Kaye [10], and others. $I\Sigma_n^-$ is the theory axiomatized over EA by the schema of induction

$$A(0) \wedge \forall x \, (A(x) \rightarrow A(x+1)) \rightarrow \forall x A(x),$$

for Σ_n formulas $A(x)$ containing no other free variables but x, and $I\Pi_n^-$ is similarly defined.[2]

It is known that the schemata $I\Sigma_n^-$ and $I\Pi_n^-$ show a very different behaviour from their parametric counterparts $I\Sigma_n$ and $I\Pi_n$. In particular, for $n \geq 1$, $I\Sigma_n^-$ and $I\Pi_n^-$ are not finitely axiomatizable, and $I\Sigma_n^-$ is strictly stronger than $I\Pi_n^-$ (in fact, stronger than $I\Sigma_{n-1} + I\Pi_n^-$). Furthermore, it is known that $I\Sigma_n$ is a conservative extension of $I\Sigma_n^-$ w.r.t. Σ_{n+2} sentences, although $I\Sigma_n^-$ itself only has a $\mathcal{B}(\Sigma_{n+1})$ axiomatization [11].

In contrast, nontrivial conservation results for $I\Pi_n^-$, for $n > 1$, seem to have been unknown. In particular, it was unknown, if the provably total recursive

---

* The research described in this publication was made possible in part by the Russian Foundation for Fundamental Research (project 96-01-01395).

[2] This definition differs from the one in [11] in that we work over EA, rather than over the weaker theories $I\Delta_0$ or PA^-. Since $I\Sigma_1^-$ in the sense of [11] obviously contains EA, the two definitions are equivalent for $n \geq 1$ in Σ case, and for $n \geq 2$ in Π case.

functions of $I\Pi_2^-$ coincide with the primitive recursive ones (communicated by R. Kaye). The case of $I\Pi_1^-$ (over PA^-) was essentially treated in [11], where the authors show that Π_2 consequences of that theory are contained in EA, cf. also [6].

In this paper we prove that the provably total recursive functions of $I\Pi_2^-$ are exactly the primitive recursive ones. Moreover, we show that $I\Pi_{n+1}^-$ is conservative over $I\Sigma_n^-$ w.r.t. boolean combinations of Σ_{n+1} sentences ($n \geq 1$).

The proofs of these results are based on a characterization of parameter free induction schemata in terms of reflection principles and (generalizations of) the conservativity results for local reflection principles obtained in [3] using methods of provability logic. In our opinion, such a relationship presents an independent interest, especially because this seems to be the first occasion when *local* reflection principles naturally arise in the study of fragments of arithmetic.

We shall also essentially rely on the results from [4] characterizing the closures of arbitrary arithmetical theories extending EA under Σ_n and Π_n induction *rules*. In fact, the results of this paper show that much of the unusual behaviour of parameter free induction schemata can be explained by their tight relationship with the theories axiomatized by induction rules.

2   Preliminaries

First, we establish some useful terminology and notation concerning rules in arithmetic (cf. also [4]). We say that a *rule* is a set of *instances*, that is, expressions of the form

$$\frac{A_1, \ldots, A_n}{B},$$

where $A_1, \ldots, A_n$ and B are formulas. Derivations using rules are defined in the standard way; $T + R$ denotes the closure of a theory T under a rule R and first order logic. $[T, R]$ denotes the closure of T under *unnested applications* of R, that is, the theory axiomatized over T by all formulas B such that, for some formulas $A_1, \ldots, A_n$ derivable in T, $\frac{A_1, \ldots, A_n}{B}$ is an instance of R. $T \equiv U$ means that theories T and U are deductively equivalent, i.e., have the same set of theorems.

A rule R_1 is *derivable* from R_2 iff, for every theory T containing EA, $T + R_1 \subseteq T + R_2$. A rule R_1 is *reducible* to R_2 iff, for every theory T containing EA, $[T, R_1] \subseteq [T, R_2]$. R_1 and R_2 are *congruent* iff they are mutually reducible (denoted $R_1 \cong R_2$). For a theory U containing EA we say that R_1 and R_2 are *congruent modulo U*, iff for every extension T of U, $[T, R_1] \equiv [T, R_2]$.

Induction rule is defined as follows:

$$\text{IR:} \quad \frac{A(0), \quad \forall x\,(A(x) \rightarrow A(x+1))}{\forall x A(x)}.$$

Whenever we impose a restriction that $A(x)$ only ranges over a certain subclass Γ of the class of arithmetical formulas, this rule is denoted Γ-IR. In general, we allow parameters to occur in A, however the following lemma holds (cf. also [5]).

Lemma 1. Π_n-IR *is reducible to parameter free* Π_n-IR. Σ_n-IR *is reducible to parameter free* Σ_n-IR.

Proof. An application of IR for a formula $A(x,a)$ can obviously be reduced to the one for $\forall z A(x,z)$, and this accounts for the Π_n case.

On the other hand, if $A(x,y,a)$ is Π_{n-1}, then an application of Σ_n-IR for the formula $\exists y A(x,y,a)$ is reducible, using the standard coding of sequences available in EA, to the one for $\exists y \forall i \leq x A((i)_0, (y)_i, (i)_1)$ (cf. Remark 4.1 in [5]).

Reflection principles, for a given r.e. theory T containing EA, are defined as follows. The *uniform* reflection principle is the schema

$$\text{RFN}_T: \qquad \forall x \, (\text{Prov}_T(\ulcorner A(\dot{x})\urcorner) \rightarrow A(x)), \quad A(x) \text{ a formula},$$

where $\text{Prov}_T(\cdot)$ denotes a canonical provability predicate for T. The *local* reflection principle is the schema

$$\text{Rfn}_T: \qquad \text{Prov}_T(\ulcorner A \urcorner) \rightarrow A, \quad A \text{ a sentence}.$$

Partial reflection principles are obtained from the above schemata by imposing a restriction that A belongs to one of the classes Γ of the arithmetic hierarchy (denoted $\text{Rfn}_T(\Gamma)$ and $\text{RFN}_T(\Gamma)$, respectively). It is known that, due to the existence of partial truthdefinitions, the schema $\text{RFN}_T(\Pi_n)$ is equivalent to a single Π_n sentence over EA. In particular, $\text{RFN}_T(\Pi_1)$ is equivalent to the consistency assertion Con_T for T. See [14, 12, 3] for some basic information about reflection principles.

We shall also consider the following *metareflection rule*:

$$RR(\Pi_n): \qquad \frac{P}{\text{RFN}_{EA+P}(\Pi_n)}.$$

We let Π_m-$RR(\Pi_n)$ denote the above rule with the restriction that P is a Π_m sentence. Main results (Theorems 1, 2 and 3) of [4] can then be reformulated as follows.

Proposition 2. 1. Π_n-IR $\cong \Pi_{n+1}$-$RR(\Pi_n)$, for $n > 1$;
2. Π_1-IR $\cong \Pi_2$-$RR(\Pi_1)$ (mod $I\Delta_0 +Supexp$).

Proposition 3. 1. Σ_1-IR $\cong \Pi_2$-$RR(\Pi_2)$;
2. Σ_n-IR $\cong \Pi_{n+1}$-$RR(\Pi_{n+1})$ (mod $I\Sigma_{n-1}$), for $n > 1$.

Since $[EA, \Sigma_n$-IR$]$ contains $I\Sigma_{n-1}$, Statement 2 implies that the rules Σ_n-IR and Π_{n+1}-$RR(\Pi_{n+1})$ are interderivable, for all $n \geq 1$.

3 Characterizing $I\Sigma_n^-$ and $I\Pi_n^-$ by reflection principles

Theorem 4. *For $n \geq 1$, over EA,*

1. $I\Sigma_n^- \equiv \{P \rightarrow \text{RFN}_{EA+P}(\Pi_{n+1}) \mid P \in \Pi_{n+1}^{st}\}$;
2. $I\Pi_{n+1}^- \equiv \{P \rightarrow \text{RFN}_{EA+P}(\Pi_{n+1}) \mid P \in \Pi_{n+2}^{st}\}$.

Proof. Both statements are proved similarly, respectively relying upon Propositions 3 and 2, so we shall only elaborate the proof of the first one. For the inclusion ($\subseteq$) we have to derive

$$A(0) \wedge \forall x \, (A(x) \rightarrow A(x+1)) \rightarrow \forall x A(x),$$

for each Σ_n formula $A(x)$ with the only free variable x. We let P denote the Π_{n+1} sentence (logically equivalent to) $A(0) \wedge \forall x \, (A(x) \rightarrow A(x+1))$. Then, by external induction on n it is easy to see that, for each n, $EA + P \vdash A(\bar{n})$. This fact is formalizable in EA, therefore

$$EA \vdash \forall x \, \text{Prov}_{EA+P}(\ulcorner A(\dot{x})\urcorner). \tag{1}$$

Denoting by T the theory axiomatized over EA by all formulas

$$Q \rightarrow \text{RFN}_{EA+Q}(\Pi_{n+1})$$

such that Q is a Π_{n+1} sentence, we conclude that

$$\begin{aligned}
T + P &\vdash \text{RFN}_{EA+P}(\Pi_{n+1}) \\
&\vdash \forall x \, (\text{Prov}_{EA+P}(\ulcorner A(\dot{x})\urcorner) \rightarrow A(x)) \\
&\vdash \forall x A(x), \quad \text{by (1).}
\end{aligned}$$

It follows that $T \vdash P \rightarrow \forall x A(x)$, as required.

For the inclusion ($\supseteq$) we observe that for any Π_{n+1} sentence P the theory $I\Sigma_n^- + P$ contains $P + \Sigma_n\text{-IR}$ by Lemma 1, and hence

$$I\Sigma_n^- + P \vdash \text{RFN}_{EA+P}(\Pi_{n+1}),$$

by Proposition 3. It follows that

$$I\Sigma_n^- \vdash P \rightarrow \text{RFN}_{EA+P}(\Pi_{n+1}).$$

Remark. Statement 2 of the above theorem also holds for $n = 0$, with a similar proof, but only over $I\Delta_0 + Supexp$ (cf. Corollary 8 below).

Remark. Essentially the same proof shows that $IB(\Sigma_n)^-$ can be axiomatized by reflection principles precisely as $I\Sigma_n^-$ in Statement 1 above. This means that $I\Sigma_n^- \vdash IB(\Sigma_n)^-$. This fact can also be directly derived from Theorem 9 below. Notice that, on the other hand, we cannot have $I\Pi_n^- \vdash IB(\Pi_n)^-$, for otherwise $I\Sigma_n^-$ would coincide with $I\Pi_n^-$. We thank the anonymous referee for this remark.

4 Relativized provability and reflection

For $n \geq 1$, $\Pi_n(\mathbf{N})$ denotes the set of all true Π_n sentences. $\mathrm{True}_{\Pi_n}(x)$ denotes a canonical truthdefinition for Π_n sentences, that is, a Π_n formula naturally defining the set of Gödel numbers of $\Pi_n(\mathbf{N})$ sentences in EA. $\mathrm{True}_{\Pi_n}(x)$ provably in EA satisfies Tarski satisfaction conditions (cf. [9]), and therefore, for every formula $A(x_1, \ldots, x_n) \in \Pi_n$,

$$EA \vdash A(x_1, \ldots, x_n) \leftrightarrow \mathrm{True}_{\Pi_n}(\ulcorner A(\dot{x}_1, \ldots, \dot{x}_n)\urcorner). \qquad (*)$$

Tarski's truth lemma $(*)$ is formalizable in EA, in particular,

$$EA \vdash \forall s \in \Pi_n^{st} \, \mathrm{Prov}_{EA}(s \dot{\leftrightarrow} \ulcorner \mathrm{True}_{\Pi_n}(\dot{s})\urcorner), \qquad (**)$$

where Π_n^{st} is a natural elementary definition of the set of Gödel numbers of Π_n sentences in EA. We also assume w.l.o.g. that

$$EA \vdash \forall x \, (\mathrm{True}_{\Pi_n}(x) \to x \in \Pi_n^{st}).$$

Let T be an r.e. theory containing EA. A provability predicate for the theory $T + \Pi_n(\mathbf{N})$ can be naturally defined, e.g., by the following Σ_{n+1} formula:

$$\mathrm{Prov}_T^{\Pi_n}(x) := \exists s \, (\mathrm{True}_{\Pi_n}(s) \wedge \mathrm{Prov}_T(s \dot{\to} x)).$$

Lemma 5. *1. For each Σ_{n+1} formula $A(x_1, \ldots, x_n)$,*

$$EA \vdash A(x_1, \ldots, x_n) \to \mathrm{Prov}_T^{\Pi_n}(\ulcorner A(\dot{x}_1, \ldots, \dot{x}_n)\urcorner).$$

2. $\mathrm{Prov}_T^{\Pi_n}(x)$ satisfies Löb's derivability conditions in T.

Proof. Statement 1 follows from $(*)$. Statement 2 follows from Statement 1, Tarski satisfaction conditions, and is essentially well-known (cf. [15], Lemma 3.7 on p. 156).

We define $\mathrm{Con}_T^{\Pi_n} := \neg\mathrm{Prov}_T^{\Pi_n}(\ulcorner 0 = 1\urcorner)$, and relativized reflection principles $\mathrm{RFN}_T^{\Pi_n}$ and $\mathrm{Rfn}_T^{\Pi_n}$ are similarly defined. For $n = 0$ all these schemata coincide, by definition, with their nonrelativized counterparts.

Lemma 6. *For all $n \geq 0$, $m \geq 1$, the following schemata are deductively equivalent over EA:*

1. $\mathrm{Con}_T^{\Pi_n} \equiv \mathrm{RFN}_T(\Pi_{n+1})$;
2. $\mathrm{Rfn}_T^{\Pi_n}(\Sigma_m) \equiv \{P \to \mathrm{RFN}_{T+P}(\Pi_{n+1}) \mid P \in \Pi_m^{st}\}$.

Proof. 1. Observe that, using $(**)$,

$$EA \vdash \neg\mathrm{Prov}_T^{\Pi_n}(\ulcorner 0 = 1\urcorner) \leftrightarrow \neg\exists s \, (\mathrm{True}_{\Pi_n}(s) \wedge \mathrm{Prov}_T(s \dot{\to} \ulcorner 0 = 1\urcorner))$$
$$\leftrightarrow \forall s \, (\mathrm{Prov}_T(\dot{\neg}s) \to \neg\mathrm{True}_{\Pi_n}(s))$$
$$\leftrightarrow \forall s \, (\mathrm{Prov}_T(\ulcorner\neg\mathrm{True}_{\Pi_n}(\dot{s})\urcorner) \to \neg\mathrm{True}_{\Pi_n}(s)).$$

The latter formula clearly follows from $\mathrm{RFN}_T(\Sigma_n)$, but it also implies $\mathrm{RFN}_T(\Sigma_n)$, and hence $\mathrm{RFN}_T(\Pi_{n+1})$, by (*).

2. By formalized Deduction theorem,

$$EA \vdash \mathrm{Con}_{T+P}^{\Pi_n} \leftrightarrow \neg\mathrm{Prov}_T^{\Pi_n}(\ulcorner\neg P\urcorner). \tag{2}$$

Hence, over EA,

$$\mathrm{Rfn}_T^{\Pi_n}(\Sigma_m) \equiv \{\mathrm{Prov}_T^{\Pi_n}(\ulcorner S\urcorner) \to S \mid S \in \Sigma_m^{st}\}$$
$$\equiv \{P \to \neg\mathrm{Prov}_T^{\Pi_n}(\ulcorner\neg P\urcorner) \mid P \in \Pi_m^{st}\}$$
$$\equiv \{P \to \mathrm{RFN}_{T+P}(\Pi_{n+1}) \mid P \in \Pi_m^{st}\}, \quad \text{by (2) and Statement 1.}$$

From this lemma and Theorem 4 we immediately obtain the following corollary.

Corollary 7. *For $n \geq 1$, the following schemata are deductively equivalent over EA:*

1. $I\Sigma_n^- \equiv \mathrm{Rfn}_{EA}^{\Pi_n}(\Sigma_{n+1})$;
2. $I\Pi_{n+1}^- \equiv \mathrm{Rfn}_{EA}^{\Pi_n}(\Sigma_{n+2})$.

Corollary 8. *Over $I\Delta_0 + Supexp$,*

$$I\Pi_1^- \equiv \mathrm{Rfn}_{EA}(\Sigma_2) \equiv \mathrm{Rfn}_{I\Delta_0+Supexp}(\Sigma_2).$$

Proof. This follows in essentially the same way from Lemma 6 and the proof of Theorem 4, where we rely upon Proposition 2 (2) rather than (1). Indeed, over $I\Delta_0 + Supexp$,

$$I\Pi_1^- \equiv \{P \to \mathrm{RFN}_{EA+P}(\Pi_1) \mid P \in \Pi_2^{st}\}$$
$$\equiv \{P \to \mathrm{Con}_{EA+P} \mid P \in \Pi_2^{st}\}$$
$$\equiv \{\mathrm{Prov}_{EA}(\ulcorner S\urcorner) \to S \mid S \in \Sigma_2^{st}\}.$$

Since the formula expressing the totality of superexponentiation function is a Π_2 sentence, the schemata $\mathrm{Rfn}_{I\Delta_0+Supexp}(\Sigma_2)$ and $\mathrm{Rfn}_{EA}(\Sigma_2)$ are also equivalent over $I\Delta_0 + Supexp$.

5 Conservation results

The following theorem is the main result of this paper.

Theorem 9. *For any $n \geq 1$, $I\Pi_{n+1}^-$ is conservative over $I\Sigma_n^-$ w.r.t. $\mathcal{B}(\Sigma_{n+1})$ sentences.*

Proof. The result follows from Corollary 7 and the following relativized version of Theorem 1 of [3].

Theorem 10. *For any r.e. theory T containing EA and any $n \geq 0$, $T + \mathrm{Rfn}_T^{\Pi_n}$ is conservative over $T + \mathrm{Rfn}_T^{\Pi_n}(\Sigma_{n+1})$ w.r.t. $\mathcal{B}(\Sigma_{n+1})$ sentences.*

Proof. The proof of this theorem makes use of a purely modal logical lemma concerning Gödel-Löb provability logic **GL** (cf. e.g. [7, 15]). Recall that **GL** is formulated in the language of propositional calculus enriched by a unary modal operator $\square$. The expressions $\Diamond\phi$ and $\square^+\phi$ are standard abbreviations for $\neg\square\neg\phi$ and $\phi \wedge \square\phi$, respectively. Axioms of **GL** are all instances of propositional tautologies in this language together with the following schemata:

L1. $\square(\phi \rightarrow \psi) \rightarrow (\square\phi \rightarrow \square\psi)$;
L2. $\square\phi \rightarrow \square\square\phi$;
L3. $\square(\square\phi \rightarrow \phi) \rightarrow \square\phi$.

Rules of **GL** are *modus ponens* and $\phi \vdash \square\phi$ (necessitation).

By an *arithmetical realization* of the language of **GL** we mean any function $(\cdot)^*$ that maps propositional variables to arithmetical sentences. For a modal formula ϕ, $(\phi)_T^*$ denotes the result of substitution for all the variables of ϕ the corresponding arithmetical sentences and of translation of $\square$ as the provability predicate $\mathrm{Prov}_T(\ulcorner \cdot \urcorner)$. Under this interpretation, axioms L1, L2 and the necessitation rule can be seen to directly correspond to the three Löb's derivability conditions, and axiom L3 is the formalization of Löb's theorem. It follows that, for each modal formula ϕ, $\mathbf{GL} \vdash \phi$ implies $T \vdash (\phi)_T^*$, for every realization $(\cdot)^*$ of the variables of ϕ. The opposite implication, for the case of a Σ_1 sound theory T, is also valid; this is the content of the important *arithmetical completeness theorem* for **GL** due to Solovay (cf. [7]).

For us it will also be essential that **GL** is sound under the interpretation of $\square$ as a *relativized* provability predicate. For an arithmetical realization $(\cdot)^*$, we let $(\phi)_{T+\Pi_n(N)}^*$ denote the result of substitution for all the variables of ϕ the corresponding arithmetical sentences and of translation of $\square$ as $\mathrm{Prov}_T^{\Pi_n}(\ulcorner \cdot \urcorner)$.

By Lemma 5 the relativized provability predicate validates the three Löb's derivability conditions. By Corollary 3.8 of [15], p. 156, the formalized analog of Löb's Theorem also holds (it follows by the standard diagonal argument from the derivability conditions). So we have the following lemma.

Lemma 11. *If $\mathbf{GL} \vdash \phi$, then $T \vdash (\phi)_{T+\Pi_n(N)}^*$, for every arithmetical realization $(\cdot)^*$ of the variables of ϕ.*

The opposite implication, that is, the arithmetical completeness of **GL** w.r.t. the relativized provability interpretation is also well-known (cf. [15]). Yet, below we do not use this fact.

The following crucial lemma is a modification of a similar lemma in [3].

Lemma 12. *Let modal formulas Q_i be defined as follows:*

$$Q_0 := p, \qquad Q_{i+1} := Q_i \vee \square Q_i,$$

where p is a propositional variable. Then, for any variables $p_0, \ldots, p_m$,

$$\mathbf{GL} \vdash \Box^+ (\bigwedge_{i=0}^{m} (\Box p_i \rightarrow p_i) \rightarrow p) \rightarrow (\bigwedge_{i=0}^{m} (\Box Q_i \rightarrow Q_i) \rightarrow p).$$

Proof. Rather than exhibiting an explicit proof of the formula above, we shall argue semantically, using a standard Kripke model characterization of **GL**.

Recall that a *Kripke model* for **GL** is a triple $(W, R, \Vdash)$, where

1. W is a finite nonempty set;
2. R is an irreflexive partial order on W;
3. $\Vdash$ is a forcing relation between elements (*nodes*) of W and modal formulas such that

$$x \Vdash \neg\phi \iff x \nVdash \phi,$$
$$x \Vdash (\phi \rightarrow \psi) \iff (x \nVdash \phi \text{ or } x \Vdash \psi),$$
$$x \Vdash \Box\phi \iff \forall y \in W \, (xRy \Rightarrow y \Vdash \phi).$$

Theorem 4 on page 95 of [7] (originally proved by Segerberg) states that a modal formula is provable in **GL**, iff it is forced at every node of any Kripke model of the above kind. This provides a useful criterion for showing provability in **GL**.

Consider any Kripke model $(W, R, \Vdash)$ in which the conclusion $(\bigwedge_{i=0}^{m}(\Box Q_i \rightarrow Q_i) \rightarrow p)$ is false at a node $x \in W$. This means that $x \nVdash p$ and $x \Vdash \Box Q_i \rightarrow Q_i$, for each $i \leq m$. An obvious induction on i then shows that $x \nVdash Q_i$ for all $i \leq m+1$, in particular, $x \nVdash Q_{m+1}$.

Unwinding the definition of Q_i we observe that in W there is a sequence of nodes

$$x = x_{m+1} R x_m R \ldots R x_0$$

such that, for all $i \leq m+1$, $x_i \nVdash Q_i$. Since R is irreflexive and transitive, all x_i's are pairwise distinct. Moreover, it is easy to see by induction on i that, for all i,

$$\mathbf{GL} \vdash p \rightarrow Q_i.$$

Hence, for each $i \leq m+1$, $x_i \nVdash p$.

Now we notice that each formula $\Box p_i \rightarrow p_i$ can be false at no more than one node of the chain $x_{m+1}, \ldots, x_0$. Therefore, by Pigeon-hole principle, there must exist a node z among the $m+2$ nodes x_i such that

$$z \Vdash \bigwedge_{i=0}^{m} (\Box p_i \rightarrow p_i) \wedge \neg p.$$

In case z coincides with $x = x_{m+1}$ we have

$$x \nVdash \bigwedge_{i=0}^{m} (\Box p_i \rightarrow p_i) \rightarrow p.$$

In case $z = x_i$, for some $i \leq m$, we have xRz by transitivity of R, and thus

$$x \nVdash \Box(\bigwedge_{i=0}^{m}(\Box p_i \rightarrow p_i) \rightarrow p).$$

This shows that the formula in question is forced at every node of any Kripke model; hence it is provable in **GL**.

Lemma 13. *For any $n \geq 0$ and any theory T, the following schemata are deductively equivalent over EA:*

$$\mathrm{Rfn}_T^{\Pi_n}(\Sigma_{n+1}) \equiv \mathrm{Rfn}_T^{\Pi_n}(\mathcal{B}(\Sigma_{n+1})).$$

Proof. As in [3], using Lemma 5.

Now we complete our proof of Theorems 9 and 10. Assume $T + \mathrm{Rfn}_T^{\Pi_n} \vdash A$, where A is a $\mathcal{B}(\Sigma_{n+1})$ sentence. Then there are finitely many instances of relativized local reflection that imply A, that is, for some arithmetical sentences $A_0, \ldots, A_m$, we have

$$T \vdash \bigwedge_{i=0}^{m}(\mathrm{Prov}_T^{\Pi_n}(\ulcorner A_i \urcorner) \rightarrow A_i) \rightarrow A.$$

Since relativized provability predicates satisfy Löb's derivability conditions, we also obtain

$$T \vdash \mathrm{Prov}_T^{\Pi_n}(\ulcorner \bigwedge_{i=0}^{m}(\mathrm{Prov}_T^{\Pi_n}(\ulcorner A_i \urcorner) \rightarrow A_i) \rightarrow A \urcorner).$$

Considering an arithmetical realization $(\cdot)^*$ that maps the variable p to the sentence A and p_i to A_i, for each i, by Lemma 12 we conclude that

$$T \vdash \bigwedge_{i=0}^{m}(\mathrm{Prov}_T^{\Pi_n}(\ulcorner B_i \urcorner) \rightarrow B_i) \rightarrow A,$$

where B_i denote the formulas $(Q_i)_{T+\Pi_n(N)}^*$. Now we observe that, if $A \in \mathcal{B}(\Sigma_{n+1})$, then for all i, $B_i \in \mathcal{B}(\Sigma_{n+1})$. Hence

$$T + \mathrm{Rfn}_T^{\Pi_n}(\mathcal{B}(\Sigma_{n+1})) \vdash A,$$

which yields Theorem 10 by Lemma 13. Theorem 9 follows from Theorem 10 and the observation that the schema $\mathrm{Rfn}_{EA}^{\Pi_n}(\Sigma_{n+2})$ corresponding to $I\Pi_{n+1}^-$ is actually weaker than the full $\mathrm{Rfn}_{EA}^{\Pi_n}$.

It is obvious, e.g., since $I\Sigma_1^-$ contains $EA + \Sigma_1\text{-}IR$, that all primitive recursive functions are provably total recursive in $I\Sigma_1^-$ and $I\Pi_2^-$. Moreover, since $I\Sigma_1^-$ is contained in $I\Sigma_1$, by a well-known result of Parsons, any provably total recursive function of $I\Sigma_1^-$ is primitive recursive. The following corollary strengthens this result and gives a positive answer to a question by R. Kaye.

Theorem 14. *Provably total recursive functions of $I\Pi_2^-$ are exactly the primitive recursive ones.*

Proof. This follows from $\mathcal{B}(\Sigma_2)$ conservativity of $I\Pi_2^-$ over $I\Sigma_1^-$.

Remark. Observe that $I\Sigma_1 + I\Pi_2^-$ (unlike each of these theories taken separately) has a wider class of provably total recursive functions than the primitive recursive ones. This follows, e.g., from the fact that

$$I\Sigma_1 + I\Pi_2^- \vdash \mathrm{RFN}_{I\Sigma_1}(\Pi_2)$$

by Theorem 4 (since $I\Sigma_1$ is equivalent to a Π_3 sentence). It follows that $I\Sigma_1 + I\Pi_2^-$ proves the totality of the Ackermann function. A direct proof of the totality of the Ackermann function is also formalizable in $I\Sigma_1 + I\Pi_2^-$. Using the methods similar to those presented in this paper we can show that the class of provably total recursive functions of $I\Sigma_1 + I\Pi_2^-$ actually coincides with the class $\mathcal{E}_{\omega^2}$ of the extended Grzegorczyk hierarchy. In contrast, it is well-known that unrestricted Π_2 induction has a yet wider class of provably recursive functions, namely $\mathcal{E}_{\omega^\omega}$.

Remark. Perhaps somewhat more naturally, conservation results for relativized local reflection principles can be stated modally within a certain bimodal system **GLB** due to Japaridze, with the operators $\square$ and $\boxdot$, that describes the joint behaviour of the usual and the relativized provability predicate (cf. [7]). Using a suitable Kripke model characterization of **GLB**, one can semantically prove that

$$\mathbf{GLB} \vdash \square(\bigwedge_{i=0}^{m}(\boxdot p_i \rightarrow p_i) \rightarrow p) \rightarrow \square(\bigwedge_{i=0}^{m}(\boxdot Q_i \rightarrow Q_i) \rightarrow p),$$

where the formulas Q_i are now understood w.r.t. the modality $\boxdot$, and this yields Theorem 10 almost directly.

References

1. Z. Adamovicz and T. Bigorajska. Functions provably total in $I^-\Sigma_1$. *Fundamenta mathematicae*, 132:189–194, 1989.
2. Z. Adamovicz and R. Kossak. A note on $B\Sigma_n$ and an intermediate induction schema. *Zeitschrift f. math. Logik und Grundlagen d. Math.*, 34:261–264, 1988.
3. L.D. Beklemishev. Notes on local reflection principles. Logic Group Preprint Series 133, University of Utrecht, 1995.
4. L.D. Beklemishev. Induction rules, reflection principles, and provably recursive functions. Logic Group Preprint Series 168, University of Utrecht, 1996. To appear in *Annals of Pure and Applied Logic*, 1997.
5. L.D. Beklemishev. A proof-theoretic analysis of collection. To appear in *Archive for Mathematical Logic*.
6. T. Bigorajska. On Σ_1-definable functions provably total in $I\Pi_1^-$. *Mathematical Logic Quarterly*, 41:135–137, 1995.
7. G. Boolos. *The Logic of Provability*. Cambridge University Press, Cambridge, 1993.

8. S. Goryachev. On interpretability of some extensions of arithmetic. *Mat. Zametki*, 40:561–572, 1986. In Russian.

9. P. Hájek and P. Pudlák. *Metamathematics of First Order Arithmetic*. Springer-Verlag, Berlin, Heidelberg, New-York, 1993.

10. R. Kaye. Parameter free universal induction. *Zeitschrift f. math. Logik und Grundlagen d. Math.*, 35(5):443–456, 1989.

11. R. Kaye, J. Paris, and C. Dimitracopoulos. On parameter free induction schemas. *Journal of Symbolic Logic*, 53(4):1082–1097, 1988.

12. G. Kreisel and A. Lévy. Reflection principles and their use for establishing the complexity of axiomatic systems. *Zeitschrift f. math. Logik und Grundlagen d. Math.*, 14:97–142, 1968.

13. Z. Ratajczyk. Functions provably total in $I^-\Sigma_n$. *Fundamenta Mathematicae*, 133:81–95, 1989.

14. C. Smoryński. The incompleteness theorems. In J. Barwise, editor, *Handbook of Mathematical Logic*, pages 821–865. North Holland, Amsterdam, 1977.

15. C. Smoryński. *Self-Reference and Modal Logic*. Springer-Verlag, Berlin-Heidelberg-New York, 1985.

16. A. Wilkie and J. Paris. On the scheme of induction for bounded arithmetic formulas. *Annals of Pure and Applied Logic*, 35:261–302, 1987.

Looking for an Analogue of Rice's Theorem in Circuit Complexity Theory

Bernd Borchert[1] and Frank Stephan[2]

[1] Mathematisches Institut, Universität Heidelberg,
Im Neuenheimer Feld 294, 69120 Heidelberg, Germany,
Email: bb@math.uni-heidelberg.de,
Phone: +49/6221/546282, Fax: +49/6221/544465.
[2] Mathematisches Institut, Universität Heidelberg,
Im Neuenheimer Feld 294, 69120 Heidelberg, Germany,
Email: fstephan@math.uni-heidelberg.de,
Phone: +49/6221/548205, Fax: +49/6221/544465.

Abstract. Rice's Theorem says that every nontrivial semantic property of programs is undecidable. It this spirit we show the following: Every nontrivial absolute (gap, relative) counting property of circuits is UP-hard with respect to polynomial-time Turing reductions.

1 Introduction

One of the nicest theorems in Recursion Theory is Rice's Theorem [17, 18]. Informally speaking it says: Any nontrivial semantic property of programs is undecidable. More formally it can be stated this way:

Theorem 1.1 [Rice 1953] *Let A be any nonempty proper subset of the partial recursive functions. Then the halting problem or its complement is many-one reducible to the following problem: Given a program p, does it compute a function from A?*

The theorem and its proof only use elementary notions of Recursion Theory. The most interesting point about Rice's Theorem is that it has messages to people in practical computing: It tells programmers that for example there is no program which finds infinite loops in given programs, or that there is no program which checks if some given program does a specified job.

Intrigued by the simple beauty of Rice's Theorem we tried to find some sister of it in Complexity Theory. We will present a concept based on Boolean circuits instead of programs, therefore our approach is different from the one of Kozen [14], Royer [19] and Case[7] who study problems on programs with given resource bounds.

The main idea behind our approach is that for circuits versus Boolean functions there is a similar syntax/semantics dichotomy like there is for programs versus partial recursive functions: A circuit is the description of a Boolean function like a program is the description of a partial recursive function. In other words, circuits (programs) are the syntactical objects whereas Boolean functions (partial recursive functions) are the corresponding semantic objects. Rice's Theorem lives on this dichotomy of syntax and semantics: It says that semantic properties are impossible to recognize on the syntactical level. Similarly we look to what extent semantic properties of Boolean functions can be recognized from the syntactical structure of their circuits. Note that for example the question whether two syntactical objects describe the same semantic object is hard in both worlds: It is Π_2-complete for programs and co-NP-complete for circuits.

By these considerations, the perfect analogue of Rice's Theorem would be the following: Let A be any nonempty proper subset of the set of Boolean functions. Then the following problem is NP-hard: Given a circuit c, does it compute a function from A? An example for which this (generally wrong) claim is true is satisfiability: it is a semantic property because it does only depend on the Boolean function, not on the way the Boolean function is represented, and the satisfiability problem for circuits is NP-complete. But unfortunately, there is the following simple counterexample for the above claim. Let A be the set of Boolean functions which have the value 0 on the all-0-assignment. For a given circuit c this question can be computed in polynomial time. Therefore, the above claim stating NP-hardness of all nontrivial semantic properties of circuits is false unless P=NP.

So we had to look for a more restrictive requirement for the set A for which a modification of the statement may be true. What we found is the (indeed very restrictive) requirement of *counting*: Let A be a set of Boolean functions for which membership in A only depends on the number of satisfying assignments, like for example the set of all Boolean functions which have at least one satisfying assignment, or the set of Boolean functions which have an odd number of satisfying assignments. We will call the corresponding sets of circuits *absolute counting problems*. In a similar fashion we will define *gap counting problems* and *relative counting problems*: they also incorporate the non-satisfying assignments in their definition. For example, the set of circuits which have more satisfying than non-satisfying assignments is a gap counting problem because it can be stated the following way: the gap (= difference) between the number of satisfying assignments and the number of non-satisfying assignments has to be greater than 0. And the same problem is a relative counting problem because it can be stated the following way: the relative number of satisfying assignments has to be greater than one half.

For each of these three types of counting problems a theorem in the fashion of Rice's Theorem can be shown:

> Any nontrivial absolute (gap, relative) counting property of circuits is
> UP-hard with respect to polynomial-time Turing reductions.

This result will be extended in terms of approximable sets and also in terms of randomized reductions.

For the following (somehow artificial) notion, which is the nonuniform counterpart of nondeterministic polynomial-time computation, a perfect analogue of Rice's Theorem could be found: Let an *existentially quantified circuit* be a circuit in which some variables are existentially quantified. Each such circuit describes in a natural way a Boolean function on its non-quantified variables. Here it can be shown: Let A be any nonempty proper subset of the set of Boolean functions which not only depends on arity. Then the following problem is NP-hard: Given an existentially quantified circuit c, does it describe a function from A?

2 Preliminaries

The standard notions of Theoretical Computer Science like words, languages, polynomial-time reductions, P, NP, etc. follow the book of Papadimitriou [16]. A central notion of this paper is the notion of a *Boolean function* which is defined to be a mapping from $\{0,1\}^n$ to $\{0,1\}$ for some integer $n \geq 0$ which is called its *arity*. *Circuits* are a standard way of representing Boolean functions. We will just assume that they are encoded as words in some standard way, for the details concerning circuits we refer for example to [16]. Remember that a given circuit can be evaluated on a given assignment in polynomial time. Each circuit c describes a Boolean function $F(x_1, \ldots, x_n)$. Note that here we have an example of the classical syntax/semantics dichotomy like we have it for programs: The circuits (programs) are the syntactical objects, which we have as finite words at our fingertips, whereas the Boolean functions (partial recursive functions) are the corresponding semantic objects far away in the mathematical sky[3]. Note that every Boolean function can be represented by a circuit, in fact it is represented by infinitely many circuits. The problem whether two circuits represent the same Boolean function is co-NP-complete.

As it was pointed out in the introduction one would like to have a theorem like: Every nontrivial semantic property of circuits is NP-hard. Unfortunately, there are counterexamples for this statement unless P=NP. First of all there is the counterexample consisting of the set of circuits with an odd number of satisfying assignments, this set is complete for $\oplus$P and not known to be NP-hard or co-NP-hard. So we have to replace in the statement above the class NP by some class which is contained in NP and $\oplus$P. The class UP would be a natural choice for that. But unfortunately there is an even stronger counterexample: the set of all circuits which evaluate to 0 on the all-0-assignment. This property of circuits is in fact a semantic property and it is nontrivial, but it can be checked with a polynomial-time algorithm. So we can only hope for an analogue of Rice's Theorem (stating UP-hardness) if we restrict ourselves to stronger

---

[3] One might argue that here we have a different situation than in the case of partial recursive functions because Boolean functions are finite objects: one can represent an n-ary Boolean function just by the length-2^n 0-1-sequence of its values on the 2^n assignments. This in fact guarantees decidability of the usual semantic questions, but we are interested in finer questions: even if semantic properties are decidable, how difficult is it to decide them? And the difficulty for deciding the semantic questions (like satisfiability) stems basically from the fact that Boolean functions are not given as the length-2^n 0-1-sequence of the function values but in a *compressed* way, namely as circuits.

semantic properties. We tried several approaches, for example by considering the equivalence relations like Boolean isomorphism presented in [1, 6]. Note that the above counterexample is also a counterexample under Boolean isomorphism. The restriction for which we could find the intended hardness result is the restriction to the counting properties which will be introduced in the next section.

3   Three Types of Counting Problems on Circuits

Given a Boolean function one can ask different question about the number of satisfying assignments: (a) what is the number of satisfying assignments? (b) what is the difference between the number of satisfying and non-satisfying assignments? (c) what is the share of the satisfying assignments compared with the total number of assignments? If the arity of the Boolean function is known then all question are equivalent. But if we do not fix the arity then the three questions are pairwise incomparable, i.e., given an answer to (a) we can not infer an answer to (b) and so on.

For a circuit c let $\#_0(c)$ and $\#_1(c)$ denote the number of non-satisfying assignments and satisfying assignments, respectively, of the Boolean function represented by c. According to the different questions (a), (b) and (c) from above we will introduce the following three types of counting problems. Note that the answer to (a) is a natural number, the answer to (b) an integer and the answer to (c) a dyadic number in the interval $[0, 1]$, we will denote this set by $\mathbf{D} = \{\frac{m}{2^n} \mid n, m \in \mathbf{N}, 0 \le m \le 2^n\}$.

Definition 3.1   (a) Let A be a subset of $\mathbf{N}$. The *absolute counting problem for A*, Absolute-Counting(A), is the set of all circuits c such that $\#_1(c) \in A$.

(b) Let A be a subset of $\mathbf{Z}$. The *gap counting problem for A*, Gap-Counting(A), is the set of all circuits c such that $\#_1(c) - \#_0(c) \in A$.

(c) Let A be a subset of $\mathbf{D}$. The *relative counting problem for A*, Relative-Counting(A), is the set of all circuits c such that the relative number of accepting assignments is in A: $c \in$ Relative-Counting(A) $\Leftrightarrow$ $\#_1(c)/(\#_0(c) + \#_1(c)) \in A$.

Examples of absolute counting problems.   The satisfiability problem for circuits, which we will denote here as SAT (though SAT traditionally refers to the satisfiability problem on CNF's), is by its definition an absolute counting problem, i.e. SAT = Absolute-Counting($\{1, 2, 3, \ldots\}$). Remember that SAT is NP-complete. Likewise, the set of unsatisfiable circuits equals the set Absolute-Counting($\{0\}$), this problem is co-NP-complete. Another example is the set of circuits with an odd number of satisfying assignments, by definition it equals Absolute-Counting($\{1, 3, 5, \ldots\}$), this problem is $\oplus$P-complete. Another example is the set 1-SAT consisting of the circuits with exactly one satisfying assignment, i.e. 1-SAT = Absolute-Counting($\{1\}$), the complexity class for which this problem is complete is usually called 1-NP.

Examples of gap counting problems.   The set C$_=$SAT of circuits which have as many satisfying as non-satisfying assingments, is a gap counting problem:

$C_=SAT = $ Gap-Counting($\{0\}$). The set PSAT of circuits which have at least as many accepting as non-accepting assignments, is a gap counting problem: PSAT = Gap-Counting($\{0, 1, 2, 3, \ldots\}$). Remember that $C_=SAT$ and PSAT are complete for the classes $C_=P$ and PP, respectively. Another example is the set Gap-2-SAT consisting of the circuits with exactly two more satisfying than non-satisfying assignments, i.e. Gap-2-SAT = Gap-Counting($\{2\}$).

Examples of relative counting problems. The two gap counting problems PSAT and $C_=SAT$ from above are also relative counting problems: $C_=SAT = $ Relative-Counting($\{\frac{1}{2}\}$), and PSAT = Relative-Counting($\{x \in \mathbf{D}; x \geq \frac{1}{2}\}$). SAT = Relative-Counting($\mathbf{D} - \{0\}$) is also a relative counting problem. The tautology problem equals Relative-Counting($\{1\}$).

Note that only relative counting has the following natural property. If we have a circuit $c(x_1, \ldots, x_m)$ and add some dummy variables $x_{m+1}, \ldots, x_n$ so that the whole new circuit $c'(x_1, \ldots, x_m, \ldots, x_n)$ represents a Boolean function on $n \geq m$ inputs then we have that $c(x_1, \ldots, x_m) \in$ Relative-Counting(A) $\iff$ $c'(x_1, \ldots, x_m, \ldots, x_n) \in$ Relative-Counting(A). In a way one would consider the Boolean functions represented by c and c' to be "basically the same". So if we identify Boolean functions modulo independent variables then only relative counting respects this natural identification.

A *counting problem* in a general sense we define the following way. Let a sequence (A_n) be given for which A_n is a subset of $\{0, \ldots, 2^n\}$. The *counting problem for* (A_n) is the set of all circuits $c(x_1, \ldots, x_n)$ such that $\#_1(c) \in A_n$, see [12] for an analogous definition of (general) counting classes. In this way, absolute, gap and relative counting problems are counting problems. It is easy to give an example of a (general) counting problem which is nontrivial but in P, for example the set of all circuits with an odd arity (it is the counting class for the sequence $\emptyset, \{0, 1, 2\}, \emptyset, \{0, 1, \ldots, 8\}, \emptyset, \{0, 1, \ldots, 32\}, \emptyset, \ldots$).

It was already mentioned that the three types of counting problems from Definition 3.1 are incomparable as mathematical sets. But the question appears if they are comparable in terms of $\leq_m^p$-complexity. For example, the tautology problem, which is a relative counting problem, is provably not an absolute counting problem, nevertheless there is an absolute counting problem, namely the non-satisfiability problem, which has the same $\leq_m^p$-complexity (both are co-NP-complete). But even for this weaker form of comparison the three types of counting problems seem to be incomparable: PSAT and $C_=SAT$ are gap and relative counting problems but do not seem to be $\leq_m^p$-equivalent to some absolute counting problem. SAT is an absolute and relative counting problem but does not seem to be $\leq_m^p$-equivalent to some gap counting problem. 1-SAT is an absolute counting problem but does not seem to be $\leq_m^p$-equivalent to some gap or relative counting problem. Gap-2-SAT is a gap counting problem but does not seem to be $\leq_m^p$-equivalent to some absolute or relative counting problem.

Remark. Analogously to the way we defined the three types of counting problems we can define three types of counting classes: for a given subset A of $\mathbf{N}$ ($\mathbf{Z}$, $\mathbf{D}$) the *absolute (gap, relative) counting class for* A consists of the languages L

for which there is a polynomial-time nondeterministic machine M such that a word x is in L iff the number of accepting paths (the difference of the number of accepting paths and non-accepting paths, the share of the accepting paths compared with the total number of paths) of M on input x is in A. This definition of *gap countable classes* equals the definition of *nice gap definable classes* in Fenner, Fortnow & Kurtz [10]. As a special case of the main result in [5, 23] it follows that an absolute (gap, relative) counting problem is $\leq_m^p$-complete for the corresponding absolute (gap, relative) counting class. In other words, absolute (gap, relative) counting problems and the corresponding absolute (gap, relative) counting classes are just two sides of the same medal. Note that the $\leq_m^p$-comparability question we discussed above is therefore equivalent to the set comparability question of the three types of counting classes.

4 Some Rice-Style Theorems for Counting Problems

In this section we will state and prove the theorems which show UP-hardness of all nontrivial counting problems of the three types defined in the previous chapter. Remember that the class UP, which was defined first in [21], is the promise class consisting of the languages L such that there is a polynomial-time nondeterministic machine M such that on every input the machine M has at most one accepting path and an input x is in L if M running on input x has an accepting path. Such machines are called *unambiguous*. By definition UP is a subset of NP. A classical result tells that UP equals P if and only if one-way functions do not exist [11]. The primality problem is a typical example of a problem in UP not known to be in P [9]. Below we use the notion of a class $\mathcal{C}$ being reducible to a language L, this is just a short way of saying that every language in the class $\mathcal{C}$ is reducible to L. The following theorem implies that any nontrivial absolute counting property of circuits is UP-hard.

Theorem 4.1 *Let A be any nonempty proper subset of $\mathbf{N}$. Then one of the following three classes is $\leq_m^p$-reducible to* Absolute-Counting(A): NP, co-NP, UP $\oplus$ co-UP.

Proof The proof distinguishes the following three cases.

First case: A has a maximum a. Then co-NP is $\leq_m^p$-reducible to Absolute-Counting(A): Let a language L in co-NP be given and let M be a machine for L in the sense that $x \in L$ iff no path of $M(x)$ is accepting. Let $m(x)$ denote the number of accepting paths of $M(x)$. Cook [8] established a method to construct in polynomial time a circuit Cook_x^M with inputs $y_1, \ldots, y_n$ (n depends on x and is bounded by a polynomial in the length of x) such that accepting computation paths of $M(x)$ and satisfying assignments of $\mathrm{Cook}_x^M(y_1, \ldots, y_n)$ correspond to each other. Therefore, $\mathrm{Cook}_x^M(y_1, \ldots, y_n)$ evaluates to 1 for exactly $m(x)$ assignments. Using a additional variables $z_1, \ldots, z_a$ the circuit given by the following specification evaluates exactly $a + m(x)$ assignments to 1:

$$d_{x,a}^M(y_1, ..., y_n, z_1, ..., z_a) = \begin{cases} \text{Cook}_x^M(y_1, ..., y_n) & \text{if } z_1 + ... + z_a = 0; \\ 1 & \text{if } z_1 + ... + z_a = 1 \\ & \text{and } y_1 + ... + y_n = 0; \\ 0 & \text{otherwise.} \end{cases}$$

Thus $d_{x,a}^M$ is in Absolute-Counting(A) iff $m(x) = 0$ iff $x \in L$. So the mapping $x \to d_{x,a}^M$ gives a $\leq_m^p$-reduction from L to Absolute-Counting(A).

Second case: $\overline{A}$ has a maximum b. Given a set L in NP recognized by the nondeterministic machine M, an analogous construction as in the previous case is used in order to obtain a circuit $d_{x,b}^M$ which evaluates exactly $b + m(x)$ assignments to 1. Now $x \in L$ iff $m(x) > 0$ iff $d_{x,b}^M \in$ Absolute-Counting(A). And so one obtains the desired $\leq_m^p$-reduction.

Third case: neither A nor $\overline{A}$ has a maximum. Then there is $a \in A$ with $a + 1 \notin A$ and $b \notin A$ with $b + 1 \in A$. For a language $L = 0L' \cup 1L''$ in UP $\oplus$ co-UP let M' and M'' be the unambiguous machines for L' and L'', respectively. The mapping which assigns to an input $0x$ the circuit $d_{x,a}^{M'}$ and to an input $1x$ the circuit $d_{x,b}^{M''}$ realizes the $\leq_m^p$-reduction from L to Absolute-Counting(A). □

Gupta [13] and Ogiwara & Hemachandra [15] introduced the class SPP as the promise class consisting of the languages L such that there is a polynomial-time nondeterministic machine M such that for every input x for the machine M either half of the computation paths are accepting of half of them plus 1 are accepting, and $x \in L$ if the second case is true. Note that the class SPP contains both UP and co-UP. Fenner, Fortnow & Kurtz [10] proved the following result which states that the class SPP is $\leq_m^p$-reducible to any gap counting problem.

Theorem 4.2 [Fenner, Fortnow & Kurtz 1994] *Let A be any nonempty proper subset of* **Z**. *Then* SPP $\leq_m^p$ Gap-Counting(A).

Now that we have UP-hardness for absolute and gap counting problems we turn to relative counting problems and will show UP-hardness for the more powerful polynomial-time Turing reducibility. Note that the classes NP, co-NP, and SPP are above UP.

Theorem 4.3 *Let A be any nonempty proper subset of* **D**. *Then* Relative-Counting(A) *is $\leq_m^p$-complete for* NP *or* co-NP, *or* SPP $\leq_T^p$ Relative-Counting(A).

Proof First consider the special case that $A(p) = A(q)$ for all dyadic numbers p, q with $0 < p < q < 1$. In this special case, A is one of the following six sets: $\{0\}, \{1\}, \{0, 1\}, \mathbf{D} - \{0\}, \mathbf{D} - \{1\}, \mathbf{D} - \{0, 1\}$. It is easy to see that the relative counting problems for first three sets are co-NP-complete, for example, Relative-Counting($\{1\}$) is the tautology problem. Similarly the relative counting problems for the last three sets are NP-complete, for example, Relative-Counting($\mathbf{D} - \{0\}$) is the satisfiability problem.

So it remains the case where there are dyadic numbers p, q such that $0 < p < q < 1$ and $A(p) \neq A(q)$. It will be shown that in this case SPP can be

$\leq_T^p$-reduced to Relative-Counting(A). Let M be a machine which witnesses that a language L is in SPP. Consider for an input x the circuit $\text{Cook}_x^M(y_1, \ldots, y_n)$ defined in the proof of Theorem 4.1. Recall that if $x \in L$ then Cook_x^M evaluates $2^{n-1} + 1$ assignments to 1 and if $x \notin L$ then Cook_x^M evaluates 2^{n-1} assignments to 1. Furthermore there is an m such that $2^{-m} \leq p < q \leq 1 - 2^{-m}$. Now the Turing-reduction works as follows:

(I): Search for p', q' with $p \leq p' < q' \leq q$, $q' - p' = 2^{-m-n}$ and $A(p') \neq A(q')$. A query to A can be translated into a query to Relative-Counting(A) as follows: Let $r = 0.a_1 \ldots a_k < 1$ be a dyadic number and let c_r denote the circuit which assigns to $(y_1, \ldots, y_k)$ the value 1 iff $0.y_1 \ldots y_k < 0.a_0 \ldots a_k$. r is in A iff $c_r \in$ Relative-Counting(A).

Using this mechanism it is possible to find p' and q' with interval search: starting with $p = p'$ and $q = q'$ one takes that $m+n$-bit dyadic number r which is nearest to $\frac{p'+q'}{2}$ and finds out whether $A(p') = A(r)$ or $A(q') = A(r)$. In the first case, p' is replaced by r, in the second, q' is replaced by r. This search is continued until the difference between p' and q' is 2^{-m-n}. Note that $A(p) = A(p') \neq A(q) = A(q')$.

(II): Now a circuit d_x is computed with Relative-Counting(A)(d_x) $= A(q')$ for $x \in L$ and Relative-Counting(A)(d_x) $= A(p')$ for $x \notin L$. Let $z = 0.z_1 \ldots z_{m+n}$ be the dyadic number determined via the binary representation of the variables.

$$d_x(z_1, \ldots, z_{m+n}) = \begin{cases} \text{Cook}_x^M(z_{m+1}, \ldots, z_{m+n}) & \text{if } z < 2^{-m}; \\ 1 & \text{if } 2^{-m} \leq z < p' + 2^{-m-1}; \\ 0 & \text{if } p' + 2^{-m-1} \leq z. \end{cases}$$

So the relative number of accepting assignments is the sum of the $p' - 2^{-m-1}$ hard wired assignments from the second line of the case-distinction plus 2^{-m-1} ($2^{-m-1} + 2^{-m-n}$) from the circuit c_x in the case of $x \notin L$ ($x \in L$). Then the whole relative number is p' for $x \notin L$ and q' for $x \in L$. It follows that $x \in L$ iff Relative-Counting(A)(d_x) $= A(q')$. So the last query whether d_x is in Relative-Counting(A) completes the decision procedure for L.

In short words: the first part of the construction uses the fact that $A \leq_m^p$ Relative-Counting(A) in order to search sufficiently close dyadic numbers p' and q' between p and q such that $A(p') \neq A(q)$. The next step produces a circuit d_x whose relative number of satisfying assignments is p' for $x \notin L$ and q' for $x \in L$. So $L(x)$ can be computed with $m + n + 1$ queries to Relative-Counting(A). Therefore, in this second case, Relative-Counting(A) is SPP-hard with respect to polynomial-time Turing reductions. □

The preceding three Theorems 4.1, 4.2, 4.3 can be summarized by the following conclusion:

Conclusion 4.4 *Any nontrivial absolute (gap, relative) counting property of circuits is UP-hard with respect to polynomial-time Turing reducibility.*

In particular a nontrivial absolute (gap, relative) counting problem on circuits is not solvable in polynomial-time unless P=UP.

5 Extensions and Limitations of the Main Results

In this section first the result is extended to randomized reductions and computations. After that it is shown that no nontrivial absolute (gap, relative) counting problem on circuits is approximable unless P=UP. Furthermore it is pointed out that it is unlikely that Theorem 4.3 holds with polynomial-time many-one-reduction in place of the polynomial-time Turing reduction. In the last part a perfect analogue of Rice's Theorem is given for the world of existentially quantified circuits.

5.1 Randomized Computations

Valiant and Vazirani [22] showed, that using randomized reductions, detecting unique solutions is as hard as solving the satisfiability problem. In particular they showed that every algorithm f which satisfies the following specification also already allows to solve the satisfiability problem in a randomized context: If the circuit x has a solutions and $a \leq 1$ then $f(x) = a$. The algorithms to reduce UP to the counting problems in Theorems 4.1, 4.2, 4.3 satisfy these requirements. So they allow to decide the set SAT = { circuits x : x has a solution} via a nondeterministic machine which has no accepting path for $x \notin$ SAT and which has more accepting than rejecting paths for $x \in$ SAT. Thus the following holds for all three counting problems, in particular for relative counting.

Theorem 5.1 *If B = Relative-Counting(A) is not trivial then* NP $\subseteq$ RP^B.

For absolute and gap counting, the result can be improved by showing that SAT is randomized polynomial-time reducible to B or to $\overline{B}$. Valiant and Vazirani [22] defined that a randomized polynomial-time reduction from some set A to another set B is given via a machine M which computes for every x and path p a circuit $M(x,p)$ such that

$$x \in A \Rightarrow M(x,p) \in B \text{ for at least } \frac{n(x)}{q(\text{length of } x)} \text{ paths } p;$$

$$x \notin A \Rightarrow M(x,p) \notin B \text{ for all paths } p.$$

where q is a suitable polynomial and $n(x)$ is the number of computational paths of M. Valiant and Vazirani [22, Theorem 1.1] constructed a randomized polynomial-time reduction from SAT to USAT$_Q$ where

$$\text{USAT}_Q(x) = \begin{cases} a & \text{if } x \text{ has } a \text{ solutions and } a \leq 1; \\ Q(x) & \text{if } x \text{ has at least two solutions;} \end{cases}$$

and where the reduction is independent of the values $Q(x)$ for all x. The result is a direct combination of this construction and the constructions for $\leq^p_m$-reducing UP to B or $\overline{B}$ in Theorems 4.1 and 4.2 which indeed are $\leq^p_m$-reductions of some suitable set USAT$_Q$ to B or $\overline{B}$, respectively.

Theorem 5.2 SAT *or its complement is randomized polynomial-time reducible to any nontrivial absolute and gap counting problem.*

5.2 Approximable Sets

A set A is approximable [4] iff there is a constant j and an algorithm which computes for each input $x_1, \ldots, x_j$ in polynomial time j bits $y_1, \ldots, y_j$ such that $A(x_h) = y_h$ for some h. Beigel [2, 3] analyzed the notion of approximable sets and showed that no NP-hard set is approximable unless P = UP. This result can be transferred to the following theorem which is an extension of Conclusion 4.4.

Theorem 5.3 *If* P $\neq$ UP *then no nontrivial absolute (gap, relative) counting problem is approximable.*

Proof The proofs are all direct combinations of Beigel's techniques with those to show the UP-hardness of these sets. Thus we restrict ourselves to show the most involved case of relative counting sets.

If a problem is NP-complete or co-NP-complete then it is not approximable under the hypothesis P $\neq$ UP [2, 3]. So one has only to adapt the main case in the proof of Theorem 4.3.

So let L be any language in UP, note that UP $\subseteq$ SPP. The first part of the Turing reduction from L to Relative-Counting(A) is exactly the same as in the proof of Theorem 4.3. In the second part it starts to differ at the definition of the circuit d_x. Based on the definition of d_x, n variants $d_{n,i}$ are defined via fixing one variable to 1 in the circuit $\mathbf{Cook}_x^M$:

$$d_{x,i}(z_1, ..., z_{m+n}) = \begin{cases} \mathbf{Cook}_x^M(z_{m+1}, ..., z_{m+n}) & \text{if } z < 2^{-m} \text{ and } z_{m+i} = 1; \\ 0 & \text{if } z < 2^{-m} \text{ and } z_{m+i} = 0 \\ & \text{or } p' + 2^{-m-1} \leq z; \\ 1 & \text{if } 2^{-m} \leq z < p' + 2^{-m-1}. \end{cases}$$

Recall that for $x \in L$ there is exactly one satisfying assignment $(a_1, \ldots, a_n)$ and for $x \notin L$ no one. So the relative number of the accepted assignments of $d_{x,i}$ is $q' = p' + 2^{-m-n}$ if $x \in L \wedge a_i = 1$ and p' otherwise.

As Beigel [3, Theorem 9] pointed out, there is an algorithm which computes for input $(d_{x,1}, \ldots, d_{x,n})$ in polynomial time $O(n^j)$ n-bit-vectors v such that one of these vectors is the characteristic function of A on the input-vector. So for each such $v = (v_1, \ldots, v_n)$ one computes the assignment $(a_1, \ldots, a_n)$ given via $a_i = 1$ for $v_i = A(q')$ and $a_i = 0$ for $v_i = A(p')$. If $\mathbf{Cook}_x^M$ has a satisfying assignment, then one of these $(a_1, \ldots, a_n)$ must be one. By evaluating $\mathbf{Cook}_x^M$ on these assignments it is found out whether $x \in L$ or $x \notin L$. Thus if Relative-Counting(A) is approximable then L is in P and P = UP. □

5.3 Relative Counting and BPP

For relative counting, we could not state UP-hardness in terms of $\leq_m^p$-reducibility. Under the hypothesis that UP is not contained in the class BPP the following Theorem 5.4 gives nontrivial relative counting problems which are not UP-hard with respect to many-one reduction.

Recall that BPP is the class of all languages L such that there is an $\epsilon > 0$ and a polynomial-time nondeterministic machine M such that for all inputs x the share of accepting paths (all of them must have the same length) is either less than $\frac{1}{2} - \epsilon$ or greater than $\frac{1}{2} + \epsilon$, and x is accepted in the second case.

Theorem 5.4 *There is a set $A \subseteq \mathbf{D}$ such that for every computable set L the following equivalence holds: $L \leq_m^p$ Relative-Counting(A) iff $L \in$ BPP.*

Proof There are uncountably many reals r between 0 and 1 and for each such real r, the set $A = \{q \in \mathbf{D} : q < r\}$ is unique. Thus, using a natural representation of dyadic numbers by words, there is a set A of this form which is not enumerable. So fix such a set A and r.

($\Rightarrow$) : Let L be a computable set which is $\leq_m^p$-reducible to Relative-Counting(A) via a function f. Let $m(x)$ denote for each x the relative number of satisfying assingments of the circuit $f(x)$, $m(x)$ can be computed from $f(x)$ using exponential time. Since $m(x)$ is in A iff x is in L, the set

$$B = \{q \in \mathbf{D} : (\exists x \in L)\,[q \leq m(x)]\}$$

is an enumerable subset of A. Since A is not enumerable, the supremum of B must be below that of A: $\sup(B) < r$. There is a dyadic number s strictly between these two numbers. So there is some $\epsilon > 0$ such that $\sup(B) < s - 2\epsilon$ and $r > s + 2\epsilon$. Let $n(x)$ be the number of inputs of circuit $f(x)$. Now the following machine M witnesses that L is in BPP.

$$M(x)(y_0, ..., y_{n(x)}) = \begin{cases} 1 & \text{if } y_0 = 0 \text{ and } f(x) \text{ evaluates } (y_1, ..., y_{n(x)}) \text{ to } 1 \\ & \text{or } y_0 = 1 \text{ and } 0.y_1...y_{n(x)} > s; \\ 0 & \text{otherwise.} \end{cases}$$

The relative number $k(x)$ of the accepting assignments is the sum of two numbers: the number $\frac{m(x)}{2}$ from the simulation of $f(x)$ and the number $\frac{1-s}{2}$ from the hard-wired paths with $y_0 = 1$. So if $x \in L$ then $m(x) \in A$, $m(a) \leq s - 2\epsilon$ and $k(x) \leq \frac{1}{2} - \epsilon$; if $x \notin L$ then $m(x) \notin A$, $m(x) \geq s + 2\epsilon$ and $k(x) \geq \frac{1}{2} + \epsilon$. Since the constant ϵ does not depend on x and $n(x)$, M witnesses that L is in BPP.

($\Leftarrow$) : Now let $L \in$ BPP. There is a real number ϵ with $|r - \frac{1}{2}| < \epsilon < \frac{1}{2}$. For L there is now a BPP-machine M which works with this ϵ, see for example [16, Chapter 11]. Let M have path length $n(x)$ on input x. Now one assigns to each x the circuit $\neg\mathbf{Cook}_x^M(y_1, \ldots, y_{n(x)})$. The relative number of the satisfying assignments for each circuit $\mathbf{Cook}_x^M$ is above $\frac{1}{2} + \epsilon > r$ for $x \notin L$ and below $\frac{1}{2} - \epsilon < r$ for $x \in L$. So the mapping $x \to \neg\mathbf{Cook}_x^M$ is an $\leq_m^p$-reduction from L to Relative-Counting(A). $\qquad \square$

Remark. The classes PP, $C_=P$ and Mod_kP could be defined as the class of all languages which are $\leq_m^p$-reducible to some suitable relative counting problem. For example, $C_=P$ is the set of languages $\leq_m^p$-reducible to Relative-Counting$(\{\frac{1}{2}\})$. In addition to the above mentioned limitation, Theorem 5.4 shows that the class BPP can be defined similarly within the class of all computable sets.

5.4   A Perfect Analogue of Rice's Theorem for Existentially Quantified Circuits

It was mentioned before that the problem whether two programs compute the same partial recursive function is complete for Π_2 in the Arithmetical Hierarchy, whereas the problem whether two circuits compute the same Boolean function is complete for co-NP $= \Pi_1^p$ in the Polynomial-Time Hierarchy. So the difficulty of the two equivalence problems is on different levels in the respective hierarchies. Therefore, we have been looking for some representation of Boolean functions such that the equivalence problem is complete for Π_2^p. Such a representation is given by *existentially quantified circuits* which are defined to be circuits in which some (but not necessarily all) variables are existentially quantified (in prenex normal form). Such an existentially quantified Boolean circuit with n free variables defines a Boolean function with arity n the following obvious way: for an assignment a to the free variables all possible assignments to the quantified variables are checked and if and only if one of them lets the circuit evaluate to 1 the final value for the assignment a is 1. For example, the existentially quantified circuit $\exists z((x \vee z) \wedge (y \vee \neg z))$ describes the same Boolean function as the circuit $x \vee y$. It is easy to see that it is a Π_2^p-complete problem to check whether two given existentially quantified circuits describe the same Boolean function.

It should be mentioned that existentially quantified circuits are equivalent (in terms of polynomial-time many-one reducibility) to the notion of *generators* studied by Yap [24] and Schöning [20]. Therefore, like generators, existentially quantified circuits circuits are a nonuniform counterpart of polynomial-time non-deterministic computation, i.e. polynomial-size existentially quantified circuits equals NP/poly, see [16, 20, 24] for the notations.

We will prove an analogue of Rice's Theorem for existentially quantified circuits, even the proof is analogous.

Theorem 5.5 *Let A be a set of Boolean functions which not only depends on arity. Then SAT or co-SAT is $\leq_m^p$-reducible to the following problem: Given an existentially quantified Boolean circuit, does the Boolean function described by it belong to A?*

Proof Because A does not only depends on arity there is an arity n such that one Boolean function of that arity is in A and another one is not in A. Assume as the first case that A does not contain the n-ary constant-0-function and let $f = f(x_1, \ldots, x_n)$ be a circuit which describes an n-ary Boolean function in A. We give an $\leq_m^p$-reduction of SAT to the set of existentially quantified Boolean circuits which describe a Boolean function in A. Let a circuit $c = c(x_1, \ldots, x_m)$ be given. Construct the existentially quantified circuit $e = \exists x_1 \ldots \exists x_n(c(x_1, \ldots, x_m) \wedge f(y_1, \ldots, y_n))$. If c is satisfiable then e describes the same function as f; otherwise e describes the n-ary constant 0-function.

If A does contain the n-ary constant-0-function then we can in an analogous fashion reduce co-SAT to the problem in question.   □

Remark.   The restriction for A of being dependent on the arity is necessary because for example for the set of Boolean functions with odd arity the decision problem in question is polynomial-time computable.

If a problem is not decidable one still can have hope that it is recursively enumerable. The following extension of Rice's Theorem has a criterion for semantic properties of programs to be not even recursively enumerable.

Theorem 5.6 [Rice 1953] *Let A be a set of partial recursive functions. If there exist two functions f, g such that f is contained in A, g is not contained in A and $f \leq g$ (i.e. if $f(n)$ terminates then $f(n) = g(n)$) then the following problem is not recursively enumerable: Given a program p, does it compute a function from A?*

We can state the analogue of the above theorem. Also the proof is analogous.

Theorem 5.7 *Let A be a set of Boolean functions with two Boolean functions $F(x_1, \ldots, x_n)$, $G(x_1, \ldots, x_n)$ such that F is in A, G is not in A and $F(a_1, \ldots, a_n) \leq G(a_1, \ldots, a_n)$ for all assignments $(a_1, \ldots, a_n)$. Then co-SAT is $\leq_m^p$-reducible to the following problem: Given an existentially quantified Boolean circuit, does the Boolean function described by it belong to A?*

Proof Let $f(x_1, , \ldots, x_n)$ be a circuit for F and let $g(x_1, , \ldots, x_n)$ be a circuit for G. We give a $\leq_m^p$-reduction of co-SAT to the set of existentially quantified Boolean circuits which describe a Boolean function in A. Let a circuit $c = c(x_1, \ldots, x_m)$ be given. Construct the existentially quantified circuit $e = \exists x_1 \ldots \exists x_n (f(y_1, , \ldots, y_n) \lor (c(x_1, \ldots, x_m) \land g(y_1, \ldots, y_n)))$. If c is not satisfiable then e describes the Boolean function F, and otherwise e describes the Boolean function G. □

6 Conclusion

We presented some results in Complexity Theory which have similarities with Rice's Theorem. Our main result of that kind is that all nontrivial absolute, gap, and relative counting properties of circuits are UP-hard with respect to polynomial-time Turing reductions.

We consider the presented results as first steps of a "Rice program" in Complexity Theory: Proving lower bounds for problems which are defined by semantic properties of subrecursive objects (like circuits).

Acknowledgements

The authors are grateful for discussions with Klaus Ambos-Spies, John Case and Andre Nies. Furthermore the authors appreciate the comments by Lance Fortnow, Ulrich Hertrampf, Heribert Vollmer, Klaus Wagner and Gerd Wechsung after a talk at Dagstuhl. The second author is supported by Deutsche Forschungsgemeinschaft (DFG) grant Am 60/9-1.

References

1. M. Agrawal, T. Thierauf. *The Boolean isomorphism problem*, 37th Symposium on Foundations of Computer Science (FOCS), 1996, 422–430.
2. R. Beigel. *A structural theorem that depends quantitatively on the complexity of SAT*, Proc. 2th Annual IEEE Conference on Computational Complexity, 1987, 28–32.
3. R. Beigel. NP-*hard sets are* P-*superterse unless* R = NP, Technical Report 88-04, John Hopkins University, Baltimore, 1988.
4. R. Beigel, M. Kummer, F. Stephan. *Approximable Sets*, Information and Computation, **120**, 1995, 304-314.
5. B. Borchert, A. Lozano. *Succinct circuit representations and leaf language classes are basically the same concept*, Information Processing Letters **58**, 1996, 211–215.
6. B. Borchert, D. Ranjan, F. Stephan. *The Computational Complexity of some Classical Equivalence Relations on Boolean Functions*, ECCC Report TR96-033, 1996.
7. J. Case. *Effectivizing Inseparability*, Zeitschrift für Mathematische Logik und Grundlagen der Mathematik (Mathematical Logic Quarterly) **37**, 1991, 97–111.
8. S. A. Cook. *The complexity of theorem proving procedures*, Proc. 3rd Annual ACM Symposium on the Theory of Computing (STOC), 1971, 151–158.
9. M. R. Fellows, N. Koblitz. *Self-witnessing polynomial-time complexity and prime factorization*, Proc. 7th Annual IEEE Conference on Computational Complexity, 1992, 107–110
10. S. A. Fenner, L. J. Fortnow, S. A. Kurtz. *Gap-definable counting Classes*, Journal of Computer and Systems Sciences **48**, 1994, 116–148.
11. J. Grollmann, A. Selman. *Complexity measures for public-key cryptosystems*, SIAM Journal on Computing **17**, 309–335
12. T. Gundermann, N. A. Nasser, G. Wechsung. *A survey on counting classes*, Proc. 5th Annual IEEE Conference on Computational Complexity, 1990, 140–153.
13. S. Gupta. *The power of witness reduction*, Proc. 6th Annual IEEE Conference on Computational Complexity, 1991, 43–59
14. Dexter Kozen. *Indexings of subrecursive classes*, Theoretical Computer Science **11**, 1980, 277–301.
15. M. Ogiwara, L. A. Hemachandra. *A complexity theory of feasible closure properties*, Proc. 6th Annual IEEE Conference on Computational Complexity, 1991, 16–29
16. Ch. Papadimitriou. *Computational Complexity*, Addison Wesley, 1994.
17. H. G. Rice. *Classes of recursively enumerable sets and their decision problems*, Trans. Amer. Math. Soc. **74**, 1953, 358–366.
18. H. G. Rice. *On completely recursively enumerable classes and their key arrays*, J. Symbolic Logic **21**, 1956, 304–341.
19. J. Royer. *A Connotational Theory of Program Structure*. Springer LNCS 273, 1987.
20. U. Schöning. *On Small Generators*, Theoretical Computer Science **34**, 1984, 337–341.
21. L. G. Valiant. *The relative complexity of checking and evaluating*, Information Processing Letters **5**, 1976, 20–23
22. L. G. Valiant, V. V. Vazirani. *NP is as easy as detecting unique solutions*, Theoretical Computer Science **47**, 1986, 85–93.
23. H. Veith. *Succinct representations, leaf languages and projection reductions*, Proc. 11th Annual IEEE Conference on Computational Complexity, 1996, 118–126.
24. C. K. Yap. *Some consequences of non-uniform conditions on uniform classes*, Theoretical Computer Science **27**, 1983, 287–300.

Two Connections Between Linear Logic and Łukasiewicz Logics

Agata Ciabattoni[1] and Duccio Luchi[2]

[1] Dipartimento di Informatica
Via Comelico, 39, Milano, Italy
ciabatto@dotto.usr.dsi.unimi.it
[2] Dipartimento di Matematica
Via Del Capitano, 15, Siena, Italy
luchi@unisi.it

Abstract. In this work we establish some syntactical and semantical links between Łukasiewicz Logics and Linear Logic. First we introduce a new sequent calculus of infinite-valued Łukasiewicz Logic by adding a new rule of inference to those of Affine Linear Logic. The only axioms of this calculus have the form $A \vdash A$. Then we compare the (provability) semantics of both logics, respectively given by MV-algebras and phase spaces. We prove that every MV-algebra can be embedded into a phase space, and every complete MV-algebra is isomorphic to some phase space. In fact, completeness is necessary and sufficient for the existence of the isomorphism. Our proof is constructive.

1 Introduction

Nonclassical logics were developed in the early twenties and thirties of this century. In this work we examine two of these: Łukasiewicz Logic, introduced in [11], and Linear Logic (LL for short), introduced in [7]. From a syntactical point of view, nonclassical logic are fragments of Classical Logic (CL for short).

In several such logics the *law of the excluded middle*, $A \vee \neg A$, fails. This fact can be expressed in various ways. For instance, let us consider Monoidal Logic, i.e., the logic whose models are integral, residuated, commutative monoids $(M, \odot, \leq)$, see [10]. In more detail, $(M, \odot, i)$ is a commutative monoid, $(M, \leq, \wedge, \vee, 0, 1)$ is a lattice with the largest element 1 and the least element 0 (with the respect to the lattice ordering $\leq$), equipped with a residuated operation "$\rightarrow$" satisfying the condition

$$a \odot b \leq c \text{ iff } a \leq b \rightarrow c$$

for every $a, b, c \in M$, and 1 is the unit element w.r.t. $\odot$. The idempotency of $\odot$ implies that the rejection of the law of the excluded middle is equivalent to refusing the *law of double negation*, $\neg\neg A \rightarrow A$. Vice versa, if one allows the law of double negation, then abandonment of the law of excluded middle forces the non idempotency of $\odot$. The first case leads to Intuitionistic Logic (IL for short), while the second case to Affine Linear Logic (ALL for short)[3]. Moreover, if we

---

[3] Affine Linear Logic stands for Linear Logic + weakening rules.

add, in the second case, the *axiom of divisibility*, i.e.,

$$A \wedge B \rightarrow (A \odot (A \rightarrow B))$$

we obtain infinite-valued Lukasiewicz Logic.

Notation

In this paper we use Girard's notation for the connectives of LL, i.e.[4]:

& "and" additive
⊕ "or" additive
⊗ "and" multiplicative
⅋ "or" multiplicative

where the meaning of "additive" and "multiplicative" will be clarified below.

This paper is devoted to investigate the relationships between Linear Logic and Lukasiewicz Logics. It is organized as follows. First we discuss Gentzen style calculi of Lukasiewicz Logics and we introduce a new sequent calculus for infinite-valued Lukasiewicz Logic whose multiplicative fragment is the same as in Affine Linear Logic. Then we establish some links between (provability) semantics of both logics, respectively given by MV-algebras and phase spaces.

2 Gentzen style Calculi of Lukasiewicz Logics

Remarkably enough, Lukasiewicz Logics lacks a well established proof theory, in the traditional sense. In particular, while there exists a sequent calculus of CL, IL and LL, only recently, see [14], Gentzen style calculi of finite and infinite-valued Lukasiewicz Logics have been introduced. But while the calculus of finite-valued Lukasiewicz Logics is very elegant and informative (albeit not cut free), the same cannot be said for the calculus of infinite-valued Lukasiewicz Logic. Indeed, in the latter calculus, the axioms are not exclusively of the form $A \vdash A$. This makes the calculus a hybrid between Hilbert and Gentzen style. In this section we introduce a new sequent calculus of infinite-valued Lukasiewicz Logic, that we call *Lukasiewicz Sequent Calculus (LSC)*. Its only axiom is the identity and this makes our calculus more in the "spirit" of Gentzen style calculi.

The *sequent calculus* was introduced by Gentzen [8] in 1934. See [15] or [16] for a detailed overview. Sequents are expressions of the form $\Gamma \vdash \Delta$ where $\Gamma(= A_1, \cdots, A_n)$ and $\Delta(= B_1, \cdots, B_m)$ are finite sequences of formulas. Γ and Δ are called, respectively, antecedent and succedent. The intended reading of $\Gamma \vdash \Delta$ is

$$A_1 \text{ and } \cdots \text{ and } A_n \quad \text{prove} \quad B_1 \text{ or } \cdots \text{ or } B_m$$

where "and" and "or", in this context, are, respectively ⊗ and ⅋.

---

[4] These connectives correspond, respectively, to $\wedge, \vee, \odot, \oplus$ on MV-algebras.

Definition 1. An *inference rule* has the form

$$\frac{S_1}{S} \quad \text{or} \quad \frac{S_1 \qquad S_2}{S}$$

where S, S_1 and S_2 are sequents.

Such a rule means that given S_1 (resp., given S_1 and S_2) we can deduce S. Rules of the calculus are divided into four groups: identity, cut, structural and logical rules. Gentzen considered the following structural rules (respectively called weakening, contraction and exchange). Each rule is actually twofold, according to the side of the sequent it modifies.

$$(w, l) \quad \frac{\Gamma \vdash \Delta}{\Gamma, B \vdash \Delta} \qquad\qquad (w, r) \quad \frac{\Gamma \vdash \Delta}{\Gamma \vdash \Delta, B}$$

$$(c, l) \quad \frac{\Gamma, B, B \vdash \Delta}{\Gamma, B \vdash \Delta} \qquad\qquad (c, r) \quad \frac{\Gamma \vdash \Delta, B, B}{\Gamma \vdash \Delta, B}$$

$$(e, l) \quad \frac{\Gamma, B, C, \Gamma' \vdash \Delta}{\Gamma, C, B, \Gamma' \vdash \Delta} \qquad\qquad (e, r) \quad \frac{\Gamma \vdash \Delta, B, C, \Delta'}{\Gamma \vdash \Delta, C, B, \Delta'}$$

Sequent calculi of Classical, Intuitionistic and Linear Logic, differ for their structural rules; indeed

- *Classical Logic* accepts weakening and contraction on both sides.
- *Intuitionistic Logic*, allows succedents to contain at most one formula; in this way it forbids weakening and contraction on the right of sequents.
- *Linear Logic* refuses both weakening and contraction. However, these two structural rules are recovered by introducing the exponential connectives "!" and "?".

Rejection of weakening and contraction in LL, entails the splitting of the usual connectives "and" and "or", of CL, into additive (or context-sharing) and multiplicative (or context-independent) version. This terminology denote, respectively, that in two-premise rules of these connectives, the contexts in both premises are either the same or disjoint (in the latter case, the premises are merged in the conclusion). So, the rules of the calculus are the following:

$$(id) \quad \frac{}{A \vdash A} \qquad\qquad (cut) \quad \frac{\Gamma \vdash \Delta, A \qquad A, \Phi \vdash \Psi}{\Gamma, \Phi \vdash \Delta, \Psi}$$

$$(\neg, l) \quad \frac{\Gamma \vdash A, \Delta}{\Gamma, \neg A \vdash \Delta} \qquad\qquad (\neg, r) \quad \frac{\Gamma, A \vdash \Delta}{\Gamma, \vdash \neg A, \Delta}$$

$$(e, l) \quad \frac{\Gamma, B, C, \Gamma' \vdash \Delta}{\Gamma, C, B, \Gamma' \vdash \Delta} \qquad\qquad (e, r) \quad \frac{\Gamma \vdash \Delta, B, C, \Delta'}{\Gamma \vdash \Delta, C, B, \Delta'}$$

Multiplicative fragment

$$(1) \quad \overline{\vdash 1} \qquad\qquad (\perp) \quad \overline{\perp \vdash}$$

$$(\wp,l) \quad \frac{\Gamma, A \vdash \Delta \qquad \Gamma', B \vdash \Delta'}{\Gamma, \Gamma', A\wp B \vdash \Delta, \Delta'} \qquad\qquad (\wp,r) \quad \frac{\Gamma \vdash A, B, \Delta}{\Gamma \vdash A\wp B, \Delta}$$

$$(\otimes,l) \quad \frac{\Gamma, A, B, \Gamma' \vdash \Delta}{\Gamma, A \otimes B, \Gamma' \vdash \Delta} \qquad\qquad (\otimes,r) \quad \frac{\Gamma \vdash \Delta, A \qquad \Gamma' \vdash B, \Phi}{\Gamma, \Gamma' \vdash \Delta, A \otimes B, \Phi}$$

Additive fragment

$$(T) \quad \overline{\Gamma \vdash \Delta, T, \Delta'} \qquad\qquad (0) \quad \overline{\Gamma, 0, \Gamma' \vdash \Delta}$$

$$(\oplus,l) \quad \frac{\Gamma, A \vdash \Delta \qquad \Gamma, B \vdash \Delta}{\Gamma, A \oplus B \vdash \Delta} \qquad\qquad (\oplus,r) \quad \frac{\Gamma \vdash A_i, \Delta}{\Gamma \vdash A_1 \oplus A_2, \Delta}$$

for $i = 1, 2$

$$(\&,l_i) \quad \frac{\Gamma, A_i, \Gamma' \vdash \Delta}{\Gamma, A_1 \& A_2, \Gamma' \vdash \Delta} \qquad\qquad (\&,r) \quad \frac{\Gamma \vdash \Delta, A, \Delta' \qquad \Gamma \vdash \Delta, B, \Delta'}{\Gamma \vdash \Delta, A \& B, \Delta'}$$

for $i = 1, 2$.

Implication can be defined as follows:

$$A \to B = \neg A \wp B$$

Remark. If we add (w, l) and (w, r) to the rules of LL we get $T = 1$ and $\perp = 0$.

The Hilbert-style axiomatization of infinite-valued Lukasiewicz Logic, conjectured by Lukasiewicz[5], see [12], and proved (unpublished) by Wajsberg, using the connectives $\neg$ and $\to$, is the following:

L1 $A \to (B \to A)$
L2 $(A \to B) \to ((B \to C) \to (A \to C))$
L3 $((A \to B) \to B) \to ((B \to A) \to A)$
L4 $(\neg A \to \neg B) \to (B \to A)$

As the only rule of inference we use *modus ponens*. Henceforth we will refer of this system as a sequent calculus whose axioms are *L1-L4* and modus ponens has the form

$$\frac{\vdash A \qquad \vdash A \to B}{\vdash B}$$

---

[5] In fact, Lukasiewicz's original axiomatization also included axiom $((A \to B) \to (B \to A)) \to (B \to A)$. Chang and Meredith, see [4, 13], proved that it is a consequence of the others.

Remark. Axioms $L1$, $L2$ and $L4$ are provable in the multiplicative fragment of Affine Linear Logic.

Let us see the role of structural rules in sequent calculi of Lukasiewicz Logics. As we will sketch at the end of this section, the sequent calculus of finite-valued Lukasiewicz Logics, introduced in [14], accepts weakening and a form of bounded contraction. On the other side, the calculus of infinite-valued Lukasiewicz Logic, introduced in the same paper, and called GW, is obtained adding axiom $L3$ to ALL. So contraction is hidden inside $L3$, while it would be interesting to explicitate it in a suitable rule. Moreover, as mentioned before, this calculus is not a "standard" sequent calculus owing to the presence of the axiom $L3$. To standardise GW we replace axiom $(A \rightarrow B) \rightarrow B \vdash (B \rightarrow A) \rightarrow A$ by a rule that allows to prove the axiom of divisibility (see p. 2). In particular, we replace axiom $L3$ by the following rule[6]:

$$(\&, l_c) \qquad \frac{\Gamma, A, B \vdash A \otimes (\neg A \,\wp\, B), \Delta}{\Gamma, A \& B \vdash A \otimes (\neg A \,\wp\, B), \Delta}$$

This establishes the existence of a sort of contraction over formulas of the kind $A \& B$ when the succedent contains $A \otimes (\neg A \,\wp\, B)$.

We name the resulting calculus the *Lukasiewicz Sequent Calculus* (LSC). Note that the only axiom of this calculus is identity. We believe that our formulation offers a different perspective than GW on infinite-valued Lukasiewicz Logic.

Notation

Henceforth we will indicate with $\vdash_{Luk}, \vdash_{ALL}$ and $\vdash_{Lsc}$ derivations in the Hilbert-style axiomatization of infinite-valued Lukasiewicz Logic, Affine Linear Logic, and LSC, respectively.

We are now going to prove that LSC is equivalent to infinite-valued Lukasiewicz Logic, i.e., $\vdash_{Lsc} \Gamma \vdash \Delta$ iff $\vdash_{Luk} \Gamma \vdash \Delta$.

Lemma 2. *The rules of Affine Linear Logic are valid in MV-algebras.*

Proof. Straightforward.

Lemma 3. *The rule $(\&, l_c)$ holds in MV-algebras.*

Proof. By the completeness theorem of MV-algebras, see [3, 5], it is sufficient to prove that if

$$\neg\Gamma \,\wp\, \neg A \,\wp\, \neg B \,\wp\, A \otimes (\neg A \,\wp\, B) \,\wp\, \Delta = 1 \text{ then } \neg\Gamma \,\wp\, \neg(A \& B) \,\wp\, A \otimes (\neg A \,\wp\, B) \,\wp\, \Delta = 1$$

in the unit real interval $[0, 1]$ where

1. $A \& B = \min\{A, B\}$
2. $\neg A = 1 - A$
3. $A \,\wp\, B = \min\{1, A + B\}$

---

[6] And the dual rule for the connective $\oplus$.

4. $A \otimes B = \max\{0, A + B - 1\}$

In order to prove the assert, note that $\neg\Gamma\,\vartheta\,\neg(A\&B)\vartheta A \otimes (\neg A\vartheta B)\vartheta\Delta = 1$ iff
$1 - \Gamma + (1 - \min\{A, B\}) + \max\{A + \min\{1 - A + B, 1\} - 1, 0\} + \Delta \geq 1$. Setting
$k = 1 - \min\{A, B\} \Rightarrow \min\{A, B\} = 1 - k$. It follows either $A = 1 - k$ or $B = 1 - k$.
We prove the case $A = 1 - k$ (case $B = 1 - k$ is symmetrical).
$A = 1 - k \Rightarrow B \geq 1 - k$. Then

$$\begin{aligned}
&\max\{A + \min\{1 - A + B, 1\} - 1, 0\} \\
&= \max\{1 - k + \min\{1 - (1 - k) + 1 - k^-, 1\} - 1, 0\} \\
&= \max\{1 - k, 0\} = 1 - k
\end{aligned}$$

So $1 - \Gamma + k + 1 - k + \Delta \geq 1$ always holds.

Theorem 4. $\vdash_{Lsc} \Gamma \vdash \Delta \Longrightarrow \vdash_{Luk} \Gamma \vdash \Delta$.

Proof. Straightforward from Lemma 2 and Lemma 3.

Theorem 5. $\vdash_{Luk} \Gamma \vdash \Delta \Longrightarrow \vdash_{Lsc} \Gamma \vdash \Delta$.

Proof. Because of remark in p. 5, it remains to prove that $\vdash_{Lsc} L3$.
Indeed $(A \rightarrow B) \rightarrow B$ is equivalent to $\neg(\neg A\vartheta B)\vartheta B$ and $(B \rightarrow A) \rightarrow A$ is
equivalent to $\neg(\neg B\vartheta A)\vartheta A$.
But

$$\neg(\neg A\&\neg B) \vdash \neg(\neg B\vartheta A)\vartheta A$$

and

$$\frac{\neg(\neg A\vartheta B)\vartheta B \vdash \neg(\neg B \otimes (B\vartheta\neg A)) \qquad \neg(\neg B \otimes (B\vartheta\neg A)) \vdash \neg(\neg B\&\neg A)}{\neg(\neg A\vartheta B)\vartheta B \vdash \neg(\neg B\&\neg A)}$$

Since

$$\neg(\neg B\&\neg A) \vdash \neg(\neg A\&\neg B)$$

using the cut rule twice we have the assert.

Remark. In the above proof, the sequent

$$\neg(\neg B \otimes (B\vartheta\neg A)) \vdash \neg(\neg B\&\neg A)$$

is provable in LSC only using the $(\&, l_c)$ rule.

Thus the Lukasiewicz Sequent Calculus is a Gentzen style formulation of infinite-
valued Lukasiewicz Logic. However, LSC (as well as GW) is not cut-free; for
example, the cut rule cannot be eliminated in the following situation:

$$\frac{\dfrac{\Gamma, A, B \vdash \Delta}{\Gamma, A\&B \vdash \Delta} \qquad \dfrac{\Gamma' \vdash A, \Delta' \qquad \Gamma' \vdash B, \Delta'}{\Gamma' \vdash A\&B, \Delta'}}{\Gamma, \Gamma' \vdash \Delta, \Delta'}$$

Where $\Delta = \Delta'', A \otimes (\neg A \,\mathfrak{B}\, B)$.

On the other hand, this would have been expected since in Lukasiewicz Logics interpolation property does not holds.

We conclude this section giving an overview of Gentzen style formulations that are equivalent to finite-valued Lukasiewicz Logics existing in literature. In [14], Prijately proposed a system, called PL_n, with $n \geq 2$, for n-valued Lukasiewicz Logics. This is given adding to ALL a bounded contraction rule, i.e.,

$$(c_n, l) \quad \frac{\Gamma, A^{n+1} \vdash \Delta}{\Gamma, A^n \vdash \Delta} \qquad\qquad (c_n, r) \quad \frac{\Gamma \vdash A^{n+1}, \Delta}{\Gamma, A^n \vdash \Delta}$$

Where $A^{(k)} = A, A, \cdots, A$, i.e., k copies of formula A.

Intuitively, these rules say that $n+1$ occurrences of a formula may be contracted to n occurrences.

Remark. (c_n, l) and (c_n, r) can be replaced by

$$(c'_n, l) \quad \frac{\Gamma, A^n, A^n \vdash \Delta}{\Gamma, A^n \vdash \Delta} \qquad\qquad (c'_n, r) \quad \frac{\Gamma \vdash A^n, A^n, \Delta}{\Gamma, A^n \vdash \Delta}$$

In fact, these rules are interderivable, using weakening rules, with those introduced above.

However, the cut rule cannot be eliminated from the system PL_n, for any $n \geq 2$. Other Gentzen style formulations, respectively for 4 and n-valued Lukasiewicz Logics are obtained by Grishin [9] and Avron [1] (see also [2]). In particular, the former, based on a calculus of hypersequents, is cut free.

3 Phase Spaces and MV-algebras

In this section we discuss some links between the semantics of Lukasiewicz Logics and LL. Henceforth, the notation used for MV-algebras will be the same as in [6] (see footnote p. 2).

We will prove that given an MV-algebra $\Omega =_{def} (M, \odot, \neg, 1)$ there is a commutative monoid $(\overline{M}, \otimes', 1')$ and a subset of $\overline{M}$, called $\bot$, such that Ω is embedded in the phase space generated by $(\overline{M}, \otimes', 1')$ and $\bot$. The embedding is an isomorphism if Ω is complete. Moreover the completeness of Ω is a necessary and sufficient condition for the existence of a phase space isomorphic to it.

MV-algebras were introduced, see [5], as an algebraic counterpart of the many-valued logics of Lukasiewicz. Phase semantics associates values to every formula of LL, in the spirit of classical model theory; therefore, unlike denotational semantics, phase semantics yields a model only for provability in LL, and not for proofs. In this section we shall establish some links between these two semantics. First we give some basic notions about MV-algebras and phase spaces.

Definition 6. An *MV-algebra* Ω is a quadruple $(\boldsymbol{M}, \odot, \neg, 1)$ where $(\boldsymbol{M}, \odot, 1)$ is a commutative monoid, satisfying the following conditions:

1. $\neg\neg\boldsymbol{x} = \boldsymbol{x}$;
2. $\boldsymbol{x} \odot \neg 1 = \neg 1$;
3. $\neg(\boldsymbol{x} \odot \neg\boldsymbol{y}) \odot \neg\boldsymbol{y} = \neg(\boldsymbol{y} \odot \neg\boldsymbol{x}) \odot \neg\boldsymbol{x}$.

(MV stands for "many-valued").

Remark. Note that Condition 3 in the previous definition corresponds to axiom L3 in the axiomatization of infinite-valued Lukasiewicz Logic presented above.

Definition 7. In any MV-algebra we can define additional operations:

- $\boldsymbol{x} \leq \boldsymbol{y}$ iff $\boldsymbol{x} \odot \neg\boldsymbol{y} = 0$.
- $0 =_{def} \neg 1$
- $\boldsymbol{x} \oplus \boldsymbol{y} =_{def} \neg(\neg\boldsymbol{x} \odot \neg\boldsymbol{y})$.
- $\boldsymbol{x} \vee \boldsymbol{y} =_{def} (\boldsymbol{x} \odot \neg\boldsymbol{y}) \oplus \boldsymbol{y}$.
- $\boldsymbol{x} \wedge \boldsymbol{y} =_{def} \neg(\neg\boldsymbol{x} \vee \neg\boldsymbol{y}) = \boldsymbol{x} \odot (\neg\boldsymbol{x} \oplus \boldsymbol{y})$.

Then $\wedge$ and $\vee$ are, respectively, the operations of inf and sup with respect to the order defined by $\leq$.

Proposition 8. *Every MV-algebra satisfies the following conditions:*

- $0 \leq \boldsymbol{x} \leq 1$
- $\boldsymbol{x} \odot \neg\boldsymbol{x} = 0$;
- $\boldsymbol{x} \leq \boldsymbol{y}$ *iff there exists z such that $\boldsymbol{x} = z \odot \boldsymbol{y}$ (divisibility).*
- *if $\boldsymbol{x} \leq \boldsymbol{y}$ then for every z we have $\boldsymbol{x} \odot z \leq \boldsymbol{y} \odot z$ (monotonicity of $\odot$).*

Proof. See [6].

Let us introduce the notion of phase spaces. These structures consist of a carrier equipped with two algebraic structures: a commutative monoid structure to interpret multiplicative connectives, and a lattice structure to interpret additive connectives.

Definition 9. Let $(\boldsymbol{M}, \odot, i)$ be a commutative monoid, $\perp$ a subset of M and let us define, for any $A, B \subseteq \boldsymbol{M}$, the operations:

- $A \to B =_{def} \{y \mid \text{for any } \boldsymbol{x} \in A, \; \boldsymbol{x} \odot y \in B\}$
- $A^{\perp} =_{def} A \to \perp = \{y \mid \text{for any } \boldsymbol{x} \in A, \; \boldsymbol{x} \odot y \in \perp\}$
- $A \otimes' B =_{def} (A \odot B)^{\perp\perp}$

The operation $(\;)^{\perp\perp} : P(M) \to P(M)$ is a preclosure, i.e., for every $A, B \subseteq \boldsymbol{M}$:

1. $A \subseteq (A)^{\perp\perp}$
2. $(A)^{\perp\perp} \subseteq ((A)^{\perp\perp})^{\perp\perp}$
3. if $A \subseteq B$ then $(A)^{\perp\perp} \subseteq (B)^{\perp\perp}$.

Now we are able to define phase spaces.

Definition 10. Let $(M, \odot, i)$ be a commutative monoid, $\perp$ a subset of M, $\overline{M} =_{def} \{A \subseteq M \mid (A)^{\perp\perp} = A\}$ and $0' =_{def} \emptyset^{\perp\perp}$. Then $(\overline{M}, 0', \perp, \otimes', ()^{\perp}, \cap)$ is a *phase space*.

Every $A \in \overline{M}$ is called a *fact*.

Definition 11. We define the following operations:

- $T =_{def} M$
- $A \cup' B =_{def} (A \cup B)^{\perp\perp}$
- $1' =_{def} \{i\}^{\perp\perp}$
- $A \otimes' B =_{def} (A^{\perp} \odot B^{\perp})^{\perp}$

Operations $\otimes', \otimes', \cap, \cup', ^{\perp}$ model, respectively, linear connectives $\otimes, \otimes, \&, \oplus, \neg$.

Proposition 12. *In every phase space we have:*

1. The intersection of a family of facts is a fact.
2. For every $A \subseteq M$, $A^{\perp}$ is a fact.
3. $(0')^{\perp} = T$, $(1')^{\perp} = \perp$.
4. $A \cap B = (A^{\perp} \cup' B^{\perp})^{\perp}$.
5. $A \cup' B = (A^{\perp} \cap B^{\perp})^{\perp}$.
6. $(\overline{M}, \cap, \cup', 0', T)$ is a lattice with minimum and maximum.
7. $(\overline{M}, \otimes', 1')$ is a commutative monoid where $\otimes'$, is the monotone tensor operation with respect to $\subseteq$.

Note that, in general, $0' \neq \perp$ and $1' \neq T$.

Remark. In MV-algebras monoidal and lattice operations are not unrelated, as in phase spaces; this causes many problems in finding a sequent calculus for Lukasiewicz Logics in which cut elimination holds (if any).

Before proving the main results, we give some examples that can help to understand the intuition behind the construction of phase spaces from particular MV-algebras.

Example 1. Let $(\{i\}, \odot, i)$ be the trivial monoid. Setting $\perp =_{def} \emptyset$, the phase space generated is isomorphic to the Boolean algebra $(\{0, 1\}, \wedge, \vee, \neg, 0, 1)$.

Example 2. Let $\Omega =_{def} (M, \odot, \neg, 1)$ be an MV-algebra where $M =_{def} \{0, \frac{1}{2}, 1\}$ and the operations are defined as follows: $x \odot y =_{def} \max\{0, x + y - 1\}$ and $\neg x =_{def} 1 - x$. Setting $\perp =_{def} \{0\}$, then the phase space generated is the following:
$$(\overline{M}, 0', \perp, \otimes', ()^{\perp}, \cap)$$
where $\overline{M} =_{def} \{\{0\}, \{0, \frac{1}{2}\}, \{0, \frac{1}{2}, 1\}\}$.
Indeed, $\{0\}^{\perp} = \{0, \frac{1}{2}, 1\}$, $\{0, \frac{1}{2}\}^{\perp} = \{0, \frac{1}{2}\}$, $\{0, \frac{1}{2}, 1\}^{\perp} = \{0\}, \{1\}^{\perp} = \{0\}, \{\frac{1}{2}\}^{\perp} = \{0, \frac{1}{2}\}, \{0, 1\}^{\perp} = \{0\}, \{\frac{1}{2}, 1\}^{\perp} = \{0\}$ so, the sets A such that $A^{\perp\perp} = A$ are all and only $\{0\}, \{0, \frac{1}{2}\}, \{0, \frac{1}{2}, 1\}$. And

$$\{0\} \otimes' x = \{0\}, \quad \{0, \frac{1}{2}\} \otimes' \{0, \frac{1}{2}\} = \{0\}, \quad \{0, \frac{1}{2}, 1\} \otimes' x = x.$$

It is easy to check that this phase space is isomorphic to Ω.

A more general example is as follows:

Example 3. Let $\Omega_n =_{def} (M_n, \odot, \neg, 1)$ be an MV-algebra, where $M_n =_{def} \{\frac{k}{n} \mid 0 \le k \le n\}$ and the operations are defined as in the previous example. Setting $\perp =_{def} \{0\}$, then in the phase space built from Ω_n a set A is a fact iff there is a $j \in M_n$ such that $A =_{def} \{y \mid y \le j\}$. Moreover the phase space resulting is isomorphic to Ω_n.

Theorem 13. *Let $\Omega =_{def} (M, \odot, \neg, 1)$ be an MV-algebra. Then there is a phase space Γ such that Ω is isomorphically embedded in Γ.*

Proof. First we note that $(M, \odot, 1)$ is a commutative monoid. We set $\perp =_{def} \{0\}$. Let $\Gamma =_{def} (\overline{M}, 0', \perp, \otimes', (\ )^{\perp}, \cap)$ be the phase space built by the usual construction (see p. 8). Now we have to define a function from Ω to Γ. Let g be the following function: $g(x) =_{def} \{y \mid y \le x\}$ where $x \in M$. To begin with we show that for every $x \in M$ the set $\{y \mid y \le x\}$ is a fact and then that this function is a homomorphism. The injectivity of g is trivial.

For every $x \in M$ let $A_x \subseteq M$ be defined as follows: $A_x =_{def} \{y \mid y \le \neg x\}$. It is easy to verify that $(A_x)^{\perp} =_{def} \{y \mid$ for any $z \in A_x \ y \odot z = 0\} = g(x)$. Indeed, let $y \in g(x)$, so $y \le x$, i.e., $y \odot \neg x = 0$. Then for every $z \in A_x$ $z \odot y \le \neg x \odot y = 0 \Leftrightarrow y \in (A_x)^{\perp}$. This implies $g(x) \subseteq (A_x)^{\perp}$. For the converse, if $y \in (A_x)^{\perp}$, then $y \odot \neg x = 0 \Leftrightarrow y \le x$, i.e., $y \in g(x)$. Thus we have $(A_x)^{\perp} = g(x)$ and for any $x \in M$, $g(x) =_{def} \{y \mid y \le x\}$ is a fact by Proposition 12. The monotonicity of g is trivial and moreover $g(0) = 0'$ and $g(1) = M$.

It remains to show:

1. $g(\neg x) = g(x)^{\perp}$
2. $g(x \odot y) = g(x) \otimes' g(y)$
3. $g(x \wedge y) = g(x) \cap g(y)$

As a matter of fact,

1. $z \in g(\neg x) \Leftrightarrow z \le \neg x \Leftrightarrow z \odot \neg\neg x = 0 \Leftrightarrow z \odot x = 0 \Leftrightarrow$ by monotonicity of $\odot$, for every $y \le x$ $z \odot y = 0 \Leftrightarrow z \in g(x)^{\perp}$.
2. $g(x) \otimes' g(y) =_{def} (g(x) \odot g(y))^{\perp\perp}$. But $g(x) \odot g(y) = g(x \odot y)$, so, being $g(x \odot y)$ a fact, it follows that $g(x \odot y) = (g(x \odot y))^{\perp\perp} = (g(x) \odot g(y))^{\perp\perp} = g(x) \otimes' g(y)$.
3. If $z \in g(x \wedge y)$, then $z \le x \wedge y$ so $z \le x$ and $z \le y$. This implies $z \in g(x)$ and $z \in g(y)$, i.e., $z \in g(x) \cap g(y)$. For the converse, if $z \in g(x) \cap g(y)$ then $z \in g(x), g(y)$, so $z \le x$ and $z \le y$. Therefore $z \le x \wedge y$, i.e., $z \in g(x \wedge y)$.

Remark. In the phase space built from an MV-algebra we necessary have $0' = \perp$ and $1' = M$. This is due to the fact that for any $x \in M$, $x \le 1$ (see Proposition 8), as one would also expect by the presence of weakening rules (see remark in p. 4).

If Ω is a complete MV-algebra then $g(x)$ is an isomorphism.

Theorem 14. *Let Ω be a complete MV-algebra. Let Γ be the phase space defined in the previous theorem. Then if A is a fact of Γ, there exists an $x \in M$ such that $A = \{y \mid y \le x\}$, i.e., $g(x)=A$.*

Proof. Let A be a fact and let $x = \sup A$, being Ω a complete MV-algebra, the sup of A exists. We prove that $A = g(x)$. Indeed, let $z \in A$ then $z \leq x$ and so $z \in g(x)$; thus $A \subseteq g(x)$. For the converse, if A is a fact then there is a $B \subseteq M$ such that $B^\perp =_{def} A$, i.e., $y \in A$ iff for every $z \in B$, $y \odot z = 0$. This implies that for every $y \in A$ and for every $z \in B$, $y \leq \neg z$. In particular, for every $z \in B$, $x = \sup A \leq \neg z$, i.e., $x \odot z = 0$. But if $k \in g(x)$ then $k \leq x$, so for every $z \in B$, $k \odot z \leq x \odot z = 0$. So $k \in B^\perp =_{def} A$ and therefore $g(x) \subseteq A$.

The following theorem establishes that the completeness of Ω is also a necessary condition for the existence of a phase space isomorphic to it.

Theorem 15. *If an MV-algebra Ω is isomorphic to a phase space Γ, then Ω is complete.*

Proof. Suppose there exists an unbounded infinite descending chain $y_1 \geq \ldots \geq y_n \geq \ldots$ (where $y_i \in M$, that is the domain of Ω). If there were a phase space Γ and an isomorphism g from Ω to Γ, then an infinite descending chain of facts $g(y_1) \supseteq \ldots \supseteq g(y_n) \supseteq \ldots$ would exist. We know that the intersection of facts is a fact and g is surjective; it follows that there exists an $x \in M$ such that

$$g(x) = \bigcap_i g(y_i).$$

As g is an isomorphism, x is a minorant of y_i, for every i, and by the monotonicity and injectivity of g it is the largest minorant. This is a contradiction, because the chain has been supposed to have no inf.

As a consequence of the previous theorem, together with Birkhoff's theorem, it follows that phase spaces do not form an equational class.

4 Conclusions

In this work we have examined Linear Logic and Lukasiewicz Logics, establishing some links between them. In particular, since the axioms of Linear Logic are a subset of those of Lukasiewicz Logics, we have proposed a sequent calculus of infinite-valued Lukasiewicz Logic, whose axioms are only in the form $A \vdash A$, and close to that of Affine Linear Logic.

From the semantical point of view, we have characterised MV-algebras as sub-models of particular phase spaces, presenting a necessary and sufficient condition for an MV-algebra to be isomorphic to a phase space. Our proof is constructive. Our future research is addressed towards finding a cut-free sequent calculus of infinite-valued Lukasiewicz Logic (if any); trying to characterise phase spaces isomorphic to MV-algebras and finally, as it was made for IL and CL, defining a translation of infinite-valued Lukasiewicz Logic into Linear Logic.

References

1. *A. Avron.* Natural 3-valued Logics. Characterization and Proof Theory. *J. of Symbolic Logic*, vol. **56**. pp. 276-294. 1991.
2. *A. Avron.* The Method of Hypersequents in the Proof Theory of Propositional Nonclassical Logics. In: **Logic: from Foundations to Applications.** W. Hodges, M. Hyland, C. Steinhorn and J. Truss eds., European Logic Colloquium. Oxford Science Pubblications. Clarendon Press. Oxford. 1996.
3. *C.C. Chang.* Algebraic analysis of many valued logics. *Trans. Amer. Math. Soc.*, vol. **88**. pp. 467-490. 1958.
4. *C.C. Chang.* Proof of an axiom of Lukasiewicz. *Trans. Am. Math. Soc.*, vol. **87**. pp. 55-56. 1958.
5. *C.C. Chang.* A new proof of the Completeness of the Lukasiewicz's axioms. *Trans. Amer. Math. Soc.*, vol. **93**. pp. 74-80. 1959.
6. *R. Cignoli, D. Mundici and I.M. D'Ottaviano.* **MV-algebras: the Mathematics of Many-Valued Logic.** In preparation.
7. *J.Y. Girard.* Linear Logic. *Theoretical Comp. Science*, vol. **50**. pp. 1-102. 1987.
8. *G. Gentzen.* Untersuchungen über das logische Schliessen I, II. *Mathematische Zeitschrift*, vol. **39**. pp. 176-210, pp. 405-431. 1934-35.
9. *V.N. Grishin.* On the Algebraic Semantics of a Logic without Contraction. In: **Studies in Set Theory and Nonclassical Logics**, D.A. Bochvar, V.N. Grishin eds., Nauka, Moskva, pp. 247-264. 1976.
10. *U. Höhle.* Commutative, residuated *l*-monoids. In: **Nonclassical logics and their applications to fuzzy subsets.** U. Höhle and P. Klement eds., Kluwer. Dordrecht. pp. 53-106. 1995.
11. *J. Lukasiewicz.* O Logice Trówartosciowej. *Ruch Filozoficzny.*, vol. **5** pp. 170-171. 1920. English Translation: On three-valued logic. In: **J. Lukasiewicz Selected Works.** North-Holland, Amsterdam. pp. 87-88. 1970.
12. *J. Lukasiewicz and A. Tarski.* Untersuchungen über den Aussagenkalkül. *Compt. Rendus de la Soc. des Sciences et des Lett. de Vars..* cl. iii vol. **23**. pp. 1-21. 1930. English translation: Investigations into the sentential calculus. Chap. 4. In: **Logic, Semantics Metamathematics.** Oxford: Clarendon Press. 1956. Reprinted Hackett. Indianapolis. 1983.
13. *C. A. Meredith.* The dependency of an axiom of Lukasiewicz. *Trans. Amer. Math. Soc.*, vol. **87**. p. 57. 1958.
14. *A. Prijately.* Bounded Contraction and Gentzen style Formulation of Lukasiewicz Logics. *Studia Logica*, vol. **57**. pp. 437-456. 1996.
15. *G. Takeuti.* **Proof Theory.** Studies in Logic and the Foundations of Math., vol. 81. North Holland. 1975.
16. *A.S. Troelstra and H. Schwichtenberg.* **Basic Proof Theory.** Cambridge University Press. 1996.

Structuring of Computer-Generated Proofs by Cut Introduction

Uwe Egly[1] and Karin Genther[2]

[1] Abt. Wissensbasierte Systeme 184/3, TU Wien
Treitlstraße 3, A–1040 Wien
e-mail: uwe@kr.tuwien.ac.at
[2] Inst. für Theoretische Informatik, Med. Universität zu Lübeck
Wallstraße 40, D–23560 Lübeck
e-mail: genther@informatik.mu-luebeck.de

Abstract. As modern Automated Deduction systems rely heavily on the use of a machine-oriented representation of a given problem, together with sophisticated redundancy-avoiding techniques, a major task in convincing human users of the correctness of automatically generated proofs is the intelligible representation of these proofs. In this paper, we propose the use of the cut-rule in the human-oriented presentation of computer-generated proofs. The intelligent application of cuts enables the integration of essential lemmata and therefore shortens and structures proof presentation. We show that many translation techniques in Automated Deduction, such as antiprenexing and some forms of normal form translations, can be described as cuts and are indeed part of the deductive solution of a problem. Furthermore, we demonstrate the connection between symmetric simplification, quantorial extension principles and the application of the cut-rule.

1 Introduction

Most of today's Automated Deduction (AD) systems use low-level formulae and proof representations. Formulae are represented as (a set of) clauses while proofs are represented as trees or sequences of clauses resulting from applications of low-level inference rules like resolution. It has been observed for a long time that such machine-oriented representations of proofs are not well suited for humans. It is widely acknowledged that the performance of AD systems strongly depends on such machine-oriented representations together with powerful redundancy-avoiding techniques, but for the acceptance of AD systems, a user-friendly interface is indispensable. The situation is similar to the case of writing and debugging programs in higher programming languages. While a compressed machine-oriented representation of the program is the input of the central unit, additional information enables the debugger to provide source code information of the high-level program and allow stepping, tracing and manipulating data in terms of the high-level program. In contrast to the debugging scenario, the generation of human-oriented presentations of automatically obtained proofs is a difficult task, mainly because a good proof structure requires ingenious lemmata and definitions. In today's AD systems, proofs are generated which are exclusively based on information provided in the given formula. Such proof systems are called *analytic* and *cut-free*. These properties correspond to the method of analyzing the structure of the given formula and to use no cuts (to be described later). If mathematicians prove theorems, they structure the theory and the proof by introducing names and new concepts by definitions and by using lemmata. The choice of appropriate lemmata is the key device to obtain readable, well-structured proofs.

A technical device how lemmata can be integrated into a calculus is the cut rule:

$$\frac{\Gamma_1 \vdash \Delta_1, F \quad F, \Gamma_2 \vdash \Delta_2}{\Gamma_1, \Gamma_2 \vdash \Delta_1, \Delta_2} \ cut$$

If we read this rule from bottom to top then this rule says that, instead of deriving $\Delta_1^\vee \vee \Delta_2^\vee$ from $\Gamma_1^\wedge \wedge \Gamma_2^\wedge$, we can derive $\Delta_1^\vee \vee F$ from $\Gamma_1^\wedge$ and $\Delta_2^\vee$ from $F \wedge \Gamma_2^\wedge$. Thereby, $\Delta^\vee$ ($\Gamma^\wedge$) denotes the disjunction (conjunction) of all elements in Δ (Γ). The formula F is the ingenious lemma which usually does not occur in the lower sequent. This is the main problem with this rule in automated proof search: how can we know what F is?

Things change if we already have a proof in some calculus (say resolution) and we want to translate it into another calculus (say NK or LK). Most of the techniques proposed so far for such translations (and many techniques in AD systems) are indeed applications of the cut rule. For systems where proofs have to be manipulated in order to fulfill other tasks, such as extraction of programs or explanation-based generalization, it is crucial not only to understand the use of literals in the proof, but also to make visible the fundamental structure of the proof.

In this paper, we examine different translation techniques and show how they are related to the cut rule. In Section 3, we show how translation schemata of computer-generated proofs into NK-proofs can be improved. Most of these translations suffer from a frequent use of proofs by contradiction which are mainly introduced if more than one substitution instance of a quantified subformula occurs in the proof. The ad-hoc solution, symmetric simplification [25], introduced to reduce the number of proofs by contradiction is identified as a specific form of the cut rule in Section 4. In Section 5, we show that antiprenexing and some translation schemata to normal forms are indeed restricted forms of the cut rule. Moreover, well-known preprocessing reductions can be described in a similar way.[3] In Section 6, we consider quantorial extension principles. It is well known [5, 11] that such extension principles are (restricted) variants of the cut rule. We show, if we already have a (cut-free) proof, how the structure (and length) of such a proof can be improved by introducing cuts representing applications of quantorial extension principles.

---

[3] It is common practice to consider such preprocessing activities not to be part of the deductive solution. Our results indicate that this viewpoint is wrong; all these techniques are indeed "inferential" and belong to the deductive solution.

2 Definitions and Notations

We consider a usual first-order language with function symbols. We use Gentzen's calculus of classical natural deduction (NK) as given in [19]. The calculus NK consists of the axiom of the excluded middle (*tertium non datur*), the rule *ex falso quodlibet* and logical and quantifier rules. Let A, B, C, and F denote formulae and F_x^t the formula which results from F by substituting all free occurrences of the variable x in F by the term t. The term t is any term not containing a bound variable.

$$\text{tertium non datur: } A \vee \neg A \qquad\qquad \text{ex falso quodlibet: } \frac{\perp}{A}\ \perp E$$

LOGICAL RULES

Negation:
$$\frac{\begin{array}{c}[A]\\ \perp\end{array}}{\neg A}\ \neg I \qquad\qquad \frac{A \quad \neg A}{\perp}\ \neg E$$

Conjunction:
$$\frac{A \quad B}{A \wedge B}\ \wedge I \qquad\qquad \frac{A \wedge B}{A}\ \wedge_1 E \qquad \frac{A \wedge B}{B}\ \wedge_2 E$$

Disjunction:
$$\frac{A}{A \vee B}\ \vee_1 I \qquad \frac{B}{A \vee B}\ \vee_2 I \qquad\qquad \frac{A \vee B \quad \begin{array}{c}[A]\\ C\end{array} \quad \begin{array}{c}[B]\\ C\end{array}}{C}\ \vee E$$

Implication:
$$\frac{\begin{array}{c}[A]\\ B\end{array}}{A \to B}\ \to I \qquad\qquad \frac{A \quad A \to B}{B}\ \to E$$

QUANTIFIER RULES

Quantifier $\forall$:
$$\frac{F(u)}{\forall x.F(x)}\ \forall I \qquad\qquad \frac{\forall x.F(x)}{F(t)}\ \forall E$$

Quantifier $\exists$:
$$\frac{F(t)}{\exists x.F(x)}\ \exists I \qquad\qquad \frac{\exists x.F(x) \quad \begin{array}{c}[F(u)]\\ C\end{array}}{C}\ \exists E$$

$\forall I$ and $\exists E$ must fulfill the *eigenvariable condition*, i.e., the variable u does not occur in $\forall x.F(x)$, $\exists x.F(x)$ or C nor in any assumption on which $\forall x.F(x)$ or C (except for $F(u)$) depends.

The *major* premise is the formula containing the occurrence of the logical operator which is eliminated in the E-rule application; all other premises are called *minor*.

It is well known (see [19]) that *tertium non datur* can be replaced by the following inference.

$$\frac{\neg\neg A}{A}\ \neg\neg E$$

For convenience, we assume to have both principles at hand, even in one proof.

The intuitionistic calculus of natural deduction NJ consists of all the rules of NK except for *tertium non datur* and $\neg\neg$E.

In a *normal* proof, each major premise of an E-rule is either an assumption or a conclusion of an E-rule different from $\vee$E, $\exists$E.

We assume familiarity with sequent calculi and use the following form of Gentzen's calculus LK. In contrast to [19], rule applications are allowed at arbitrary places in a sequent, thus the exchange rule can be omitted. Let Γ, Δ, Π and Λ denote sequences of formulae and let A and B denote arbitrary formulae. Initial sequents (axioms) are of the form $A \vdash A$ for a formula A.

LOGICAL RULES

$$\frac{\Gamma \vdash \Delta, A}{\neg A, \Gamma \vdash \Delta} \; \neg l \qquad\qquad \frac{A, \Gamma \vdash \Delta}{\Gamma \vdash \Delta, \neg A} \; \neg r$$

$$\frac{A, \Gamma \vdash \Delta}{A \wedge B, \Gamma \vdash \Delta} \; \wedge_1 l \qquad \frac{B, \Gamma \vdash \Delta}{A \wedge B, \Gamma \vdash \Delta} \; \wedge_2 l \qquad \frac{\Gamma \vdash \Delta, A \qquad \Gamma \vdash \Delta, B}{\Gamma \vdash \Delta, A \wedge B} \; \wedge r$$

$$\frac{A, \Gamma \vdash \Delta \qquad B, \Gamma \vdash \Delta}{A \vee B, \Gamma \vdash \Delta} \; \vee l \qquad \frac{\Gamma \vdash \Delta, A}{\Gamma \vdash \Delta, A \vee B} \; \vee_1 r \qquad \frac{\Gamma \vdash \Delta, B}{\Gamma \vdash \Delta, A \vee B} \; \vee_2 r$$

$$\frac{\Gamma \vdash \Delta, A \qquad B, \Gamma \vdash \Delta}{A \to B, \Gamma \vdash \Delta} \; \to l \qquad \frac{A, \Gamma \vdash \Delta, B}{\Gamma \vdash \Delta, A \to B} \; \to r$$

QUANTIFIER RULES

$$\frac{A(t), \Gamma \vdash \Delta}{\forall x.A(x), \Gamma \vdash \Delta} \; \forall l \qquad\qquad \frac{\Gamma \vdash \Delta, A(u)}{\Gamma \vdash \Delta, \forall x.A(x)} \; \forall r$$

$$\frac{A(u), \Gamma \vdash \Delta}{\exists x.A(x), \Gamma \vdash \Delta} \; \exists l \qquad\qquad \frac{\Gamma \vdash \Delta, A(t)}{\Gamma \vdash \Delta, \exists x.A(x)} \; \exists r$$

$\forall r$ and $\exists l$ must fulfill the eigenvariable condition, i.e., the variable u does not occur in the lower sequent. Again, the term t is any term not containing a bound variable.

STRUCTURAL RULES

Cut:
$$\frac{\Gamma \vdash \Delta, A \qquad A, \Pi \vdash \Lambda}{\Gamma, \Pi \vdash \Delta, \Lambda} \; cut$$

Weakening:
$$\frac{\Gamma \vdash \Delta}{A, \Gamma \vdash \Delta} \; wl \qquad\qquad \frac{\Gamma \vdash \Delta}{\Gamma \vdash \Delta, A} \; wr$$

Contraction:
$$\frac{A, A, \Gamma \vdash \Delta}{A, \Gamma \vdash \Delta} \; cl \qquad\qquad \frac{\Gamma \vdash \Delta, A, A}{\Gamma \vdash \Delta, A} \; cr$$

3 Transformation of Computer-Generated Proofs

A major problem in Automated Deduction is the presentation of proofs found by sophisticated and efficient deduction systems in a form more suitable for humans to read. Calculi which are often considered (more) adequate for representing proofs are Gentzen's NK and LK (see [19]). A number of publications develop procedures for transforming automatically generated proofs stated in various deduction formalisms (such as expansion-tree proofs or refutation graphs) into variants of NK and LK (e.g. see [2], [7], [17], [21], [22], [23], and [24]).

Many of the transformations to NK suffer from a frequent use of proofs by contradiction, which seem to be not "intuitive" for the reader. For instance, such proofs are introduced if more than one substitution instance of a subformula is required in the automatically generated proof. Consider the following NK-proof "by contradiction" of the formula $p(a) \lor p(b) \to \exists x.p(x)$ as an example.

$$
\cfrac{[p(a) \lor p(b)]_1 \qquad \cfrac{\cfrac{\cfrac{[\neg \exists x.p(x)]_2}{\neg p(a)} \neg\exists E \quad \cfrac{[\neg \exists x.p(x)]_2}{\neg p(b)} \neg\exists E}{\cfrac{\neg p(a) \land \neg p(b)}{\neg(p(a) \lor p(b))} \neg\land\text{-taut}} \land I}{\cfrac{\bot}{\cfrac{\neg\neg\exists x.p(x)}{\exists x.p(x)} \boxed{\neg\neg E}} \neg I_2} \neg E}{(p(a) \lor p(b)) \to \exists x.p(x)} \to I_1
$$

The rules $\neg\exists E$, $\neg\land$-taut and $\neg\neg E$ are derived rules in NK and are used here to keep the size of the example manageable. The framed application of $\neg\neg E$ is not necessary for the proof of the formula because it can be proven intuitionistically, i.e., without the use of *tertium non datur*. The proof structure is very obscure because of the use of this classical inference and does not reflect the simple structure of e.g. a connection proof.

In [18], a transformation procedure from an extension procedure to NK is suggested that avoids the use of proofs by contradiction in case that more than one substitution instance of a quantified subformula occurs in the proof. The underlying principle of the transformation rule used in such a case is first to generalize each of the substitution instances and then to use a "contracting" rule in order to derive the quantified subformula. An example for the case of two substitution instances of the quantified subformula is the following short proof of the formula $p(a) \lor p(b) \to \exists x.p(x)$.

$$
\cfrac{[p(a) \lor p(b)]_1 \qquad \cfrac{\cfrac{[p(a)]_2}{\exists x.p(x)} \exists I \quad \cfrac{[p(b)]_2}{\exists x.p(x)} \exists I}{\exists x.p(x)} \lor E_2}{(p(a) \lor p(b)) \to \exists x.p(x)} \to I_1
$$

Obviously, this direct proof is shorter, more readable, and no derived rules are necessary. In the next section, we consider a technique, symmetric simplification, which was introduced in order to remove "obscure" parts of translated proofs. We show that this technique is a special form of cut introduction.

4 Symmetric Simplification

In [25], a simplification and restructuring method for NK-proofs is proposed. The proofs which are restructured have been obtained by a translation of expansion-tree proofs into NK-proofs. In almost all translations of expansion-tree proofs or resolution proofs into NK-proofs [21, 23, 24], an overwhelming number of indirect proofs are present. The main reasons for this are the generation of $\bot$ and the use of *ex falso quodlibet* if more than one substitution instance of a quantified (sub)formula occurs in the proof. We first recapitulate symmetric simplification and then we show the connection to the cut rule.

The algorithm introduced in [25] uses the fact that, instead of the rule of indirect proof, the law of excluded middle ($A \vee \neg A$) may be included in the logical calculus (i.e., NK). Given a set Γ of assumptions and a conclusion C to be deduced from Γ, there are three typical situations where a classical rule of inference has to be applied:

1. $C = F' \vee F''$ and neither F' nor F'' alone follow from Γ,
2. $C = \exists x.F'(x)$ and there exists no single instance t such that $F'(t)$ follows from Γ,
3. C is atomic[4]

and no progress can be made by applying an intuitionistic rule to formulae of Γ. These situations are referred to as *translation impasses*: a standard translation procedure can no longer make progress by applying intuitionistic rules of inference and hence uses the rule of indirect proof (e.g., the proof of $(p(a) \vee p(b)) \rightarrow \exists x.p(x)$ has to cope with an impasse of the second type). The following illustrates the general pattern for an NK-proof using the law of excluded middle:

$$
\cfrac{A \vee \neg A \qquad \cfrac{[A]_1 \\ \vdots\ \mathcal{D}_1 \\ C} \qquad \cfrac{[\neg A]_1 \\ \vdots\ \mathcal{D}_2 \\ C}}{C} \vee E_1
$$

In case one of the impasses occurs in the translation process, the symmetric simplification procedure uses heuristics for choosing a formula A (see [25] for detailed examples). The algorithm then tries to simplify A and $\neg A$ simultaneously in such a way that both subproof obligations ($\mathcal{D}_1$ and $\mathcal{D}_2$) can still be fulfilled. Since A and $\neg A$ remain identical apart from the negation sign the method is called "symmetric". The simplification of A consists of three steps:

1. Single Instantiation: Subformulae of A of the form $\forall x.A'(x)$ are replaced by $A'(t)$, if the term t instantiates x somewhere in the deduction $\mathcal{D}_1$ (dual for $\neg A$ and $\mathcal{D}_2$).

---

[4] In this case, there is at least one assumption of the form $F' \rightarrow F''$ or $\neg F'$ and for any such an assumption, C does not follow from the remaining assumptions.

2. Single Deletion: If there exists a positive (negative) subformula occurrence $A' \wedge A''$ ($A' \vee A''$) in A and none of the instances of A' is used in $\mathcal{D}_1$, erase A' from A and $\neg A$ (dual for positive subformulae $\neg A' \vee \neg A''$ in $\neg A$ and $\mathcal{D}_2$).

3. Propositional Restructure (Single Mating Change): Roughly speaking, this step tries to change $\mathcal{D}_1$ in such a way as to avoid the use of an assumption A' (as in the scenario for *Single Deletion*), i.e., redundancies introduced when creating $A \vee \neg A$ can be deleted. This step may enable further *Single Instantiations* and *Single Deletions*.

The following illustrates that symmetric simplification is indeed cut introduction.[5]

$$
\frac{A \vee \neg A \quad \overset{[A]_1}{\underset{C}{\vdots}} \quad \overset{[\neg A]_1}{\underset{C}{\vdots}}}{C} \vee E_1
\qquad
\frac{\vdash A \vee \neg A \qquad \dfrac{\dfrac{A, \Gamma \vdash C}{A, \Gamma, \Delta \vdash C}\, wl \quad \dfrac{\neg A, \Delta \vdash C}{\neg A, \Gamma, \Delta \vdash C}\, wl}{\dfrac{A \vee \neg A, \Gamma, \Delta \vdash C}{}}\, \vee l}{\Gamma, \Delta \vdash C}\, cut
$$

The formula $A \vee \neg A$ and the corresponding $\vee E$-inference are only needed in the NK-proof to connect the two subderivations of C and to close the indicated occurrences of A and $\neg A$. If the formula A is a quantified formula, for instance, of the form $\forall x.B(x)$ then symmetric simplification essentially coincides with an introduction of a classical tautology of the form $\forall x.(B(x) \vee \neg B(x))$ and a *functional extension* (F-extension)[6] step resulting in the formula $\forall x.B(x) \vee \exists x.\neg B(x)$ (see [13, 14] and [5] for a detailed discussion). We will come back to this issue in Section 6.

5 Preprocessing and Cuts

Early books on first-order automated theorem proving favored prenex forms because of the simplicity of describing skolemization. From a logical point of view, this favor does not matter because a first-order formula is logically equivalent to any of its prenex forms. From a computational (proof search) point of view, there is a significant distinction between proof search for prenex forms and proof search for non-prenex forms. Baaz and Leitsch [6] showed that there is a class of formulae $(F_i)_{i \in \mathbb{N}}$ [26] for which the following holds. There exist triple-exponential analytic proofs of F_i, but there exists a prenex form of F_i such that any analytic proof of this prenex form is not elementary. Since the search space is elementarily related to the shortest analytic cut-free proof, a similar relation holds for the different search spaces.

---

[5] The relation between symmetric simplification and cut is even more obvious in the context of tableaux with cut, because there, cut *directly* encodes the principle of bivalence.

[6] F-extension is introduced in Section 6.

The important observation is that, given a prenex form, a non-prenex form can be obtained by an application of cut. Hence, antiprenexing [7, 15] is a form of cut introduction.

$$\frac{\overset{\alpha}{\vdash F'} \qquad \overset{\beta}{F' \vdash F}}{\vdash F} \ cut \tag{1}$$

The proof of the antiprenexed formula F' is denoted by α. The proof β of the sequent $F' \vdash F$ is an "instance" of the correctness proof of antiprenexing. In a presentation of the proof, this proof is not visible by default, but can be generated easily in a systematic way without search.

In many cases, antiprenexing does not have such an extreme (non-elementary) impact on proof length and search space. From a practical point of view, however, antiprenexing can simplify proof search even if there is not a considerable length reduction of the proof. For illustration, consider a normal NK-proof of

$$G = \exists x \forall y.(p(x) \rightarrow p(y)). \tag{2}$$

This formula has a simple proof in LK (even without cut), but any normal NK-proof is rather complex.[7]

If antiprenexing is applied, we get $H = (\forall x.p(x)) \rightarrow \forall y.p(y)$. The cut introduced by antiprenexing is encoded as an application of $\vee$E in the NK-proof. Again, β is an instance of the correctness proof of antiprenexing.

$$
\frac{G \vee \neg H \quad [G]_1 \qquad \dfrac{[\neg H]_1 \qquad \dfrac{\dfrac{\dfrac{[\forall x.p(x)]_2}{p(u)} \, \forall E}{\forall y.p(y)} \, \forall I}{H} \to I_2}{\dfrac{\dfrac{\bot}{G} \, \bot E}{}} \neg E}{G} \vee E_1
$$

Skolemization *cannot* be described as a cut in this simple manner. The problem is that a sequent of the form

$$F' \vdash F$$

is required where F' is a skolemized form of F. If we can derive a sequent of the form $\vdash F'$ then an application of cut would suffice to derive $\vdash F$. However, the right upper sequent in (1) is not derivable. Let $F := \forall x.(p(x) \vee \neg p(x))$. Then F' is of the form $p(c) \vee \neg p(c)$ for a Skolem constant c. But

$$p(c) \vee \neg p(c) \vdash \forall x.(p(x) \vee \neg p(x))$$

---

[7] The interested reader is invited to search for a normal proof of G. We use G because it is identified as a "hard" formula for symmetric simplification (see [25]).

is not derivable because of the eigenvariable restriction of $\forall r$. But since skolemization is a validity-preserving operation, we can allow nonlogical axioms of the form $F' \vdash F$. Then skolemization can be described as an application of cut as in (1).

Various skolemization techniques are available [10, 1, 3, 20] which fit into the above scheme (by modifying F'). Usually, skolemization is a computationally inexpensive operation but some extended and more expensive forms are available. In general, these extended skolemization techniques produce much "better" skolemized forms [20].

Translations to normal forms (e.g., disjunctive normal form (DNF) or negation normal form (NNF)) can be described in an analogous way. For some normal forms, translations can be very harmful with respect to the minimal proof length of the resulting normal form (e.g., for DNF or for some kinds of structure-preserving translations) [4, 16]. Assume we want to translate a given formula F into DNF F''. As we already know, skolemization can be encoded as an application of the cut rule. Consider the following derivation.

$$\frac{\dfrac{\vdash F'' \quad F'' \vdash F'}{\vdash F'} \; cut \; (DNF) \qquad F' \vdash F}{\vdash F} \; cut \; (Sk)$$

The lower cut encodes skolemization, the upper cut the translation of a skolemized form of F into DNF. Other forms of translations to a normal form can be described similarly. In the same manner, preprocessing reduction steps [8] (like, e.g., purity reduction) can be encoded as cuts. As the examples suggest, any translation of a formula into a normal form as well as many simplifications of the resulting normal form can be described as cuts. Hence, these operations always belong to the deductive solution of a problem.

6 Quantorial Extension Principles

In this section, we explain the basic ideas about quantorial extension principles. We first start with an example and introduce the definition afterwards.

Let us consider a formula of the form

$$\forall x \forall y.(p(x,y) \lor q(x,y)). \tag{3}$$

The variables x and y occur in both literals; for this reason, no antiprenexing is possible. By the valid formula

$$(\forall x \forall y.(p(x,y) \lor q(x,y))) \rightarrow \forall x.(\forall y.p(x,y) \lor \exists y.q(x,y)) \tag{4}$$

together with (3), the consequent of (4) can be derived by modus ponens. Obviously, a formula with a newly introduced $\exists$-quantifier is derived. If this formula is skolemized then $p(x,y) \lor q(x,g(x))$ is obtained where g is a new Skolem function symbol. Now, the name *quantorial extension* (and *functional extension* in case of skolemization) becomes clear; the $\exists$-quantifier (or the Skolem function symbol) extends (the language of) the formula.

Definition 1. Let $A^- = \forall x_1 \ldots \forall x_m \, [\forall \mathbf{u} \, B_1 \vee \forall \mathbf{v} \, B_2]$ be a negative occurrence of a formula and let $A^+ = \exists x_1 \ldots \exists x_m \, [\exists \mathbf{u} \, B_1 \wedge \exists \mathbf{v} \, B_2]$ be a positive occurrence of a formula. Let $\mathbf{u}$ denote the free variables occurring only in B_1 and let $\mathbf{v}$ denote the free variables occurring only in B_2. Let $\{y_1, \ldots, y_k\} \subseteq \{x_1, \ldots, x_m\}$, $\{z_1, \ldots, z_l\} = \{x_1, \ldots, x_m\} \setminus \{y_1, \ldots, y_k\}$, $Q_i \in \{\forall, \exists\}$, and let Q_i^d be the quantifier dual to Q_i $(1 \leq k \leq m, 1 \leq i \leq k)$. Then

$$\forall z_1 \ldots \forall z_l \, [Q_1 y_1 \ldots Q_k y_k \forall \mathbf{u} \, B_1 \vee Q_1^d y_1 \ldots Q_k^d y_k \forall \mathbf{v} \, B_2] \tag{5}$$

$$\exists z_1 \ldots \exists z_l \, [Q_1 y_1 \ldots Q_k y_k \exists \mathbf{u} \, B_1 \wedge Q_1^d y_1 \ldots Q_k^d y_k \exists \mathbf{v} \, B_2] \tag{6}$$

are called *quantorial extensions* of A^- and A^+, respectively. If the strong quantifiers (i.e., negative occurrences of $\exists$ and positive occurrences of $\forall$) are removed by skolemization, then we call the result an *F-extension* of A^- and A^+, respectively.[8]

Quantifier extension is considered as an extension of calculi used for automated proof search. Since such calculi usually rely on a reduced syntax type (e.g., conjunctive normal form), F-extension is usually defined for clauses only, and skolemization is used in order to remove the newly introduced $\exists$-quantifiers.

Since $(\forall \, (B_1 \vee B_2)) \to$ (5) and (6) $\to \exists \, (B_1 \wedge B_2)$, quantorial extension (and F-extension) can be considered as restricted variants of the cut rule. Results how F-extension can speed-up proofs can be found in [5, 12, 6, 14]. We now come back to symmetric simplification and show how some kinds of quantifier extension can be "simulated" by symmetric simplification.

In what follows, we assume that we have an initial proof whose structure should be improved by applying quantorial extension. Let us reconsider (2) from Section 5. There, we used antiprenexing to derive $(\forall x.p(x)) \to \forall y.p(y)$ in order to simplify the proof. As an alternate, the lemma $\forall x.p(x) \vee \neg \forall x.p(x)$ can be chosen and (2) can be proven easily as follows.

$$
\cfrac{\forall x.p(x) \vee \neg\forall x.p(x) \qquad \cfrac{[p(t)]_3 \qquad \cfrac{[\forall x.p(x)]_1}{p(u)} \, \forall \mathrm{E}}{\cfrac{\cfrac{p(t) \to p(u)}{\forall y.(p(t) \to p(y))} \, \forall \mathrm{I}}{\exists x \, \forall y.(p(x) \to p(y))} \, \exists \mathrm{I}} \to \mathrm{I}_3 \qquad \alpha}{\exists x \, \forall y.(p(x) \to p(y))} \, \forall \mathrm{E}_{1,2}
$$

α is as follows.[9]

---

[8] Strictly speaking, the whole formula has to be considered as the input for skolemization, because one has to assure that *globally* new Skolem function symbols are introduced.

[9] We use derived rules $\neg \exists \mathrm{E}$, $\neg \forall \mathrm{E}$, and $\neg \to \mathrm{E}$ in order to keep the example manageable.

$$\frac{\dfrac{[\neg(\exists x\,\forall y.(p(x)\to p(y)))]_4}{\neg(\forall y.(p(u)\to p(y)))}\ \neg\exists E \qquad \dfrac{[\neg(p(u)\to p(v))]_5}{p(u)}\ \neg\to E}{p(u)}\ \neg\forall E_5$$

$$\cfrac{[\neg\forall x.p(x)]_2 \qquad \cfrac{\cfrac{p(u)}{\forall x.p(x)}\ \forall I}{}}{\cfrac{\cfrac{\bot}{\neg\neg\exists x\,\forall y.(p(x)\to p(y))}\ \neg I_4}{\exists x\,\forall y.(p(x)\to p(y))}\ \neg\neg E}\ \neg E$$

If we apply quantifier extension to $\forall x.(p(x)\vee\neg p(x))$, we derive $\forall x.p(x)\vee\exists x.\neg p(x)$. Using this formula in the $\vee E$-inference, α becomes even simpler and no derived rule has to be used.

$$\cfrac{[\exists x.\neg p(x)]_2 \qquad \cfrac{\cfrac{\cfrac{\cfrac{\cfrac{[\neg p(u)]_4 \qquad [p(u)]_5}{\bot}\ \neg E}{p(v)}\ \bot E}{p(u)\to p(v)}\ \to I_5}{\forall y.(p(u)\to p(y))}\ \forall I}{\exists x\,\forall y.(p(x)\to p(y))}\ \exists I}{\exists x\,\forall y.(p(x)\to p(y))}\ \exists E_4$$

In [18], *logical flow graphs* [9] are used to decide whether an application of quantorial extension (in an LK-derivation) yields a correct derivation of the same end sequent. Usually, proofs with quantorial extensions tend to be shorter and have more structure, because a cut is introduced.

7 Conclusion

We have shown that many AD techniques can be considered as applications of the cut rule. Usually, these cuts are not part of an automatically generated proof because computationally advantageous calculi like resolution or connection calculi rely on reduced syntax classes like conjunctive normal form and simple inference rules. All these calculi are basically cut-free and, therefore, the use of essential lemmata is restricted. In contrast, for a presentation of an automatically generated proof, cuts are highly desirable, because their intelligent application shortens and structures proofs.

We identified one source of "unintuitive" proofs in current translations, namely if different instances of a subformula are required in a proof, and showed how to improve the translation in order to avoid these parts in the resulting proof. Another solution, symmetric simplification, uses the unintuitive proofs and tries to replace a proof by contradiction by a proof with *tertium non datur*. The relation of the latter technique to the cut rule and to quantorial extension is explained in detail. It turns out that symmetric simplification is a weak form of cut introduction. Moreover, some forms of quantorial extension are closely related to symmetric simplification.

References

1. P. B. Andrews. Resolution in Type Theory. *J. Symbolic Logic*, 36:414–432, 1971.
2. P. B. Andrews. Transforming Matings into Natural Deduction Proofs. In W.Bibel and R. Kowalski, editors, *Proceedings of the 5ᵗʰ Conference on Automated Deduction*, volume 87 of *Lecture Notes in Computer Science*, pages 281–292. Springer Verlag, 1980.
3. P. B. Andrews. Theorem Proving via General Matings. *Journal of the ACM*, 28(2):193–214, 1981.
4. M. Baaz, C. Fermüller, and A. Leitsch. A Non-Elementary Speed Up in Proof Length by Structural Clause Form Transformation. In *Proceedings of the Logic in Computer Science Conference*, pages 213–219, 1994.
5. M. Baaz and A. Leitsch. Complexity of Resolution Proofs and Function Introduction. *Annals of Pure and Applied Logic*, 57:181–215, 1992.
6. M. Baaz and A. Leitsch. On Skolemization and Proof Complexity. *Fundamenta Informaticae*, 20:353–379, 1994.
7. W. Bibel. *Automated Theorem Proving*. Vieweg, Braunschweig, second edition, 1987.
8. W. Bibel. *Deduction: Automated Logic*. Academic Press, London, 1993.
9. S. R. Buss. The Undecidability of k-Provability. *Annals of Pure and Applied Logic*, 53:75–102, 1991.
10. C. L. Chang and R. C. Lee. *Symbolic Logic and Mechanical Theorem Proving*. Academic Press, New York, 1973.
11. E. Eder. *Relative Complexities of First Order Calculi*. Vieweg, Braunschweig, 1992.
12. U. Egly. Shortening Proofs by Quantifier Introduction. In A. Voronkov, editor, *Proceedings of the International Conference on Logic Programming and Automated Reasoning*, pages 148–159. Springer Verlag, 1992.
13. U. Egly. On Different Concepts of Function Introduction. In G. Gottlob, A. Leitsch, and D. Mundici, editors, *Proceedings of the Kurt Gödel Colloquium*, pages 172–183. Springer Verlag, 1993.
14. U. Egly. *On Methods of Function Introduction and Related Concepts*. PhD thesis, TH Darmstadt, Alexanderstr. 10, D–64283 Darmstadt, 1994.
15. U. Egly. On the Value of Antiprenexing. In F. Pfenning, editor, *Proceedings of the International Conference on Logic Programming and Automated Reasoning*, pages 69–83. Springer Verlag, 1994.
16. U. Egly. On Different Structure-preserving Translations to Normal Form. *J. Symbolic Computation*, 22:121–142, 1996.
17. A. P. Felty. Using Extended Tactics to do Proof Transformations. Technical Report MS-CIS-86-89 LINC LAB 48, Department of Computer and Information Science, Moore School, University of Pennsylvania, Philadelphia, PA 19104, 1986.
18. K. Genther. Repräsentation von Konnektionsbeweisen in Gentzen-Kalkülen durch Transformation und Strukturierung. Master's thesis, TH Darmstadt, 1995.
19. G. Gentzen. Untersuchungen über das logische Schließen. *Mathematische Zeitschrift*, 39:176–210, 405–431, 1935. English translation: "Investigations into Logical Deduction" in [27], pp. 68-131.
20. J. Goubault. A BDD-Based Simplification and Skolemization Procedure. *J. of the IGPL*, 3(6):827–855, 1995.
21. C. Lingenfelder. Transformation and Structuring of Computer Generated Proofs. Technical Report SR-90-26, Universität Kaiserslautern, 1990.

22. D. Miller and A. Felty. An Integration of Resolution and Natural Deduction Theorem Proving. In T. Kehler, S. Rosenschein, R. Folman, and P. F. Patel-Schneider, editors, *Proceedings of the 5th AAAI National Conference on Artificial Intelligence*, pages 198–202. Morgan Kaufmann Publishers, 1986.

23. D. A. Miller. Proofs in Higher-Order Logic. Technical Report MS-CIS-83-37, Department of Computer and Information Science, Moore School, University of Pennsylvania, Philadelphia, PA 19104, 1983.

24. D. A. Miller. Expansion Tree Proofs and their Conversion to Natural Deduction Proofs. In R. E. Shostak, editor, *Proceedings of the 7th Conference on Automated Deduction*, volume 170 of *Lecture Notes in Computer Science*, pages 375–393. Springer Verlag, 1984.

25. F. Pfenning and D. Nesmith. Presenting Intuitive Deductions via Symmetric Simplification. In M. E. Stickel, editor, *Proceedings of the 10th Conference on Automated Deduction*, volume 449 of *Lecture Notes in Computer Science*, pages 226–350. Springer Verlag, 1990.

26. R. Statman. Lower Bounds on Herbrand's Theorem. In *Proc. AMS 75*, pages 104–107, 1979.

27. M. E. Szabo, editor. *The Collected Papers of Gerhard Gentzen*. North–Holland, Amsterdam, 1969.

NaDSyL and some Applications

Paul C. Gilmore

Dept of Computer Science, University of B.C.,Vancouver, B.C. Canada V6T 1Z4
e-mail: gilmore@cs.ubc.ca

Abstract. NaDSyL, a Natural Deduction based Symbolic Logic, and some of its applications are briefly described. The semantics of NaDSyL is based on the term models of the lambda calculus and is motivated by the belief that a confusion of use and mention is the source of the paradoxes. Proofs of the soundness, completeness and the eliminability of cut are sketched along with three applications: The foundations for recursive definitions of well-founded and non-well founded predicates, classical and intuitionistic second order arithmetic, and a study of Cantor's diagonal argument and paradox.

1 Introduction

This paper provides a brief description of the logic NaDSyL and some of its applications. A full introduction with a semantic proof of cut-elimination is provided in [10]. [11] is an extended description of the logic and some of its applications.

NaDSyL, a Natural Deduction based Symbolic Logic, like the logics described in [7,8,9], is motivated by the belief that a confusion of use and mention is the source of the set theoretic paradoxes, a view also expressed in [18,19]. That NaDSyL is second order is a consequence of the need to maintain the distinction between the use and mention of predicate names. In [16] the same distinction is made: Used occurrences of predicate variables are said to be in "extensional" positions while mentioned occurrences are said to be in "intensional" positions.

NaDSyL differs from the earlier logics [7,8,9] in several important respects. "Truth gaps", as they were called in [15], are essential to the consistency of the earlier logics, but are absent from NaDSyL; the law of the excluded middle is derivable for all the sentences of NaDSyL. But the logic has an undecidable elementary syntax, a departure from tradition that is of little importance since the semantic tree presentation of the proof theory incorporates the decision process for sentence.

The notation of the lambda calculus, rather than of set theory, is used in NaDSyL. But the semantics of NaDSyL is based on the term models of the lambda calculus and not on the Scott models [2]. These term models are the "natural" intepretation of the lambda calculus for the naive nominalist view of the logic sketched in [7,8] that justifies the belief in the source of the paradoxes.

The elementary syntax is described in §2, the semantics in §3, and the proof theory in §4; sketches of semantic proofs of soundness, completeness and of the eliminability of cut are also provided there. The notations of NaDSyL are extended in §5 with definitions of identity, extensional identity, ordered pair, and partial first order functions. Least and greatest set definitions provides the foundations for recursive definitions of well-founded and non-well-founded predicates in §6; classical and intuitionistic second order arithmetic are also formalized here. §7 describes a study of Cantor's diagonal argument that illustrates well the manner in which NaDSyL avoids the paradoxes.

2 Elementary Syntax

It is necessary to distinguish between seven different denumerable sets of strings used in the elementary syntax of NaDSyL, first and second order constants, variables, and parameters, as well as the abstraction variables used in λ-abstraction terms. It is convenient, but not necessary, to have available in the metalanguage in which the logic is described the following "typing" notations for the seven sets of strings:

1. 1C, 1V, and 1P are the sets of first order constants, [quantification] variables, and parameters.
2. [n.1]C, [n.1]V, and [n.1]P are the sets of second order constants, [quantification] variables, and parameters of arity n, where $n \geq 0$.
3. av is the set of abstraction variables.

The particular strings of characters used in this paper as members of these sets will become evident as examples are provided.

The notation ':' is used in the metalanguage to denote the membership relation between a string and one of these given types, or between a string and other types to be defined below. For example, B:[2.1]C expresses that B is a second order constant of arity 2. The notation [n.1] is to be understood here and later as expressing that a string of this type accepts n first order order arguments. Thus for example, [] is the same as [0.1], [1] as [1.1], and [1,1] as [2.1].

Additional primitive symbols used in the basic elementary syntax are '(', ')', '.', '[', ']', '$\downarrow$', 'λ', and '$\forall$'. '$\downarrow$' is the joint denial logical connective that is 'true' only when both its arguments are 'false'. However all the usual logical connectives '$\neg$', '$\rightarrow$', '$\wedge$', '$\vee$', and '$\leftrightarrow$', and the existential quantifier '$\exists$' will be freely used in examples and applications of the logic without further explanation.

Definition of Stg
1. $1C \cup 1V \cup 1P \subset Stg$; $[n.1]C \cup [n.1]V \cup [n.1]P \subset Stg$ for $n \geq 0$; and $av \subset Stg$.
2. R, S:Stg => (R.S):Stg
3. R, S:Stg => [R$\downarrow$S]:Stg
4. R:Stg & v:av => (λv.R):Stg
5. x:1V or x:[n.1]V with $n \geq 0$ & R:Stg => ($\forall$x.R):Stg

Free and bound variables are defined in the usual way along with a substitution operator: [R/v]S is the result of replacing free occurrences of a variable v in S by R with appropriate changes of bound variables if necessary. The following abbreviating definitions introduce customary notations:

R(S) <df> (R.S)
$R(S_1, S_2, \ldots, S_n, S)$ <df> $R(S_1, S_2, \ldots, S_n)(S)$
$\lambda u_1, u_2, \ldots, u_n.S$ <df> $(\lambda u_1.(\lambda u_2. \ldots .(\lambda u_n.S) \ldots))$
$\forall x_1, x_2, \ldots, x_n.S$ <df> $(\forall x_1.(\forall x_2. \ldots .(\forall x_n.S) \ldots))$

2.1 First Order Terms, Formulas, Degrees & Higher OrderTerms

A *first order term* is a member of Stg in which no second order parameter has an occurrence and no second order variable has a free occurrence. The type notation for first order terms is simply 1; thus t:1 asserts that t is a first order term. Since first order terms are understood to be *mentioned*, and thus to be implicitly within single quotes, the restriction on first order terms is necessary to avoid the error described in footnote 136 of [5], which is "to use in the role of a syntactical variable the expression obtained by enclosing a variable of an object language in quotation marks".

A member of Stg of the form $T(t_1, \dots , t_n)$, where $T:[n.1]C \cup [n.1]V \cup [n.1]P$ and $t_1, \dots , t_n:1$, $n \geq 0$, is an *atomic formula*, the type notation of which is []A.

Definition of []: The Formulas
1. $[]A \subset []$.
2. $F, G:[] \Rightarrow [F \downarrow G]:[]$.
3. $R:Stg, [R/v]T(S_1, \dots , S_n):[]$ & v:av $\Rightarrow (\lambda v.T)(R, S_1, \dots , S_n):[]$, for $n \geq 0$.
4. $F:[]$ & x:1V or x:[n.1]V for $n \geq 0 \Rightarrow (\forall x.F):[]$.

Note that clause (3) ensures that the set [] of formulas is undecidable because of its relationship to head normal form reductions in the lambda calculus [2]. That this is of little consequence will be demonstrated in §4.1.

Definition of deg(F): The Degree of a Formula F
1. $F:[]A \Rightarrow \deg(F) = 0$.
2. $\deg(F) = d1$ & $\deg(G) = d2 \Rightarrow \deg([F \downarrow G]) = d1 + d2 + 1$.
3. $\deg([R/v]T(S_1, \dots , S_n)) = d \Rightarrow \deg((\lambda v.T)(R, S_1, \dots , S_n)) = d+1$.
4. $\deg([p/x]F) = d \Rightarrow \deg(\forall x.F) = d+1$, where p is a parameter of the same order and arity as x that does not occur in F.

cStg is the set of *closed* strings of Stg, namely those in which no variable has a free occurrence. For any type τ, $c\tau$ is the set $\{S \mid S:\tau \ \& \ S:cStg\}$. *Sentences* are members of c[].

An Extended Type Notation
For $k \geq 0$, $R(p):[\tau_1, \dots , \tau_k]$ & p:τP $\Rightarrow R:[\tau, \tau_1, \dots , \tau_k]$, where τ is 1 or [n.1], $n \geq 0$.

Examples
Let Pb:[1]P, Qr:[2.1]P, P:[1]P, p:1P and x:1V . Then Pb(p):c[]A, P(x):[]A, and Qr(x,p):[]A, so that $[Pb(p) \lor \exists x.[P(x) \land Qr(x,p)]]:c[]$. Using clause (3) of the definition of formula, $(\lambda u.[Pb(u) \lor \exists x.[P(x) \land Qr(x,u)]])(p):c[]$, and
$(\lambda w,u.[Pb(u) \lor \exists x.[w(x) \land Qr(x,u)]])(P,p):c[]$. Thus
$(\lambda w,u.[Pb(u) \lor \exists x.[w(x) \land Qr(x,u)]])(P):c[1]$, and hence
$(\lambda w,u.[Pb(u) \lor \exists x.[w(x) \land Qr(x,u)]]):c[[1],1]$
   Consider the following abbreviating definitions:

$$G <df> \ (\lambda wb,wr,w,u.[wb(u) \lor [\exists x:w].wr(x,u)])$$
$$Lt <df> \ (\lambda w,u. \forall Z.[\forall x.[w(Z,x) \to Z(x)] \to Z(u)])$$

G:c[[1],[2.1],[1],1] since its 4 arguments are of type [1], [2.1], [1], and 1 respectively. Thus G(Pb,Qr):c[[1],1]. Since the first argument of Lt is of type [[1],1] and the second of type 1, a further extension of the types [10] results in Lt:c[[[1],1],1].

Bound abstractions and quantifiers can be defined in the usual way:

$[\lambda u:T].F <df> \lambda u.[T(u) \land F]$ and $[\forall x:T].F <df> \forall x.[T(x) \to F]$, for T:[1].

3. Semantics

3.1 An Equivalence Relation on Stg

The term models of the pure λ-calculus [2] are extended here for Stg to provide the basis for the semantics of NaDSyL.

Definitions of the Relations ⊳ and ⊳⊳ on Stg

1. R ⊳ S, when S is an α-, β-, or η-contractum of R, where
    .1. $(\lambda u.[u/v]S)$ is an α-contractum of $(\lambda v.S)$, provided u is free to replace v in S, and $(\forall y.[y/x]S)$ is an α-contractum of $(\forall x.S)$, provided y is free to replace x in S.
    .2. $[R/v]S$ is a β-contractum of $(\lambda v.S)(R)$.
    .3. S is an η-contractum of $(\lambda v.S(v))$, provided v has no free occurrence in S.
2. Let R ⊳ S:
    .1. T:Stg => (T.R) ⊳ (T.S) & (R.T) ⊳ (S.T).
    .2. T:Stg => [R↓T] ⊳ [S↓T] & [T↓R] ⊳ [T↓S].
    .3. $(\lambda v.R)$ ⊳ $(\lambda v.S)$.
    .4. $(\forall x.R)$ ⊳ $(\forall x.S)$

The ⊳⊳ relation on Stg is the reflexive and transitve closure of ⊳.

Theorem (Church-Rosser). For R, S, T:Stg, R ⊳⊳ S & R ⊳⊳ T => there is an R' for which S ⊳⊳ R' and T ⊳⊳ R'.

Any proof of the theorem for the pure λ-calculus, for example the ones provided in [2], can be adapted for the strings Stg. Define $R \approx S$ to mean that there exists a T:Stg for which R ⊳⊳ T & S ⊳⊳ T. A corollary of the theorem is that the relation $\approx$ is an equivalence relation on Stg.

3.2.   Interpretations of NaDSyL

A *signed* sentence is a sentence prefixed with one of the signs $\pm$. Preliminary to the definition of the models of NaDSyL is the definition of an interpretation $\mathbb{I}$, and the definition of a set $\Omega[\mathbb{I}]$ of signed sentences recording the sentences that are true and false in $\mathbb{I}$. As with the semantics described in [14], not all interpretations are models.

The domain *dom* of an interpretation is the set of t:c1 in which no parameter occurs. Thus the members of (c1–dom) are the members of c1 in which first order parameters occur. Clearly dom is closed under ⊳; that is, d:dom & d ⊳ d' => d':dom.

The set D(0) has as its members the two truth values 'true' and 'false'. For $n > 0$, D(n) is the set of functions $f: \text{dom}^n \to D(0)$ for which $f(d_1, \dots, d_i, \dots, d_n)$ is $f(d_1, \dots, d'_i, \dots, d_n)$, whenever $d_i \approx d'_i$ and $d_1, \dots, d_n, d'_i$:dom. The effect of this is to make each f a function of the equivalence classes of dom under $\approx$.

A *base* Bse is a set sets $B(n)$ for which $B(0)$ is $D(0)$ and $B(n) \subseteq D(n)$, $B(n)$ nonempty for $n > 0$. An *interpretation* with base Bse is defined by two functions Φ_1 and Φ_2 where Φ_1: $dom \cup 1P \rightarrow dom$, with $\Phi_1[d]$ being d for d:dom., and $\Phi_2[S] \in B(n)$, for $n \geq 0$ and $S:[n.1]C \cup [n.1]P$.

The domain of Φ_1 is extended by definition to include all of c1: For t:c1, $\Phi_1[t]$ is the result of replacing each occurrence of a first order parameter p in t by $\Phi_1[p]$. Note that $t \doteq t' => \Phi_1[t] \doteq \Phi_1[t']$, for t, t':c1. Using the extended Φ_1, the domain of Φ_2 is extended by definition to include all of c[]A: $\Phi_2[R(t_1, ..., t_n)]$ is $\Phi_2[R](\Phi_1[t_1],$ $..., \Phi_1[t_n]$) when $R:[n.1]C \cup [n.1]P$ and $t_1, ..., t_n$:c1. A proof of the following lemma is immediate:

Lemma 1. Let $R:[n.1]C \cup [n.1]P$, $t_1, ..., t_n$, t_i':c1, and $t_i \doteq t_i'$ for some i, $1 \leq i \leq n$. Then $\Phi_2[R(t_1, ..., t_i, ..., t_n)]$ is $\Phi_2[R(t_1, ..., t_i', ..., t_n)]$.

Let $\mathbb{I}$ be an interpretation with functions Φ_1 and Φ_2, and let p be a parameter, first or second order. An interpretation $\mathbb{I}^*$ is a *p variant* of $\mathbb{I}$ if it has the same base as $\mathbb{I}$ and its functions Φ_1^* and Φ_2^* are such that when p is first order Φ_2^* is Φ_2 and $\Phi_1^*[q]$ differs from $\Phi_1[q]$ only if q is p, and when p is second order Φ_1^* is Φ_1 and $\Phi_2^*[q]$ differs from $\Phi_2[q]$ only if q is p.

An interpretation $\mathbb{I}$ assigns a single truth value $\Phi_2[F]$ to each F:c[]A. This assignment will be extended to an assignment of a single truth value $\Phi_2[F]$ to each F:c[]. This assignment of truth values is recorded as a set $\Omega[\mathbb{I}]$ of signed sentences: $+F \in \Omega[\mathbb{I}]$ records that $\Phi_2[F]$ is 'true' and $-F \in \Omega[\mathbb{I}]$ that $\Phi_2[F]$ is 'false'.

Definition of the Set $\Omega[\mathbb{I}]$
The set $\Omega[\mathbb{I}]$ is defined to be $\cup\{\Omega_k[\mathbb{I}] \mid k \geq 0\}$ where $\Omega_k[\mathbb{I}]$ is defined for $k \geq 0$:
1. $\Omega_0[\mathbb{I}]$ is the set of $\pm F$, F:c[]A, for which $\Phi_2[F]$ is 'true', respectively 'false'.
2. Assuming $\Omega_k[\mathbb{I}]$ is defined for all interpretations $\mathbb{I}$, $\Omega_{k+1}[\mathbb{I}]$ consists of all members of $\Omega_k[\mathbb{I}]$ together with the sentences
    .1. $+[F \downarrow G]$ for which $-F \in \Omega_k[\mathbb{I}]$ and $-G \in \Omega_k[\mathbb{I}]$; and
        $-[F \downarrow G]$ for which $+F \in \Omega_k[\mathbb{I}]$ or $+G \in \Omega_k[\mathbb{I}]$.
    .2. $\pm(\lambda v.T)(R, S_1, ..., S_n)$ for which $\pm[R/v]T(S_1, ..., S_n) \in \Omega_k[\Phi]$, $n \geq 0$ and
        v:av.
    .3. $+\forall x.F$ for which $+[p/x]F \in \Omega_k[\mathbb{I}^*]$ for every p variant $\mathbb{I}^*$ of $\mathbb{I}$; and
        $-\forall x.F$ for which $-[p/x]F \in \Omega_k[\mathbb{I}^*]$ for some p variant $\mathbb{I}^*$ of $\mathbb{I}$,
        where [p/x]F:c[], and p is a parameter that does not occur in F and is of
        the same order and arity as x.

Lemma 2. For each interpretation $\mathbb{I}$ and sentence F,
1. $+F \notin \Omega[\mathbb{I}]$ or $-F \notin \Omega[\mathbb{I}]$, and
2. $+F \in \Omega[\mathbb{I}]$ or $-F \in \Omega[\mathbb{I}]$.

Proof: That $+F \notin \Omega_k[\mathbb{I}]$ or $-F \notin \Omega_k[\mathbb{I}]$ follows by induction on k, as does $\deg(F) \leq k$ $\Rightarrow +F \in \Omega_k[\mathbb{I}]$ or $-F \in \Omega_k[\mathbb{I}]$, from which (2) can be concluded.

Definition of $\Phi_2[T]$

Let T:c[n.1], $n \geq 0$, and $\mathbb{I}$ be an interpretation with functions Φ_1 and Φ_2. Let $d_1, \ldots$ d_n:dom so that $T(d_1, \ldots d_n)$:c[]. By lemma 2, exactly one of $+T(d_1, \ldots d_n)$ and $-T(d_1, \ldots d_n)$ is in $\Omega[\mathbb{I}]$. Define $\Phi_2[T]$ to be the function $f \in D(n)$, for which $f(d_1, \ldots d_n)$ is 'true', respectively 'false', if $+T(d_1, \ldots d_n)$, respectively $-T(d_1, \ldots d_n)$, is in $\Omega[\mathbb{I}]$.

Lemma 3. Let H be a formula in which at most the quantification variable x has a free occurrence and let p be a parameter of the same order and arity as x not occurring in H. Let t:c1 if x:1V, and t:c[n.1] if x:[n.1]V. Let $\mathbb{I}^*$ be a p variant of an interpretation $\mathbb{I}$ for which $\Phi_i^*[p]$ is $\Phi_i[t]$, for i=1 or i=2. Then for $k \geq 0$,
a) $\pm[p/x]H \in \Omega_k[\mathbb{I}^*] \Rightarrow \pm[t/x]H \in \Omega_k[\mathbb{I}]$.

Proof: The proof of the lemma will be by induction on k and is illustrated for two cases. Let k be 0. If x:1V, then H takes the form $R(r_1, \ldots, r_n)$, where $R:[n.1]C \cup [n.1]P$, and x may have a free occurrence in any of $r_1, \ldots, r_n:1$. In this case $\Phi_2^*[[p/x]H]$ is $\Phi_2[R](\Phi_1^*[[p/x]r_1], \ldots, \Phi_1^*[[p/x]r_n])$. But $\Phi_1^*[[p/x]r_i]$ is $[\Phi_1^*[p]/x]\Phi_1[r_i]$, since p does not occur in r_i, and this is $[\Phi_1[t]/x]\Phi_1[r_i]$, which is $\Phi_1[[t/x]r_i]$. Thus $\Phi_2^*[[p/x]H]$ is $\Phi_2[[t/x]H]$. If x:[n.1]V, then H takes the form $x(t_1, \ldots, t_n)$ where $t_1, \ldots, t_n$:c1, and t is some $R:[n.1]P \cup n.1]C$. Thus $\Phi_2^*[p]$ is $\Phi_2[R]$ and $\Phi_1^*[t_i]$ is $\Phi_1[t_i]$ from which (a) follows immediately.

Assume now that (a) holds and let H be $\forall y.F$, where it may be assumed that x is distinct from y and has a free occurrence in F, so that $[p/x]\forall y.F$ is $\forall y.[p/x]F$ and $[t/x]\forall y.F$ is $\forall y.[t/x]F$. Let q be of the same order and arity as y and not occur in $[p/x]F$. Then $+\forall y.[p/x]F \in \Omega_{k+1}[\mathbb{I}^*] \Rightarrow +[q/y][p/x]F \in \Omega_k[\mathbb{I}^{**}]$ for every q variant $\mathbb{I}^{**}$ of $\mathbb{I}^*$. For each $\mathbb{I}^{**}$ there is a q variant $\mathbb{I}^{*'}$ of $\mathbb{I}$ for which $\mathbb{I}^{**}$ is a p variant of $\mathbb{I}^{*'}$. Thus by the induction assumption $+\forall y.[p/x]F \in \Omega_{k+1}[\mathbb{I}^*] \Rightarrow +[q/y][t/x]F \in \Omega_k[\mathbb{I}^{*'}]$ for every q variant $\mathbb{I}^{*'}$ of $\mathbb{I} \Rightarrow +\forall y.[t/x]F \in \Omega_{k+1}[\mathbb{I}]$. The $-\forall$ case can be similarly argued.

3.3.   Models

Let $\mathbb{I}$ be an interpretation with base $\{ B(n) \mid n \geq 0\}$ and functions Φ_1 and Φ_2. For every term t:c1, $\Phi_1[t]$:dom. But although $\Phi_2[T] \in D(n)$, for T:c[n.1], it does not follow that $\Phi_2[T] \in B(n)$. $\mathbb{I}$ is a *model* if $\Phi_2[T] \in B(n)$ for every T:c[n.1], $n \geq 1$. Thus every interpretation for which B(n) is D(n) for $n \geq 0$ is a model.

A Gentzen sequent $\Gamma \vdash \Theta$ is said to be *satisfied* by a model $\mathbb{M}$ if there is a sentence F for which $F \in \Gamma$ and $-F \in \Omega[\mathbb{M}]$, or $F \in \Theta$ and $+F \in \Omega[\mathbb{M}]$. The sequent is *valid* if it is satisfied by every model. A *counter-example* for a sequent $\Gamma \vdash \Theta$ is a model $\mathbb{M}$ which does not satisfy the sequent; that is a model for which for each $F \in \Gamma$, $+F \in \Omega[\mathbb{M}]$, and for each $F \in \Theta$, $-F \in \Omega[\mathbb{M}]$.

4. Logical Syntax

NaDSyL is formalized in a semantic tree version of the semantic tableaux of [3]. The *initial* nodes of a semantic tree, *based on* a Gentzen sequent $\Gamma \vdash \Theta$ of sentences, consist of +F for which $F \in \Gamma$, and –G for which $G \in \Theta$; not all members of Γ and Θ need appear signed as initial nodes. The remainder of a semantic tree based on a sequent can be seen to be a systematic search for a counter-example to the sequent. The semantic rules used to extend a semantic tree beyond its initial nodes are:

$+\downarrow$	$+[F\downarrow G]$	$+[F\downarrow G]$	$-\downarrow$	$-[F\downarrow G]$
	$-F$	$-G$		$+F$ $\quad$ $+G$

$+\lambda 1$	$+R(t_1, \ldots, t_i, \ldots, t_n)$	$-\lambda 1$	$-R(t_1, \ldots, t_i, \ldots, t_n)$
	$+R(t_1, \ldots, t_i', \ldots, t_n)$		$-R(t_1, \ldots, t_i', \ldots, t_n)$

where $R:[n.1]P \cup [n.1]C, t_1, \ldots, t_i, \ldots, t_n, t_i':c1$, and $t_i \vdots t_i'$ with $1 \le i \le n$.

$+\lambda 2$	$+(\lambda v.T)(R, S_1, \ldots, S_n)$	$-\lambda 2$	$-(\lambda v.T)(R, S_1, \ldots, S_n)$
	$+[R/v]T(S_1, \ldots, S_n)$		$-[R/v]T(S_1, \ldots, S_n))$

$+\forall$	$+\forall x.F$	$-\forall$	$-\forall x.F$
	$+[t/x]F$		$-[p/x]F$

Here x is the only free variable of F:[], and either x:1V, p:1P & t:c1, or x:[n.1]V, p:[n.1]P & t:c[n.1]; p may not occur in any node above the conclusion of $-\forall$.

Cut	
	$+F$ $\quad$ $-F$

Each of the rules, with the exception of $-\downarrow$ and cut, has a single conclusion that may be added as a node to any branch on which the premiss occurs as a node. The two conclusion rules split a branch into two branches with each of the conclusions appearing as a node immediately below the split. The cut rule may be applied at any time; the $-\downarrow$ rule requires that its premiss is a node on the branch that its conclusions split. An example of a derivation is given in §4.1.

A branch of a semantic tree is *closed* if there is an *atomic* sentence F for which both ±F are nodes on the branch. Such a pair is called a *closing pair* of the branch. A *derivation* of a sequent is a semantic tree based on the sequent in which every branch is closed.

The cut rule of deduction is theoretically not needed since it is *eliminable* in the following sense: Any derivation of a sequent in which cut is used can be replaced by a derivation of the same sequent in which it is not used. A sketch of the proof of this result is provided in §4.4 and a full proof in [10]. Cut is nevertheless retained as a rule because it can justify the reuse of given derivations in the construction of new derivations.

For theoretical purposes it is convenient to have the $\pm\!\downarrow$ rules as the only propositional rules and $\pm\forall$ as the only quantifier rules. But for applications of the logic it is better to use all of the logical connectives $\neg$, $\rightarrow$, $\wedge$, $\vee$, and $\leftrightarrow$, and to have the existential quantifier $\exists$ available as well. It is assumed that the reader can derive all the appropriate rules for these connectives and quantifier.

As proved in [10], the $\pm\lambda 1$ rules can be generalized to the following eliminable $\pm$ rules:

$$+\qquad \frac{+[r/x]F}{+[t/x]F} \qquad\qquad\qquad -\qquad \frac{-[r/x]F}{-[t/x]F}$$

where $x{:}1V$, $r,t{:}c1$, and $r \triangleright t$.

4.1. Properties of Derivations

A node ηc of a semantic tree is a *descendant* of a node η if ηc is η, or if there is a descendant ηp of η which is the premiss of a rule of deduction with conclusion ηc. Given a derivation of a sequent, the derivation may be transformed to one of the same sequent for which the same signed sentence does not occur twice as a node of the same branch; each branch of the derivation has exactly one closing pair; and each node of the derivation has a closing node as descendant.

It has been claimed that the undecidablity of the elementary syntax of NaDSyL is of little consequence. This statements is justified by the fact that no creativity is needed to construct a derivation for S $\vdash$ S if S:cStg $\cap$ []: The choice of the t for an application of the $+\forall$ rule can always be a p previously introduced in an application of the $-\forall$ rule.

Since F $\vdash$ F is derivable for all F:c[], the requirement that a closing pair of a branch consist of signed atomic sentences $\pm$F can be relaxed. A branch with nodes $\pm$F, where F is a sentence that is not atomic, can be closed by attaching to the leaf node of the branch a derivation of F $\vdash$ F from which the initial nodes $\pm$F have been dropped.

Rules $\pm\lambda$ that can replace the rules $\pm\lambda 1$ and $\pm\lambda 2$ of NaDSyL are described and proved to be eliminable in [10] but are not used here.

Example Derivation
A derivation of $\vdash\forall Y.[\forall x.[RG(Y,x)\rightarrow Y(x)]\rightarrow[\forall x{:}Lt(RG)].Y(x)]$, where RG:c[[1],1]:
$-\forall Y.[\forall x.[RG(Y,x) \rightarrow Y(x)] \rightarrow [\forall x{:}Lt(RG)].Y(x)]$
$-[\forall x.[RG(P,x) \rightarrow P(x)] \rightarrow [\forall x{:}Lt(RG)].P(x)]$
$+\forall x.[RG(P,x) \rightarrow P(x)]$
$-[\forall x{:}Lt(RG)].P(x)$
$-[Lt(RG)(p) \rightarrow P(p)]$
$+Lt(RG)(p)$
$-P(p)$
$+(\lambda w,u.\forall Z.[\forall x.[w(Z,x) \rightarrow Z(x)] \rightarrow Z(u)])(RG,p)$ *defn of Lt*
$+(\lambda u.\forall Z.[\forall x.[RG(Z,x) \rightarrow Z(x)] \rightarrow Z(u)])(p)$
$+\forall Z.[\forall x.[RG(Z,x) \rightarrow Z(x)] \rightarrow Z(p)]$
$+[\forall x.[RG(P,x) \rightarrow P(x)] \rightarrow P(p)]$

$-\forall x.[RG(P,x) \rightarrow P(x)]$ $+P(p)$
$================$ $====$

The double lines indicate that the branch is closed. The left branch is closed because $\forall x.[RG(P,x) \rightarrow P(x)]$ is a sentence when $RG:c[[1],1]$.

4.2.   A Sequent Calculus Formulation of NaDSyL

A semantic tree derivation can be seen to be an abbreviated version of a Gentzen sequent calculus derivation. Consider a semantic tree derivation for a sequent $\Gamma \vdash \Theta$. Let η be any node of the derivation which does not have an initial node below it. Define $\Gamma[\eta]$ and $\Theta[\eta]$ to be the sets of sentences F for which $+F$, respectively $-F$, is η itself or is a node above η. Thus if η is the last of the initial nodes of the derivation, $\Gamma[\eta]$ is Γ and $\Theta[\eta]$ is Θ.

Consider $F,G \in \Gamma[\eta]\cup\Theta[\eta]$. F is said to *entail* G if $F \in \Gamma[\eta]$ and $+F$ is the premiss of a rule of deduction with a conclusion $\pm G$, where $G \in \Gamma[\eta]$ respectively G $\in \Theta[\eta]$; or $F \in \Theta[\eta]$ and $-F$ is the premiss of a rule of deduction with a conclusion $\pm G$, where $G \in \Gamma[\eta]$ respectively $G \in \Theta[\eta]$. Sets $\Gamma[\eta]*$ and $\Theta[\eta]*$ are obtained from $\Gamma[\eta]$ and $\Theta[\eta]$ by repeatedly removing from $\Gamma[\eta]\cup\Theta[\eta]$ any sentence that entails a member of $\Gamma[\eta]\cup\Theta[\eta]$ but is not itself entailed by a member of $\Gamma[\eta]\cup\Theta[\eta]$.

A Gentzen sequent calculus derivation of $\Gamma \vdash \Theta$ can be constructed from the semantic tree derivation by turning the latter upside down and replacing each node η by the sequent $\Gamma[\eta]* \vdash \Theta[\eta]*$. The rules of deduction of the sequent calculus are obtained from the semantic tree rules in the obvious way. An intuitionistic formulation results when the sequents of the derivation are required to have at most one sentence in the succecedent.

4.3.   Soundness, Completeness & Cut-Elimination

Define the *height* of a node η on a given branch of the derivation to be the number of descendants of η, other than η itself, on the branch; the *height* $h(\eta)$ of η in a tree is the maximum of its heights on the branches on which it occurs. By induction on $h(\eta)$, $\Gamma[\eta] \vdash \Theta[\eta]$ can be can be shown to be valid; the soundness of NadSyL is a consequence.

Theorem (Completeness & Cut-Elimination). The logical syntax of NaDSyL is complete without the cut rule.

A proof of the theorem is provided in [10] by adapting the proof of [17] to NaDSyL. That cut is an eliminable rule is a corollary. Here a sketch of the proof is provided.

A function Φ_1 is defined: $\Phi_1[p_i]$ is c_{2i-1} for all $i \geq 1$, where c_1, c_2, ... and p_1, p_2, ... are enumerations of 1C and 1P. Define $T^\dagger$ for any term or formula T as follows: Each occurrence of c_i in T is replaced by c_{2i}, and then each occurrence of p_i is replaced by c_{2i-1}, for $i \geq 1$. Thus for $t:c1$, $t^\dagger:dom$ and $\Phi_1[t]$ is $t^\dagger$. Φ_1 is the first order function of all the interpretations to be defined.

The set c[] of sentences has been defined in terms of given sets [n.1]P of second order parameters for $n \geq 0$. Let $[*.1]P$ denote the union of all these sets. The proof makes use of a transfinite sequence of sets $[*.1]P_\alpha$ of second order parameters, where α is an ordinal less than the first ordinal with a cardinality greater than the cardinality of D(2). $[*.1]P_0$ is $[*.1]P$, and $[*.1]P_{\alpha+1}$ is $[*.1]P_\alpha \cup \{P\}$, where P is a second order parameter not in $[*.1]P_\alpha$ of an arity to be determined. For a limit ordinal β, $[*.1]P_\beta$ is

$\cup\{[*.1]P_\alpha \mid \alpha < \beta\}$. For each set $[*.1]P_\alpha$ of parameters, $c[]_\alpha$ and $c[*.1]_\alpha$ are respectively the set of sentences and set of closed second order terms in which only second order parameters from $[*.1]P_\alpha$ are used.

Let $\Gamma \vdash \Theta$ be a sequent of sentences from $c[]_0$ which has no cut free derivation in which the eliminable rules $\pm \cdot$ are used in place of the $\pm\lambda 1$ rules. Thus there is a set $\mathbb{K}_0$ of signed sentences, including $+F$ for each $F \in \Gamma$ and $-F$ for each $F \in \Theta$, that are the signed sentences of a branch of a semantic tree that cannot be closed. From $\mathbb{K}_0$ the base $\{B_0(n) \mid n \geq 0\}$ and second order function $\Phi_{2,0}$ of an interpretation $\mathbb{I}_0$ are constructed.

Let $[*.1]P_\alpha$, $\mathbb{K}_\alpha$, and $\mathbb{I}_\alpha$ be defined for a given α, where $\{B_\alpha(n) \mid n \geq 0\}$ and $\Phi_{2,\alpha}$ have been obtained from $\mathbb{K}_\alpha$. These are defined for $\alpha+1$ as follows. If $\mathbb{I}_\alpha$ is not a model, then for some $k \geq 1$ there exists a $T:c[k.1]_\alpha$ for which $\Phi_{2,\alpha}[T] \notin B_\alpha(k)$. The arity of the new P to be added to $[*.1]P_\alpha$ is the arity k of T. $\mathbb{K}_{\alpha+1}$ is obtained from $\mathbb{K}_\alpha$ by adding to $\mathbb{K}_\alpha$ all signed sentences obtained from members of $\mathbb{K}_\alpha$ by replacing occurrences of T by P. $B_{\alpha+1}(k)$ is $B_\alpha(k)\cup\{f\}$, where $f \in D(k)-B_\alpha(k)$ and is defined from T and $\mathbb{K}_\alpha$. For a limit ordinal β, $\mathbb{K}_\beta$ is $\cup\{\mathbb{K}_\alpha \mid \alpha < \beta\}$, and $B_\beta(n)$ is $\cup\{B_\alpha(n) \mid \alpha < \beta\}$ for $n \geq 1$.

The following result can be established by transfinite induction: For each α, and each sentence F, if $\pm F \in \mathbb{K}_\alpha$ then $\pm F^\dagger \in \Omega[\mathbb{I}_\alpha]$. Thus in particular $\pm F^\dagger \in \Omega[\mathbb{I}_\alpha]$ when $F \in \Gamma$, respectively $F \in \Theta$. Since for each α there is a k and a f for which $B_{\alpha+1}(k)$ is $B_\alpha(k)\cup\{f\}$, where $f \in D(k)-B_\alpha(k)$, there is an ordinal β for which $\mathbb{I}_\beta$ is a model that is a counter-example for $\Gamma \vdash \Theta$. From the model $\mathbb{I}_\beta$ a counter-example $\mathbb{M}$ can be constructed that interprets only sentences of the original set $c[]$.

5. Additional Notations

Notations are defined respectively for intensional and extensional identity, and ordered pairs:

$$= <df> \ (\lambda u,v.\forall Z.[Z(u) \rightarrow Z(v)]),$$
$$=_{en} \ <df> \ \lambda u,v.\forall x_1, \ldots , x_n.[u(x_1, \ldots , x_n) \leftrightarrow v(x_1, \ldots , x_n)]$$
$$\langle\rangle \ <df> \ (\lambda u,v,w.w(u,v))$$

The definition of ordered pair originated in [Church41]. The usual infix notation for these notations can be introduced by definition schemes:

$$s=t \ <df> \ =(s,t), \text{ where } s,t:1. \qquad R=_{en}S \ <df> \ =_{en}(R,S), \text{ where } R,S:[n.1]$$
$$\langle R, S \rangle \ <df> \ \langle\rangle(R, S), \text{ where } R, S:1\cup[*.1]$$

Eliminable rules of deduction can be introduced for these notations. First and second order unique existance quantifiers $\exists!y.F$ and $\exists!Y.F$ can be defined using intensional and extensional identity. The head and tail predicates for ordered pair can be defined in the usual way.

A notation fT for partial first order functions can be added to NaDSyL. If $T:c[(n+1).1]$, $n \geq 0$, and $t_1, \ldots , t_n:c1$, then $fT(t_1, \ldots , t_n):c1$. The term $fT(t_1, \ldots , t_n)$ is treated in the logic as a definite description and may not occur in an atomic sentence. Associated with each function fT is a domain or "type" δT on which the function is defined: δT <df> $\lambda u_1, \dots , u_n.\exists!y.T(u_1, \dots , u_n, y)$. The $\pm f$ rules for removing occurrences of terms $fT(t_1, \dots , t_n)$ from nodes of a semantic tree have two conclusions one of which is a checking of "type": all branches under $-\delta T(t_1, \dots , t_n)$ must close. NaDSyL with the functional notation is a conservative extension of NaDSyL without. Details are provided in [10].

6. Recursive Definitions

One of the advantages of a logic like NaDSyL that admits impredicative definitions is illustrated here with the definition of well-founded and non-well-founded predicates.

A *recursion generator* is any string RG for which RG:c[[1],1]; an example is G(Pb,Qr) defined in §2.1. Lt was also defined in §2.1; Gt is defined here:

$$\text{Gt <df> } (\lambda w,u.\exists Z.[\forall x.[Z(x) \rightarrow w(Z,x)] \wedge Z(u)])$$

Consider the sequents:

Lt.1) $\vdash \forall Y.[\forall x.[RG(Y,x) \rightarrow Y(x)] \rightarrow [\forall x:Lt(RG)].Y(x)]$
Gt.1) $\vdash \forall Y.[\forall x.[Y(x) \rightarrow RG(Y,x)] \rightarrow [\forall x:Y].Gt(RG)(x)]$

A derivation of (Lt.1) appears in §4.1; (Gt.1) is also derivable. They are generalizations of mathematical induction for the natural numbers and justify the choice of the name "recursion generator". The predicates Lt(RG) and Gt(RG) are of arity 1. Predicates of greater arity can be defined by appropriate generalizations of Lt, Gt, and RG.

Definition of Positive Sentences and Recursion Generators
A *positive* sentence is atomic, of the form [G∧H] or [G∨H] where G and H are positive, of the form $(\lambda v.T)(R, S_1, \dots , S_n)$ where $[R/v]T(S_1, \dots , S_n)$ is positive, or of the form $\exists x.F$ where [p/x]F is positive for a parameter p of the same order and arity as x. A recursion generator RG is *positive* if RG(P,p) is positive for P:[1]P and p:1P. The recursion generator G(Pb,Qr), for example, is positive. A recursion generator RG is said to be *monotonic* if $\vdash$ Mon(RG) is derivable where

$$\text{Mon <df> } \lambda w.\forall X,Y.[\forall x.[X(x) \rightarrow Y(x)] \rightarrow \forall x.[w(X, x) \rightarrow w(Y, x)]]$$

By induction on the definition of positive sentence it can be proved that a positive recursion generator is monotonic.
   The following sequents are derivable for positive recursion generators RG:

Lt.2) $\vdash \forall x.[RG(Lt(RG),x) \rightarrow Lt(RG)(x)]$
Gt.2) $\vdash \forall x.[Gt(RG)(x) \rightarrow RG(Gt(RG),x)]$
Lt.3) $\vdash \forall x.[Lt(RG)(x) \rightarrow RG(Lt(RG),x)]$
Gt.3) $\vdash \forall x.[RG(Gt(RG),x) \rightarrow Gt(RG)(x)]$
LtGt) $\vdash \forall x.[Lt(RG)(x) \rightarrow Gt(RG)(x)]$

Lt(RG) and Gt(RG) are said to be *fixed points* of RG, respectively the *least* and the *greatest*, because the following sequents are also derivable:

FixLt) ⊢ FixPt(RG,Lt(RG))
FixGt) ⊢ FixPt(RG,Gt(RG))

FxPt is defined using extensional, not intensional identity: FixPt <df> λw,v.w(v)=$_e$v.

Semantics for Programming Languages and Category Theory

A semantics for programming languages was illustrated in [13] using an example from [20] in which two flow diagram programs are proved equivalent. Although the inconsistent logic NaDSet was used, the results can be more easily be developed in NaDSyL [11]. On the other hand the results of [12], in which category theory was formalized within NaDSet, can only be established in a form that maintains the distinction between small and large categories [11].

Arithmetic

Representation for the natural numbers is provided by the predicate N where

$\quad$ N <df> Lt(G(BN,RN)) $\quad$ BN <df> λu.u=0 $\quad$ RN <df> λu,v.v=S(u)
$\quad$ 0 <df> λu.$\neg$u=u $\quad\quad\quad$ S <df> λu,v.v=u

Since BN:c[1] and RN:c[2.1], G(BN,RN):c[[1],1] and is a positive recursion generator. The first two of Peano's axioms follow from (Lt2), the next two follow from the definitions of '0' and 'S', and mathematical induction follows from (Lt.1). Addition and multiplication predicates can be defined using the arity 3 form of Lt and appropriate recursion generators. All of the usual axioms for these predicates can be derived within NaDSyL.

Iteration and Based Recursion Generators

A recursion generator of the form G(B,R), where B:c[1] and R:c[2.1], is *based*; G(BN,RN), for example, is based. The predicate Lt(G(B,R)) can be characterized as λv.[$\exists$x:N].It(B,R)(x,v), where λv.It(B,R)(n,v) is the set obtained by iterating R n times beginning with B. The predicate Gt(G(B,R)) can be characterised under an assumption on R as λv.[Lt(G(B,R))(v) $\vee$ [$\forall$x:N].It(V,R)(x,v)], where V is the universal set λu.u=u. These characterizations justify calling predicates Lt(G(B,R)) *well-founded* and predicates Gt(G(B,R)) *non-well-founded*.

$\quad$ The predicate It is defined in terms of the arity 2 form of Lt and a recursion generator GIt appropriate for it:

$\quad\quad$ It <df> λwb,wr.Lt2(GIt(wb,wr)) and GIt <df>
$\quad\quad$ λwb,wr,w,u1,u2.[[u1=0$\wedge$wb(u2)]$\vee\exists$x1,x2.[w(x1,x2)$\wedge$u1=S(x1)$\wedge$wr(x2,u2)]]

The assumption on R is that ⊢ FP(R) is derivable where FP expresses that R has at most two predecessors:

$\quad\quad$ FP <df> λw.$\forall$y.$\exists$x1,x2.$\forall$x.[w(x,y) $\rightarrow$ x=x1 $\vee$ x=x2]

$\quad$ The significance of FP can be seen in the following derivable sequent:

FPIt) ⊢ [$\forall$Y:FP].$\forall$z.[[$\forall$x:N].It(V,Y)(x,z)$\rightarrow\exists$y.[$\forall$x:N].[It(V,Y),V)(x,y)$\wedge$Y(y,z)]]

from which a fundamental property of trees, namely that a finitely branching infinite tree has an infinite branch, can be concluded. The derivation of this sequent in NaDSyL, and the sequents expressing the characterizations, involve inductive arguments on a variety of least set defined predicates [11].

A Formalization of Second Order Arithmetic

A sentence of second order arithmetic is transformed to a sentence of NaDSyL by replacing each first order quantifier $\forall x$ by $[\forall x:N]$ and each second order quantifier $\forall X$, where $X:[n.1]V$, by

$$[\forall X:(\lambda w.[\forall x_1, \ldots, x_n:N].[w(x_1, \ldots, x_n) \rightarrow w(x_1, \ldots, x_n)])]$$

This transforms the axioms of second order arithmetic into derivable sequents of NaDSyL. By additionally requiring derivations to satisfy the intuitionistic restriction described in §4.2, an intuitionistic second order arithmetic is obtained [11].

7. Cantor's Diagonal Argument

Cantor first used the argument which now bears his name to prove that no mapping of the natural numbers into the reals could have all the reals as its range; this conclusion is sometimes called *Cantor's Lemma*. He subsequently recognized that the argument in its most general form could lead to what is called *Cantor's paradox*; namely, that the set of all sets has a cardinal not exceeded by any other cardinal, and yet the cardinal of its power set must be greater [6]. In this century the argument has found many incontrovertible applications; for example, Turing's use of the argument to prove that the computable reals cannot be enumerated by a Turing machine. Here the validity of Cantor's diagonal argument in NaDSyL is examined.

A real number can be represented by a sequence of 0's and 1's; that is a total single valued function with arguments from N1, and values that are 0 or 1 where:

Sq <df> $\lambda u.[\forall x:N1].[\exists !y:B].u(x,y)$ N1 <df> $\lambda u.[N(u) \wedge \neg u=0]$
B <df> $\lambda u.[u=0 \vee u=1]$ 1 <df> $S(0)$

Identity between sequences is necessarily extensional identity $=_{e2}$. A mapping w of the numbers N1 onto sequences is then a member of Map as defined here:

Map <df> $\lambda w.[\forall x:N1].[\exists !X:Sq].w(x,X)$

Cantor's lemma can then be formalized as the sequent scheme

CLs) $Map(Mp) \vdash [\exists Y:Sq].[\forall x:N1].\neg Mp(x,Y)$, where $Mp:c[1,[2.1]]$

Each instance of (CLs) is derivable in NaDSyL [11]. Turing's result follows from an instance. But Cantor's paradox requires deriving a universally quantified form of (CLs) in which Mp is replaced with a second order variable X; that is not possible because $X(x,Y)$ is not a formula.

Acknowledgements The evolution of NaDSyL from the earlier logics [7,8,9] began with a letter from J.Y. Girard in early 1994 reporting on the inconsistency of the logic NaDSet described in [12,13]; I am much indebted to Girard for reporting this discovery to me. Conversations with Eric Borm and George Tsiknis were helpful during the early development of the logic. More recently my research associate Jamie Andrews, who is adapting for NaDSyL his theorem-prover for Prolog based on [1], has been very helpful in clarifying details. The financial support of the Natural Science and Engineering Council of Canada is gratefully acknowledged.

REFERENCES

1. Jamie Andrews. *Logic Programming: Operational Semantics and Proof Theory.* Cambridge University Press Distinguished Dissertation Series, 1992.
2. H.P. Barendregt. *The Lambda Calculus, Its Syntax and Semantics.* Revised Edition, North-Holland, 1984.
3. E.W. Beth. Semantic Entailment and Formal Derivability, *Mededelingen de Koninklijke Nederlandse Akademie der Wetenschappen, Afdeeling Letterkunde, Nieuwe Reeks,* 18(13):309-342, 1955.
4. Alonzo Church. *The Calculi of Lambda Conversion.* Princeton University Press, 1941.
5. Alonzo Church. *Introduction to Mathematical Logic I.* Princeton University Press, 1956.
6. Joseph Warren Dauben, *Georg Cantor, His Mathematics and Philosphy of the Infinite.* Harvard University Press, 1979.
7. Paul C. Gilmore. A Consistent Naive Set Theory: Foundations for a Formal Theory of Computation, IBM Research Report RC 3413, June 22, 1971.
8. Paul C. Gilmore. Combining Unrestricted Abstraction with Universal Quantification. A revised version of [7]. *To H.B. Curry: Essays on Combinatorial Logic, Lambda Calculus and Formalism*, Editors J.P. Seldin, J.R. Hindley, Academic Press, 99-123, 1980.
9. Paul C. Gilmore. Natural Deduction Based Set Theories: A New Resolution of the Old Paradoxes, *Journal of Symbolic Logic*, 51:393-411, 1986.
10. Paul C. Gilmore. Soundness and Cut-Elimination for NaDSyL. Dept. of Computer Science Technical Report TR97-1, 27 pages.
11. Paul C. Gilmore. A Symbolic Logic and Some Applications. A monograph on NaDSyL in preparation.
12. Paul C. Gilmore and George K. Tsiknis. A logic for category theory, *Theoretical Computer Science*, 111:211-252, 1993.
13. Paul C. Gilmore and George K. Tsiknis. Logical Foundations for Programming Semantics, *Theoretical Computer Science*, 111:253-290, 1993.
14. Leon Henkin. Completeness in the Theory of Types, *J. Symbolic Logic*, 15:81-91, 1950.
15. Saul Kripke. Outline of a Theory of Truth, *Journal of Philosophy*, November 6, 690-716, 1975.
16. Gopalan Nadathur and Dale Miller. Higher-Order Logic Programming, CS-1994-38, Dept of Computer Science, Duke University, 1994.
17. Dag Prawitz. Completeness and Hauptsatz for second order logic, *Theoria*, 33:246-258, 1967.
18. Wilfred Sellars. Abstract Entities, *Rev. of Metaphysics*, 16:625-671, 1963.
19. Wilfred Sellars. Classes as Abstract Entities and the Russell Paradox, *Rev. of Metaphysics*, 17:67-90, 1963
20. J.E. Stoy. *Denotational Semantics: The Scott-Strachey Approach to Programming Language Theory.* MIT Press, 1977.

Markov's Rule Is Admissible in the Set Theory with Intuitionistic Logic

Khakhanian V.Kh *

Moscow State University of railway communications, Chair of Applied
Mathematics-2, Obraztsova str. 15, Moscow 101475, Russia
e-mail: post@miit.msk.su

Abstract. We prove that the strong Markov's rule with only set parameters is admissible in the full set theory with intuitionistic logic

0. It is well known that Markov's Rule (MR) is admissible in such predicative theories as HA and HAS [1]. The first proofs were very complicated and
used higher order functional constructions or the standartization (or the normalization) of deductions. In 1977 H.Friedman [2] and independently in 1979
A.Dragalin [3] suggested the very elegant method of MR-proving for predicative
or not predicative systems, for example a set theory. But in the last case the
axiom of extensionality and set parameters are omitted. The MR-admissibility
is a good constructive test for a formal theory and it is natural to verify the full
set theory by such test.

In the present article the admissibility of the strong MR with set parameters
(but without natural parameters) is proved for the set theory with all standard
set theoretical axioms and two kinds of variables (numerical and set-theoretical).
The double negative complement principle of sets (double complement of sets
or DCS) is added and so our theory is equiconsistent with classical ZF [4]. Also
the examined theory has the disjunction and existension properties [5],[6] (see
Appendix for details).

1. Now we give the description of the set theory ZFI2+DCS. The language of
ZFI2+DCS contains two kinds of variables (numerical and set-theoretical), binary predicate symbols of the natural equality, belonging of a natural number to
a set and a set to a set, symbols of primitive recursive functions, the symbols for
conjuction, disjunction, implication, negation and the quantifiers of universality
and existence for all kinds of variables. An axioms list includes the intuitionistic
predicate logic, the arithmetic and the axioms of a set theory. The formal notations of set-theoretical axioms are given below. Letters m,n,k are used for the
notation of natural variables, x,y,z,u,v for set variables, t,p,r,q for terms.

---

* The work was supported by Reseach Scientific Foundation of Russian Ministry of
Transport

2. We consider the following variant of the strong MR with set parameters only:

$$(*) \qquad \frac{\forall n(\phi(n,x) \vee \neg\phi(n,x)), \neg\forall n\neg\phi(n,x)}{\exists n\phi(n,x)}$$

Markov's rule claims that if a property $\phi(n)$ is decidable and if we proved $\neg\neg\exists n\phi(n)$ then we can find such n that $\phi(n)$ is true.

3. THEOREM 1. The rule $(*)$ is admissible in ZFI2+DCS.

4. Let us extend ZFI2+DCS to ZFI2+DCS* so that if ZFI2+DCS $\vdash \exists x \forall y(y \in x \leftrightarrow \phi(y))$ then we add a constant C_ϕ and the axiom $\forall y(y \in C_\phi \leftrightarrow \phi(y))$ (see details in [5] or Appendix). Such extension is conservative. Then we extend the last language by splitting each of its terms to a set of other terms and we construct from new terms a universe (it is assumed that every term C_ϕ has a rank and we construct the universe by induction on term rank, see also [5] or Appendix). Let A be a some metastatment and rk(t) be the rank of a term t.

5. Construction of the universe DA.

Let ω be the set of natural numbers. We define now.

$\alpha = 0; D_\alpha = \{C_{\phi,x}|(rk(C_\phi) = 0) \wedge (X \subseteq \omega)\};$

$\alpha = \beta + 1; D_\alpha = \{C_{\phi,x}|(rk(C_\phi) = \alpha) \wedge (X \subseteq \omega \cup \bigcup\{D_\gamma|\gamma < \alpha\}) \wedge [(C_{\phi,Y} \overset{\alpha}{\sim} C_{\eta,Z}) \wedge (C_{\phi,Y} \in X) \Rightarrow (C_{\eta,Z} \in X)] \wedge (\bigcup\{rk(t) + 1|t \in X\} = \alpha\};$ if t=$C_{\phi,X}$ then $t^- = C_\phi$ and $t^+ = X;$

$C_{\phi,Y} \overset{\alpha}{\sim} C_{\eta,Z} \rightleftharpoons (rk(C_\phi) = rk(C_\eta) \leq \beta) \wedge \forall n(n \in Y \vee A \leftrightarrow n \in Z \vee A) \wedge \forall q(rk(q) < \alpha \rightarrow (q \in Y \vee A \leftrightarrow q \in Z \vee A)).$

The limit case: $D_\alpha = \bigcup\{D_\beta|\beta < \alpha\}$. At last DA=$\bigcup\{D_\alpha|\alpha \in On\}$.

6. The constants of ZFI2+DCS** are the elements of DA and we add the axiom $\forall y(y \in C_{\phi,X} \leftrightarrow \phi(y))$ for every constant $C_{\phi,X}$.

LEMMA 1. ZFI2+DCS** is a conservative extension of ZFI2+DCS*.

It is evident that ZFI2+DCS* $\vdash \phi^-$ iff ZFI2+DCS** $\vdash \phi$, where ϕ^- is obtained from ϕ by the simultaneuos substitution of t^- instead of t for all terms t from ϕ.

7. We define the satisfiability of a closed formula ϕ of the language of ZFI2+DCS theory: $\models \phi$.

a) $(\models t = r) \rightleftharpoons (t = r \vee A)$ is true, where t,r are arithmetic terms;

b) $(\models t \in q) \rightleftharpoons (t \in q^+ \vee A)$ is true, where t is an arithmetic term or t,q belong to DA;

c) $(\models \perp) \rightleftharpoons A$ is true;

d) $\models (\phi * \psi) \rightleftharpoons (\models \phi) * (\models \psi)$, where $*$ is one of the logical connectives;

e) $\models \forall n\phi(n) \rightleftharpoons \forall n. \models \phi(n);$

f) $\models \forall x\phi(x) \rightleftharpoons \forall t \in DA. \models \phi(t);$

g) $\models \exists n\phi(n) \rightleftharpoons \exists n. \models \phi(n);$

h) $\models \exists\phi(x) \rightleftharpoons \exists t \in DA. \models \phi(t).$

8 THEOREM 2. If ZFI2+DCS $\vdash \phi$, then $\models \phi$.

We develop the proof of the THEOREM 2 by induction on ϕ-deduction.

9. The predicate logic is verified as usual. We notice that the axiom $\perp \rightarrow \phi$ is verified by induction on ϕ-construction.

10. The verification of the satisfiability of arithmetic axioms is also trivial.

11. The verification of set-theoretic axioms.

a) extensionality $\forall xyz[\forall n(n \in x \leftrightarrow n \in y) \wedge \forall u(u \in x \leftrightarrow u \in y) \wedge x \in z \rightarrow y \in z]$.
Let t,q,p be from DA and $\models t \in p, \forall n(\models n \in t \Leftrightarrow \models n \in q), \forall r \in DA(\models r \in t \Leftrightarrow r \in q)$; as $\models t \in p \rightleftharpoons t \in p^{+} \vee A$ then we examine two cases: if A, then $q \in p^{+} \vee A$ and $\models q \in p$; if $t \in p^{+}$, then (from the second and the third premises) $t \overset{\alpha}{\sim} q$, where $\alpha = rk(t)$ and $q \in p^{+}$, i.e. $\models q \in p$.

b) pair $\forall mnyz\exists x(n \in x \wedge m \in x \wedge y \in x \wedge z \in x)$.

Let $\beta = \max(rk(t), rk(q))$ and $rk(C_\phi)$ equals $\beta + 1$ and $X = \{r \in DA | r \overset{\beta}{\sim} t \vee r \overset{\beta}{\sim} q\}$; then $C_{\phi,X} = p \in DA, rk(p) = \beta + 1$. We have now $t \in p^{+}, q \in p^{+}, m \in p^{+}, n \in p^{+}$.

c) the axioms of union and power are verified in the same way.

d) infinity $\exists x \forall n(n \in x)$. If $rk(C_\phi) = 0$, then $C_{\phi,\omega}$ is the required constant.

e) scheme of the replacement $\exists x[\forall y(y \in x \leftrightarrow y \in a \wedge \phi(y)) \wedge \forall n(n \in x \leftrightarrow n \in a \wedge \eta(n))]$, with the usual boundary on ϕ, η.
We must prove that $\forall q \in DA \exists t \in DA[\forall r \in DA(\models r \in t \Leftrightarrow \models r \in q \wedge \models \phi(r)) \wedge \forall n(\models n \in t \Leftrightarrow \models n \in q \wedge \models \eta(n))]$.

Let $X = \{r \in DA | \models r \in q \wedge \models \phi(r)\} \cup \{n| \models n \in q \wedge \models \eta(n)\}$ and $rk(C_\psi) = \alpha$, where $\alpha = \bigcup\{rk(r) + 1 | r \in X\}$. Then $C_{\psi,X}$ is the required constant t.

LEMMA 2. If $t \overset{\alpha}{\sim} q$, then $(\models \phi(t)) \Leftrightarrow (\models \phi(q))$.
The proof of LEMMA 2 is given on ϕ-construction.
Therefore $t \in DA$. Now, if $\models n \in q \wedge \models \eta(n)$, then $n \in t^{+}$, i.e. $\models n \in t$ and vice versa, as $A \Rightarrow \models \phi$ for every formula ϕ. The set part of the scheme is proved similarly.

f) scheme of the induction $\forall x[\forall y(y \in x \rightarrow \phi(y)) \rightarrow \phi(x)] \rightarrow \forall x \phi(x)$.
We use the external induction on a term rank and will prove that $\forall q \in DA. \models \phi(q)$. As $[\models \forall y(y \in q \rightarrow \phi(y))] \Rightarrow \models \phi(q)$, we presuppose that for every $p \in DA$, such as $rk(p) < rk(q) \models \phi(p)$ is true; then $\models \forall y(y \in q \rightarrow \phi(y))$, i.e. $\models \phi(q)$ is true.

g) scheme of the collection $[\forall x(x \in a \rightarrow \exists y \phi(x,y)) \wedge \forall n(n \in a \rightarrow \exists y \psi(n,y))] \rightarrow \exists B[\forall x(x \in a \rightarrow \exists y(\phi(x,y) \wedge y \in B)) \wedge \forall n(n \in a \rightarrow \exists y(\psi(n,y) \wedge n \in B))]$.
We consider only the set part of the scheme. Let $q \in DA$. If $\models q \in t$, then $\exists r \in DA. \models \phi(q,r)$, i.e. $\forall q(q \in t^{+} \Rightarrow \exists r. \models \phi(q,r))$. Then $\exists B \forall q \in t^{+} \exists r \in B \models \phi(q,r)$. We suppose $\alpha = \sup\{rk(r) + 1 | r \in B\}$ and let C_η has the rank α. We claim that $C_{\eta,B}$ is the required constant, of course, for the set part of our scheme. Really, let $q \in t^{+} \vee A$. If A then $\models \exists y(y \in C_{\eta,B} \wedge \phi(q,y))$; if $q \in t^{+}$, then $\exists r \in DA. \models (r \in C_{\eta,B} \wedge \phi(q,r))$.

h) double complement of sets (DCS) $\forall a \exists x[\forall y(y \in x \leftrightarrow \neg\neg y \in a) \wedge \forall n(n \in x \leftrightarrow \neg\neg n \in a)]$. Let $t \in DA$, then t is the required term. We treat only set part of the DCS. Let $\models \neg\neg q \in t$, that is $(q \in t^{+} \vee A \Rightarrow A) \Rightarrow A$. It is necessary to prove $q \in t^{+} \vee A$. We analyse two cases: if A then $q \in t^{+} \vee A$; if A is false, then (as $(q \in t^{+} \vee A \Rightarrow A) \Rightarrow A$ is true) $(q \in t^{+} \vee A) \Rightarrow A$ is false, i.e. $q \in t^{+} \vee A$ is true. And vice versa: let $q \in t^{+} \vee A$ is true: if A, then $\models \neg\neg q \in t$; if A is false, then $(q \in t^{+} \vee A)] \Rightarrow A$ is false and again $\models \neg\neg q \in t$.

The THEOREM 2 is proved.

12. The proof of the THEOREM 1 from the point 3.

Let ZFI2+DCS $\vdash \forall n(\phi(n) \vee \neg\phi(n)) \wedge \neg\forall n\neg\phi(n)$ and let A be the following metastatement: ZFI2+DCS $\vdash \exists n\phi(n)$ (set parameters now are absent). We use the THEOREM 2 and have $(\models \forall n\neg\phi(n)) \Rightarrow A$. We have also, that ZFI2+DCS $\vdash phi(n)$ or ZFI2+DCS$\vdash \neg\phi(n)$. If ZFI2+DCS $\vdash \phi(n)$, then A and $\models \neg\phi(n)$; if ZFI2+DCS $\vdash \neg\phi(n)$, then, from the THEOREM 2, $\models \neg\phi(n)$, that is $\models \forall n\neg\phi(n)$ and therefore A is true and ZFI2+DCS $\vdash \exists n\phi(n)$. The THEOREM 1 is proved.

13. Appendix: some notes and details from [5],[6] and [7].

a) If Markov's rule $(*)$ has set parameters, then we introduce in the language of ZFI2+DCS theory a new constant C and we take $rk(C) = 0$ later on. Let us denote this theory by ZFI2+DCS$_C$. As in [5] we have, that ZFI2+DCS $\vdash \forall x\phi(x)$ iff ZFI2+DCS$_C \vdash \phi(C)$ and now we begin our construction of the universe DA in this case from ZFI2+DCS$_C$.

b) Every formula ϕ is satisfiable in our model (see [3] or the point 9 of the article and the proof of the THEOREM 1 from the point 12).

c) Y.Gavrilenko used Dragalin's result (Dragalin gave a new simple proof of the admissibility of the weak Markov's rule in HA) and proved in [7] the admissibility (it is also a new proof with a very simple metamathematics) of the strong Markov's rule with parameters in HA. He used the realizability with bounded predicate of deducibility (see [1], p.181) and the reflection principle (see also [1],§5) for the bounded predicate of deducibility.

d) J.Myhill proved in [5], that a set theory with intuitionistic logic and the replasement scheme (he denoted it by ZFI) has the disjunction and existension properties. The author extended the result to ZFI+DCS set theory in [6] as this theory is equiconsistent with classical ZF and the same problem for ZFI set theory is open. Here, we reproduce (very briefly) the all important places from [5] and [6].

The first, we construct the conservative extension of ZFI (let us denote it by ZFI*), which is formed by adding comprehension terms to ZFI, i.e. by adjoining the rule: if ZFI $\vdash \exists x\forall y(y \in x \leftrightarrow \phi(y))$, where ϕ has no parameters except y, add a new constant C_ϕ with defining axiom $\forall y(y \in C_\phi \leftrightarrow \phi(y))$.

The second, we construct the conservative extension of ZFI* (let us denote it by ZFI**), which is obtained by splitting each term C_ϕ of ZFI* into many terms $C_{\phi,X}$, where X is a set of terms of ZFI**. To do this, we define for each term t of ZFI* a rank $rk(t) \in On$, using standard interpretation, so that if ZFI$^* \vdash t \in q$, then $rk(t) < rk(q)$. Now we can construct a universe of terms of ZFI** by induction on rank of a term. We add an every constant $C_{\phi,X}$ with defining axiom $\forall y(y \in C_{\phi-,X} \leftrightarrow \phi(y))$, where ϕ^- is obtained from ϕ by replacing each constant t by t^- (if t=$C_{\phi,X}$, then $t^- = C_\phi$ and $t^+ = X$).

ZFI** is the conservative extension of ZFI*. We must note, that for each term $C_{\phi,X}$ the set X is extensional, i.e. for any terms t and q, if t$\sim$q and t$\in$X, then q$\in$X and t$\sim$q$\rightleftharpoons (ZFI^* \vdash \forall x(x \in t^- \leftrightarrow x \in q^-)) \wedge (t^+ = q^+)$.

The third, we define our model for an atomic formula by $t \in C_{\phi,X}$ iff $t \in X$. At last, we define the notion of realizability by induction on ϕ-construction in the same way as Kleene's realizability for arithmetic.

The author noted in [6], that it is sufficient to use DCS in the metamathematics for the proving of the satisfiability of DCS in our realizability model.

e) Let ZFI+DCS be the intuitionistic set theory as above (i.e. with only set variables). Markov's rule has the following form in this case:

$$(**) \qquad \frac{\forall x \in \omega(A(x,y) \vee \neg A(x,y)), \neg\forall x \in \omega\neg A(x,y)}{\exists x \in \omega A(x,y)}$$

THEOREM 3. The rule $(**)$ is admissible in ZFI+DCS

To prove this we use a double formula traslation from ZFI+DCS to ZFI2+DCS and vice versa and the THEOREM 1.

f) Open problem. Let us consider the following form of Markov's rule:

$$(\text{GMR}) \qquad \frac{\forall x(B(x) \to A(x) \vee \neg A(x)), \neg\forall x(B(x) \to \neg A(x))}{\exists x(B(x) \wedge A(x))}$$

For what kinds of formula B GMR is admissible in ZFI+DCS?

References

1. Troelstra A. Metamathematical investigations of intuitionistic arithmetic and analysis. Lecture Notes in Mathematics, 344, (1973), chapter Y.
2. Friedman H. Classically and intuitionistically provably recursive functions. Lecture Notes in Mathematics, 669, (1977), 21–27.
3. Dragalin A. New kinds of realizability and Markov's Rule. Soviet Mathematical Dokl.,v.251, 3, (1980), 534–537
4. Powell W. Extending Gödel's negative interpretation to ZF. The journal of Symbolic Logic, v.40, 2, (1975), 221–229.
5. Myhill J. Some properties of intuitionistic Zermelo-Fraenkel set theory. Lecture Notes in Mathematics, 337, (1973), 206–231.
6. Khakhanian V. The independence of collection from DCS-principle in the intuitionistic set theory. Proceedings of the Higher Academic Establishments, series Mathematics, 2, (1993), 81–83 (in Russian).
7. Gavrilenko Y. About admissibility of Markov's rule in the intuitionistic arithmetic. In book: II Soviet-Finnish colloquium, (1979), 12–15 (in Russin).

Bounded Hyperset Theory and Web-like Data Bases

Alexei Lisitsa and Vladimir Sazonov *

Program Systems Institute of Russian Academy of Sciences,
Pereslavl-Zalessky, 152140, Russia
e-mail: {sazonov,lisitsa}@logic.botik.ru

1 Introduction

We present in this paper rather abstract, "static" set-theoretic view on the World-Wide Web (WWW) or, more generally, on Web-*like* Data Bases (WDB) and on the corresponding querying to WDB. Let us stress that it is not only about databases with an access via Web. The database itself should be organized in the same way as Web. I.e. it must consist of hyperlinked pages distributed among the computers participating either in global network like Internet or in some local, isolated from the outside world specific network based essentially on the same principles, except globality, and called also Intranet [15].

This approach is based on a work on Bounded Set Theory (BST) and on its Δ-language [27]–[33], [19, 20] considered as a query language for data bases with complex or nested data. The notion of PTIME and LOGSPACE computability of set-theoretic operations was defined there in terms of graph transformers. It was used representation of hereditarily-finite *well-founded* sets (of sets of sets, etc.) by vertices of finite graphs (possibly labelled [30] and acyclic, except [31, 20]). Another source is P. Aczel's *non-well-founded* set theory [2, 4, 9, 24, 31] which allows considering "cyclic" sets and data represented by cyclic graphs.

Here we reintroduce in a natural way non-well-founded version of BST and Δ [31, 33] starting from motivations in WWW or WDB, rather than in abstract sets. We consider WDB as a (finitely branching) graph with labelled edges in a way analogous to [1, 22, 25], however, with an essential difference based on the above mentioned approach to consider graph vertices as (denoting) *antifounded sets*, called also *hypersets* [4][2]. To this end, the well-known notion of graph bisimulation due to D. Park (cf. [23]) is used crucially. In this respect our approach is closer to that of [18, 5] also based on bisimulation of graphs representing possibly cyclic "unstructured" databases. (Cf. more comments in Sect. 5)

---

* Both authors are supported by the grants RBRF 96-01-01717 and INTAS 93-0972. The second author was also supported by Swedish Royal Academy of Sciences.

[2] We will see that the abstract notion of hyperset occasionally and happily correlates also with that of *HyperText*: "Web documents are created by authors using a language called HTML (HyperText Markup Language) that offers short codes (also called tags) to designate graphical elements and links. Clicking on links brings documents located on a server to a browser, irrespective of the server's geographic location." (From the Netscape Handbook.)

On the other hand, this paper aims to give a natural interpretation of abstract hypersets and Δ, as a corresponding query language, in more realistic terms of WWW and WDB and to demonstrate both to theoreticians and to more practically oriented researchers possible applied value of hypersets. (Cf. [4] for other applications of hypersets to Computer Science and Philosophy.) This also dictates some "intermediate" form of presentation of the material. Properly speaking, it is given not a final set-theoretic description of the Web as it is, but rather some concise approach relating the corresponding ideas.

Mathematical results (Sect. 6) consist (i) in defining in terms of Δ (and also FO + IFP) a linear ordering on hereditarily-finite hypersets (respectively, on finite strongly extensional graphs) and (ii) in using this fact to characterize Δ-definable operations over hypersets exactly as PTIME-computable ones. The proofs for a pure (label-free) version of Δ are given in [20] on the base of [31].

2 Web-like Data Bases

We define a (state or instance of) World-Wide Web or a *Web-like Data Base* (WDB) as a labelled graph (abbreviated $\mathcal{LG}$; it is also called *labelled transition system* [4]). See a motivation below. Formally, $\mathcal{LG}$ is a map $g : |g| \to \mathcal{P}(L \times |g|)$ with $|g|$ a set of *vertices*, $L = L^g$ a set of *labels* or *attributes* (from some given large class of labels $\mathcal{L}$) and $\mathcal{P}(|g|)$ the powerset. We consider vertices and labels quite abstractly, but *represent* or *encode* them as finite strings in a fixed finite alphabet. All the computations over a WDB will be *relativised* to its graph g considered as an "oracle" set-valued function. Three important subclasses of $\mathcal{LG}$ are *finite, finitely-branching* and *countably branching* labelled graphs ($\mathcal{FLG}$, $\mathcal{FBLG}$ and $\mathcal{CBLG}$). We write $l : u \,\epsilon^g\, v$ or $g \models l : u \in v$ instead of $\langle l, u \rangle \in gv$ and say this as "u is an l-element of v in the sense of g".

Considering vertices u, v of a graph g and the relation ϵ^g, respectively, as *formal names of sets* (consisting of labelled elements) and as the *membership relation* between these sets might seem rather artificial for arbitrary graphs, however we will give in Sect. 3 more arguments for such way of thinking.

Motivations for "WDB as Graph". Which way such a graph g is related to the Web? Let us think that vertices $u, v, \ldots$ are formal expressions like http://www.botik.ru/PSI/AIReC/logic/, i.e. addresses (Uniform Resource Locators, URLs, in a computer network) of WWW-pages. Then a *page*, which an address v refers to, is just the corresponding set of *references* $gv = \{l_1 : u_1, l_2 : u_2, \ldots\} \subseteq L \times |g|$. Formally "$l_i : u_i$" is an ordered pair, a vertex u_i labelled by l_i. In a real *finitely-branching* WDB this page may be represented as a finite file (or as a part of a file containing many such pages) which consists of the lines "$l_i : u_i$" listed in *any* order. This file is saved in one of the computers participating in the distributed WDB network. (Of course, there might be just one computer for the whole WDB.) The computer and file may be located by the address v. In the *infinitely-branching* case an "infinite page" would be generated line-by-line by a program. All these details are implicit in the program, a *browser*, which computes g on the base of the current state of all the computers in the Internet (or Intranet) participating eventually in the computation of g.

As in the real Web, it is reasonable if any page looks like a list of lines "l_i" consisting of labels only so that the corresponding hidden addresses "u_i", whose precise shape is not so relevant to a human user, are *invisible* on the screen display. *This idea of visibility is crucial for the future considerations.* We consider that labels play the role of atomic or primitive data or attributes, as in a "flat" relational DB, whereas addresses only organize these data into a complex/nested/graphical structure, possibly with cycles. By "clicking" mouse button on some occurrence of a label "l_i", we invoke the function g applied to the corresponding hidden address "u_i" to get visible a new page gu_i, etc. This process of "browsing" through the Web also suggests the "inverse" direction of (labelled) edges in the graph g defined as $u \xleftarrow{l} v$ iff $l : u \, \epsilon^g \, v$.

According to this point of view on the nature of a Web page $p = \{l_1 : u_1, l_2 : u_2, \ldots\}$, its "visible part" is considered as *unordered multiset of words* $[\![p]\!] \rightleftharpoons \{l_1, l_2, \ldots\}$, i.e. just of labels l_i, *rather than a text*, i.e. ordered sequence of words. If a label l_i, as a word (or a picture), contains "enough" information then it is "non-clickable" ($g(u_i) = \emptyset$). Of course, any text may be represented as a very long word. E.g. one label on a page may be a longer text in which other shorter labels are mentioned in some way so that you can decide *informally* for yourself which of these labels must be clicked next. So, this representing data only via labels (which are appropriately organized via addresses) is sufficiently adequate and flexible abstract approach to Web-like Data Bases.[3]

If all the labels l_i in $gv = \{l_1 : u_1, l_2 : u_2, \ldots\}$ coincide then v actually defines a "uniform" set of elements $\{u_1, u_2, \ldots\}$. If all l_i are different then the page gv is a *tuple* or *record* with the *attributes* or *fields* l_i. (So, in a sense we need no special type of records.) Otherwise it could be called *quasituple* or *quasirecord* with u_i being l_ith *projection* of this quasituple. Let in general $v.l \rightleftharpoons \{u_i | l_i : u_i \in gv \& l_i = l\} = \{u_{i_1}, u_{i_2}, \ldots\}$. In the case of tuples this gives a *singleton set* $v.l_i = \{u_i\}$ (what is slightly different from the tradition to define $v.l_i = u_i$). If all gu_i are (quasi)tuples then gv is considered as *(quasi)relation* (a possibly heterogeneous set of (quasi)tuples). We see that the ordinary *relational* approach to Databases is easily absorbed by this WDB approach, and the same for the case of *nested relations* and *complex objects* (cf. also [30]) even having *cyclic structure*.

The World-Wide Web seems so widely known that there is no strong reason to present any illustrating examples. Rather, WWW is serving as a nice illustration and possible application of Hyper Set Theory whose "bounded" version we describe below. Say, there is nothing unusual in the possibility of cycles in the graph g representing a current instance of WWW (WDB).

---

[3] "Isn't this like reasoning about a city by its tramway plan?"—asked one reader skeptically. Yes, almost such. Extend this plan by pedestrian schema, plans of some buildings, etc. The real Web demonstrates that this is no problem at all.

3   Denotational Semantics of WDB via Hypersets

Now, the key point of the present approach to WWW or WDB is in defining *denotational semantics* $[v]^g$ for any *vertex* v of a graph g in terms of visible part of a *page* $[p]$ considered above. (Such a semantics [-] was also called *decoration* in [2] for the case of arbitrary graphs g; cf. also [4, 9, 31]. For well-founded g it is just A. Mostowski's *collapsing* [3] which we used in [27]–[30], [33]; cf. also [6]). Different vertices u and v may have the *same meaning*:

$$u \approx^g v \rightleftharpoons [u]^g = [v]^g. \tag{1}$$

For example, it may be the case that $gu = gv$. (Just create or copy on your server a Web page which already exists somewhere else in the Web and assign it the corresponding new address.) These two addresses may be reasonably considered as equivalent by referring to exactly the same page. More generally, the pages (as sets of pairs) gu and gv may be different, but *visibly coinciding* (as multisets of labels): $[gu] = [gv]$. Actually, $u \approx^g v$ means that it is impossible to find any difference between the addresses u and v if to *take into account only visible part* of corresponding arising pages during the multiple repeated process of browsing started respectively from u and v. However this seems intuitively rather clear, a formal definition is necessary; cf. (4). This approach of *neglecting the precise form of addresses* of the Web pages is quite natural because they play only the role of references or *hyperlinks*—no less, but also no more.

Alternatively and equivalently, membership-like notation ϵ^g suggests to consider $[v]^g$ as a *set of sets of sets, etc.* defined recursively by the identity

$$[v]^g = \{l : [u]^g \,|\, l : u \,\epsilon^g\, v\}. \tag{2}$$

For example, if g contains exactly five (non-labelled) edges $u \,\epsilon^g\, v$, $u \,\epsilon^g\, v'$, $u \,\epsilon^g\, w$, $v \,\epsilon^g\, w$ and $v' \,\epsilon^g\, w$ then $[u]^g = \emptyset$, $[v]^g = [v']^g = \{\emptyset\}$, and $[w]^g = \{\{\emptyset\}, \emptyset\}$. If g is any *well-founded* $C\mathcal{BLG}$, i.e. has no infinite chains of vertices satisfying $\ldots \epsilon^g v_{i+1} \epsilon^g v_i \epsilon^g \ldots \epsilon^g v_1 \epsilon^g v_0$ or, equivalently, has no (directed) *cycles*, for the case of finite g, then $[v]^g$ is "calculated" similarly for any $v \in |g|$ as a hereditarily-countable well-founded set. However, in non-well-founded case, such as in the graph $\circlearrowleft$, there is a problem. We need sets like $\Omega = \{\Omega\}$. This leads us to rather unusual non-well-founded or even *antifounded* or *hyper* set theory where some sets may serve as their own members, etc. [2, 4].

We confine ourselves to a subuniverse $\mathrm{HCA}_{\mathcal{L}}$ (for some supply of labels $\mathcal{L}$) of this set theory containing only *hereditarily-countable* (or finite) *hypersets*. Besides countability of each set $x \in \mathrm{HCA}$, it must satisfy the following *Anti-Foundation Axiom* (AFA) which is just a *finality property* in the category $C\mathcal{BLG}$:

> *For any $C\mathcal{BLG}$ g there exists a unique denotational semantics map $[-]^g$: $|g| \to \mathrm{HCA}$ satisfying the identity (2) with respect to the membership relation $\in$ understood in the sense of HCA (i.e. as $\in^{\mathrm{HCA}}$).*

Evidently, HCA (if it exists) is determined by this axiom uniquely, up to isomorphism. According to the tradition of non-well-founded set theory, we will use $\in$

with no superscript HCA because we "believe" (or just behave as believing) in "real existence" of such a universe in the same way as for the "ordinary" well-founded universe, say, HC or HF of hereditarily-countable, respectively, -finite) sets. If g ranges only over finite, respectively, finitely branching graphs then $[\![v]\!]^g$ for $v \in |g|$ will range over the corresponding subuniverses HFA $\subset$ HFA$^\infty$ of HCA.[4] While HFA consists of hereditarily-finite sets x whose *transitive closure*

$$\mathsf{TC}(x) \rightleftharpoons \{l : z | \exists n \geq 0 \exists \bar{l}\bar{z}.(l : z \in z_1, l_1 : z_1 \in \ldots \in z_n, l_n : z_n \in x)\} \qquad (3)$$

is finite, HFA$^\infty$ consists also of hereditarily-finite sets, however, their transitive closure may be infinite (just countable). Also sets from the subuniverses HF $\subseteq$ HFA and HC $\subseteq$ HCA correspond to the vertices of finitely and, respectively, countably branching *well-founded* graphs. It can be proved in the line of [2]

Theorem 1. *There exists the unique, up to isomorphism, universe HCA satisfying Anti-Foundation Axiom, and analogously also for* HFA$^\infty$ *and* HFA.

We only show how to define alternatively the corresponding equivalence relation $\approx^g \subseteq |g|^2$ (1) in terms of the graph g only, without any mentioning the semantic map $[\![\text{-}]\!]$ and the universe HCA (or HFA, HFA$^\infty$).

First note that according to the above recursive description (2) of $[\![\text{-}]\!]^g$ in the universe HCA the relation $\approx^g$ must evidently satisfy the equivalence

$$u \approx^g v \Leftrightarrow \& \begin{cases} \forall l : x \,\epsilon^g\, u \exists m : y \,\epsilon^g\, v(l = m \& x \approx^g y) \\ \forall m : y \,\epsilon^g\, v \exists l : x \,\epsilon^g\, u(l = m \& x \approx^g y). \end{cases} \qquad (4)$$

Actually, (4) have been informally discussed above in terms of visible indistinguishability of u and v. Any binary relation $R \subseteq |g|^2$ satisfying (4) with $\approx^g$ replaced by R and $\Leftrightarrow$ replaced by $\Rightarrow$ is called a *bisimulation* on g. Evidently, bisimulations are closed under unions so that there exists a *largest* one on g. It can be proved also (essentially, by using only AFA, (1) and (2)) that the latter *coincides* with $\approx^g$. So, the reader may completely forget (if he prefers this) about the universes HFA, HFA$^\infty$ and HCA and work directly in terms of graphs and bisimulations. However, it would be rather unreasonable to ignore our everyday and fruitful set-theoretic experience also applicable to hypersets. On the other hand, graph representation of such hypersets is very useful because this allows to consider computability notions over these abstract universes of sets in terms of graph transformers.

Moreover, in finitely branching case $\approx^g$ may be obtained as intersection[5]

$$\approx^g = \bigcap_{i=0}^{\infty} \approx_i^g \quad \text{where} \qquad (5)$$

---

[4] There are used also alternative denotations HF$_0$ for HF, HF$_{1/2}$ for HFA and HF$_1$ for HFA$^\infty$ [4] (actually, in the case of *pure* sets which do not involve labels). Also, in [30] HFA denoted *well-founded* universe with 'A' staying for 'attributes' which are called labels here. In this paper 'A' means 'Anti-Foundation Axiom' or 'Aczel'.

[5] In general case i must range over *ordinals*, instead of natural numbers.

$$u \approx_0^g v \ \rightleftharpoons \ \text{true} \quad \text{and}$$
$$u \approx_{i+1}^g v \ \rightleftharpoons \ \& \begin{cases} \forall l : x \ \epsilon^g \ u \exists m : y \ \epsilon^g \ v(l = m \& x \approx_i^g y) \\ \forall m : y \ \epsilon^g \ v \exists l : x \ \epsilon^g \ u(l = m \& x \approx_i^g y) \end{cases} \tag{6}$$

so that $\approx_0^g \supseteq \approx_1^g \supseteq \dots \supseteq \approx^g$. For the case of a finite graph g this sequence evidently stabilizes: $\approx_i^g = \approx^g$ for some i, so that $g \mapsto \approx^g$ is *computable* (in poly-time wrt cardinality of $|g|$). For finitely branching graphs with infinite number of vertices we have only *semidecidability* of $\not\approx^g$ (wrt the "oracle" g). By considering graphs *up to bisimulation* (in contrast to, say, [1] devoted to infinite finitely branching graphs as WWW instances), this is the main technical reason why we need some additional efforts for developing corresponding approach for this infinite case (i.e. for the case of HFA$^\infty$; corresponding paper is in progress.) So, we concentrate here mainly on the finite case (HFA).

Now, it is natural to consider the "membership" relation ϵ^g *up to bisimulation* $\approx^g$ so that $\approx^g$ proves to be a *congruence* relation on $|g|$:

$$l : u \in^g v \rightleftharpoons \exists l' : u' \ \epsilon^g \ v(l = l' \& u \approx^g u'), \tag{7}$$
$$u \approx^g u' \& v \approx^g v' \Rightarrow (l : u \in^g v \Leftrightarrow l : u' \in^g v').$$

Also the ordinary *extensionality axiom* written in the form (like (4))

$$u = v \Leftrightarrow \& \begin{cases} \forall l : x \in u \exists m : y \in v(l = m \& x = y) \\ \forall m : y \in v \exists l : x \in u(l = m \& x = y) \end{cases} \tag{8}$$

holds in any g if symbols $=$ and $\in$ are interpreted as $\approx^g$ and, respectively, as $\in^g$ (i.e. as ϵ^g considered up to $\approx^g$). Actually, the following *axiom of strong extensionality* holds in the quotient $g/\approx^g$ or in HCA, HFA$^\infty$, HFA:

$$xRy \Rightarrow x = y \quad \text{for arbitrary bisimulation relation R wrt } \in.$$

Graph g is called *strongly extensional* ($\mathcal{SE}$) or *canonical* if $\approx^g = =^g$. It follows also that the vertices of any (countable, $\mathcal{SE}$) WDB g and the relation $\in^g$ corresponding to the edges of g (up to $\approx^g$) may be identified with corresponding HCA-sets and, respectively, with the membership relation $\in^{\text{HCA}}$ for these sets.

We conclude that the Web or any WDB (considered statically) may be naturally interpreted as an "initial" (*transitive*) part of the universe HCA or HFA$^\infty$ or HFA, depending on its type: countable, finitely branching, or just finite WDB.

WDB Schemas. Different instances g, g' of WDB may have the same *schema* or a *type s* which is also presented as a (desirably $\mathcal{SE}$) graph $s : |s| \to \mathcal{P}(\mathcal{L} \times |s|)$. Given arbitrary binary relation on labels $T \subseteq \mathcal{L}^2$, consider corresponding largest, *relativized to T, bisimulation relation* (now denoted as $::_T^{g,s}$; cf. [30])

$$u ::_T^{g,s} v \Leftrightarrow \& \begin{cases} \forall l : x \ \epsilon^g \ u \exists m : y \ \epsilon^s \ v(lTm \& x ::_T^{g,s} y) \\ \forall m : y \ \epsilon^s \ v \exists l : x \ \epsilon^g \ u(lTm \& x ::_T^{g,s} y), \end{cases}$$
$$g ::_T s \rightleftharpoons \forall u \in |g| \exists v \in |s|(u ::_T v) \& \forall v \in |s| \exists u \in |g|(u ::_T v).$$

(Alternatively, we could consider, as in [30], that only the first conjunct is present in these formulas. In this case we should rather call $::$ *simulation* relation.)

Evidently, $\approx \; = \; ::=_{\mathcal{L}}$, $::_{T_1 \circ T_2} = (::_{T_1}) \circ (::_{T_2})$ and $::_{T^{-1}} = (::_T)^{-1}$. If $T \subseteq \mathcal{L}^2$ is transitive (respectively, reflexive, symmetric) then such is $::_T$.

Let now lTm denote some typing relation "l *is of type m*" between labels like: "`Pereslavl-Zalessky` *is of type* `city`", "`Moscow` *is of type* `capital`", "`Moscow` *is of type* `word`", "`word` *is of type* `word`", "`5` *is of type* `integer`", etc. Then $g ::_T s$ also means that g *is of type* s or that s is a *schema* of the database (state) g. If $T \subseteq \mathcal{L} \times \mathcal{L}$ is a (partial) map $\mathcal{L} \to \mathcal{L}$ then such s is defined uniquely, up to $\approx$, and may be easily obtained from g by replacing in g all labels by their T-images.

Any given typing relation $T \subseteq \mathcal{L}^2$ also induces corresponding typing relation $x ::_T \alpha$ between arbitrary HFA-sets x and α. Any $\alpha \in$ HFA may serve as a type. However, for T a map, a type α is usually simpler than x, and it properly reflects the "nested" structure of x. Cf. also [30] for corresponding tabular representations of (well-founded) HF-sets according to their types. This makes the present approach related with (nested) relational databases. Note, that now both sets and their types may be circular.

Isomorphism of WDBs up to Bisimulation. Let T, now denoted as $\simeq$, be a bijection $L \to L'$ between labels of two WDB graphs g and g'. Then corresponding relation between vertices $::_T^{g,g'}$, denoted as $\simeq^{g,g'}$, proves to be an *isomorphism, up to bisimulation*, between "initial parts" of g and g' because

$$u \simeq^{g,g'} u' \;\& \; v \simeq^{g,g'} v' \;\& \; l \simeq l' \Rightarrow$$
$$(l : u \in^g v \Leftrightarrow l' : u' \in^{g'} v') \& (u \approx^g v \Leftrightarrow u' \approx^{g'} v')$$

holds for all $u, v \in |g|$, $l \in L$ and $u', v' \in |g'|$, $l' \in L'$. We usually write $u \simeq u'$ instead of $u \simeq^{g,g'} u'$ to simplify notation and consider this as *three* place relation between u, u' *and* a bijection $\simeq \subseteq L \times L'$. If there is some predefined relational structure on labels such as a linear order $\prec_{\mathcal{L}}$, a typing relation, etc. (considered as primitive predicates of the Δ-language below) then we assume additionally that $\simeq$ is the ordinary *first-order isomorphism* between L and $L' \subseteq \mathcal{L}$.

4 Δ-Language for Hypersets

It is widely recognized high expressive power and flexibility of the classical set theory. Let us present the syntax of Δ-*language* for some its *bounded* version:

Δ-formulas $::= l = m \mid P(\bar{t}) \mid \varphi \& \psi \mid \varphi \vee \psi \mid \neg \varphi \mid \forall l : x \in s.\varphi \mid \exists l : x \in s.\varphi \mid$
$\qquad\qquad$ [the-least $P.(P(\bar{x}) \Leftrightarrow \theta(\bar{x}, P))](\bar{t})$

Δ-terms $::=$ set-variables $x, y, \ldots \mid \emptyset \mid \{l_1 : t_1, \ldots, l_k : t_k\} \mid$
$\qquad\qquad \bigcup \{t(l, x) \mid l : x \in s \& \varphi(l, x)\} \mid \mathsf{TC}(s) \mid \mathsf{D}(s, t)$

Here $t, \bar{t}, s$ are Δ-terms (or lists of terms); φ, ψ are Δ-formulas; l, m, l_i are label variables which formally are not Δ-terms; P is a predicate variable (only of set arguments); variables x and l are not free in s; θ is Δ-formula which involves (at any depth) only atomic terms; and all occurrences of the predicate variable P in θ are *positive*. Also there may be some additional primitive predicates over labels, besides the *equality* $l =_{\mathcal{L}} m$, such as some given *linear order* $l \prec_{\mathcal{L}} m$, or the predicate "l *is a substring of* m" or discussed above some typing relation.

The meaning of the most of these constructs in the universe HFA (or even in HCA) is straightforward. $\bigcup \{t(l,x)|l : x \in s \& \varphi(x,l)\}$ is the *union* of the family of sets $t(l,x)$ where x and l satisfy the condition after "$|$". The *transitive closure* TC has been defined in (3). [the-least $P.(P(\bar{x}) \Leftrightarrow \ldots)$] denotes the least predicate on the universe HFA such that the corresponding equivalence holds for all $\bar{x} \in$ HFA. It evidently exists by the positivity requirement on P and may be obtained as *union* $\bigcup_i P_i$ with $P_0 \rightleftharpoons$ false and $P_{i+1}(\bar{x}) \rightleftharpoons \theta(\bar{x}, P_i)$. Note, that this union is actually "locally finite" due to $\theta(\bar{x}, \bar{y}, P) \Leftrightarrow \theta(\bar{x}, \bar{y}, P \upharpoonright \mathsf{TC}(\{\bar{x}, \bar{y}\}))$ where $\bar{x}, \bar{y}$ are all free set variables of θ (which has no complex terms!). This construct is actually equivalent in HFA (essentially as in [14]) to its *inflationary* version [inflationary $P.(P(\bar{x}) \Leftrightarrow (P(\bar{x}) \vee \theta(\bar{x}, P)))](\bar{t})$ with no requirement of positivity on P in θ.

Unordered sets may encode *ordered triples* of the kind

$$\langle l, u, v \rangle \rightleftharpoons \{l : \{l : u\}, l : \{l : u, l : v\}\}. \tag{9}$$

Then arbitrary set $s \in$ HFA serves also as a finite graph by taking all ordered triples $\langle l, u, v \rangle \in s$ (if such exist) as the labelled edges of this graph. Therefore for *decoration operation* D we may postulate the axiom analogous to (2) for [-]:

$$\mathsf{D}(s, v) = \{l : \mathsf{D}(s, u)|l : \langle l, u, v \rangle \in s\}.$$

Decoration is the only construct of this language which has not a meaning in general for the ordinary well-founded universe HF. However, this is a natural operation which, given arbitrary (finite) graphical, possibly cyclic "plan" $s \in$ HFA, constructs a set $\mathsf{D}(s, v) \in$ HFA according to this plan.

Note that strong extensionality allows to define in Δ equality relation $u = v$ between HFA-sets (missing in the present version of Δ) recursively, as the largest fixed point via (4) or (8) (easily reducible to the-least construct) by using primitive equality $l =_{\mathcal{L}} m$ on labels. Therefore $\in^{\mathrm{HFA}}$ is also Δ-definable by using bounded quantification as $l : u \in v \rightleftharpoons \exists m : w \in v(l = m \& u = w)$.

The origins of Δ-language go back to the well-known notion of Δ_0-formulas of A. Lévy and in so called *basic = rudimentary* set-theoretic operations of R.O. Gandy [10] and R.B. Jensen [17]. Note also somewhat related theoretical approach to SETL in [6, 7] and set-theoretic programming language STARSET [11]. The latter, in comparison with SETL and Δ, does not use deeply nested sets. So called semantic or Σ-programming [12] is based mainly on considering unbounded, unlike Δ, positive existential quantification.

Some label-free version of this Δ-language for pure universe HFA was considered in [27], as well as labelled one for the well-founded case in [30]. Natural *axioms* describing the meaning of Δ-definable operations and predicates are presented in [27, 28, 33] and [31, 33] for pure well-founded and, respectively, non-well-founded case, as well as corresponding proof-theoretic considerations, so that we actually have a *Bounded Set Theory with Anti-Foundation Axiom*, BSTA. It may be used, say, for proving correctness/equivalence of Δ-queries.

5   Querying and Computing over WDB

Any Δ-term $t(\bar{l}, \bar{x})$ or Δ-formula $\varphi(\bar{l}, \bar{x})$ defines a map, called also *query*, $\text{HFA}^k \to \text{HFA}$ or, respectively, $\text{HFA}^k \to \{\text{true}, \text{false}\}$. (We represent any label argument $l_i \in \mathcal{L}$ by the set $\{l_i : \emptyset\} \in \text{HFA}_{\mathcal{L}}$.) Such a map $q : \text{HFA} \to \text{HFA}$ is called (PTIME) *computable* by a graph transformer $Q : \langle g, v \rangle \mapsto \langle g', v' \rangle = Q(g, v)$ if

$$q([\![\langle g, v \rangle]\!]) = [\![Q(g, v)]\!] \quad (\text{where} \quad [\![\langle g, v \rangle]\!] \rightleftharpoons [\![v]\!]^g) \tag{10}$$

holds for all finite g and $v \in g$, and Q is (PTIME) computable. We may also assume that g' extends the graph g by some "temporary" vertices and edges used to "calculate" the result $[\![\langle g', v' \rangle]\!] = q([\![\langle g, v \rangle]\!])$ without changing the meaning of the old vertices in g. We also abbreviate $\langle g', v' \rangle = Q(g, v)$ as $v' = Q(v)$.

It is natural to call q and Q *abstract* and *concrete* versions, respectively, of a query to WDB g, or as its *denotational* and *operational semantics*. The specific way of calculating a query Q which guarantees the identity (10) consists in

- multiple repeated *browsing*, i.e. applying the function g, starting from an input address v of a page,
- *searching* the labels on a current page satisfying some (also computable) condition (e.g. on containing by a text-label some substring(s) of symbols),
- *(re)organizing* this information according to the query Q by *creating new pages* with the labels found above and *assigning them some addresses*,
- taking into account during the computation *only visible part* of pages,
- *outputting* as the final result (a *page* with) some *address* v' which, together with all (possibly new) pages accessible from it via hyperlinks will constitute the whole answer to the query.

In particular, this is the case for arbitrary Δ-definable queries/updates $Q(\bar{v})$ to WDB g where now set variables $\bar{v}$ range over addresses (i.e. graph vertices) rather than over corresponding HFA-sets (cf. also [31]). Here are examples.

1. Given labels $l_1, \ldots, l_n$ and addresses $v_1, \ldots, v_n$ $(n \geq 0)$, create a new page $\{l_1 : v_1, \ldots, l_n : v_n\}$ (i.e. a new file on a computer in the network with the corresponding address) consisting of the lines $l_1 : v_1, \ldots, l_n : v_n$.
2. Take the *union* $v_1 \cup v_2 = \bigcup\{v_1, v_2\}$ of two pages given by their addresses v_1 and v_2 by concatenating corresponding files. The same for more general union construct $\bigcup\{t(l, x) | l : x \in s\}$, assuming that we can compute $t(l, x)$.[6]

---

[6] Let us postulate additionally that during all the subcomputations $t(l, x)$ for $l : x \in gs$ the sets of newly created addresses (in all the intermediate steps) for different pairs $l : x$ do not intersect. This allows to accomplish all these subcomputations independently/concurrently, with no conflicts between new addresses, and actually guarantees polytime complexity of the resulting query. Cf. Theorem 3 (a) below.

3. Bounded quantifiers $\forall l : x \in s.\varphi$ and $\exists l : x \in s.\varphi$ and the general union $\bigcup\{t(l, x)| l : x \in s\&\varphi(l, x)\}$ are also based on looking through pages having references $l : x$ in gs and satisfying given (computable) condition $\varphi(l, x)$.

4. To compute **the-least** construct use its local property (cf. Sect. 4).

5. To compute decoration $D(s, w)$ by any two given page addresses s and w
   (a) start depth 3 browsing from s to gather (actually taking into account bisimulation which is itself computable by **the-least** construct) all the "triples" $l : \langle l, u, v \rangle$ (9) from the page gs,
   (b) consider each this triple as a l-labelled edge of a graph γ,
   (c) create new (i.e. not used in the WDB instance g) different addresses $\text{NewAddr}(u)$, $\text{NewAddr}(v)$, $\text{NewAddr}(w)$, ... for all the vertices of γ,
   (d) extend g by creating new pages with these new addresses to imitate γ,
   (e) output $\text{NewAddr}(w)$ as the final result.
   (We omit some more details.)

We see that Δ is not only a *pure* set-theoretic language over HFA whose meaning may be understood quite abstractly, but simultaneously a "graph" language interpreted solely in terms of the computation over the Web or WDB.

Unlike HFA, when working in terms of a WDB state g the ordinary equality and g-membership relations between addresses $u = v$ and $l : u \in^g v$ *cannot* be considered here as legal Δ-queries. Otherwise, we would be able to distinguish visibly non-distinguishable addresses. However, bisimulation $u \approx^g v$ (4) and another membership relation of the kind $l : u \in^g v$ (7) have been actually Δ-defined. Fortunately, "membership" relation of the form $l : u \in^g v$ *is allowed* in the context of bounded quantification which is effectively computable by looking through the finite page gv.

Database Integrity Constraints also may be expressed in Δ-language. A trivial example: for $x \in$ HFA to be a tuple (vs. quasituple, i.e. an arbitrary HFA-set) is formulated as the constraint $\forall l : y, l' : y' \in x(l = l' \Rightarrow y = y')$. We may describe that some relations (sets of tuples) are of a specified arity, or postulate some functional dependencies or some interleaving between tuples and relations when going deeper and deeper into a (possibly cyclic) set, etc.

Important Properties of Δ-Definable Queries. It can be shown by induction on the definition of Δ-language that (in terms of its operational semantics in WDB terms sketched above) all Δ-definable queries *preserve bisimulation relation $u \approx^g v$ between (addresses of) pages* (cf. [31] for more details). This means that any page query t and boolean query φ should not distinguish between pages with the same meaning:

$$u \approx^g v \Rightarrow t(u) \approx^g t(v)\&(\varphi(u) \Leftrightarrow \varphi(v)).$$

Evidently, all Δ-queries *preserve (do not create new) labels* in the sense that any label participating in $\mathsf{TC}(t(\bar{l}, \bar{u}))$ must participate also in $\{\bar{l}\} \cup \mathsf{TC}(\{\bar{u}\})$.

A very important property of Δ-definable queries consists in their *genericity* (or *abstractness*), **up to bisimulation**. We define this notion following [30] where it was applied directly to HF-sets rather than to representing (finite acyclic) graphs. Cf. also [1] where the notion of genericity is considered only for boolean queries and without any connection to bisimulation, unlike this paper and [30].

A query $Q(u)$ is called *generic (up to bisimulation)* if it preserves (i) bisimulation, (ii) labels and also (iii) isomorphism, up to bisimulation, of its inputs. I.e. for any two graphs g and g' and an isomorphism $\simeq$ between corresponding sets of labels L and L' (and therefore between some vertices in $|g|$ and $|g'|$)

$$u \in |g|, \ u' \in |g'|, \ u \simeq u' \quad \text{implies} \quad Q(u) \simeq Q(u')$$

(according to the *same* bijection $\simeq$ between labels!).

These properties of queries appear as quite natural, even independing on the above Δ-language of queries. Moreover, considering properly the *computability* *notion* of a query as a graph transformer $Q : \langle g, v \rangle \mapsto \langle g', v' \rangle = Q(g, v)$ satisfying (10) (reasonably defined by using encoding of *abstract* labelled graphs; cf. also [30]) leads *inevitably* to genericity of Q.

Some Comparisons. Of course, it is not strictly necessary to support set-theoretic (based on bisimulation) level of abstraction if paying attention to the precise syntax of addresses; cf. [1, 22, 25]. E.g. in [1] a general Turing machine model of browsing and searching is discussed where genericity of the underlying finitely-branching graph is understood in terms of the ordinary isomorphism of graphs, what corresponds to understanding addresses of the Web pages essentially more literally, than here. Bisimulation and corresponding approach to querying for the graphs representing data has been discussed in

- [27]–[30],[33]: the acyclic case, via decoration restricted to collapsing, including a more general approach to encoding of HF-sets and to PTIME-computability over HF; cf. especially labelled case [30],
- [32, 19]: approaching to LOGSPACE-computability over HF,
- [31, 33]: cyclic label-free case on a more abstract theoretical level than here, including some proof theoretic considerations on BST, AFA, and a Logic of Inductive Definitions,
- and also in the papers of A. Kosky [18] and P. Buneman, et al [5].

The main informal difference of our approach with [18, 5] is the set-theoretical point of view based on the observation that the graphs and operations over them considered up to bisimulation are essentially (non-well-founded) sets of sets of sets, etc., and well-known set-theoretical operations, respectively. Also note, that query languages UnQL and UnCAL discussed in [5] and formulated in terms of graphs may be essentially (up to some unclear places in syntax and semantics) embedded in Δ (if it is considered as a language of graph transformers). However, the decoration operation D of Δ is not definable there and also in SRI [18]. Any operation of these languages preserves the well-foundeness of input data, i.e. being applied to the acyclic data structure results also in the acyclic one, while D obviously does not. (An argument to D may be cyclic graph—something like a syntactic specification of a data—*represented* by an HF-set, but all HF-sets themselves *are* of "acyclic" nature.) It follows that these languages do not define all PTIME computable queries, unlike Δ (cf. Theorem 3 below).

6 Linear Ordering of HFA and PTIME Computability

Let, for a while, a finitely branching graph g (say, = HFA or HFA^∞) contain no labels. Then we can define iteratively as in [20] (by appropriate reconstructing a formula from [30] with using an idea from [8]) strict linear preorders $\prec_k$ on g (which are decidable uniformly on k with respect to the oracle g):

$$x \prec_0 y \rightleftharpoons \text{false},$$

$$x \prec_{k+1} y \rightleftharpoons x \prec_k y \vee$$

$$\{x \approx_k y \;\&\; \exists u \in y[\forall v \in x(u \not\approx_k v) \;\&\;$$

$$\forall w \in x \cup y(u \prec_k w \rightarrow \exists p \in x \exists q \in y(p \approx_k w \;\&\; q \approx_k w))]\}.$$

This definition may be easily adapted for the case of labelled graphs with some predefined strict linear order $\prec^L$ on labels. Note, that expressions like $x \approx_k y$ may be replaced by actually equivalent formulas $x \not\prec_k y \;\&\; y \not\prec_k x$. It follows that

Theorem 2. *The "limit" strict linear order $\prec \rightleftharpoons \bigcup_i \prec_i$ on HFA (and even on HFA^∞) is definable in Δ relative to $\prec_{\mathcal{L}}$ (both by the inflationary and positive version of the-least construct of Δ). The same holds for vertices of any finite or finitely-branching $\mathcal{SE}$ graph with linear ordered labels and FO+LFP (or IFP).*

Remember that FO+LFP (or IFP) is First-Order Logic + the corresponding logical analogue of the-least (respectively, inflationary) fixed point construct of Δ. Both versions are equivalent over finite structures [14]. Definability in Δ over HFA was reduced in [31] to definability of graph transformers in FO+LFP. It is well-known that such kind of recursion in *linearly ordered* finite domains, like a finite segment of natural numbers, is equivalent to PTIME-computability [26, 16, 21, 34, 13]. Therefore, Theorem 2 (cf. also [20]) implies

Theorem 3. *(a) Definability in FO + IFP (i.e. recursion) over finite strongly extensional labelled graphs with predefined linear order only on the labels is equivalent to PTIME-computability over these graphs.*
(b) Assuming a linear order $\prec_{\mathcal{L}}$ on labels, Δ-definable operations over $\text{HFA}_{\mathcal{L}}$ = all (generic) PTIME-computable queries HFA $\rightarrow$ HFA (actually, relative to oracles corresponding to free predicate variables in Δ-query) under graph representations of sets according to the semantics $[\![\cdot]\!] : \mathcal{FLG} \rightarrow \text{HFA}_{\mathcal{L}}$.

We conclude that set-theoretic and WDB query language Δ is *sufficiently complete and natural*. Also for the case of "small" finite WDB all the definable queries are, in a sense, "feasibly" computable. However, the latter note should not be overestimated. So, bisimulation/non-bisimulation is practically rather expensive procedure, however being PTIME computable. (It is PTIME-complete problem even for acyclic graphs [6].) On the other hand, it can be evaluated permanently, by some program, so that we could use partial results ot its work, when needed. For the case of finitely branching graphs (and HFA^∞, unlike HFA) we can assert only that $\prec$ and $\not\approx$ (or $\neq$ in HFA^∞) are *semicomputable*.

Anyway, the present work is conceptual rather than practical one. We just attempted to present a uniform, concise set-theoretic view on querying to hyperlinked data.

References

1. Abiteboul, S., Vianu, V.: Queries and computation on the Web. Database Theory—ICDT'97, 6th International Conference, Delphi, Greece, January 1997, Proceedings. Lecture Notes in Computer Science **1186** Springer (1995) 262–275
2. Aczel, P.: Non-well-founded sets. CSLI Lecture Notes No. 14, 1988
3. Barwise, J.: Admissible sets and structures. Springer, Berlin, 1975
4. Barwise, J., Moss, L.: Vicious Circles: on the mathematics of circular phenomena, CSLI Lecture Notes, 1996
5. Buneman, P., Davidson, S., Hillebrand, G., Suciu, D.: A query Language and Optimization Techniques for Unstructured Data. Proc. of SIGMOD, San Diego, 1996
6. Dahlhaus, E.: Is SETL a suitable language for parallel programming?—a theoretical approach. Proc. of CSL'87, Lecture Notes in Computer Science **329** (1987) 56–63
7. Dahlhaus, E., Makowsky, J.: The Choice of programming Primitives in SETL-like Languages. Proc. of ESOP'86, Lecture Notes in Computer Science **213** (1986) 160–172
8. Dawar A., Lindell S., Weinstein, S.: Infinitary Logic and Inductive Definability over Finite Structures. Information and Computation, **119**, No. 2 (1995) 160-175
9. Fernando, T.: A Primitive recursive set theory and AFA: on the logical complexity of the largest bisimulation. Report CS-R9213 ISSN 0169-118XCWI P.O.Box 4079, 1009 AB Amsterdam, Netherlands
10. Gandy, R.O.: Set-theoretic functions for elementary syntax. Proc. Symp. in Pure Math. **Vol. 13, Part II** (1974) 103–126
11. Gilula, M.M.: The Set Model For Database and Information Systems. Addison Wesley, 1994
12. Goncharov, S.S., Ershov, Ju.L., Sviridenko, D.I.: Semantic programming, Kugler H.J., ed. Information Processing'86 , Elsevier, North Holland (1986) 1093–1100
13. Gurevich, Y.: Algebras of feasible functions. Proc. 24th IEEE Conf. on Foundations of Computer Science (1983) 210–214
14. Gurevich, Y., Shelah, S. : Fixed-point extensions of first-order logic. Annals of Pure and Applied Logic **32** (1986) 265-280
15. Cf. http://www.intranetjournal.com/.
16. Immerman, N.: Relational queries computable in polynomial time. Proc. 14th. ACM Symp. on Theory of Computing (1982) 147–152; cf. also Information and Control **68** (1986), 86–104

17. Jensen, R.B.: The fine structure of the constructible hierarchy. Ann. Math. Logic **4** (1972) 229-308
18. Kosky, A. : Observational properties of databases with object identity. Technical Report MS-CIS-95-20. Dept. of Computer and Information Sciience, University of Pennsylvania (1995)
19. Lisitsa, A.P., Sazonov, V.Yu.: Δ-languages for sets and LOGSPACE-computable graph transformers, Theoretical Computer Science **175** (1997) 183–222
20. Lisitsa, A.P., Sazonov, V.Yu.: On linear ordering of strongly extensional finitely-branching graphs and non-well-founded sets. DIMACS Technical Report 97-05. Rutgers University, February 1997. (A shorter version of this paper to appear in Proc. of LFCS'97, Springer LNCS).
21. Livchak, A.B.: Languages of polynomial queries. In: Raschet i optimizacija teplo-tehnicheskih ob'ektov s pomosh'ju EVM, Sverdlovsk (1982) 41 (in Russian)
22. Mendelzon, A. O., Mihaila, G.A., Milo, T. : Querying the World Wide Web. Draft, available by ftp: milo@math.tau.ac.il (1996)
23. Milner, R. : Operational and algebraic semantics of concurrent processes. In: Handbook of Theoretical Computer Science, Ed. J. van Leeuwen, Elsevier Science Publishers, B.V. (1990) 1201–1242
24. Mislove, M., Moss, L., Oles, F.: Non-Well-Founded Sets Modeled as Ideal Fixed Points. Information and Computation, **93** (1991) 16-54
25. Quass, Y., Rajaraman, A., Sagiv, Y., Ullman, J., Widom, J.: Querying semistructured heterogeneous information. Technical Report, Stanford University, December 1995. Available by anonymous ftp from db.stanford.edu.
26. Sazonov, V.Yu.: Polynomial computability and recursivity in finite domains. Elektronische Informationsverarbeitung und Kybernetik **16**, No. 7 (1980) 319–323
27. Sazonov, V.Yu.: Bounded set theory and polynomial computability. Proc. of All Union Conf. on Applied Logic, Novosibirsk (1985) 188–191 (In Russian).
28. Sazonov, V.Yu.: Bounded set theory, polynomial computability and Δ-programming. Application aspects of mathematical logic. Computing systems **122** (1987) 110–132 (In Russian). Cf. also a short English version of this paper in: Lect. Not. Comput. Sci. **278**, Springer (1987) 391–397
29. Sazonov, V.Yu.: Bounded set theory and inductive definability. Abstracts of Logic Colloquium'90. JSL **56** Nu.3 (1991) 1141–1142
30. Sazonov, V.Yu.: Hereditarily-finite sets, data bases and polynomial-time computability. Theoretical Computer Science, **119** (1993) 187-214.
31. Sazonov, V.Yu.: A bounded set theory with anti-foundation axiom and inductive definability. Computer Science Logic, 8th Workshop, CSL'94 Kazimierz, Poland, September 1994, Selected Papers. Lecture Notes in Computer Science **933** Springer (1995) 527–541
32. Sazonov, V.Yu., Lisitsa, A.P.: Δ-languages for sets and sub-PTIME graph transformers. Database Theory – ICDT'95, 5th International Conference, Prague, Czech Republic, January 1995, Proceedings. Lecture Notes in Computer Science **893** Springer (1995) 125–138
33. Sazonov, V.Yu.: On Bounded Set Theory (10th International Congress on Logic, Methodology and Philosophy of Sciences, Florence, 1995), M.L.Dalla Chiara et al. (eds.), Logic and Scientific Methods, Kluwer Academic Publishers, 1997, 85–103
34. Vardi, M.: Complexity of relational query languages. Proceedings of 14th Symposium on Theory of Computation (1982) 137–146

Invariant Definability

(Extended Abstract)

J.A. Makowsky*

Department of Computer Science
Technion—Israel Institute of Technology
IL-32000 Haifa, Israel
janos@cs.technion.ac.il

Abstract. We define formally the notion of *invariant definability* in a logic $\mathcal{L}$ and study it systematically. We relate it to other notions of definability (implicit definability, Δ-definability and definability with built-in relations) and establish connections between them. In descriptive complexity theory, invariant definability is mostly used with a linear order (or a successor relation) as the auxiliary relation. We formulate a conjecture which spells out the special role linear order plays in capturing complexity classes with logics and prove two special cases.

1 Introduction

This paper initiates an analysis of the notion of *invariant definability* special cases of which have been used in various contexts of finite model theory and descriptive complexity theory. We assume the reader is familiar with the basics of complexity theory as given in [Joh90, BS90] and of descriptive complexity theory as given in [EF95]. We want to explore the impact of invariant definability on descriptive complexity, so we are mostly interested in the complexity classes **P** (deterministic polynomial time), **NP** (non-deterministic polynomial time), **L** (deterministic logarithmic space), **NL** (non-deterministic logarithmic space) and **PSpace** (deterministic polynomial space).

Informally, a global relation R over a class of τ-structures is invariantly definable using another auxiliary relation S if there is a formula, containing symbols from τ and a symbol for S, which defines R using S but the interpretation of this definition is independent of the interpretation of S. A formal definition is given below in section 2. The by now classical results in descriptive complexity theory, cf. the monograph [EF95], show that various complexity classes such a polynomial time (space) can be captured by logics in the presence of a linear order or labeling of the underlying structures. Furthermore, it was observed that capturing complexity classes with logics does not depend on the particular labeling chosen on the structures, in other words, if a logic $\mathcal{L}$ captured a complexity

* Supported by the German Israeli Foundation (GIF) and the Technion Foundation for Promotion of Research in the framework of the project *Definability and Computability*

class **C** over linearly ordered structures then very often one could also prove that every query in **C** was inviariantly definable in $\mathcal{L}$ using a linear order.

The role of the linear order is intriguing especially in capturing complexity classes contained in deterministic polynomial time. Gurevich conjectured that no reasonable logic (defined precisely in [Gur88]) captures determinsitic polynomial time on all finite structures, unless, say, $\mathbf{P} = \mathbf{NP}$. But the presence of a linear order can be less explicit. For most results, i.e. when the logic allows us to define the transitive closure of a relation, a successor relation is enough. If the order is not total, but omits only a bounded number of elements, it can be extended to a definable total linear order using parameters.

In this paper we want to investigate to what extent invariant definability using linear orders is special in capturing complexity classes. For this purpose we first study, in sections 3 and 4 the notion of invariant definability in general, and associate with each logic $\mathcal{L}$ its closure under invariantly definable relations, denoted by $INV(\mathcal{L})$. We shall also look at a more restricted version, denoted by $INV_{K_0}(\mathcal{L})$, where the auxiliary relation S ranges over K_0, a fixed class of structures with one relation S. As it turns out $INV(\mathcal{L})$ is a sublogic of $\Delta(\mathcal{L})$, the extension of $\mathcal{L}$ consisting of the implicitly Σ_1-definable relations. On the other hand we shall see, in section 4.3, that it is incomparable to the extension $IMP(\mathcal{L})$ of $\mathcal{L}$ which consists of the implicitly definable relations. To see this we shall exploit a close relationship between the notion of invariant definability and a notion of definability previously studied by Ajtai [Ajt83, Ajt89]. Ajtai's notion of definability was introduced to prove negative results in circuit complexity and is stronger than invariant definability, cf. [BS90], and even non-recursive languages may be Ajtai-definable (as well as weakly invariantly definable). The reason we introduce invariant definability for its own sake lies in the fact that it gives rise to well defined logics with its definable classes of structures lying well within the polynomial hierarchy (in $\mathbf{NP} \cap \mathbf{CoNP}$ in the case of first order logic).

In section 5 we examine more closely the role of invariant definability in descriptive complexity theory. We first look at fixed point logics PFP and LFP (partial fixed point and least fixed point) and observeq that $INV(PFP)$ captures exactly deterministic polynomial space, but that $INV(LFP)$ is more expressive than deterministic polynomial time. In the light of this it is natural to ask for which K_0 is it true that $INV_{K_0}(LFP)$ contains all **P**-computable relations ? Our main contribution here is the exact formulation of a conjecture: $INV_{K_0}(LFP)$ contains all **P**-computable rrelations iff in K_0 linear order is parametrically definable using an LFP-formula.

In section 6 we sketch a proof of a particular case of our conjecture due to E. Rosen. It deals with K_0 closed under substructures and having exactly one model in each finite cardinality (f-categorical). Both of these properties are quite natural in the context of invariant definability. Closure under substructures assures that the logic INV_{K_0} is closed under relativization and f-categoricity makes invariant definability equivalent to Ajtai's notion of definability.

Acknowledgments

I am indebted to Eric Rosen with whom I have discussed this work extensively during his stay at the Technion as a Postdoctoral fellow. The partial solutions to my conjecture are his and he also pointed out to me the relevance of Ajtai's work to invariant definability. Unfortunately he left the Technion before we could write up all his results in detail. We plan to publish an expanded version of our results jointly as [RM97] and I would like to thank Eric Rosen for allowing me to include his results and contributions in this extended abstract. I also wish to thank A. Dawar, Y. Gurevich, L. Hella, M. Otto and E. Ravve for various insightful comments on earlier drafts of this paper.

2 Definitions and Summary of Results

We denote vocabularies by ρ, σ, τ, possibly augmented with indices. Unless stated otherwise, all vocabularies are purely relational. If $\rho = \{R_0, \ldots, R_m\} = \bar{R}$ and ϕ is a ρ-sentence, we also write $\phi(\bar{R})$ to indicate the relation symbols. Similarly for $\sigma = \bar{S}$ and $\tau = \bar{T}$. For two sets X, Y we denote by $X \sqcup Y$ the disjoint union of X and Y. Again, unless stated otherwise, all structures are finite. A logic $\mathcal{L}$, as defined in [Ebb85, EF95], is called *weakly regular*, if it is closed under boolean operations, existential quantification, substitution and renaming. Following [MP94], it is called 1-regular, if it is additionally closed under relativization, and regular, if it is also closed under vectorization. Note that 1-regular is called regular in [Ebb85, EF95]. We denote by LFP and PFP the extensions of FOL by the least fixed point operator and the (partial) fixed point operator respectively, and by $FOL[DTC]$ and $FOL[TC]$ the extensions of FOL by the deterministic transitive and transitive closure operators respectively, as defined for example in [EF95]. All these logics are regular.

2.1 Invariant definability and descriptive complexity

Definition 1. Let $\mathcal{L}$ be a weakly regular logic and K_0 be a class of ρ-structures which has models in every cardinality and is closed under isomorphisms. A class of σ-structures K is *weakly K_0-invariantly $\mathcal{L}$-definable* if there is a $\mathcal{L}(\rho \sqcup \sigma)$-sentence ϕ such that

(i) $\mathcal{A} \in K$ iff there is an expansion $\bar{\mathcal{A}}$ to a $\rho \sqcup \sigma$-structure such that $\bar{\mathcal{A}} \models \phi$;
(ii) if $\mathcal{B}$ is a σ-structure with universe B and $< B, \bar{X}_0 >$ and $< B, \bar{X}_1 >$ are two expansions to $\rho \sqcup \sigma$-structures such that $\langle B, \bar{X}_0 \rangle$ and $\langle B, \bar{X}_1 \rangle$ are both in K_0 then
$$\langle \mathcal{B}, \bar{X}_0 \rangle \models \phi \text{ iff } \langle \mathcal{B}, \bar{X}_1 \rangle \models \phi(\bar{R}, \bar{S}).$$

We say that K *strictly K_0-invariantly $\mathcal{L}$-definable* if additionally there is a $\mathcal{L}(\rho)$-sentence θ such that $K_0 = Mod(\theta)$.

The following variant of the definition is easily seen to be equivalent.

Lemma 1. *Let K_0 be a class of ρ-structures which has models in every cardinality and is closed under isomorphisms.*
A class of σ-structures K is K_0-invariantly $\mathcal{L}$-definable iff for every σ-structure $\langle A, S_1^A, \ldots, S_m^A \rangle$ and for every ρ-structure $\langle A, R_1^A, \ldots, R_n^A \rangle \in K_0$ with the same universe we have that

$$\langle A, S_1^A, \ldots, S_m^A, R_1^A, \ldots R_n^A \rangle \models \phi \text{ iff } \langle A, S_1^A, \ldots, S_m^A \rangle \in K$$

Example 1. Let $EVEN$ be the class of structures over the empty vocabulary which are finite and of even cardinality. $EVEN$ is K_0–invariantly FOL-definable with the following K_0:

(i) $K_0 = ALTSUCC$, the class of all finite $\{R, P\}$-structures such that the interpretation of R is a successor relation with a last element, and the interpretation of P is given by $P(x)$ iff $\exists y R(y, x) \land \neg P(x)$.

(ii) $K_0 = 2EQUI$, the class of all equivalence relations where all equivalence classes have cardinality at most 2 and there is at most one equivalence class of cardinality 1.

2.2 Capturing complexity classes

The notion of invariant definability emerged from work of Immerman [Imm87], Sazonov [Saz80] and Vardi [Var82]. Vardi and Immerman study the case where K_0 is the class of finite linear orders ORD and the logics are LFP over finite structures or First Order Logic FOL augmented with transitive closure operators.

A global relation over a class of finite τ-structures K is a function Q which maps each τ-structure $\mathcal{A}$ into a k-ary relation over its universe such that the function is isomorphisms invariant. Let us assume that τ is ordered, i.e. it has two components, a finite set of relation symbol $\bar{R}$ and a distinguihsed relation symbol $R_<$ which is always interpreted as a linear order. Now $Q(\mathcal{A})$ depends on the interpretation $\mathcal{A}(\bar{R})$ of $\bar{R}$ and on the linear order given by $\mathcal{A}(R_<)$, so its appropriate to write $Q(\mathcal{A}(\bar{R}), \mathcal{A}(R_<))$ instead of $Q(\mathcal{A})$. The linear order can be viewed as a labeling of the $\bar{R}$-structure.

What Immerman and Vardi showed is that each global relation over some ordered K computable in **P** is definable in LFP. This is not true in general if $R_<$ is not present. For instance, if $\bar{R} = \emptyset$ and $Q(\mathcal{A}(P))$ is a relation of size $\frac{n}{2}$ if the size of $\mathcal{A}$ is even and is empty otherwise, Q is global unary relation in **P** which is not definable in $LFP(\emptyset)$ but it is definable in $LFP(R_<)$. Furthermore, there is a formula $\phi(x) \in LFP(R_<)$ such that for any two ordered τ-structures $\mathcal{A}_1 = \langle A, <_1 \rangle$ and $\mathcal{A}_2 = \langle A, <_2 \rangle$ with the same universe the structures $\langle A, \phi[\mathcal{A}_1] \rangle$ and $\langle A, \phi[\mathcal{A}_2] \rangle$ are isomorphic, i.e. the global relation defined by ϕ does not depend on the order (labeling) on $\mathcal{A}$.

To see the connection with our invariant definability, let K_0 be the class of linearly ordered finite structures and K be the graph of Q, i.e. the pairs $\langle A, Q(\mathcal{A}_1) \rangle$. Then this shows that K is (strongly) invariantly definable over K_0. In other words, the Immerman-Sazonov-Vardi result can be restated as follows:

Theorem 2 (Immerman, Sazonov and Vardi).
A class K of finite σ-structures is (strictly) ORD-invariantly LFP-definable iff membership in K is recognizable in polynomial time.

Question 1 (For finite Structures). How does the notion of K_0-invariant $\mathcal{L}$-definability vary with different choices of K_0 keeping $\mathcal{L}$ fixed? In particular, what is so special about the use of ORD in the Immerman-Vardi Theorem?

This question was the main motivation for this paper. Maybe our main contribution consists in the exact formulation of a conjecture which asserts that, in a certain sense, ORD is the only class of structures for which the Immerman-Sazonov-Vardi theorem holds. To make this more precise we need first a definition.

Definition 2. Let K_0 is a class of ρ–structures. We say that ORD *is parametrically $\mathcal{L}$-definable in K_0* if there is an $\mathcal{L}(\rho)$ formula (with parameters $\bar{y}$) $\chi(x_1, x_2, \bar{y})$ such that for every $\mathcal{A} \in K_0$ there are $\bar{a} \in A$ such that

$$\{\langle a_1, a_2 \rangle \in A^2 : \mathcal{A} \models \chi(a_1, a_2, \bar{a})\}$$

is a linear order on A.

By modifying the proof of the Immerman-Vardi Theorem it is easy to see that if ORD is parametrically definable in K_0 then every class $K \in \mathbf{P}$ is still K_0-invariantly LFP-definable. We conjecture the following converse.

> **Conjecture:** If K_0 is a class of ρ–structures such that every $K \in \mathbf{P}$ is K_0-invariantly LFP-definable then ORD is parametrically LFP-definable in K_0.

In section 6 we shall sketch a proof of the following special case of this conjecture due to E. Rosen. It is purely model theoretic, using Ramsey's theorem and indiscernible sequences. We say that K_0 is f-categorical if it has exactly one model in every (sufficiently large) finite cardinality. Clearly, ORD is f-categorical and also closed under substructures. These two properties also play a role in our study of invariant definability in general, cf. section 4.3, and they are also enough to establish another case of our conjecture.

Result 1 (E. Rosen). *If K_0 is f-categorical and closed under substructures and every $K \in \mathbf{P}$ is K_0-invariantly LFP-definable then ORD is parametrically LFP-definable (actually even FOL-definable) in K_0.*

2.3 Implicit definability and definability using built in relations

Once the notion of invariant definability is introduced it is natural to explore its meaning for logics in general.

Question 2 (For finite or arbitrary structures). How is (strict) invariant definability related to other notions of definability such as *implicit definability* or Δ-*definability*, as discussed in [Kol90] for finite model theory, or more systematically in [Mak85]?

In section 4.3 we give all the necessary definitions and show

Result 2. *For First Order Logic over finite structures, invariant definability is strictly weaker than Δ-definability (corollary 17) and it is incomparable to implicit definability (proposition 20).*

Invariant definability is closely related (but not equivalent) to *definability using built in relations*. To make this notion precise we choose a similar form as in the previous definition. The definition is basically due to Ajtai [Ajt83]. Ajtai-definability was studied for FOL in [Ajt83, Ajt89] and for Monadic Second Order Logic in [AF90, FSV93, Sch96].

Definition 3. Let $\mathcal{L}$ be a weakly regular logic and K_0 be a class of ρ-structures which has models in every cardinality and is closed under isomorphisms. A class of σ-structures K is K_0-*Ajtai-$\mathcal{L}$-definable* if there is a $\mathcal{L}(\rho \sqcup \sigma)$-sentence ϕ such that for each n there is a ρ-structure $\mathcal{A} \in K_0$ with cardinality exactly n such that for every expansion $\bar{\mathcal{A}}$ of $\mathcal{A}$ to an $\rho \sqcup \sigma$-structure we have that

$$\bar{\mathcal{A}} \mid_\rho \in K \text{ iff } \bar{\mathcal{A}} \models \phi.$$

If K_0 is the class of all ρ-structures we say that K is Ajtai $\mathcal{L}$-definable (over ρ).

Remark. Ajtai only considers the case where K_0 is the class of all finite ρ-structures $Str_{fin}(\rho)$. We have introduced K_0 as an additional parameter to facilitate the transition to invariant definability. Ajtai also requires only that there are arbitrarily large structures in K_0. He proves mostly negative results stating that some specific K is not $Str_{fin}(\rho)$-Ajtai FOL-definable for any ρ.

Using lemma 1 we get immediately:

Proposition 3. *Let K_0 be a class of ρ-structures and K be a class of σ-structures. K is Ajtai $\mathcal{L}$-definable iff there is a class of ρ-structures K_1 such that K is weakly K_1-invariantly $\mathcal{L}$-definable.*

The connection between invariant and Ajtai-definability is rather simple. A more general relationship is given in section 4.2 proposition 13.

Result 3. *Assume K_0 is ($\mathcal{L}$-definable and) f-categorical. Then K is weakly (strictly) K_0-invariantly $\mathcal{L}$-definable iff K is K_0-Ajtai-$\mathcal{L}$-definable.*

Ajtai-definability was used by Immerman [Imm87, BS90] to give the following characterization of non-uniform $\mathbf{AC_0}$, the class of problems recognizable by polynomial size, constant depth unbounded fan-in circuits.

Theorem 4 (Immerman). *A class K of finite σ-structures is Ajtai FOL-definable iff membership in K is recognizable in non-uniform* **AC₀**.

Related work on uniform circuit complexity classes appears in [BI94].

Question 3 (For finite structures). Can one characterize other non-uniform circuit complexity classes similarly, replacing FOL by other suitable logics? Can one characterize uniform circuit complexity classes using strict $\mathcal{L}$-invariant definability for suitable $\mathcal{L}$?

In section 4.4 we discuss briefly the case of infinite structures. We shall see that for logics satisfying Craig's Interpolation Theorem, every strictly invariantly $\mathcal{L}$-definable K is already $\mathcal{L}$-definable.

Question 4 (For infinite structures). What is the expressive power of invariant definitions over infinite structures for logics which do not satisfy Craig's Interpolation Theorem?

We leave these last two questions open for further research.

3   The Logics $INV(\mathcal{L})$ and $\Delta(\mathcal{L})$

In this section we deal with weakly regular logics $\mathcal{L}$ on finite structures and prove some general facts concerning invariant definability. Background on logics and logical reductions can be found in [EFT94, EF95, MP95]. The proofs of all the statements in this section are easy and left to the reader.

We first introduce our notion of parametrically definable classes of structures (or global relations). It will serve as a tool to obtain more invariantly definable classes once we have some at our disposal.

Definition 4. Let K_0 and K_1 be $\mathcal{L}$-definable classes of τ and σ-structures respectively and let $\sigma = \{S_1, \ldots, S_m\}$ with j_i the arity of S_i.
For a τ-structure $\mathcal{A}$, $\bar{a} \in A$ and $\mathcal{L}(\tau)$-formulas $\psi_i(\bar{x}_i, \bar{y})$ we denote by $\psi_i(\bar{x}_i, \bar{a})[\mathcal{A}]$ the set

$$\{\bar{a} \in A : \mathcal{A} \models \psi_i(\bar{x}_i, \bar{a})\}$$

K_1 is *parametrically definable* in K_0 if there are $\mathcal{L}(\tau)$-formulas $\chi_i(\bar{x}_i, \bar{y})$ with parameters $\bar{y}$ and $\bar{x}_i$ a vector of j_i free variables such that for every $\mathcal{A} \in K_0$ there is $\bar{a} \in A$ such that

$$\mathcal{A}_{\bar{a}}^* = \langle A, \chi_i(\bar{x}_1, \bar{a})[\mathcal{A}], \ldots, \chi_i(\bar{x}_m, \bar{a})[\mathcal{A}]\rangle \in K_1.$$

K_0 is *parametrically reducible* to K_1 if additionally $\mathcal{A} \in K_0$ iff there is $\bar{a} \in A$ with $\mathcal{A}_{\bar{a}}^* \in K_1$.

Proposition 5. *Let K_0, K_1 be classes of ρ_0 (ρ_1)-structures respectively.*

(i) If K_0 is parametrically reducible to K_1 and K is strictly K_1-invariantly definable in $\mathcal{L}$, then K is also strictly K_0-invariantly definable in $\mathcal{L}$.

(ii) If K_0 is parametrically definable in K and K is strictly K_0-invariantly definable in $\mathcal{L}$, then K is parametrically definable in $\mathcal{L}$.

Using the above we get that invariant definability is trivial if K_0 is parametrically definable in K in the following sense:

Example 2. For the following K_0 every K which is strictly K_0 invariantly $\mathcal{L}$-definable is already $\mathcal{L}$-definable:

(i) K_0 the class of all (finite) ρ-structures.
   Proof: Let K be strictly K_0-invariantly definable in $\mathcal{L}$ by a formula $\theta(\bar{R}, \bar{S})$. As the definition is invariant and the empty (full) relations are definable, we can substitute their definitions for $\bar{R}$ in θ.

(ii) K_0 the class of all equivalence relations (or all partial orders) with ρ consisting of one binary relation symbol.
   Proof: Similar as above noting that equality is trvially definable and that equality is both an equivalence relation and a partial order (with all elements incomparable).

More generally, if there is a global relation Q on K such that Q is (parametrically) $\mathcal{L}$-definable and for each $\mathcal{A} \in K$, $\langle A, Q(A) \rangle \in K_0$ then K is K_0-invariantly $\mathcal{L}$-definable iff K is $\mathcal{L}$-definable.

Given a weakly regular logic $\mathcal{L}$ we can associate with it new logics by closing $\mathcal{L}$ under some definability notion under which $\mathcal{L}$ is not closed such as relativized definability, vectorized definability, implicit definability etc. One way of doing this is using projective (Σ_1^1) classes. We recall the following standard definitions, cf. [BF85]

Definition 5. Let K be a class of σ-structures.

(i) K is a *$\mathcal{L}$-projective class* if there is a vocabulary ρ and a sentence ϕ in $\mathcal{L}(\rho \sqcup \sigma)$ such that $\mathcal{A} \in K$ iff there is an expansion $\bar{A}$ such that $\bar{A} \models \phi$.

(ii) $\Delta(\mathcal{L})$ is the family of classes K of σ-structures such that both K and its complement are $\mathcal{L}$-projective classes.

First we note that our strict invariantly definable classes are special cases of projective classes.

Lemma 6. *Let K_0 be $\mathcal{L}(\rho)$-definable by a sentence θ. If a class of σ-structures K is strictly K_0-invariantly $\mathcal{L}$-definable by the $\mathcal{L}(\rho \sqcup \sigma)$-sentence ϕ then*

(i) $\mathcal{A} \in K$ iff $\mathcal{A} \models \exists \bar{R}(\theta(\bar{R}) \wedge \phi(\bar{R}, \bar{S}))$
(ii) $\mathcal{A} \in K$ iff $\mathcal{A} \models \forall \bar{R}(\theta(\bar{R}) \rightarrow \phi(\bar{R}, \bar{S}))$

In other words, if K is strictly K_0-invariantly $\mathcal{L}$-definable, then both K and its complement are $\mathcal{L}$-projective classes.

Next we introduce some notation for invariantly definable classes.

Definition 6. Let $\mathcal{L}$ be a logic.

(i) Let K_0 be $\mathcal{L}$-definable. $INV_{K_0}(\mathcal{L})$ $(wINV_{K_0}(\mathcal{L})$) is the family of classes of σ-structures K such that K is strictly (weakly) K_0-invariantly $\mathcal{L}$-definable.
(ii) $INV(\mathcal{L})$ $(wINV(\mathcal{L}))$ is the family of classes of σ-structures K such that K is strictly (weakly) K_0-invariantly $\mathcal{L}$-definable for some $\mathcal{L}$-definable K_0.

The following are easily verified:

Lemma 7. (i) If K_0 is $\mathcal{L}$-projective and $K \in wINV_{K_0}(\mathcal{L})$ then there is a K_1 which is $\mathcal{L}$-definable and $K \in INV_{K_1}(\mathcal{L})$.
(ii) Let K_0, K_1 be $\mathcal{L}$-definable. If $K \in INV_{K_1}(INV_{K_0}(\mathcal{L}))$ there is a $\mathcal{L}$-definable K_2 such that $K \in INV_{K_2}(\mathcal{L})$.
(iii) $INV_{K_0}(INV_{K_0}(\mathcal{L})) = INV_{K_0}(\mathcal{L})$.
(iv) $INV(INV(\mathcal{L})) = INV(\mathcal{L})$.
(v) $INV_{K_0}(\mathcal{L}) \subseteq INV(\mathcal{L}) \subseteq \Delta(\mathcal{L})$.

Both $\Delta(\mathcal{L})$ and $INV(\mathcal{L})$ are closure operators yielding weakly regular logics. More precisely:

Proposition 8. Let $\mathcal{L}$ be a weakly regular logic and K_0 be $\mathcal{L}$-projective.

(i) ([MSS76]:) $\Delta(\mathcal{L})$ is a weakly regular logic and $\Delta(\Delta(\mathcal{L})) = \Delta(\mathcal{L})$.
(ii) $INV_{K_0}(\mathcal{L})$ and $INV(\mathcal{L})$ are weakly regular logics and $INV_{K_0}(INV_{K_0}(\mathcal{L})) = INV_{K_0}(\mathcal{L}) \subseteq \Delta(\mathcal{L})$, respectively $INV(INV(\mathcal{L})) = INV(\mathcal{L}) \subseteq \Delta(\mathcal{L})$.
(iii) If, additionally, $\mathcal{L}$ is 1-regular and K_0 is closed under substructures $INV_{K_0}(\mathcal{L})$ is a 1-regular logic.

For closure under relativization and vectorization (the regular case) one has to be more cautious and add the closure conditions to a modified definition of $\Delta(\mathcal{L})$ and $INV(\mathcal{L})$ respectively, cf. [BF85].

4 Invariant Definability, Δ–Closure and Implicit Definability

In this section we relate invariant definability over finite structures to other notions of definability. We shall discuss the case for arbitrary finite and infinite structures as well as some open problems in section 4.4.

4.1 EVEN, S-T-CONN and Spectra

First we introduce some classes of finite structures whose definability status we shall determine. Recall that EVEN(τ) the class of finite τ-structures of even cardinality.

Definition 7. (i) We denote by RELEVEN(τ) the class of finite $\tau \sqcup \{P\}$-structures $\mathcal{A}$ where P is unary and $P^{\mathcal{A}}$ is of even cardinality.

(ii) We denote by S-T-CONN the class of $\tau = \{E, c_s, c_t\}$ structures $\mathcal{A}$ such that E^A is a binary relation, and such that there is a directed E-path from c_s^A to c_t^A.

(iii) Let $f : \mathbf{N} \to \mathbf{N}$ by a function over the natural numbers, which tends to infinity arbitrarily slowly. We denote by S-T-CONN$_f$ the class of $\tau = \{E, c_s, c_t\}$ structures $\mathcal{A}$ such that E^A is a binary relation, and such that there is an E-path between c_s^A and c_t^A of length $\leq f(card(A))$.
Note that for $f(n) = n - 1$ S-T-CONN$_f$ coincides with S-T-CONN.

Fact 9. EVEN(τ) and RELEVEN(τ) both are $\Delta(FOL)$-definable (over finite structures) but not LFP-definable, and, a fortiori, not FOL-definable.
S-T-CONN$_f$ (for f tending to infinity) is not FOL-definable. Its definability in $\Delta(FOL)$ depends on the particular choice of f and is more intricate.

In example 1 we have observed that EVEN is in $INV(FOL)$. This can be generalized in the following way:

Definition 8. The *spectrum of a $\mathcal{L}$-sentence ϕ, spec(ϕ)* is the set of reducts of finite models of ϕ to the empty vocabulary.

Proposition 10 (E. Rosen). *(i) For every set of numbers, $S \subseteq \omega$ there is a class of structures K_S (not necessarily definable in $\mathcal{L}$), such that, for any vocabulary τ, $K(\tau) = \{\mathcal{A} : \mathcal{A} \text{ is a } \tau\text{-structure with } card(A) \in S\}$ is (weakly) K_S-invariantly $\mathcal{L}$-definable. Note that this includes non-recursive sets S.*
(ii) If, furthermore, the class F_S of structures over the empty vocabulary whose cardinality is in S is in $\Delta(\mathcal{L})$, K_S can be chosen to be $\mathcal{L}$-definable and $K(\tau)$ is strictly K_S-invariantly $\mathcal{L}$-definable. Note that here S is in $\mathbf{NP} \cap \mathbf{CoNP}$

Proof. We only prove (ii), leaving (i) to the reader. By our assumption, viewing spec(ϕ) as a set of models over the empty vocabulary, let ϕ_1 [ϕ_2] be a $\mathcal{L}$-sentence such that $spec(\phi_1) = S$ [$spec(\phi_2) = \omega - S$]. Let $\phi = (\phi_1 \vee \phi_2)$. Let $K_S = Mod(\phi)$, and observe that it contains models of every cardinality. Let $\psi = \phi_1$. Then over any vocabulary τ, $Mod_\tau(\psi)$ consists of exactly those finite τ-structures $\mathcal{A}$ with $card(A) \in S$ $Mod_\tau(\psi) = K(\tau)$ is K_S invariantly FOL-definable. ∎

4.2 Ajtai definability

In this section we study the exact relationship between Ajtai-definability, introduced in section 1, and invariant definability. Ajtai definability is used mostly to prove negative results. Indeed in the sequel, we shall use it to prove that certain properties are not weakly K_0-invariantly FOL-definable.

Notation 11. Let $\mathcal{L}$ be a weakly regular logic. We denote by $AJTAI(\mathcal{L})$ the class of classes of structures K which are K_0-Ajtai$\mathcal{L}$-definable for some K_0.

We first make an observation concerning monotonicity properties.

Observation 12. Let $K_0 \subseteq K_1$ be classes of ρ-structures which have models in all finite cardinalities.

(i) If a class of σ-structures K is K_0-Ajtai-$\mathcal{L}$-definable then it is also K_1-Ajtai-$\mathcal{L}$-definable.
(ii) However, if a class of σ-structures K is (strictly) K_1-invariantly -$\mathcal{L}$-definable then it is also (strictly) K_0-invariantly -$\mathcal{L}$-definable.

In other words, Ajtai–definability is upwardly monotone, whereas invariant definability is downwardly monotone.

Proposition 13. *Let K_0 be a class of ρ-structures and K be a class of σ-structures.*

(i) Assume K_0 is $\mathcal{L}$-definable. If K is (strictly) K_0-invariantly $\mathcal{L}$-definable then K is K_0-Ajtai $\mathcal{L}$-definable. In other words $INV(\mathcal{L}) \subseteq AJTAI(\mathcal{L})$.
(ii) Assume K_0 is f-categorical. Then, for all K, K is (strictly) K_0-invariantly $\mathcal{L}$-definable iff (K_0 is $\mathcal{L}$-definable and) K is K_0-Ajtai-$\mathcal{L}$-definable.
(iii) K is K_0-Ajtai $\mathcal{L}$-definable iff there is a (f-categorical) $K_1 \subseteq K_0$ such that K is weakly K_1-invariantly $\mathcal{L}$-definable. In other words $wINV(\mathcal{L}) = AJTAI(\mathcal{L})$.

Proof: (i): Let K_0 be $\mathcal{L}$-definable class of finite ρ-structures and let K be K_0-invariantly $\mathcal{L}$-definable. Let $K_1 \subseteq K_0$ be f-categorical subfamily of K_0. By the monotonicity we have that K be K_1-invariantly $\mathcal{L}$-definable. Using lemma 1 and the f-categoricity of K_1 we get that K is K_1-Ajtai $\mathcal{L}$-definable. Using monotonicity again we get that K is K_0-Ajtai $\mathcal{L}$-definable.
(ii): One direction follows from (i). Conversely, if K is K_0-Ajtai $\mathcal{L}$-definable (and K_0 is $\mathcal{L}$-definable) and f-categorical, K is (strictly) K_0-invariantly $\mathcal{L}$-definable. Here we use again lemma 1 and the f-categoricity of K_0
(iii) just combines the proofs of (i) and (ii). ∎

Using proposition 13 and 10(i) we get

Corollary 14. *There are non-recursive classes of finite structures in $AJTAI(FOL)$.*

We now can state and use Ajtai's main results [Ajt89, Ajt83].

Theorem 15 (Ajtai). *The following are not Ajtai-FOL-definable, and therefore not in $INV(FOL)$.*

(i) RELEVEN, by [Ajt83].
(ii) S-T-CONN$_f$ for f tending to infinity, by [Ajt89].
(iii) S-T-CONN, by fact 9 above.

This shows that in theorem 8 closure under relativization is not guaranteed even if $\mathcal{L}$ is regular. More precisely:

Corollary 16. *Although FOL is regular $INV(FOL)$ is only weakly regular.*

Proof: $EVEN \in INV(FOL)$ but $RELEVEN \notin INV(FOL)$. But RELEVEN is the relativized version of $EVEN$. ∎

We also see that invariant $\mathcal{L}$–definability is, over finite structures, strictly weaker than definability in $\Delta(\mathcal{L})$.

Corollary 17. $FOL \subset INV(FOL) \subset \Delta(FOL)$, *hence there are logics* $\mathcal{L}$ *with* $INV(\mathcal{L}) \subset \Delta(\mathcal{L})$ *(i.e. the inclusion are proper).*

Proof: The inclusions were proven in lemma 7. To see that the inclusion is proper we note that $RELEVEN \in \Delta(FOL)$ by fact 9. On the other hand $RELEVEN \notin INV(FOL)$ by theorem 15. ∎

4.3 Implicit definability

Now we want to show that invariant definability is incomparable with implicit definability. Given a logic $\mathcal{L}$ the logic $IMP(\mathcal{L}) \subseteq \Delta(\mathcal{L})$ is defined in [EF95] as follows:

Definition 9 (The logic $IMP(\mathcal{L})$). An $IMP(\mathcal{L})(\tau)$-formula $\phi(\bar{x})$ with free variables among $\bar{x}$ is a tuple

$$(\psi_1(R_1), \ldots, \psi_m(R_m), \psi(\bar{x}, \bar{R}))$$

where each ψ_i is an $\mathcal{L}(\tau \cup \{R_i\})$-sentence and ψ is an $\mathcal{L}(\tau \cup \{\bar{R}\})$-formula with free variables among $\bar{x}$. Furthermore, we require that every finite τ-structure $\mathcal{A}$ has exactly one expansion to an $(\tau \cup \{\bar{R}\})$-structure $\bar{\mathcal{A}}$ satisfying the conjunction $\psi_1(R_1) \wedge \ldots \wedge \psi_m(R_m)$. Now the meaning of ϕ is given by $\mathcal{A} \models \phi(\bar{a})$ iff $\bar{\mathcal{A}} \models \psi(\bar{a}, \bar{R})$ where the R_i in $\mathcal{A}$ are interpreted to satisfy $\psi_i(R_i)$.

Definition 10. A τ–structure $\mathcal{A}$ is *trivial* if every permutation of the universe of $\mathcal{A}$ is a τ–automorphism. A relation R_A on A is trivial if the structure $< A, R^A >$ is trivial.

The following is well known and can be proved using pebble games:

Lemma 18. *Assume the vocabulary* τ *is empty and* $\mathcal{L}$ *is a sublogic of finite variable infinitary logic* $\mathcal{L}^\omega_{\omega_1, \omega}$. *Then the implicit definitions just define trivial relations. Furthermore, every implictly FOL-definable relation is already FOL-definable.*

The following theorem was proved by Kolaitis [Kol90], cf. also [EF95][corollary 7.5.9].

Theorem 19 (Kolaitis). $LFP \subseteq IMP(FOL)$ *over finite structures, i.e. every LFP-definable class of finite structures* K *is also* $IMP(FOL)$-*definable.*

Remark. Theorem 19 is false if we allow infinite structures. To see this we note that LFP on infinite structures properly extends FOL and is not compact, whereas $IMP(FOL) = FOL$.

Proposition 20. *INV(FOL) and IMP(FOL) are incomparable.*

Proof: By proposition 10, $EVEN \in INV(FOL)$. But by lemma 18, $EVEN \notin IMP(FOL)$, as $EVEN$ is not FOL-definable by fact 9.

On the other hand, S-T-CONN $\in IMP(FOL)$, by theorem 19 and the fact that S-T-CONN is LFP-definable. But by theorem 15, S-T-CONN $\notin INV(FOL)$. ∎

4.4 Invariant Definability over Infinite Structures

The notion of invariant definability is not interesting for traditional First Order Logic (FOL) over arbitrary τ-structures, as the following shows:

Proposition 21. *If a class of σ-structures K is strictly K_0-invariantly FOL-definable then K is already FOL-definable. Moreover, If $\mathcal{L}$ is a Δ-closed logic, i.e. if $\Delta(\mathcal{L}) = \mathcal{L}$, then every class $K \in INV(\mathcal{L})$ is already $\mathcal{L}$-definable.*

Proof: By lemma 6, both K and its complement are projective classes in FOL, and therefore, by Craig's Interpolation Theorem for FOL, K is FOL-definable. ∎

Example 3. Let $FOL(Q_0)$ be FOL augmented with the quantifier Q_0, which says 'there exist infinitely many x'. $FOL(Q_0)$ is not Δ-closed, [BF85]. Furthermore, the class of finite structures over the empty vocabulary is definable in $FOL(Q_0)$ and over finite structures the quantifier Q_0 can be eliminated. We conclude that the class $EVEN$ of finite structures of even cardinality is not $FOL(Q_0)$-definable.

However, $EVEN$ is $ALTSUCC$-invariantly $FOL(Q_0)$-definable.

Question 5. In [MSS76, Vä85] it is shown that $\Delta(FOL(Q_0))$ has a nice description in terms of infinitary logic. In fact $\Delta(FOL(Q_0)) = L^{CK}_{\omega_1,\omega}$ i.e. First Order Logic augmented with countable conjuctions and disjunction which are in the least admissible set containing ω.

Is there a similarly appealing description of $INV(FOL(Q_0))$?

5 Capturing Complexity Classes with Invariant Definability

In this section we finally study the relationship of invariant definability to complexity classes. Fagin and Christen proved, [Fag74, Chr76] that **NP** coincides with the FOL-projective classes. Extensions to other complexity classes are proven using the explicit use of order. As already discussed in section 1, looking at the proofs, cf. [EF95], one sees that the theorems can be also proven for $INV_{ORD}(\mathcal{L})$.

Theorem 22 (Immerman, Sazonov, Vardi). *The following capturing results hold:*

$$INV_{ORD}(FOL[DTC]) = \mathbf{L}, INV_{ORD}(FOL[TC]) = \mathbf{NL},$$
$$INV_{ORD}(LFP) = \mathbf{P}, INV_{ORD}(PFP) = \mathbf{PSpace}.$$

Our main question concerns the role of ORD in this theorem. Clearly, ORD can be replaced by the finite successor relation $SUCC$, but in the presence of deterministic transitive closure ORD and $SUCC$ are intertranslatable. From our previous analysis we have

Proposition 23. *Assuming that* $\mathbf{P} \neq \mathbf{NP} \cap \mathbf{CoNP}$ *we have*

$$INV(FOL[DTC]) \nsubseteq \mathbf{L}, INV(FOL[TC]) \nsubseteq \mathbf{NL}, INV(LFP) \nsubseteq \mathbf{P}.$$

This is in contrast to

$$INV(PFP) = \mathbf{PSpace}.$$

Proof: Use proposition 10 and the Fagin-Christen characterization of **NP** for the first part. For the last statement we observe that $\Delta(PFP) \subseteq \mathbf{PSpace}$ because deterministic and non-deterministic polynomial space coincide by Savitch's theorem, cf. [GJ79]. For the other inclusion we use that $\mathbf{PSpace} = INV_{ORD}(PFP)$. ∎

This suggests the general question, for what K does theorem 22 remain true, or more generally, for what K does $\mathbf{P} \subseteq INV_K(LFP)$? From proposition 5 we already know

Corollary 24. *If* ORD *is parametrically definable in* K *then*

$$\mathbf{L} \subseteq INV_K(FOL[DTC]), \mathbf{NL} \subseteq INV_K(FOL[TC]),$$
$$\mathbf{P} \subseteq INV_K(LFP), \mathbf{PSpace} \subseteq INV_K(PFP).$$

Observe that in the corollary we require only that ORD is parametrically definable in K, whereas in theorem 22 $K = ORD$, so that the corollary applies in a more general setting. On the other hand, the conclusion has been weakened to a claim of containment rather than equality. This has been done in the light of proposition 23, as there are FOL-definable K with $INV_K(FOL) \nsubseteq \mathbf{P}$, provided $\mathbf{P} \neq \mathbf{NP} \cap \mathbf{CoNP}$.

It is easy to show that the requirement parametrically definable in the conditions of corollary 24 over a class K can not be weakened to just definable (without parameters). The following two examples witness this fact.

Example 4. Let $\sigma = \{S\}$ consist of a binary relation symbol, and let $K_1 = \{\mathcal{A} : \exists A' \subseteq A$ such that $card(A') = card(A) - 2$ and A' is linearly ordered by $S^A\}$. Minor changes in the proof of the Immerman-Vardi result establish that $INV_{K_1}(LFP) = \mathbf{P}$. On the other hand, since K_1 contains non-rigid models, i.e. models having non-trivial automorphism groups, no logical formula, in any language, can define a linear order on all of K_1.

Example 5. Let $\sigma = \{Exy\}$, and let $K_2 = \{A \mid A \text{ is a chain}\}$. Here, A is a *chain* iff it is a connected simple undirected graph such that every vertex has degree exactly 2. Observe that, up to isomorphism, there is a unique chain of each finite cardinality, and that it is non-rigid with automorphism group of size $= 2 \cdot card(A)$.

Observe that a linear order is FOL–definable with parameters over K_1 and LFP–definable with parameters over K_2. We conjecture that this definability property provides both a necessary and sufficient condition for $INV_K(LFP)$ to contain P, (the sufficiency having just been established). More precisely, as stated in the introduction

Conjecture: For any class K, $\mathbf{P} \subseteq INV_K(LFP)$ iff a linear order is parametrically definable on K.

This is the major conjecture of this note, though one can also ask many similar questions concerning invariant definability over other languages, especially other fixed-point logics.

6 Partial Cases of the Conjecture

Recall definition 10 of trivial structures from section 4.3. E. Rosen observed the following:

Theorem 25 (E. Rosen). *If K_0 is f-categorical and closed under substructures then either*

(i) ORD is parametrically FOL-definable in K_0 or every $A \in K_0$ is trivial
(ii) If every $A \in K_0$ is trivial, K_0 is definable by a first order sentence and every global relation on K_0 is explicitly definable with equality only.

Sketch of proof: The proof uses standard model theoretic tools from [CK90]. First, one uses Ramsey's theorem to show that under the two hypotheses all finite structures in K_0 admit an indiscernible ordering. If not every permutation is an automorphism we use the basic diagram of the structure $A \in K_0$ of size n, where n is the arity of a relation S^A (from σ) that is not trivial on A. This gives us the required formula defining a linear order with $n - 2$ parameters. Otherwise we use the basic equality diagram of a small structure in K_0 to define the relations. This definition is good for all structures in K_0 due to the substructure property. This establishes (ii). ∎

The two requirements on K_0 arose naturally in our discussion and have a broader significance. Closure under substructures guarantees that $INV_{K_0}(\mathcal{L})$ is regular, provided that $\mathcal{L}$ is, cf. our discussion in section 3, theorem 8. On the other hand, f-categoricity assures us that the isomorphism type of members of K_0 depends only on the cardinality. It also fits the intuition behind Ajtai's notion of definability using built-in relations.

To prove the conjecture for this case we observe that if K_0 is FOL–definable with equality only, then $\mathcal{L} = INV_{K_0}(\mathcal{L})$ by proposition 5, and hence $EVEN$ is not K_0-invariantly LFP-definable. Hence we have

Corollary 26. *If K_0 is f-categorical and closed under substructures and $\mathcal{L} \neq INV_{K_0}(\mathcal{L})$ then a linear order is definable with parameters on K_0. In particular, this is the case if $EVEN \in INV_{K_0}(LFP)$, $\mathbf{L} \subseteq INV_{K_0}(LFP)$ or $\mathbf{P} \subseteq INV_{K_0}(LFP)$.*

The general case of the conjecture seems difficult, as K can be rather complicated and the assumption that there is no parametrically definable order is hard to handle. The latter reminds us of Shelah's characterization of stability, cf. [She90], although the situation is not quite the same: Our models are finite and the order, if present, has to be total. On the other side, we do not deal with complete theories.

E. Rosen has announced that the conjecture is true if K consists of a class of pre-orders, i.e linear orders of equivalence classes. More precisely:

Theorem 27 (E. Rosen). *If K is a class of pre-orders such that every $K_1 \in \mathbf{P}$ is K-invariantly LFP-definable then ORD is parametrically LFP-definable (actually even FOL-definable) in K.*

The proof will be published in [RM97].

References

[AF90] M. Ajtai and R. Fagin. Reachability is harder for directed than for undirected finite graphs. *Journal of Symbolic Logic*, 55.1:113–150, 1990.

[Ajt83] M. Ajtai. Σ_1^1–formulae on finite structures. *Annals of Pure and Applied Logic*, 24:1–48, 1983.

[Ajt89] M. Ajtai. First-order definability on finite structures. *Annals of Pure and Applied Logic*, 45:211–225, 1989.

[BF85] J. Barwise and S. Feferman, editors. *Model-Theoretic Logics*. Perspectives in Mathematical Logic. Springer Verlag, 1985.

[BI94] D.A.M. Barrington and N. Immerman. Time, hardware and uniformity. In *Proccedings of the 9th Structure in Complexity Theory*, pages 176–185. IEEE Computer Society, 1994.

[BS90] R.B. Boppana and M. Sipser. The complexity of finite functions. In J. van Leeuwen, editor, *Handbook of Theoretical Computer Science*, volume 1, chapter 14. Elsevier Science Publishers, 1990.

[Chr76] C.A. Christen. Spektralproblem und Komplexitätstherie. In E. Specker and V. Strassen, editors, *Komplexität von Entscheidungsproblemen*, volume 43 of *Lecture Notes in Computer Science*, pages 102–126. Springer, 1976.

[CK90] C.C. Chang and H.J. Keisler. *Model Theory*. Studies in Logic, vol 73. North–Holland, 3rd edition, 1990.

[Ebb85] H.D. Ebbinghaus. Extended logics: The general framework. In *Model-Theoretic Logics*, Perspectives in Mathematical Logic, chapter 2. Springer Verlag, 1985.

[EF95] H.D. Ebbinghaus and J. Flum. *Finite Model Theory*. Perspectives in Mathematical Logic. Springer, 1995.

[EFT94] H.D. Ebbinghaus, J. Flum, and W. Thomas. *Mathematical Logic, 2nd edition*. Undergraduate Texts in Mathematics. Springer-Verlag, 1994.

[Fag74] R. Fagin. Generalized first-order spectra and polynomial time recognizable sets. In R. Karp, editor, *Complexity of Computation*, volume 7 of *American Mathematical Society Proc*, pages 27–41. Society for Industrial and Applied Mathematics, 1974.

[FSV93] R. Fagin, L. Stockmeyer, and M. Vardi. On monadic NP vs. monadic co-NP. In *STOC'93*, pages 19–30. ACM, 1993.

[GJ79] M.G. Garey and D.S. Johnson. *Computers and Intractability*. Mathematical Series. W.H. Freeman and Company, 1979.

[Gur88] Y. Gurevich. Logic and the challenge of computer science. In E. Börger, editor, *Trends in Theoretical Computer Science*, Principles of Computer Science Series, chapter 1. Computer Science Press, 1988.

[Imm87] N. Immerman. Languages that capture complexity classes. *SIAM Journal on Computing*, 16(4):760–778, Aug 1987.

[Joh90] D.S. Johnson. A catalog of complexity classes. In J. van Leeuwen, editor, *Handbook of Theoretical Computer Science*, volume 1, chapter 2. Elsevier Science Publishers, 1990.

[Kol90] P.G. Kolaitis. Implicit definability on finite structures and unambiguous computations. In *LiCS'90*, pages 168–180. IEEE, 1990.

[Mak85] J.A. Makowsky. Compactness, embeddings and definability. In *Model-Theoretic Logics*, Perspectives in Mathematical Logic, chapter 18. Springer Verlag, 1985.

[MP94] J.A. Makowsky and Y.B. Pnueli. Oracles and quantifiers. In *CSL'93*, volume 832 of *Lecture Notes in Computer Science*, pages 189–222. Springer, 1994.

[MP95] J.A. Makowsky and Y.B. Pnueli. Computable quantifiers and logics over finite structures. In M. Krynicki, M. Mostowski, and L.W. Szczerba, editors, *Quantifiers: Generalizations, extensions and and variants of elementary logic*, volume I, pages 313–357. Kluwer Academic Publishers, 1995.

[MSS76] J.A. Makowsky, S. Shelah, and J. Stavi. Δ–logics and generalized quantifiers. *Annals of Mathematical Logic*, 10:155–192, 1976.

[RM97] E. Rosen and J.A. Makowsky. Invariant definability. In preparation, 1997.

[Saz80] V. Sazonov. Polynomial computability and recursivity in finite domains. *Elektronische Informationsverarbeitung und Kybernetik*, 16:319–323, 1980.

[Sch96] T. Schwentick. On winning Ehrenfeucht games and monadic NP. *Annals of Pure and Applied Logic*, 79:61–92, 1996.

[She90] S. Shelah. *Classification Theory and the number of non–isomorphic models*. Studies in Logic and the Foundations of Mathematics. North Holland, 2nd edition, 1990.

[Vä85] J. Väänanen. Set theoretic definability of logics. In *Model-Theoretic Logics*, Perspectives in Mathematical Logic, chapter 18. Springer Verlag, 1985.

[Var82] M. Vardi. The complexity of relational query languages. In *STOC'82*, pages 137–146. ACM, 1982.

Comparing Computational Representations of Herbrand Models

Robert Matzinger*

Technische Universität Wien

Abstract. Finding computationally valuable representations of models of predicate logic formulas is an important issue in the field of automated theorem proving, e.g. for automated model building or semantic resolution. In this article we treat the problem of representing single models independently of building them and discuss the power of different mechanisms for this purpose. We start with investigating context-free languages for representing single Herbrand models. We show their computational feasibility and prove their expressive power to be exactly the finite models. We show an equivalence with "ground atoms and ground equations" concluding equal expressive power. Finally we indicate how various other well known techniques could be used for representing essentially infinite models (i.e. models of not finitely controllable formulas), thus motivating our interest in relating model properties with syntactical properties of corresponding Herbrand models and in investigating connections between formal language theory, term schematizations and automated model building.

1 Introduction

Representing single models of predicate logic formulas plays an important role in various subfields of automated theorem proving, e.g.

Automated model building not only tries to decide the satisfiability of some PL formula, but also attempts to generate a model in case satisfiability is found; see e.g. [Tam91,CZ92,Sla92,FL93,BCP94,CP95,FL96b]. Because a model of a formula $\neg A$ is in fact nothing else than a counterexample for A[1], knowing this model may be of enormous help for a human engineer who tries to find the reason for A not being a tautology (imagine A to be e.g. a faulty system description).

Model based resolution refinements restrict the number of clauses generated by a resolution theorem prover by evaluating the clauses in a given model and resolving true against false clauses only (see [Lov78,CL73,Sla93]).

---

* Institut für Informationssysteme, Abteilung für Wissensbasierte Systeme; Treitlstr. 3/E184-3, A-1040 Wien/Austria/Europe; email: matzi@kr.tuwien.ac.at; URL: http://www.kr.tuwien.ac.at/~matzi

[†] This work was partially supported by the Austrian Science Foundation under FWF grant P11624–MAT.

[1] As usual in the resolution literature we refute the negation of a given formula A.

For this purpose the availability of computationally feasible model representations is an important issue.

However, as soon as we do not restrict ourselves to finite models, we have to ask what "representing a model" actually should mean. But even if we cope with finite models a concise symbolic representation may be a lot easier to deal with than the explicit representation of a finite model with, say, some thousand domain elements. Still techniques for representing potentially infinite models symbolically are rarely explored. That's why we considered it worthwhile to investigate model representation techniques independently from actually building the models, thus accompanying the effort that is going on in the automated model building field. In this paper we are going to survey some model representation techniques known from the literature and focus particularly on the

Characterization Question, i.e. to characterize the models we can represent or the formulas we can satisfy with a representable model (which leads to investigating model properties in terms of syntactical properties of the set of its true ground atoms), and the

Comparison Question, i.e. to compare different representation mechanisms with respect to their expressive power and their computational feasibility.

From a theoretical point of view it makes sense to restrict ourselves to Herbrand models, but this is also justified from a practical viewpoint to utilize the intuitive requirement of

Understandability. The model representation should be intuitively readable by human engineers,

because in Herbrand models both the domain and the interpretation of the function symbols are clear, fixed and intuitive. Still we know we can find a Herbrand model for any satisfiable formula.

To describe a certain Herbrand model over a fixed signature we just have to find a method for specifying potentially infinite sets of ground atoms, which we can arbitrarily read as terms or as strings themselves. Clearly there is a lot of knowledge around about representing sets of terms or strings, so it's a basic idea of this article to take some key representation mechanisms and discuss their ability to serve as a model representation mechanism:

- *Formal language theory* offers a great tradition of representing infinite sets of strings. Context-free grammars (see e.g. [Sal73]) or regular tree automata (see [GS84]) in particular will serve as a simple and understandable device for representing models. They appear to be very usual in computer science (e.g. for specifying the syntax of programming languages) so we expect them to be intuitively readable for many engineers in the field.
- There is considerable experience in representing infinite sets of terms in the field called *term schematizations* or *recurrent schematizations* (see e.g. [CHK90,Sal92,Com95,GH97]). But as these ideas focus on different objectives than we do (like e.g. unification algorithms for term representations which we do not need here), adaptations will be necessary to utilize model representations.

- *Constrained clauses* introduced in [CZ92,BCP94,CP95] (see also [CL89]) evolved directly from the automated model building field. However their actual power for representing models is rarely discussed.
- Finally there are a lot of simple representation mechanisms like *atom representations, linear atom representations* (see [FL93,FL96b]), *ground atoms plus ground equations* (see [FL96a]), that evolved directly from the construction of model building algorithms, so knowing the expressive power of these mechanisms is of special interest.

Clearly there are a lot more important mechanisms around, which we are going to miss unfortunately. However to call a mechanism a computational model representation we want to require (as proposed in [FL96b]) with descending importance the availability of:

Ground Atom Test. For a given model representation it should be (computationally) easy to decide whether a given ground atom is true or false in the represented model.

Clause Test. There should be an algorithm to decide whether a given clause is true or false in the represented model.

Equivalence Test. It should be possible to decide whether two given representations represent the same model or not.

2 Preliminaries

We assume the reader to be familiar with the terms first order logic, clause form, model, Herbrand model, resolution, etc. (see e.g. [CL73]) as well as with formal languages, grammars (context-free, regular, ...) and so on (see e.g. [Sal73]). Basic Knowledge about regular tree automata (see [GS84]), equational systems, rewrite systems and the Knuth-Bendix algorithm (see [KB70]) is also helpful. Unless stated different (all the symbols may appear with arbitrary indices) $u, v, w, x, y, z, \ldots$ denote variables in formulas, $f, g, h, \ldots$ denote function symbols (of arbitrary arity > 0), $a, b, c, \ldots$ denote constant symbols, $r, s, t, \ldots$ denote (ground) terms, $A, B, C, \ldots, M, N, O, P, Q, \ldots$ denote clauses, predicate symbols, or nonterminal symbols, $\mathcal{A}, \mathcal{B}, \mathcal{C}, \ldots$ denote sets of clauses, formulas, etc. $\tau(t)$ shall denote the term depth of a term t.

Throughout this article we talk about first order logic with arbitrary function symbols but we do not deal with equality (except explicitly mentioned). Without loss of generality[2] we can assume all formulas to be in negation normal form[3]. We call any such formula to be in *skolemized form* iff it does not contain existential quantifiers. By Herbrand's theorem we know that there is a sat-equivalent[4] skolemized formula for any predicate logic formula (in negation normal form). Note that for model building we cannot be satisfied with pure sat-equivalence

---

[2] We will use the abbreviation *w.l.o.g.* in the sequel.

[3] Formulas with quantifiers $\forall$ and $\exists$, connectives $\lor$ and $\land$, where $\neg$ is only allowed in front of atoms.

[4] Two formulas are called *sat-equivalent* if one is satisfiable iff the other is satisfiable.

among the formulas. Additionally we have to require a translation of the models, i.e. we need a simple method to obtain a model of the original formula from a model of its sat-equivalent pendant. For skolemization this is trivial as by the standard method of introducing function symbols for existential quantifiers a Herbrand model of the skolemized formula is always a model of the original formula itself (simply by interpreting the existential quantifiers according to the skolem functions we introduced). Therefore it makes sense to restrict ourselves to building and representing Herbrand models of skolemized formulas or – equivalently – of clause sets.

For a skolemized formula $\mathcal{F}$ we call $\Sigma_{\mathcal{F}} = (\mathcal{P}_{\mathcal{F}}, F_{\mathcal{F}}, C_{\mathcal{F}})$ the *signature* of $\mathcal{F}$ where $\mathcal{P}_{\mathcal{F}}$ is the set of predicate symbols, $F_{\mathcal{F}}$ is the set of function symbols (with arity > 0) and $C_{\mathcal{F}}$ is the set of constant symbols occurring in $\mathcal{F}$. W.l.o.g. we always assume $C_{\mathcal{F}} \neq \emptyset$. $HD_{\mathcal{F}}$ denotes the set of terms that can be built from constant symbols $C_{\mathcal{F}}$ and function symbols $F_{\mathcal{F}}$, i.e. the Herbrand domain of $\mathcal{F}$. Analogously we will use HD_{Σ} to denote the set of terms built from a signature Σ. Throughout this article we consider an arbitrary, but fixed signature $\Sigma = (\mathcal{P}, F, C)$ and abbreviate $\mathcal{A}_{\Sigma} := F \cup C \cup \{ '(' , ')' , ',' \}$.

Since we want to describe Herbrand models by describing the ground terms for which a certain predicate is true, it is often convenient to restrict ourselves to one-place (monadic) predicate letters. As we allow full functional structure in terms, we do not lose generality because we can translate every skolemized formula to a sat-equivalent one that contains monadic predicates only. This can be done in a way that allows easy translation of the models in the required direction (but not in the other one, this would require a slightly more complicated translation): Assume for a skolemized formula $\mathcal{F}$ we obtain a formula $\mathcal{F}'$ by replacing every atom $P(t_1, \ldots, t_n)$ in $\mathcal{F}$ by an atom $T(p(t_1, \ldots, t_n))$ (p being a new n-place function symbol for every predicate P)

Lemma 1. $\mathcal{F}'$ and $\mathcal{F}$ are sat-equivalent formulas. Moreover any model of $\mathcal{F}'$ can trivially be translated to a model of $\mathcal{F}$.

Proof. Since we work on skolemized formulas (i.e. there are no existential quantifiers), it is easy to verify that we get a Herbrand model $\mathcal{H}$ of $\mathcal{F}$ if we take an arbitrary Herbrand model $\mathcal{H}'$ of $\mathcal{F}'$, strip every term from its domain that contains one of the new function symbols p and define $P(t_1, \ldots, t_n)$ to be true in $\mathcal{H}$ iff $T(p(t_1, \ldots, t_n))$ is true in $\mathcal{H}'$ (on ground terms $t_1, \ldots, t_n$ that are in both domains). Thus $\mathcal{F}$ has a model whenever $\mathcal{F}'$ has one and the translation of the models is indeed trivial. On the other hand if $\mathcal{F}'$ does not have a model, there must be a resolution refutation of the clause form of $\mathcal{F}'$. But as all the p occur as leading term symbols only and every leading term symbol is one of the p, unification cannot introduce a p somewhere deeper in a term. Thus we can rewrite every $T(p(\ldots))$ to $P(\ldots)$, getting a refutation of $\mathcal{F}$ which proves its unsatisfiability. $\square$

Note that a translation to monadic formulas is nowhere essential. We can always as well represent the (nonmonadic) true atoms themselves. However restricting ourselves to monadic predicate symbols is often technically convenient for developing or representing a method.

It may sound strange to discuss the kind of models we can represent and to restrict ourselves to Herbrand models at the same time. However properties of models are clearly reflected by the syntactical structure of the set of true ground atoms in the model, so we define

Definition 2. A *Herbrand image* of an interpretation Ω of a skolemized formula is a Herbrand interpretation Ω_H such that there exists a (not necessarily surjective) homomorphism $h : \Omega_H \to \Omega$ (with respect to both functions and predicates in the obvious sense). We call Ω and Ω_H *Herbrand-equivalent*, if h is surjective.

Note that although the mapping works from the Herbrand interpretation to the original one, the Herbrand image is still uniquely fixed by the interpretation of the constant symbols. Furthermore the interpretation of ground atoms is always the same in a model and its Herbrand image. Thus Herbrand-equivalent models are truth-equivalent for arbitrary skolemized formulas, so we may call them just *equivalent*. Note that every model of a skolemized formula must contain a submodel that is equivalent to the Herbrand image.

3 Formulating Herbrand Models as Context Free Languages

If we restrict ourselves to monadic skolemized formulas (with arbitrary function symbols), we can fully describe a Herbrand model Ω of a formula $\mathcal{F}$ by assigning a set of ground terms L_P to every predicate symbol P and stating that $P(t)$ should be interpreted as true iff $t \in L_P$. The basic idea of this section is to describe each L_P as a formal language $\subseteq \mathrm{HD}_{\mathcal{F}}$.

Example 3. $\mathcal{C}_{ex} := \{ \{P(x) \vee P(f(x))\}, \{\neg P(x) \vee \neg P(f(x))\} \}$. It is easy to see that $\mathcal{C}_{ex}$ has exactly two Herbrand models: $\{P\left(f^{2i}(a)\right) | i \geq 0\}$ and $\{P\left(f^{2i+1}(a)\right) | i \geq 0\}$ (with the convention $f^0(a) = a$). The first one could be expressed by the grammar $G_1 := (\mathcal{A}_{\Sigma}, \{P, Q\}, \Pi_1, P)$ with $\Pi_1 := \{ P \longrightarrow f(Q), P \longrightarrow a, Q \longrightarrow f(P) \}$

When dealing with grammars we will often omit the starting symbol, writing it as an index of the generated language instead, i.e. L_P denotes the language produced via the starting symbol P.

Definition 4. A context-free grammar $G = (\mathcal{A}_{\Sigma}, \mathcal{P} \cup \mathcal{N}, \Pi)$ ($\mathcal{N}$ being a finite set of additional nonterminal symbols disjoint with $\mathcal{P}$, F and C of our signature $\Sigma = (\mathcal{P}, F, C)$) is said to be a *term grammar*[5] iff for each production $(N \longrightarrow t) \in \Pi$, t is a term over the extended signature $\Sigma' = (\mathcal{P}, F, C \cup \mathcal{P} \cup \mathcal{N})$. It is called *flat* iff each production is either of the form $N \longrightarrow a$ with a being a constant or is of the form $N \longrightarrow f(N_1, \ldots, N_n)$ with f being a function symbol of arity n and $N, N_1, \ldots, N_n \in \mathcal{P} \cup \mathcal{N}$ being nonterminals.

---

[5] If we treat the difference between the representation of terms as trees or as strings inessential, term grammars are clearly equivalent to regular tree automata as introduced in [GS84].

Note that if we stick to our notational conventions, it suffices to specify the production set of a term grammar. In the remainder of this article we can restrict ourselves to flat term grammars as we clearly do not loose expressiveness: we can always rewrite arbitrary term grammars to flat term grammars by introducing new nonterminals.

Clearly if a language L_P is produced by a term grammar G, $P \in \mathcal{P}$, then $L_P \subseteq$ HD_Σ, i.e. the grammar produces Herbrand terms only. However in [Mat97b] we could prove that also the opposite is true, i.e. every context-free subset of HD_Σ can be generated by a term grammar, which justifies the name *context free model* in the following definition:

Definition 5. The Herbrand interpretation (or Herbrand model) Ω_G *defined via the term grammar* G is defined by: $P(t)$ is interpreted as true iff $t \in L_P$ where L_P is generated by the grammar G (P being a predicate symbol as well as a nonterminal symbol in the term grammar). Any Herbrand model of a (skolemized) monadic formula (with arbitrary function symbols) which can be defined via a term grammar is called a *context-free model*.

Representing Herbrand models by term grammars well fulfills our requirement of **understandability** as we already mentioned. We also get an efficient **atom test** as evaluating $P(t)$ (for a ground term t) is nothing else than a syntax check for t with respect to the term grammar (a context-free grammar), which can be done in polynomial time. Based on [GS84] it can even be shown that:

Theorem 6. *The ground atom test can be solved in linear time (w.r.t the length of the atom).*

Before tackling clause test and equivalence test we asked ourselves what kind of models we are able to express as context free models. We are especially interested in the question whether (essentially) infinite models can be expressed or not. Although the answer is no, which we show by transforming an arbitrary context-free model to a finite one, we find it interesting that also the opposite is possible, i.e. to transform any finite model to a context-free one (in some sense). For the first direction we are going to produce a grammar that has the property that every ground term $t \in HD_\Sigma$ is generated by exactly one nonterminal symbol, so we can read the nonterminal symbols as elements of a finite domain of a model which indeed is equivalent to the original Herbrand model. Thus we proceed by showing that we can transform any term grammar G to an (in a certain sense) equivalent grammar G^0_{unique} that satisfies the above property.

Definition 7. For a set of nonterminals $\mathcal{N} = \{N_1, \ldots, N_n\}$, $\mathcal{N}^{\neg \wedge}$ denotes the equivalence classes of the terms that can be built from $\mathcal{N}$ and $\neg \mathcal{N} = \{ \neg N_1, \ldots, \neg N_n \}$ with a binary function symbol $\wedge$ with respect to idempotency, commutativity and associativity of $\wedge$. An element of $\mathcal{N}^{\neg \wedge}$ is called *complete*, if it has the form $(\neg) N_1 \wedge \ldots \wedge (\neg) N_n$. The set of complete elements in $\mathcal{N}^{\neg \wedge}$ is denoted as $\mathcal{N}^{\neg \wedge}_{\text{complete}}$.

Clearly $\mathcal{N}^{\neg \wedge}$ is isomorphic to the powerset of $\mathcal{N} \cup \neg \mathcal{N}$, thus it's a finite set. We naturally associate a language with every nonterminal in $\mathcal{N}^{\neg \wedge}$:

Definition 8. Let $\mathcal{N}$ be a set of nonterminals of a grammar. The *language corresponding to*
- $N \in \mathcal{N}$ is L_N, i.e. the language produced by the grammar from the nonterminal N,
- $\neg N$ is $\mathrm{HD}_\Sigma \setminus L_N$ for $N \in \mathcal{N}$,
- $N_1 \wedge N_2$ is $L_{N_1} \cap L_{N_2}$ for $N_1, N_2 \in \mathcal{N}^{\neg\wedge}$.

We just mentioned that if we treat the difference of representing terms as trees or as strings inessential, term grammars and regular tree automata (see [GS84]) are actually just a different notation of the same thing. Regular tree languages are well known to be closed under intersection and under inversion[6]. Finally for any regular tree automaton there exists an equivalent deterministic one[7] (its state-space is the powerset of the states of the indeterministic automaton). By applying these lemmas and translating the ideas to our terminology it is possible to prove:

Lemma 9. *For a term grammar* $G = (\mathcal{A}_\Sigma, \mathcal{N}, \Pi)$ *with* $\mathcal{N} = \{N_1, \ldots, N_n\}$ *there exists a grammar* G_{unique} *such that:*

- *The nonterminal set of* G_{unique} *is* $\mathcal{N}^{\neg\wedge}_{\mathrm{complete}}$.
- *Every nonterminal in* G_{unique} *produces its corresponding language.*

In [Mat97a] we gave a constructive proof for this lemma and showed that G_{unique} can indeed be effectively computed. Let's finally obtain G^0_{unique} by stripping all nonterminals and all productions from G_{unique} that do not produce any ground terms at all (note that this can be done in time linear with the number of nonterminals in G_{unique}, so we don't add significant complexity here). A number of interesting facts can be concluded for G^0_{unique}:

Lemma 10.
- *All languages generated by* G^0_{unique} *are pairwise disjoint and their union is* HD_Σ, *i.e. all ground terms.*
- *Every ground term of* HD_Σ *can be derived from exactly one nonterminal.*
- *If* t_i *are ground terms derivable from nonterminals* $\overline{N_i}$ $(1 \leq i \leq n)$ *in* G^0_{unique}, *then in* G^0_{unique} *there must be exactly one production* $\overline{X} \longrightarrow f(\overline{N_1}, \ldots, \overline{N_n})$ *for every function symbol* f.

Definition 11. Now let Γ_G be the Herbrand interpretation of a skolemized monadic formula defined via the grammar G. We define a finite interpretation $\Gamma_G^{\mathrm{finite}}$: For the domain we take the nonterminal symbols of G^0_{unique}. Every ground term is interpreted as the unique nonterminal it is derivable from. Every ground atom $P(t)$ is interpreted as true iff t can be derived from a nonterminal in G^0_{unique} that contains P positively (i.e. not as $\neg P$).

Lemma 12. Γ_G *and* $\Gamma_G^{\mathrm{finite}}$ *are equivalent interpretations,*

---

[6] with respect to terms, i.e. the inversion of L is $\mathrm{HD}_\Sigma \setminus L$.

[7] In our notation this is the property of knowing there is at most one production $\overline{P} \longrightarrow f(\overline{N_1}, \ldots, \overline{N_k})$ for any choice of nonterminals $\overline{N_i}$.

because Γ_G^{finite} is a homomorphic image of Γ_G according to the definition. Thus we have succeeded to construct a finite model for every context free model of a monadic skolemized formula (with arbitrary function symbols).

On the other hand any (finite) model must have a submodel that is equivalent to some Herbrand model (i.e. that is the image of a homomorphic mapping of a Herbrand model). We can read this finite submodel as a grammar in the following way: The elements of the domain are the nonterminals of the grammar and if ground terms t_i are interpreted to the domain elements N_i ($1 \leq i \leq n$) and $f(t_1, \ldots, t_n)$ is interpreted to N, then we put the production $N \longrightarrow f(N_1, \ldots, N_n)$ into the grammar. Clearly all ground terms must be derivable from exactly one nonterminal in this grammar. According to whether a predicate $P \in \mathcal{P}$ is true or false on a certain element of the finite domain, we can denote the nonterminals as complete nonterminals $\in (\mathcal{P} \cup \mathcal{Q})_{\text{complete}}^{\neg \wedge}$ (for some appropriate set $\mathcal{Q}$). Thus we have constructed a grammar with the same properties as G_{unique}^0. We can conclude the next theorem:

Theorem 13. *For any context-free model of a skolemized monadic formula (with arbitrary function symbols) there is an equivalent finite model and every finite model contains a submodel that is equivalent to a context-free model.*

Corollary 14. *For any finite model of a skolemized formula (with arbitrary function symbols) the set of true ground atoms for a particular predicate is a context-free language.*

Corollary 15. *A (skolemized monadic) formula has a finite model iff it has a context-free one.*

Note that the translation between Herbrand models of monadic skolemized formulas and models of arbitrary formulas we already discussed in Section 2 can easily be taken over to (finite) homomorphic images. Thus the restriction to skolemized monadic formulas in Corollary 15 is inessential and we can say that we can indeed represent models of exactly those formulas having a finite model.

Example 16. For the grammar G_1 in Example 3 we obtain
$G_{1\,\text{unique}}^0 = (\ A_{\Sigma}, \{P \wedge \neg Q, \neg P \wedge Q\}, \{P \wedge \neg Q \longrightarrow a, P \wedge \neg Q \longrightarrow f(\neg P \wedge Q), \neg P \wedge Q \longrightarrow f(P \wedge \neg Q)\}\ )$. As Γ_{G_1} is a Herbrand model of $\mathcal{C}_{ex}$ of Example 3 we obtain the finite model $\Gamma_{G_1}^{\text{finite}}$ of $\mathcal{C}_{ex}$: The domain is $\{P \wedge \neg Q, \neg P \wedge Q\}$, applying the function f makes us toggle between the two elements of the domain and on the domain element $P \wedge \neg Q$, P is interpreted as true, Q as false (and vice-versa for $\neg P \wedge Q$).

The fact that every context-free model is (truth-)equivalent to a finite one immediately settles the problem of **clause evaluation** as we can always evaluate over a finite domain. (However it will still be interesting to investigate methods for speeding up clause evaluation.) Knowing that the equivalence problem for regular tree languages is decidable (see [GS84]), we can conclude the decidability of the equivalence problem for context-free models[8].

---

[8] In [Mat97a] we used automated theorem proving technology to construct an efficient decision algorithm for the equivalence problem. We translated each grammar to a

4 Atom Representations

In [FL93,FL96b] a model representation technique called *atom representation* is introduced (along with a model building algorithm for obtaining linear[9] atom representations for formulas in interesting clause classes). Atom representations define a Herbrand model by stating that all ground instances of a finite set of atoms should be true in the model, the rest should be false. E.g. the atom representation $\{P(a, f(x)), Q(g(x, y))\}$ represents the Herbrand model in which $P(a, f(t_1))$ and $Q(g(t_2, t_3))$ are true for any ground terms t_1, t_2, t_3 and all other ground atoms are false. Straightforward algorithms for atom test, clause evaluation and equivalence on these model representations are presented in [FL93] and [FL96b]. Unfortunately there are very simple satisfiable sets of clauses that do not have a model with an atom representation, e.g. Example 3.

Obviously term grammars are a strict extension of linear atom representations: For every atom $P(t)$ in the atom representation consider a production $P \longrightarrow t$ and replace every variable in t by a nonterminal T that is ensured to derive just any term by appropriate productions. With atom representations in general, context-free models appear to be incomparable as the model represented by the atom representation $\{P(f(x, x))\}$ could not be achieved with term grammars; on the other hand the model generated by the grammar $\{P \longrightarrow f(f(P)), P \longrightarrow a\}$ has no atom representation.

An interesting variant of atom representations is presented in [FL96a] as a result of a model building procedure dealing with equality too: *Ground atoms plus ground equations*. They define a Herbrand model by stating that those ground atoms that can be derived from the given ones with help of the equations shall be considered true, the others false.

Example 17. $(\{P(a)\}, \{a = f(f(a))\})$ defines a model of Example 3, i.e. $\{P\left(f^{2i}(a)\right) | i \geq 0\}$

Clearly the ground atom test can be solved by completing the set of equations to a terminating and confluent set of rewrite rules with the Knuth-Bendix algorithm (which is always possible for ground equations), but evaluating clauses and deciding equivalence is also shown to be possible. So we asked ourselves about the expressive power of "ground atoms with ground equations". We found that it is equal to the expressive power of the context free models (and is thus exactly the finite models). We will spend the remainder of this section proving this fact by presenting translations between term grammars and "ground atoms with ground equations".

Let's call any Herbrand model that can be represented by "ground atoms with ground equations" a *GAE-model*. We need some notational conventions: Recall

---

clause set that has exactly the one Herbrand model defined via the grammar. Thus the equivalence test can be solved by a satisfiability test of the union of two such clause sets which we solved by applying an appropriate resolution refinement that guarantees quick termination.

[9] An expression is called linear iff every variable occurs at most once within the expression. A clause is called linear iff all of its literals are linear.

that we use $P \longrightarrow t$ to denote a production in a grammar. We will use $s \vdash^1 t$ if the term t can be derived from s with the help of the productions in the grammar (s, t may contain nonterminals) in one step and $\vdash$ for its reflexive and transitive closure. While $\equiv$ denotes syntactical equality, $t \doteq s$ denotes an equation in the equational system and $t = s$ denotes equality with respect to the equations. For rewrite systems, $t \rightarrow s$ indicates a particular rewrite rule, while $t \mapsto^1 s$ denotes that s can be derived from t with exactly one application of a rewrite rule; the symbol $\mapsto$ is used for the reflexive and transitive closure of $\mapsto^1$. t^0 denotes an (the) irreducible form of t.

We will use the Knuth-Bendix algorithm (see [KB70]) to normalize an equational system to a complete[10] set of rewrite rules. Recall that this is always possible for sets of ground equations. Moreover by using an appropriate ordering within the Knuth-Bendix algorithm, this is always possible in a way such that every rewrite rule $s \rightarrow t$ is oriented according to the term depth, i.e. $\tau(s) \geq \tau(t)$.

Lemma 18. *For any GAE-model there is a term grammar that represents this model.*

Proof. W.l.o.g. we restrict ourselves to monadic predicate symbols. Let $(\mathcal{A}, \mathcal{E})$ be a "ground atoms with ground equations"-representation of a Herbrand model (i.e. $\mathcal{A}$ is a set of ground atoms and $\mathcal{E}$ is a set of ground equations). First of all we use the Knuth-Bendix algorithm to obtain a complete set of rewrite rules $\mathcal{R}$ equivalent to $\mathcal{E}$ (in the sense that two ground terms are equal in $\mathcal{E}$ iff they rewrite to the same unique normal form under $\mathcal{R}$). We normalize[11] all atoms in $\mathcal{A}$ and denote the result as $\mathcal{A}^0$. Clearly the set of ground atoms derivable from $\mathcal{A}$ with help of $\mathcal{E}$ is exactly the same as the set of ground atoms normalizing to some atom in $\mathcal{A}^0$ with respect to the rewrite rules $\mathcal{R}$. Note that after applying the Knuth-Bendix algorithm, any rule in $\mathcal{R}$ is normalized with respect to all the other rules. As we only deal with ground terms this means that every right-hand side of a rewrite rule is normalized under $\mathcal{R}$ and every proper subterm of a left-hand side of a rewrite rule is normalized under $\mathcal{R}$ (this follows easily from the term-depth property of the rewrite rules which we mentioned earlier). Clearly any subterm of a normalized term is normalized. We need the following

Lemma 19. *If t is a ground term and $t \equiv f(t_1, \ldots, t_n)$ then there exists a rewrite derivation $t \mapsto f(t_1^0, \ldots, t_n^0) \mapsto t^0$ (with respect to $\mathcal{R}$).*

Proof. t^0 is the unique normal form of t and $f(t_1^0, \ldots, t_n^0)$ (because the latter are equal with respect to $\mathcal{R}$). We know that $t_i \mapsto t_i^0$ for all i, thus $t \equiv f(t_1, \ldots, t_n) \mapsto f(t_1^0, \ldots, t_n^0)$. Therefore we get $f(t_1^0, \ldots, t_n^0) \mapsto t^0$. □

We construct a grammar G that defines the same model as $(\mathcal{A}, \mathcal{E})$ as follows: $\mathcal{A}_\Sigma$ is the set of terminal symbols (as always). If $\{t_1, \ldots, t_k\}$ is the set of all syntactically different terms and subterms of right-hand sides of rules in $\mathcal{R}$, we define

---

[10] A set of rewrite rules is *complete* iff every derivation of a term leads to the same irreducible normal form.

[11] I.e. we replace every term by its unique irreducible form.

$\mathcal{N} = \{N_{t_1}, \ldots, N_{t_k}\}$. We define $\mathcal{N} \cup \mathcal{P}$ ($\mathcal{P}$ being the set of predicate symbols in our signature) to be the nonterminal set of G. For defining the production set of G we need:

Definition 20. The operator GP, mapping each **normalized** term (i.e. irreducible w.r.t. $\mathcal{R}$) to some nonterminal-containing term is defined as follows. Let t^0 be some normalized term.

- If t^0 is the right-hand side of some rewrite rule $s \dot{\to} t^0 \in \mathcal{R}$, then $\mathrm{GP}(t^0) \equiv N_{t^0}$
- if not and $t^0 \equiv f(t_1^0, \ldots, t_n^0)$, then $\mathrm{GP}(t^0) \equiv f(\mathrm{GP}(t_1^0), \ldots, \mathrm{GP}(t_n^0))$
- else it must hold that $t^0 \equiv a$ for some constant a, then $\mathrm{GP}(t^0) \equiv a$

We now define the production set of G to contain the following rules:

- $N_{f(t_1, \ldots, t_n)} \longrightarrow f(N_{t_1}, \ldots, N_{t_n})$ for any $N_{f(t_1, \ldots, t_n)} \in \mathcal{N}$,
- $N_a \longrightarrow a$ for any $N_a \in \mathcal{N}$ (a being some constant),
- $\mathrm{GP}(t) \longrightarrow f(\mathrm{GP}(t_1), \ldots, \mathrm{GP}(t_n))$ for any rewrite rule $f(t_1, \ldots, t_n) \dot{\to} t \in \mathcal{R}$ (note that $\mathrm{GP}(t)$ must be a nonterminal in this case),
- $\mathrm{GP}(a) \longrightarrow b$ for any rewrite rule $b \dot{\to} a \in \mathcal{R}$ (note that $\mathrm{GP}(a) \equiv N_a$).

Clearly $\mathrm{GP}(t) \vdash t$ for any irreducible term t. Furthermore we get:

Lemma 21. *If $t_1^0, \ldots, t_n^0$ are normalized and $f(t_1^0, \ldots, t_n^0) \mapsto t^0$ (t^0 normalized), then either $f(t_1^0, \ldots, t_n^0) \equiv t^0$ or $\mathrm{GP}(t^0) \vdash f(\mathrm{GP}(t_1^0), \ldots, \mathrm{GP}(t_n^0))$.*

Proof. If $f(t_1^0, \ldots, t_n^0)$ is normalized, clearly $f(t_1^0, \ldots, t_n^0) \equiv t^0$. Otherwise $f(t_1^0, \ldots, t_n^0)$ must be the left-hand side of a rewrite rule $f(t_1^0, \ldots, t_n^0) \dot{\to} t^0 \in \mathcal{R}$, because otherwise one of the t_i^0 must be reducible. Thus the production $N_{t^0} \longrightarrow f(\mathrm{GP}(t_1^0), \ldots, \mathrm{GP}(t_n^0))$ is in the production set of our grammar (because $N_{t^0} \equiv \mathrm{GP}(t^0)$) and the lemma follows. $\square$

By using induction on term depth and Lemmas 19 and 21 it is not hard to prove that if $t \mapsto t^0$, then the derivation $\mathrm{GP}(t_0) \vdash t$ is possible w.r.t. G. On the other hand it is easily verified that every grammar production turns into an equation valid under $\mathcal{E}$, when replacing $\longrightarrow$ by $\dot{=}$ and any nonterminal N_t by t. Thus any derivation of $N_t \vdash t'$ can be turned into an equational proof of $t = t'$. So let's finally add productions $P \longrightarrow \mathrm{GP}(t^0)$ for any atom $P(t^0)$ in $\mathcal{A}^0$ (recall that any grammar can be transformed to an equivalent grammar without chain-productions[12]) which ensures that $P \vdash t$ iff t^0 is the normal form of t and $P(t^0) \in \mathcal{A}^0$. This concludes the proof of Lemma 18. $\square$

Note that there is some structural similarity of this proof to the congruence closure algorithm presented in [NO80], but this algorithm deals with building congruence classes of finite sets of (ground) terms, whereas we are interested in (the number and the structure of) the congruence classes of the whole (infinite) Herbrand domain w.r.t. some ground equations.

Lemma 22. *Every context-free model can be represented by some "ground atoms with ground equations".*

---

[12] I.e. productions of the form $N \longrightarrow M$.

Proof. Let Γ_G be a Herbrand model that is defined via a flat term grammar $G = (\mathcal{A}_\Sigma, \mathcal{N}, \mathcal{P})$. Let's consider G^0_{unique} as defined in Section 3, i.e. the grammar that produces corresponding languages on the nonterminal set $\mathcal{N}^{\neg\wedge}_{\text{complete}}$. For $\overline{N} \in \mathcal{N}^{\neg\wedge}_{\text{complete}}$, let $L_{\overline{N}}$ be the language generated by G^0_{unique} with starting symbol $\overline{N}$. Recall that the Knuth-Bendix algorithm uses an ordering $>_K$ that is total on ground terms (and can be adjusted such that if $t >_K s$, then $\tau(t) \geq \tau(s)$, which we choose to do). Thus if $L_{\overline{N}}$ is nonempty (which can effectively be tested), it contains a unique minimal element w.r.t. $>_K$. As this element is of minimal term depth in $L_{\overline{N}}$ (and there exist only finitely many terms of that depth) this element can be found algorithmically; let's call it $t_{\overline{N}}$. Now we obtain a rewrite system $\mathcal{R}$ from $\mathcal{P}$ by replacing every production $\overline{N} \longrightarrow f(\overline{N_1}, \ldots, \overline{N_n})$ by a rewrite rule $f(t_{\overline{N_1}}, \ldots, t_{\overline{N_n}}) \rightarrow t_{\overline{N}}$ and finally adding $t \rightarrow t_{\overline{N}}$ for any $t \in L_{\overline{N}}$ that has the same term depth as $t_{\overline{N}}$. It can be verified that the resulting rewrite system $\mathcal{R}$ is indeed complete, because turning all rewrite rules into equations and applying the Knuth-Bendix algorithm will just yield $\mathcal{R}$ itself. The fact that $t \mapsto t_{\overline{N}}$ iff $\overline{N} \vdash t$ can now be proven by induction on term depth. Thus turning all rewrite rules in $\mathcal{R}$ into equations and considering for every $\overline{N} \in \mathcal{N}^{\neg\wedge}_{\text{complete}}$ all ground atoms $P(t_{\overline{N}})$ for which P occurs positively in $\overline{N}$, indeed yields a "ground atoms plus ground equations" system defining the same model as the grammar G, which concludes the proof of Lemma 22. $\qquad\square$

By summarizing Lemma 18 and 22 we conclude

Theorem 23. *Every context-free model is a GAE-model and vice-versa.*

Corollary 24. *Theorem 13 and the Corollaries 14 and 15 apply to GAE-models too.*

Thus we have shown that with "ground atom with ground equations" we can once again represent models of exactly those formulas having a finite model.

Example 25. We have already seen that the Herbrand model $\{P(f^{2i}(a)) \mid i \geq 0\}$ can be represented by $(\{P(a)\}, \{a = f(f(a))\})$. The corresponding canonical rewrite system is $f(f(a)) \rightarrow a$ and $P(a)$ is already normalized. a is the only (normalized) term appearing as right-hand side of a rewrite rule, so the nonterminal set of the grammar is $\{N_a, P\}$, $\text{GP}(a) = N_a$ and $\text{GP}(f(a)) = f(N_a)$. Thus the production set of our grammar is $\{N_a \longrightarrow a, N_a \longrightarrow f(f(N_a)), P \longrightarrow N_a\}$ which obviously represents the same model.
On the other hand if we start with the flat term grammar given by the productions $\{P \longrightarrow a, P \longrightarrow f(Q), Q \longrightarrow f(P)\}$ (which is already deterministic and all $f^i(a)$ can be derived from exactly one terminal), we get $t_P \equiv a$ and $t_Q \equiv f(a)$. The resulting rewrite system is $f(f(a)) \rightarrow a$ (i.e. $f(t_Q) \rightarrow t_P$) as the other grammar productions lead to redundant rewrite rules. So the result is $(\{P(a)\}, \{f(f(a)) \doteq a\})$, which obviously represents the same model as the grammar.

5 Some Notes on Overcoming Finiteness

The results of Sections 3 and 4 raise the question of how to overcome finiteness, or which kind of representations may yield the power of representing models of formulas that are not finitely controllable (i.e. that do not have a finite model). We already encountered nonlinear atom representations which do have the power of representing some (essentially) infinite models, but are not yet powerful enough to represent all finite models.

Example 26. A Herbrand model of the clause set $C_1 = \{\{\neg P(a, f(x))\}, \{P(x, x)\}, \{\neg P(f(x), f(y)) \vee P(x, y)\}\}$ (taken from [FL96b]) can be represented by the atom representation $\{P(x, x)\}$ (w.r.t. the signature $\{f, a\}$). It can be verified that C_1 does not have a finite model.

Term schematizations as introduced in [CHK90,Sal92,Com95,GH97]) also offer a considerable power of representing infinite sets of terms (which we want to use for representing Herbrand models). They are mainly designed to represent sets of derivations which is utilized by the existence of an unification algorithm. As matching of ground terms could be seen as unification, this yields an atom test for all of them. Let's very briefly survey R-terms as described in [Sal92]. An R-term is either an ordinary term or an expression $f(t_1, \ldots, t_k)$ where $t_1, \ldots, t_k$ are R-terms or expressions $\Phi(\hat{t}, N, s)$, where $\hat{t}$ is some term that may contain the special symbol $\Diamond$ wherever it normally is allowed to contain a variable; N is a linear expression in the so-called counter variables and s is an R-term itself. Intuitively an R-term $\Phi(\hat{t}, N, s)$ represents the set of terms that can be obtained by substituting arbitrary integers for counter-variables in N, substituting $\hat{t}$ for $\Diamond$ in $\hat{t}$ N times and finally substituting some term represented by s for $\Diamond$.

Example 27. $\Phi(f(\Diamond), N, a)$ represents $\{f^n(a) \mid n \geq 0\}$.
$P(\Phi(f(\Diamond), N, a), \Phi(f(\Diamond), N + M + 1, a))$ represents $\{P(f^n(a), f^{n+m+1}(a))\} \mid n, m \geq 0\}$.

R-terms can indeed be used to represent essentially infinite models, e.g. the clause set $C_2 = \{\{P(x, f(x))\}, \{\neg P(x, x)\}, \{\neg P(x, y) \vee P(x, f(y))\}\}$ does not have a finite model. It's only Herbrand model is $\{P(f^n(a), f^{n+m+1}(a)) \mid n, m \geq 0\}$ that can be expressed by R-terms as we have just seen. Note that this Herbrand model cannot be expressed by an atom representation. However we cannot hope that recurrent schematization mechanisms like R-terms can be used to represent all the finite models, as allowing alternatives in the application of the substitutions which are a natural thing in grammars, spoil decidability of unification. This indicates incomparability with methods representing the finite models. Still if we allow (free) variables in R-terms (stating that they can be ground-substituted arbitrarily), representing models by sets of R-terms is strictly more powerful than atom representations. Note that the ground atom test can still be performed by using the R-term unification algorithm.

Another interesting method to represent models that directly evolved from an effort to design model building algorithms are sets of constrained atoms; see e.g. [CZ92,BCP94,CP95]. A constrained clause is a pair $[C; P]$ where C is an ordinary clause and P is a formula $\exists \overline{w} P'$ such that P' is a quantifier free boolean formula[13] (using $\wedge$, $\vee$ and $\neg$ in the natural way) over equations of terms. If C contains one atom only $[C; P]$ is called a constrained atom. Such P can be solved by transforming to certain normal forms (see [CL89] for an exhaustive survey) which can be used to solve the ground atom test. A ground atom A is said to be represented by $[C; P]$, if $A \equiv C\theta$ and the substitution θ is a solution of the free variables of P over the free term algebra.

Example 28. $[P(x, y); x \neq y]$ is a constrained atom. It represents all ground atoms $P(t_1, t_2)$, such that $t_1 \not\equiv t_2$. It thus represents a model of $\mathcal{C}_3 = \{\{P(a, f(x))\}, \{\neg P(x, x)\}, \{P(f(x), f(y)) \vee \neg P(x, y)\}\}$, which can also be verified not to have a finite model.

Clearly sets of constrained atoms are a strict extension of atom representations. Still they do not cover all the finite models, as it appears to be impossible to represent e.g. $\{P(f^{2i}(a))|i \geq 0\}$. With term schematizations there seems to be incomparability, because obviously term sets like $\{P(f^n(g^n(a))) \mid n \geq 0\}$ or $\{P(f^n(a), f^{n+m+1}(a)) \mid n, m \geq 0\}$ cannot be expressed by constrained atoms whereas alternatives and arbitrary inequations seem not to be representable by R-terms.

6 Summary and Future Work

We investigated the use of context-free grammars (or regular tree automata) for representing Herbrand models. Context-free grammars are a simple and understandable device for representing infinite sets of ground atoms, i.e. Herbrand models. We argued that we do not lose generality when restricting ourselves to monadic predicate symbols (but allowing arbitrary function symbols) and showed sufficiency of building models for those formulas. We found the expressive power of context-free grammars to be exactly the (essentially) finite models. We showed the existence of computational *ground atom test* (which works in linear time), *equivalence test* and *clause evaluation*, which we stated to be the key requirements for model representations. As we could prove the equivalence of context-free grammars to "ground atoms with ground equations" by giving effective translations in both directions, we could also settle the key requirements for this representation mechanism and conclude the same expressive power (i.e. exactly the finite models). We briefly surveyed various candidates for model representation mechanisms that go beyond finite controllability and compared their expressive power. It will be an interesting task for the future to give precise characterizations of the representable models (or of the formulas that are satisfiable by such a model). We hope that some hierarchy of infinite models will evolve from this. Clearly many other interesting representations are still not

---

[13] Actually we could allow universal quantifiers in P', because they can be eliminated (this is the case, because P' is interpreted in the free term algebra over a fixed signature; see [CL89]).

investigated. E.g. it would be interesting to add equations to the grammars or to switch to other language classes like some super classes of regular tree languages which allow certain well–restricted non–linearities (see [BT92,CCD93,Wei96], (certain) context sensitive languages, L-systems, etc. (see e.g. [Sal73,RS92]). Finally we think that on the base of our investigations it should be possible to construct model building algorithms for interesting classes of formulas. However we believe we proved it worthwhile to think of model representations independently of model building and to investigate model properties in terms of syntactical properties of the corresponding Herbrand models

References

[BCP94] C. Bourely, R. Caferra, and N. Peltier. A method for building models automatically. Experiments with an extension of otter. In Alan Bundy, editor, 12th International Conference on Automated Deduction, pages 72–86, Nancy/France, June 1994. Springer Verlag, LNAI 814.

[BT92] B. Bogaert and Sophie Tison. Equality and disequality constraints on direct subterms in tree automata. In A. Finkel, editor, Proc. 9th Annual Symposium on Theoretical Aspects of Computer Science, volume 577 of LNCS, pages 161–172, Cachan, 1992. Springer.

[CCD93] Anne-Cécile Caron, Jean-Luc Coquidé, and Max Dauchet. Encompassment properties and automata with constraints. In C. Kirchner, editor, Proc. 5th RTA, volume 690 of LNCS, pages 328–342. Springer, June 1993.

[CHK90] H. Chen, J. Hsiang, and H.-C. Kong. On finite representations of infinite sequences of terms. In S. Kaplan and M. Okada, editors, Proceedings 2nd International Workshop on Conditional and Typed Rewriting Systems, volume 516 of LNCS, pages 100–114, Montreal, Canada, June 1990. Springer.

[CL73] C.-L. Chang and R. C. T. Lee. Symbolic Logic an Mechanical Theorem Proving. Academic Press, New York, 1973.

[CL89] H. Comon and P. Lescanne. Equational Problems and Disunification. Journal of Symbolic Computation, 7(34):371–426, March/April 1989.

[Com95] H. Comon. On unification of terms with integer exponents. Mathematical Systems Theory, 28(1):67–88, 1995.

[CP95] R. Caferra and N. Peltier. Decision procedures using model building techniques. In Computer Science Logic (9th Int. Workshop CSL'95), pages 131–144, Paderborn, Germany, 1995. Springer Verlag. LNCS 1092.

[CZ92] R. Caferra and N. Zabel. A method for simultanous search for refutations and models by equational constraint solving. Journal of Symbolic Computation, 13(6):613–641, June 1992.

[FL93] C. Fermüller and A. Leitsch. Model building by resolution. In Computer Science Logic (CSL'92), pages 134–148, San Miniato, Italy, 1993. Springer Verlag. LNCS 702.

[FL96a] C. Fermüller and A. Leitsch. Decision procedures and model building in equational clause logic. Technical Report TR-CS-FL-96-1, TU Wien, Vienna/Austria, 1996.

[FL96b] C. Fermüller and A. Leitsch. Hyperresolution and automated model building. J. of Logic and Computation, 6(2):173–203, 1996.

[GH97] R. Galbavý and M. Hermann. Unification of infinite sets of terms schematized by primal grammars. Theoretical Computer Science, 176, April 1997.

[GS84]   F. Gécseg and M. Steinby. *Tree Automata*. Akadémiai Kiadó, Budapest, 1984.

[KB70]   D.E. Knuth and P.B. Bendix. Simple word problems in universal algebras. In J. Leech, editor, *Computational Problems in Abstract Algebra*, pages 263–297. Pergamon Press, Oxford, 1970.

[Lov78]   D. Loveland. *Automated Theorem Proving — A Logical Basis*. North Holland, 1978.

[Mat97a]   R. Matzinger. Computational representations of Herbrand models using grammars. In D.v. Dalen, editor, *Proceedings of the CSL'96*, LNCS. Springer, 1997. To appear.

[Mat97b]   R. Matzinger. Context free term sets are regular - and some applications to logic. Technical Report TR-WB-Mat-97-2, TU Wien, Vienna/Austria, 1997.

[NO80]   G. Nelson and D.C. Oppen. Fast decision algorithms based on congruence closure. *JACM*, 23(2):356–364, 1980.

[RS92]   G. Rozenberg and A. Salomaa, editors. *Lindenmayer Systems*. Springer, 1992.

[Sal73]   A. Salomaa. *Formal languages*. Academic Press, Orlando, 1973.

[Sal92]   G. Salzer. The unification of infinite sets of terms and its applications. In A. Voronkov, editor, *Proceedings Logic Programming and Automated Reasoning, St. Petersburg (Russia)*, volume 624 of *Lecture Notes in Computer Science (in Art. Intelligence)*. Springer, 1992.

[Sla92]   J. Slaney. FINDER (finite domain enumerator): Notes and guide. Technical Report TR-ARP-1/92, Australien National University Automated Reasoning Project, Canberra, 1992.

[Sla93]   J. Slaney. SCOTT: A model-guided theorem prover. In *Proceedings of the 13th international joint conference on artificial intelligence (IJCAI '93)*, volume 1, pages 109–114. Morgan Kaufmann Publishers, 1993.

[Tam91]   T. Tammet. Using resolution for deciding solvable classes and building finite models. In *Baltic Computer Science*, pages 33–64. Springer Verlag, 1991. LNCS 502.

[Wei96]   C. Weidenbach. Unification in pseudo-linear sort theories is decidable. In M.A. McRobbie and J.K. Slaney, editors, *Automated Deduction - Cade-13*, LNAI 1104, pages 343–357, New Brunswick, NJ, USA, July 1996. Springer-Verlag.

Restart Tableaux with Selection Function

Christian Pape and Reiner Hähnle

Universität Karlsruhe
Institut für Logik, Komplexität und Deduktionssysteme
76128 Karlsruhe, Germany
{pape,reiner}@ira.uka.de

Abstract. Recently, several different sound and complete tableau calculi were introduced, all sharing the idea to use a selection function and so-called restart clauses: A-ordered tableaux, tableaux with selection function, and strict restart model elimination. We present two new sound and complete abstract tableau calculi which generalize these on the ground level. This makes differences and similarities between the calculi clearer and, in addition, gives insight into how properties of the calculi can be transferred among them. In particular, a precise borderline separating proof confluent from non-proof confluent variants is exhibited.

1 Introduction

In this paper we introduce two new ground[1] tableau calculi, called *restart tableaux* and *strict restart tableaux*. Restart tableaux generalize the recently developed A-ordered tableaux [4] and tableaux with selection function [5], whereas strict restart tableaux subsume strict restart model elimination [1].

All of these calculi can be uniformly described by restricting the usual extension rule of clause tableaux:

1. a selection function (i.e., a mapping from clauses to subsets of their literals) restricts possible connections of clauses used for extension steps;
2. extension steps are either weakly connected (i.e. to any branch literal) or strongly connected (i.e. to a leaf);
3. so-called restart steps (extension steps with certain unconnected clauses) are used to continue proof search, whenever it is not possible to employ connected extension steps.

The idea of restarting a tableau proof is not new and often used to extend logic programming to full first-order logic. Restart steps appear, for example, as certain applications of the cut rule in the *simplified problem reduction format* [10]. They are used more explicit in the *near-Horn PROLOG* system [8] to treat non-Horn clauses: If a goal matches one (of several) head literals of a non-Horn clause then all other head literals are deferred and handled later with a

---

[1] A thorough treatment of first-order logic would have doubled the size of the paper, although there are no principal obstacles. See the last section for a few hints to lifting.

restart step. In [1] it is pointed out that both calculi — the simplified problem reduction format and near-Horn PROLOG — can be seen as instances of certain variants of restart model elimination.

The above abstract features — selection functions, extension steps and restarts — are used by the currently known calculi in differing ways. Our notions (see Section 3.1 below) generalize them all.

Further ingredients of tableau calculi specify to which extent branches are regular (i.e. are free of repetitions) and which literals are permitted for closing branches. It turns out that these latter conditions are determined by those on restart steps and on the amount of strong connections. In fact there is a direct trade-off between restrictive restarts and preference of strong connections (i.e. high connectivity) and regularity/reduction steps (at least if complete calculi are desired).

In this paper we define abstract tableau calculi whose properties can be adjusted within a wide range while the completeness proof is fully generic. This leaves open the possibility to fine-tune a calculus to each theorem one wants to prove with it.

An important difference between A-ordered tableaux and tableaux with selection function on the one side and restart model elimination on the other is the lack of proof confluency[2] in the latter. We investigate the reasons for this and show that already a very slight liberalization of restart model elimination gives proof confluency. Thus we exhibit a precise proof theoretical borderline separating calculi that are proof confluent from those that are not. Moreover, restart tableaux turn out to be a proof confluent procedure which is extremely close to restart model elimination.

Section 2 states basic notions and Section 3 briefly reviews existing calculi. In Section 4 we define restart tableaux and prove their completeness. A-ordered tableaux and tableaux with selection function are obtained as instances. In Section 5 restart tableaux are modified to strict restart tableaux which are also proven to be complete. Strict restart model elimination is obtained as an instance. At the end we make some brief remarks about lifting to first-order logic.

2 Notation

From a signature Σ (predicate, constant and function symbols) **atoms** and **literals** are constructed using the negation sign $\neg$ as usual. The set of all literals over Σ is denoted by $\mathcal{L}_\Sigma$. We omit the index Σ if no confusion can arise. In this paper we only deal with *ground clauses* that is all atoms are variable free.

A **clause** is a sequence $L_1 \vee \ldots \vee L_n, n \geq 1$ of disjunctively connected literals. $\mathcal{C}$ is the set of all clauses. We write $L \in C$ for short if a literal L occurs in a clause C. $\overline{L}$ is the **complement** of a literal L, i.e. $\overline{A} = \neg A$ and $\overline{\neg A} = A$ if A is an atom.

---

[2] A calculus is *proof confluent* if each partial proof for a provable theorem can be completed. Typical well-known examples of non-proof confluent calculi are model elimination and linear resolution.

A **tableau** $\mathfrak{T}$ is an ordered tree where the root node is labeled with *true* or a literal and all other nodes are labeled with literals. For a node n of $\mathfrak{T}$ the clause clauseof(n) is constructed from the literals of the children of n in the order from left to right. A path from the root node to a leaf literal of $\mathfrak{T}$ is called a **branch** of $\mathfrak{T}$. A tableau is **closed** if every branch contains (at least) two complementary literals. We sometimes describe a branch as a finite set of literals. We also often identify branches with the set of literals on them.

A branch B is said to be **regular** if (i) every literal of a node of B occurs only once on B and (ii) clauseof(n) is no tautology for every node n of B. A tableau $\mathfrak{T}$ is **regular** if all branches of $\mathfrak{T}$ are regular.

Partial interpretations are associated with consistent sets of ground literals. An interpretation I **satisfies** a ground clause C iff there is an $L \in C$ with $L \in I$. I is a **model** for a set S of clauses iff I satisfies all clauses of S.

3 Tableaux with Restarts and Selection Function

In this section we rehash definitions of tableaux with selection function [5], of A-ordered tableaux [4], and of restart model elimination [1]. For motivation and more examples, see the papers cited. Completeness of these calculi is obtained later from more general results and not re-stated here. We only give ground versions. Note that in the following the various notions of restart and of selection function slightly differ among the calculi. We unify them later.

3.1 Tableaux with Selection Function

Definition 1. A **selection function** is a total function f from ground clauses to sets of literals such that $\emptyset \neq f(C) \subseteq C$ for all $C \in \mathcal{C}$.

f is used to restrict connections between clauses to literals selected by f. Unrestricted extension steps, so called restarts, are only allowed with clauses that have at least a connection to another clause.

Definition 2. Let $S \subset \mathcal{C}$, f a selection function, B a tableau branch. C is a **restart clause (of S)** iff there is a $D \in S$ and a literal L such that $L \in f(D)$ and $\overline{L} \in f(C)$. C has a **weak connection via f and L into B** iff there is an $L \in B$ such that $\overline{L} \in f(C)$.

Definition 3. Let $S \subset \mathcal{C}$, f a selection function, then a **tableau with selection function f for S** is a regular clause tableau $\mathfrak{T}$ such that for every node n of $\mathfrak{T}$ the clause clauseof(n) (i) has a weak connection via f into the branch ending in n or (ii) it is a restart clause.

Example 1. Fig. 1 shows a closed tableau with selection function f for the set of ground clauses $S = \{\underline{\neg A} \vee \neg B, \underline{A} \vee B, \underline{A} \vee \neg C, \underline{B} \vee C, \underline{B} \vee \neg C\}$ (f selects the underlined literals). The first three clauses of S can be used for restarts. The solid lines correspond to extension steps (weak connection) and the dashed lines to other closures.

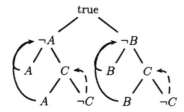

Fig. 1. Tableau with selection function

3.2 A-ordered tableaux

Definition 4. An **A-ordering** is an irreflexive, transitive relation $\prec$ on atoms, such that for all atoms A, B: $A \prec B$ implies $A\sigma \prec B\sigma$ for all substitutions σ (**stability wrt substitutions**).

A-orderings can be extended to literal orderings via $L \prec L'$ iff atom$(L) \prec$ atom(L') (where atom(A) = atom$(\neg A)$ = A). As in ordered resolution connections between clauses are restricted to literals that occur $\prec$-maximally in these clauses.

Definition 5. A literal L_j occurs $\prec$-**maximally** in a clause $L_1 \vee \ldots \vee L_n$ iff $L_j \not\prec L_i$ for all $i = 1, \ldots, n$.

A clause $C = L_1 \vee \ldots \vee L_n$ has a $\prec$-**maximal connection** to a clause $C' = L_1' \vee \ldots \vee L_{n'}'$ iff $L_i = \overline{L_j'}$ for some $1 \le i \le n, 1 \le j \le n'$, L_i occurs maximally in C, and L_j' occurs maximally in C'. If both $C, C' \in S$ then C is a **restart clause** of S.

A clause C has a **maximal connection into a set of literals** B iff C has a maximal connection to a clause consisting of a single literal of B.

A-ordered tableaux are regular clause tableaux with the restriction that extension steps are only possible with clauses that have a maximal connection into the branch they extend or to another clause of S:

Definition 6. Let $\prec$ be an A-ordering and S a set of ground clauses. A $\prec$-**ordered clause tableau for** S is a regular clause tableau $\mathfrak{T}$ for S such that for every node n of $\mathfrak{T}$ the clause clauseof(n) (i) has a $\prec$-maximal connection into the branch ending in n or (ii) it is a restart clause.

Every A-ordering $\prec$ induces a selection function $f_\prec$ on literals (such that $f_\prec(C)$ are exactly the $\prec$-maximal literals of C), hence A-ordered tableaux can be seen as an instance of tableaux with selection function, see [5].

Example 2. Let S be as in Ex. 1. For the A-ordering $A > B > C$ obviously $f_< = f$ on S, where f is as in Ex. 1. Thus, Fig. 1 constitutes as well an A-ordered tableaux for S and $<$.

3.3 (Strict) Restart Model Elimination

In contrast to tableaux with selection function in restart model elimination (RME) the selection function f (i) applies only to non-negative clauses and (ii) selects exactly one positive literal. As a consequence, a clause is never connected via f to another clause and, thus, there are no restart clauses in the sense of tableaux with selection function and A-ordered tableaux. Instead, the role of restart clauses is taken on by negative clauses.

Definition 7. A **selection function** is a total function f from non-negative ground clauses to literals, such that $f(C)$ occurs in C and is positive. Every negative clause is a **restart clause**.

RME is a refinement of model elimination [7], hence all connections are restricted to the leaf literals of a branch.

Definition 8. Let f be a selection function, B a tableau branch, $C \in \mathcal{C}$. If L is the leaf literal of B and $\overline{L} = f(C)$, then C has a **strong connection via f and L into** B.

A clause that has a connection (either weak or strong) into a branch B, where it is used gives rise to immediate closure of at least one of the new branches, namely the one with a selected (or maximal) literal L whose complement occurs on B. In this case we say that L is a **connection literal**. Closed branches not closed by a connection literal are said to contain a **reduction step**.

Most implementations of theorem provers based on model elimination and on RME are using Prolog Technology Theorem Proving (PTTP) [11]. In PTTP a set of clauses is compiled into a Prolog program such that every literal of a clause is the head of a Prolog clause (so called contrapositives). In RME only one contrapositive of each non-negative clause needs to be used in a PTTP implementation. This can lead to a significant reduction of the search space during the proof.

Unfortunately, the above restriction in combination with regularity [6] leads to an incomplete calculus (see Ex. 3). To restore completeness, the regularity condition of RME has to be relaxed:

Definition 9. Given a tableau $\mathfrak{T}$ and a subbranch[3] $\langle L_{n_0}, \ldots, L_{n_i} \rangle$ of $\mathfrak{T}$. Then $\langle L_{n_1}, \ldots, L_{n_i} \rangle$ is called a **block (corresponding to a restart clause C)** iff clauseof$(n_0) = C$ is a restart clause, for no $j = 1, \ldots, i - 1$ is clauseof(n_j) a restart clause, and

1. either n_i is a leaf of $\mathfrak{T}$ or
2. clauseof(n_i) is a restart clause.

A branch B is **blockwise regular** iff every block of B is regular.

---

[3] The sequence of labels of a contiguous subset of the nodes in a branch.

Definition 10. A restart model elimination (RME) tableau for a set S of ground clauses is a clause tableau $\mathfrak{T}$ for S such that:

1. For every node n of $\mathfrak{T}$ (i) clauseof(n) has a strong connection into the branch ending in n, or (ii) clauseof(n) is a restart clause and the label of n is a positive literal.
2. Every branch is (i) regular wrt positive literals (**positive regular**) and (ii) blockwise regular.

Example 3. Let S be as in Ex. 1. As f selects exactly one positive literal in each non-negative clause f is a suitable selection function for a restart model elimination tableau, if we ignore it on the only negative clause. This clause is by Def. 7 also the only restart clause. Fig. 2 shows an RME tableau for S. Due to symmetry in S the open branch on the right can be closed in a similar way as the left one. Note that for completeness of the calculus it is essential to permit multiple occurrences of negative literals on a tableau branch (condition 2.(i) in Def. 10).

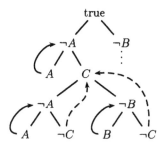

true

Fig. 2. A restart model elimination tableau

In a PTTP implementation reduction steps usually slow down the speed of the theorem prover, because they involve search among the literals on the current branch. The following modification of RME restricts reduction steps to negative leaf literals which gains some benefit in a PTTP implementation.

Definition 11. A branch B in a **strict restart model elimination tableau** for a set S of ground clauses is **closed** if its leaf literal L is (i) a connection literal or (ii) is negative and $\overline{L} \in B$.

4 Restart Tableaux with Selection Function

In RME the selection function selects exactly one positive literal, while in tableaux with selection function all literals can be selected. In restart tableaux we use a notion of selection function which generalizes Defs. 1, 7. This version of selection function is meant in the remainder of the paper if not explicitly said otherwise.

Definition 12. **A selection function** is a total function f from ground clauses to sets of literals such that $f(C) \subseteq C$ for all $C \in \mathcal{C}$.

Next we unify the various notions of restart clause.

Definition 13. Let $S \subset \mathcal{C}$, f a selection function, B a tableau branch. C is a **restart clause (of S)** iff (i) there is a $D \in S$ and a literal L with $L \in f(D)$ and $\overline{L} \in f(C)$ or (ii) $f(C) = \emptyset$.

In tableaux with selection function a non-restart clause C can extend a tableau branch B only if a selected literal of C is complementary to a literal on B, whereas in restart model elimination C has to be complementary to the leaf literal on B. Therefore, a unifying calculus has to deal with both kinds of connections. We merely repeat (parts of) Defs. 2, 8:

Definition 14. Given a clause C, a tableau branch B, and a selection function f. Then C has a **weak connection via f and L into** B iff there exists $L \in B$ with $\overline{L} \in f(C)$. It has a **strong connection via f and L into** B iff $\overline{L} \in f(C)$ for the leaf literal L of B. In both cases we say that C is **connected to B via f (and L)**.

For a given set S of clauses we fix in advance whether a clause of S can extend a tableau branch with a weak or with a strong connection. This leads to a partition of S into two disjoint subsets S_w and S_s, such that

- the clauses of S_w can extend a tableau branch B if they have a weak connection into B,[4] and
- the clauses of S_s can extend B only if they have a strong connection into B.

It will become necessary to compare blocks (cf. Def. 9) wrt to their literal sets. In the following definition and elsewhere we handle blocks (which are defined as sequences of literals) as sets of literals without the danger of confusion.

Definition 15. Two blocks b and b' are **distinct** iff neither $b \subseteq b'$ nor $b' \subseteq b$.

In the next definition a somewhat complicated notion of regularity is used which is motivated as follows: A clause $C \in S_w$ can be used to extend a branch B whenever a selected literal of C is complementary to a literal on B. The complement of the connection literal can be anywhere in the current block. Now, in the case of a strong connection two different clause $C, D \in S_s$ may be required to extend a branch via the same selected literal (in different blocks) which, therefore, occurs twice on this branch, cf. Figure 2. On the other hand, if

---

[4] Note that every strong connection is also a weak connection.

B contains a literal L such that its complement $\overline{L}$ is never selected in S_s, then a clause of S_s can never extend a branch with leaf literal L and branch-wise regularity can be enforced wrt such literals. Formally, these literals are defined as $L_{S,f} = \mathcal{L} - \bigcup_{C \in S_s} \{\overline{L} | L \in f(C)\}$.

If we choose, for example, $S_s = \emptyset$ then $L_S = \mathcal{L}$ which implies the usual regularity condition as of tableaux with selection function. If $S_w = \emptyset$ and the selection function only selects positive literals, then L_S contains at least all positive literals. This is the regularity condition of restart model elimination, see Def. 10 2.(i).

The above considerations are summarized and formally expressed in the following definition.

Definition 16. $S_w \cup S_s$ is any partition of $S \subseteq \mathcal{C}$ and f a selection function. A **restart tableau (RT) for S and f** is a clause tableau $\mathfrak{T}$ for S such that:

1. For every node n of $\mathfrak{T}$ one of the following holds:
   (a) clauseof$(n) \in S_s$ has a strong connection via f into the block ending in n *(strong extension step)*;
   (b) clauseof$(n) \in S_w$ has a weak connection via f into the block ending in n *(weak extension step)*;
   (c) clauseof(n) is a restart clause and it is not possible to extend the block above n with a strong or a weak extension step.
2. For every branch B of $\mathfrak{T}$ all of the following hold:
   (a) B is blockwise regular;
   (b) B is regular wrt $L_{S,f}$;
   (c) B contains only distinct blocks.

The definition of closure of RT is as usual (i.e. not as in Def. 11).

Definition 17. A RT $\mathfrak{T}$ for a ground clause set S is called **saturated** iff there exists no RT $\mathfrak{T}'$ for S such that $\mathfrak{T}$ is a proper subtree of $\mathfrak{T}'$.

Lemma 18. *Let S be a finite set of ground clauses. Then every branch of a saturated RT $\mathfrak{T}$ for S is finite.*

Proof. First note that each block of a branch B of $\mathfrak{T}$ is finite by Def. 16.2a. By Def. 16.2c there is a finite number of possible different blocks in B which proves the claim. □

Theorem 19. *Given a finite unsatisfiable set S of ground clauses and a selection function f. Then there exists a closed RT for S and f.*

Proof. By regularity we can assume that S contains no tautologies.

Assume there is no closed RT for S. Then, by Lemma 18, for every saturated RT $\mathfrak{T}$ for S an open and finite branch B exists. The literals of B constitute a partial interpretation I of S (via $I = B$).

Let S' be the set of all clauses from S not satisfied by I. As $\mathfrak{T}$ is saturated there is no restart clause in S', otherwise such a clause could be used to extend B by Def. 16.1c.

J is the partial interpretation that satisfies the selected literals of clauses in S'. J is well-defined, otherwise two clauses of S' would be connected via f and hence restart clauses.

$I \cup J$ is well-defined, too: If not, there are literals $L \in I$ and $\overline{L} \in J$. We distinguish two cases: either (i) $C \in S_w$ or (ii) $C \in S_s$, where $C = L \vee L_1 \vee \ldots \vee L_n$ and $\overline{L} \in f(C)$. We show that in both cases $\mathfrak{T}$ is not saturated.

case (i) here we have $C \in S_w \cap S'$ and C has a weak connection into a block b of B. Any weak extension step (Def. 16.1b) with C produces several new branches with leaf literals not already on B (otherwise C would not be in S'). Therefore, regularity is maintained.

case (ii) We show how to initiate a restart on the leaf of B and to extend B with a new block, such that regularity still holds.

First select a block $\langle L'_{n_1}, \ldots, L'_{n_m} \rangle$ of B with nodes $n_1, \ldots, n_m$ and beginning with a restart clause D such that L occurs in this block, say, $L = L'_{n_i}$, $1 \le i \le m$. Then extend B with D and clauseof(n_1) up to clauseof(n_{i-1}) and call the resulting block b. This is possible, simply because it was possible earlier in B, but we must take care to make b distinct from all other blocks. This is done by extending b with C (recall that in the present case $C \in S_s \cap S'$). This strong extension step (Def. 16.1b) generates n new distinct blocks $b_1, \ldots, b_n$ (see right part of Fig. 3) each of which satisfies Def. 16.2a–16.2c:

(a) Every block b_i is regular up to L_i because it is a prefix of a regular block and B does not contain L_i, otherwise I satisfies D.

(b) The b_i do not contain a literal $L' \in L_{S,f}$: Such a literal causes a restart and L' would have to be the last node of b which then cannot be extended by C. Thus, the new branches are regular wrt L_S.

(c) each b_i is a new distinct block on B, because it contains a literal L_i not occurring on B (otherwise C would not be in S').

We conclude that $I \cup J$ is a model of S contradicting its unsatisfiability. $\qquad \square$

Note that, as a consequence of our proof, RT are proof confluent.

Ground completeness of tableaux with selection function follows easily from Theorem 19:

Corollary 20 [5]. *For each finite unsatisfiable set S of ground clauses and selection function f (in the sense of Def. 1) there exists a closed tableau with selection function f for S.*

Proof. Set $S_w = S$. Then L_S contains all literals from S and, hence, every branch of a restart tableau has to be regular wrt to all literals, which holds also for tableaux with selection function. $\qquad \square$

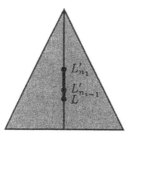 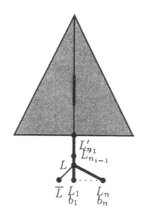

Fig. 3. Initiate a new restart (see text)

Restart tableaux are impossible to instantiate to restart model elimination, because restart model elimination is known not to be proof confluent.[5] On the other hand, if we allow restarts on negative leaf literals as well (and call the resulting calculus **unrestricted restart model elimination**), then we obtain:

Corollary 21. *For each finite unsatisfiable set S of ground clauses and selection function f (in the sense of Def. 7) there exists an unrestricted restart model elimination tableau for S and f.*

Proof. In RME all non-restart clauses must be part of strong extension steps, therefore set $S_s = S$. f selects only positive literals from a clause or none if the clause is negative, hence $L_{S,f}$ contains at least all positive literals. For restart tableaux these settings imply that each branch is blockwise regular and regular wrt to positive literals. □

It is remarkable that merely admitting positive literals in restart steps decides whether restart model elimination is proof confluent or not.

5 Strict Restart Tableaux

Restart tableaux, although very close in spirit to restart model elimination, bear a small, but crucial, difference to the latter: restarts are restricted to positive leaf literals. In addition, in strict restart model elimination reduction steps with negative leaf literals are excluded.

In this section we modify restart tableaux to a calculus of which (strict) restart model elimination is an instance.

---

[5] Consider the unsatisfiable set $\{\neg A, A \vee \neg B, A \vee \neg C, C\}$ and the partial tableau generated by the first two clauses. The open branch with negative leaf $\neg B$ cannot be closed or extended.

Recall the definition of the literal set $L_{S,f}$ in Section 4 which controls regularity. Let $L_R \subset L_{S,f}$ be any set that does not contain complementary literals.

Definition 22. $S_w \cup S_s$ is any partition of $S \subset C$ and f a selection function. A **strict restart tableau (SRT) for S and f** is a clause tableau $\mathfrak{T}$ for S such that:

1. For every node n of $\mathfrak{T}$ one of the following holds:
   (a) clauseof$(n) \in S_s$ has a strong connection via f into the block ending in n *(strong extension step)*;
   (b) clauseof$(n) \in S_w$ has a weak connection via f into the *branch* ending in n *(weak extension step)*;
   (c) clauseof(n) is a restart clause and the literal of n is not in $\{\overline{L} \mid L \in L_R\}$.
2. For every branch B of $\mathfrak{T}$ both of the following hold:
   (a) B is blockwise regular;
   (b) B is regular wrt L_R.

Definition 23. A branch B of an SRT is **closed** iff its leaf literal L is (i) a connection literal or (ii) $L \notin L_R$ and there is $\overline{L} \in B$. The latter is called a *strict reduction step*.

In comparison to restart tableaux one notes two important relaxations: weak connections need not be local to a block anymore and blocks may be identical. Moreover, restarts must be permitted even when extensions steps are still possible. For this reason RT and SRT are not instances of each other.

Some instances of strict restart tableaux are proof confluent, others are not, so there is no way to obtain a completeness proof based on saturation as for restart tableaux. One way to view strict restart tableau proofs is as a kind of normal form for restart tableau proofs. The proof below is by a tableau transformation that computes exactly this normal form, thus establishing completeness of strict restart tableaux. Note that this does not impose any assumption on proof confluency of strict restart tableaux as we start out with a closed tableau.

Unfortunately, the transformation destroys regularity conditions Def. 16.2b and 16.2c, because it copies parts of the proof tree. The following lemma shows that at least Def. 22.2b can be regained. Details of proofs had to be left out due to space restrictions, but the proofs are rather tedious than deep, anyway.

Lemma 24. *For each closed strict restart tableau $\mathfrak{T}$ for S which is blockwise regular but not necessarily regular wrt L_R there exists a closed strict restart tableau $\mathfrak{T}'$ for S which is also regular wrt L_R.*

Proof. The proof is by a careful analysis showing that critical occurrences of duplicate literals can be deleted without changing the rest. $\square$

Theorem 25. *For any finite unsatisfiable set $S \subset C$ and selection function f there exists a closed strict restart tableau for S and f.*

Proof. By Theorem 19 there exists a closed restart tableau $\mathfrak{T}$ for S. From $\mathfrak{T}$ we construct in two steps a closed strict restart tableau $\mathfrak{T}''$ for S.

First, we transform $\mathfrak{T}$ into closed restart tableau such that for all leaf literals that are involved in a reduction step $L \notin L_R$ holds, thereby satisfying Def. 23(ii). We proceed by induction on the size of the set $Red(\mathfrak{T})$ of all leaf literals $L \in L_R$ of $\mathfrak{T}$ that are involved in a reduction step.

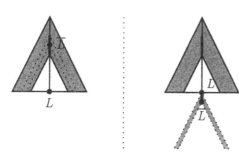

If $|Red(\mathfrak{T})| = 0$ we are done, so assume $|Red(\mathfrak{T})| = n > 0$ and choose an arbitrary literal $L \in Red(\mathfrak{T})$.

Consider a branch B such that L is used in a reduction step with $\overline{L}$. The latter is contained in a block b. Now copy the part of b above and up to $\overline{L}$ of (including all subtableaux) below B (see Fig. 4). The same is done with all remaining leaf literals L if there are any. The

Fig. 4. Trans. for invalid reduction steps

resulting tableau is handled by the induction hypothesis. We obtain a closed tableau $\mathfrak{T}'$ in which no leaf literal from L_R is used in any reduction step.

In a second step we remove from $\mathfrak{T}'$ all restarts below a literal $L \in \overline{L_R} = \{\overline{L} | L \in L_R\}$, thereby establishing Def. 22.1c. We proceed by induction on the size of the set $Res(\mathfrak{T}')$ of all literals $L \in \overline{L_R}$ which terminate a block and are no leaf literals.

If $|Res(\mathfrak{T}')| = 0$ we are done, so assume $|Res(\mathfrak{T}')| = n > 0$ and choose an arbitrary literal $L \in Res(\mathfrak{T}')$.

Fig. 5. Transformation for invalid restarts

Consider an occurrence in $\mathfrak{T}'$ of such an L. L marks the end of a block and is not a leaf literal.

1. If L is not used in a reduction step then it is possible to remove L (and the corresponding subtableaux) from $\mathfrak{T}'$. Note that L is not a connection literal, because all connections are local to blocks.
2. If L is used in a reduction step on a branch B, then we copy the block (with its subtableaux) in which L occurs below B (see Fig. 5). This is a proper restart, because $\overline{L} \notin L_R$ holds.

We continue this procedure with remaining occurrences of L until the resulting tableau contains no literal L before a restart. The rest is handled by the induction hypothesis.

The constructions above preserves blockwise regularity of branches, but (i) they may introduce identical copies of blocks and (ii) they may destroy global regularity of branches. By Lemma 24, however, we can assume that $\mathfrak{T}'$ is regular wrt L_R. □

Corollary 26 [1]. *For any finite unsatisfiable set $S \subset C$ and selection function f (in the sense of Def. 7) there exists a strict restart model elimination tableau for S and f.*

Proof. Set $S_w = \emptyset$ and $S_s = S$. As f selects only positive literals in clauses of S, $L_{S,f}$ consists of at least all positive literals, so take as $L_R \subset L_{S,f}$ the set of all positive literals. With these settings a strict restart tableau is blockwise regular, positive regular, and reduction steps are only allowed on negative literals; thus it is a strict restart model elimination tableau. □

As noted in [1] completeness of RME can be derived from this result as well.

6 Outlook

We introduced two new abstract, sound and complete tableau calculi that generalize other calculi using restart clauses: A-ordered tableaux, tableaux with selection function, and restart model elimination. This gives a whole spectrum of new proof procedures that can be fine-tuned at the selection function they use and at the desired amount of connectivity in proofs. We show explicitly how regularity and restrictions on reduction steps are influenced by the choice of these parameters. Such knowledge can in the future provide the basis for computing an instance of a proof procedure optimized for solving a given problem.

Our framework also helps to determine proof confluency of restart calculi. In particular, restart tableaux which are proof confluent can be instantiated such that they are extremely close to restart model elimination which is not.

Lifting of (strict) restart tableaux to first-order logic can be done as usual if the selection function, resp., the A-ordering is stable wrt substitutions, cf. Def. 4. In addition, a similar property has to be stated for the partition of S into weakly and strongly connected clauses. In an optimal implementation this approach leads to constraints for the regularity condition of restart tableaux. Like in [5] for tableaux with selection function one can make a compromise and get rid of the constraints in a slightly less restrictive proof procedure which admits a much weaker assumption than stability wrt substitutions: stability wrt variable renaming.

It is well known that connected tableaux can be simulated by resolution and vice versa (where the size of the tableau proof can be exponential in the size of the resolution proof).[6] Therefore, completeness of, for instance, A-ordered tableaux can be proven using a transformation of A-ordered resolution proofs into a corresponding A-ordered tableaux. In [2] it is proven that lifting of A-ordered

---

[6] For instance, this relation has been investigated in [3] for the connection method.

resolution is possible with a weaker condition for the A-ordering than stability wrt substitutions. Thus, if the transformation maps first-order resolution proofs to first-order tableau proofs we obtain as an additional result that A-ordered tableaux are liftable with these orderings as well — even if we use constraints for a maximal restriction of the calculus. Unfortunately, because of the different types of variables used in the calculi — universally quantified variables in resolution and free variables in tableaux — such a transformation based proof would be quite complex, and using this method proof confluence of the tableau caclulus cannot be proven, which is a major drawback.

In contrast to that transformation the proof confluence of restart tableaux with selection function follows immediately from our proof of Theorem 19 which, as an additional advantage, can be extended to arbitrary first-order formulas as well, in the same way as it has been done in [9] for A-ordered tableaux.

References

1. P. Baumgartner and U. Furbach. Model Elimination without Contrapositives and its Application to PTTP. *Journal of Automated Reasoning*, 13:339–359, 1994.
2. H. de Nivelle. Resolution Games and Non-Liftable Resolution Orderings. In *Collegium Logicum. Annals of the Kurt-Gödel-Society*, volume 2, pages 1–20. Springer-Verlag, Wien New York, 1996.
3. E. Eder. A Comparison of the Resolution Calculus and the Connection Method, and a new Calculus Generalizing both Methods. In E. Börger, H. K. Büning, and M. M. Richter, editors, *Proceedings of the 2nd Workshop on Computer Science Logic*, volume 385 of *LNCS*, pages 80–98, Berlin, Oct. 1989. Springer.
4. R. Hähnle and S. Klingenbeck. A-Ordered Tableaux. *Journal of Logic and Computation*, 6(6):819–834, 1996.
5. R. Hähnle and C. Pape. Ordered Tableaux: Extensions and Applications. In D. Galmiche, editor, *Proc. International Conference on Automated Reasoning with Analytic Tableaux and Related Methods, Pont-à-Mousson, France*, volume 1227 of *LNAI*, pages 173–187. Springer, 1997.
6. R. Letz, J. Schumann, S. Bayerl, and W. Bibel. SETHEO: A High-Perfomance Theorem Prover. *Journal of Automated Reasoning*, 8(2):183–212, 1992.
7. D. Loveland. Mechanical Theorem Proving by Model Elimination. *Journal of the ACM*, 15(2), 1968.
8. D. W. Loveland. Near-Horn Prolog and Beyond. *Journal of Automated Reasoning*, 7(1):1–26, Mar. 1991.
9. C. Pape. Vergleich und Analyse von Ordnungseinschränkungen für freie Variablen Tableau. Diplomarbeit, Fakultät für Informatik, Universität Karlsruhe, May 1996. In German. Available via anonymous ftp to ftp.ira.uka.de under pub/uni-karlsruhe/papers/techreports/1996/1996-30.ps.gz.
10. D. A. Plaisted. A Simplified Problem Reduction Format. *Artificial Intelligence*, 18:227–261, 1982.
11. M. E. Stickel. A Prolog Technology Theorem Prover. In E. Lusk and R. Overbeek, editors, *9th International Conference on Automated Deduction (CADE)*, LNCS, pages 752–753, Argonne, Ill, 1988. Springer.

Two Semantics and Logics Based on the Gödel Interpretation

V. Plisko*

Department of Mathematical Logic and Theory of Algorithms, Faculty of Mechanics and Mathematics, Moscow State University, Moscow, 119899, Russia

Abstract. Two arithmetic constructive theories based on Dialectica interpretation are introduced and studied in the paper. For an arithmetic formula A let A^D be its Dialectica interpretation in the language of arithmetic in all finite types. The translation $(A^D)^\circ$ of A^D back into the language of first-order arithmetic using the system **HRO** of hereditary recursive operations is considered. The theories T_1 and T_2 consist of arithmetic sentences A such that $(A^D)^\circ$ is true in the standard model and provable in the intuitionistic arithmetic respectively. Using the author's recent results on the arithmetic complexity of the predicate logics of constructive arithmetic theories it is proved that the logic of T_1 is not recursively enumerable and the logic of T_2 is $\mathit{\Pi}_2$-complete.

1 Introduction

Constructive semantics of formal languages are widely used in intuitionistic proof theory. Now the interest in such semantics is growing because of their applications in theoretical computer science, especially in extracting algorithms from constructive proofs. Historically, the first precise constructive semantics of the language of formal arithmetic LA was recursive realizability introduced by S. C. Kleene [3] in 1945. Later, in 1958, K. Gödel [2] defined a translation from LA into the language of arithmetic in all finite types LA^ω expressing the constructive point of view in another way. This translation is usually called the Gödel interpretation or Dialectica interpretation (see [7]). The Gödel interpretation gives a spectrum of semantics of arithmetic sentences depending on the choice of models of the language LA^ω.

For any semantics it is of interest to study the corresponding logic as the set of predicate formulas being schemata of arithmetic sentences true in the given semantics. In this paper we consider two semantics of arithmetic sentences and corresponding predicate logics based on the Gödel interpretation in the case when the system of hereditary recursive operations **HRO** is chosen as a model of the language LA^ω. The main results concern arithmetic complexity of considered predicate logics. These results are obtained by applying the author's recent metamathematical results on the logics of constructive theories (see [4], [5]).

* Partially supported by RFBR grant 96-01-01470 and by INTAS–RFBR grant 95-0095.

2 The Language of Arithmetic in All Finite Types

Let HA be the formal system of intuitionistic arithmetic in the language LA containing symbols for all primitive recursive functions. Many facts concerning partial recursive functions are formalized in HA. For an appropriate enumeration of the class of partial recursive functions there exists a quantifier-free formula $T(x, y, z, u)$ such that for every natural numbers l, m, n, $HA \vdash \exists u T(l, m, n, u)$ if and only if $\{l\}(m) = n$ ($\{l\}$ denotes the partial recursive function with number l). In what follows we shall write $\{t_1\}(t_2) = t_3$ instead of the formula $\exists u T(t_1, t_2, t_3, u)$, t_1, t_2, t_3 being terms.

Types are defined inductively by the following two clauses:

- 0 is a type;
- if σ, τ are types, then $(\sigma \to \tau)$ is a type.

The language of arithmetic in all finite types LA^ω is obtained by adding to the language LA an infinite set of variables $x^\tau, y^\tau, z^\tau, \ldots$ for each type $\tau \neq 0$, the variables of LA being variables of type 0. *Terms of type τ* are defined inductively as follows:

- terms of the language LA are terms of type 0;
- variables of type τ are terms of type τ;
- if t is a term of type $(\sigma \to \tau)$, s is a term of type σ, then $t(s)$ is a term of type τ.

Prime formulas of the language LA^ω are of the form $s = t$, s and t being terms of type 0. More complicated *formulas* are constructed from the prime formulas in the usual way by using connectives $\neg, \&, \vee, \supset$ and quantifiers $\forall, \exists$ over variables of any type.

A *model* **M** of the language LA^ω is given by specifying a non-empty domain M_τ for each type τ, M_0 being the set of natural numbers $N = \{0, 1, 2 \ldots\}$. The set M_τ is the range of the variables of type τ. The elements in M_τ are called *objects of type τ* in the model **M**. For every pair of types σ, τ an *application map* $app_{\sigma,\tau} : M_{(\sigma \to \tau)} \times M_\sigma \longrightarrow M_\tau$ is given too. In this case, if t is a term of type $(\sigma \to \tau)$, $a \in M_{(\sigma \to \tau)}$ is its value in the model **M**, s is a term of type σ, and $b \in M_\sigma$ is its value in **M**, then $app_{\sigma,\tau}(a, b) \in M_\tau$ is the value of the term $t(s)$ in the model **M**.

Let us agree on the following notations. We abbreviate $t_1(t_2) \ldots (t_n)$ as $t_1 t_2 \ldots t_n$. If **t** is a finite string of terms $t_1, t_2, \ldots, t_n$, t_i being a term of type τ_i $(i = 1, \ldots, n)$, then **t** will be called a list of type $\boldsymbol{\tau}$, $\boldsymbol{\tau}$ being the string of types $\tau_1, \tau_2, \ldots, \tau_n$. For a type σ, $(\boldsymbol{\tau} \to \sigma)$ will denote the type

$$(\tau_1 \to (\tau_2 \to \ldots (\tau_n \to \sigma) \ldots)).$$

If **s** is another string of terms $s_1, \ldots, s_m$, s_j being of type $(\boldsymbol{\tau} \to \sigma_j)$ $(j = 1, \ldots, m)$, then **s**(**t**) will denote the list of terms $s_1 t_1 t_2 \ldots t_n, \ldots s_m t_1 t_2 \ldots t_n$. If the list **s** is empty, then the list **s**(**t**) is considered to be empty too, and if the list **t** is empty, then the list **s**(**t**) is **s** by definition. Finally, $(\boldsymbol{\sigma} \to \tau)$ will denote the list of types

$(\sigma \to \tau_1), \ldots, (\sigma \to \tau_n)$. If one of the lists σ, τ is empty, let us assume the same agreement on the list $(\sigma \to \tau)$ as in the case of the list of terms $s(t)$. If $\mathbf{x}$ is a string of variables $x_1, \ldots, x_n$ and $\mathbf{Q}$ is a quantifier $\forall$ or $\exists$, then $\mathbf{Qx}$ will denote $\mathbf{Q}x_1 \ldots \mathbf{Q}x_n$.

3 Hereditary Recursive Operations

The system of hereditary recursive operations **HRO** is a model of the language LA^ω described in [7, 2.4.8]. In this model the objects of each type are natural numbers. The set HRO_τ of the objects of type τ is defined as follows:

$$HRO_0 \rightleftharpoons N; HRO_{(\sigma \to \tau)} \rightleftharpoons \{n | (\forall x \in HRO_\sigma)\{n\}(x) \in HRO_\tau\}.$$

The application map $app_{\sigma,\tau}$ is defined as partial recursive function application: $app_{\sigma,\tau}(x, y) = \{x\}(y)$. The definition of the set $HRO_{(\sigma \to \tau)}$ implies that the map $app_{\sigma,\tau}$ is defined everywhere on $M_{(\sigma \to \tau)} \times M_\sigma$ and its range is M_τ. Obviously, for each type σ the set M_σ can be defined by an arithmetic formula $V_\sigma(x)$:
$V_0(x) \rightleftharpoons x = x; V_{(\sigma \to \tau)}(x) \rightleftharpoons \forall y(V_\sigma(y) \supset \exists z(\{x\}(y) = z \,\&\, V_\tau(z)))$.

Following [1] let us define the notion of *a partial recursive term* (p.r.t.). It will be a metamathematical expression allowing more intuitive notation for arithmetic sentences. Namely, any arithmetic term, i.e. a term of the language LA is a p.r.t. Further, if t_1, t_2 are p.r.t., then the expression $\{t_1\}(t_2)$ is a p.r.t. too. For a p.r.t. t and a variable y an arithmetic formula $t = y$ is defined inductively by the following convention:

$$\{t_1\}(t_2) = y \rightleftharpoons \exists u \exists v(t_1 = u \,\&\, t_2 = v \,\&\, \{u\}(v) = y).$$

Let us assign a p.r.t. t° to any term t of the language LA^ω as follows:

- if t is an arithmetic term, then t° is t;
- $(s(t))^\circ$ is $\{s^\circ\}(t^\circ)$.

The operation $^\circ$ is extended to the formulas in the following way:

- $(s = t)^\circ \rightleftharpoons \exists y(s^\circ = y \,\&\, t^\circ = y)$, y being a variable not occuring in s and t;
- $(\neg A)^\circ \rightleftharpoons \neg A^\circ$;
- $(A\lambda B)^\circ \rightleftharpoons (A^\circ \lambda B^\circ)$, λ being a connective $\&$, $\vee$ or $\supset$;
- $(\forall x A)^\circ \rightleftharpoons \forall x(V_\sigma(x) \supset A^\circ)$ (here and in the next case x is a variable of type σ);
- $(\exists x A)^\circ \rightleftharpoons \exists x(V_\sigma(x) \,\&\, A^\circ)$.

All the variables in A° are considered as natural variables, therefore A° is an arithmetic formula.

4 The Gödel Translation

In [2] K. Gödel defined a translation from LA into LA^ω that assigns a formula A^D of the form $\exists \mathbf{u} \forall \mathbf{v} A_D(\mathbf{u}, \mathbf{v})$ to every arithmetic formula A (here $\mathbf{u}$ and $\mathbf{v}$ are lists of variables, A_D is a quantifier-free formula). Formulas A^D and A_D are defined inductively in the following way:

- if A is a prime formula, then A^D and A_D coincide with A.

In the next cases we assume that the formula B^D of the form $\exists \mathbf{x} \forall \mathbf{y} B_D(\mathbf{x}, \mathbf{y})$ and the formula C^D of the form $\exists \mathbf{u} \forall \mathbf{v} C_D(\mathbf{u}, \mathbf{v})$ are defined, $\mathbf{x}$, $\mathbf{y}$, $\mathbf{u}$, $\mathbf{v}$ being mutually disjoint lists of variables of types π, ρ, σ, τ respectively.

- $(B \,\&\, C)^D \rightleftharpoons \exists \mathbf{x}, \mathbf{u} \forall \mathbf{y}, \mathbf{v} (B \,\&\, C)_D$, where $(B \,\&\, C)_D$ is $B_D(\mathbf{x}, \mathbf{y}) \,\&\, C_D(\mathbf{u}, \mathbf{v})$;
- $(B \lor C)^D \rightleftharpoons \exists z, \mathbf{x}, \mathbf{u} \forall \mathbf{y}, \mathbf{v} (B \lor C)_D$, where z is a variable of type 0, $(B \lor C)_D$ is $(z = 0 \supset B_D(\mathbf{x}, \mathbf{y})) \,\&\, (z \neq 0 \supset C_D(\mathbf{u}, \mathbf{v}))$;
- $(B \supset C)^D \rightleftharpoons \exists \mathbf{U}, \mathbf{Y} \forall \mathbf{x}, \mathbf{v} (B \supset C)_D$, where $\mathbf{U}$ is a list of variables of type $(\pi \to \sigma)$, $\mathbf{Y}$ is a list of variables of type $(\pi, \tau \to \rho)$, $(B \supset C)_D$ is $B_D(\mathbf{x}, \mathbf{Yxv}) \supset C_D(\mathbf{Ux}, \mathbf{v})$;
- $(\neg B)^D \rightleftharpoons \exists \mathbf{Y} \forall \mathbf{x} (\neg B)_D$, where $\mathbf{Y}$ is a list of variables of type $(\pi \to \rho)$, $(\neg B)_D$ is $\neg B_D(\mathbf{x}, \mathbf{Yx})$;
- $(\exists z B(z))^D \rightleftharpoons \exists z, \mathbf{x} \forall \mathbf{y} (\exists z B(z))_D$, where $(\exists z B(z))_D$ is $B_D(\mathbf{x}, \mathbf{y}, z)$;
- $(\forall z B(z))^D \rightleftharpoons \exists \mathbf{X} \forall z, \mathbf{y} (\forall z B(z))_D$, where $\mathbf{X}$ is a list of variables of type $(0 \to \pi)$, $(\forall z B(z))_D$ is $B_D(\mathbf{X}z, \mathbf{y}, z)$.

We shall say that an arithmetic sentence A is *true in the first sense* (or 1-true) if the formula $(A^D)^\circ$ is (classically) true in the standard model of arithmetic. Let T_1 be the set of all 1-true arithmetic sentences.

We say that an arithmetic sentence A is *true in the second sense* (or 2-true) if $HA \vdash (A^D)^\circ$. Let T_2 be the set of all 2-true arithmetic sentences. It follows from the definition that $T_2 \subseteq T_1$. One can say that T_2 consists of sentences whose membership in T_1 is provable in HA.

5 Constructive Arithmetic Theories

In [4] we introduced the notion of constructive arithmetic theory as an arbitrary set of arithmetic sentences closed under derivability in the theory HA+CT+M obtained by adding to HA the Church thesis, i.e. the schema

$$\text{CT}: \qquad \forall x \exists y A \supset \exists z \forall x \exists y(\{z\}(x) = y \,\&\, A)$$

and the Markov principle, i.e. the schema

$$\text{M}: \qquad \forall x(A \lor \neg A) \,\&\, \neg\neg\exists x A \supset \exists x A.$$

Theorem 1. T_1 *and* T_2 *are constructive theories.*

Proof. To prove this theorem let us recall some well-known facts concerning systems of intuitionistic arithmetic in all finite types, hereditary recursive operations, and the Gödel interpretation. In [7, 2.4.10] the formal system HRO^- in the language obtained from LA^ω by adding some constants is described. The precise formulation of this theory is not important for our purposes. We shall use only some facts about it stated in [7]. Let us formulate them not in the most general form but extracting only the content which is needed for us.

Fact 1. HRO^- is a conservative extension of HA [7, 3.6.2]. It means that any arithmetic formula F is deducible in HRO^- if and only if it is deducible in HA.

Fact 2. The system of hereditary recursive operations is a model of HRO^- and this is provable in HA [7, 2.4.10, 3.6.2]. It means that for any arithmetic sentence F provable in HRO^- the formula F° is provable in HA. Moreover, if F is deducible in HRO^- from $F_1, F_2, \ldots$, then F° is deducible in HA from $F_1^\circ, F_2^\circ, \ldots$.

Fact 3. If A is a sentence in the language LA^ω and A^D is its Gödel translation, then $HRO^- \vdash A$ implies $HRO^- \vdash A^D$ [7, 3.5.4]. Moreover, if A is deducible in HRO^- from $A_1, A_2, \ldots$, then A^D is deducible in HRO^- from $A_1^D, A_2^D, \ldots$.

Fact 4. Each instance of the schema

$$\text{M}' : \qquad \neg\neg\exists x A(x) \supset \exists x A(x) \qquad (A(x) \text{ quantifier-free})$$

is Dialectica interpretable in HRO^- [7, 3.5.9]. It means that the Gödel translation of each instance of the schema M' is deducible in HRO^-.

Fact 5. The Church thesis is Dialectica interpretable in HRO^- [7, 3.5.16]. It means that the Gödel translation of each instance of the schema CT is deducible in HRO^-.

Let us return now to the proof of the theorem. We must show that T_1 and T_2 are constructive theories, i.e. closed under derivability in $HA+CT+M$. It follows from Fact 1 that any sentence A provable in HA is deducible in HRO^-. If we combine this with Fact 3, we get $HRO^- \vdash A^D$. By Fact 2, $HA \vdash (A^D)^\circ$, so that A is in T_2 according to the definition of T_2. Since $T_2 \subseteq T_1$, it follows that $A \in T_1$.

If A is any closed instance of the Church thesis CT, then, by Fact 5, A^D is provable in HRO^-. The application of Fact 2 yields $HA \vdash (A^D)^\circ$, so that $A \in T_2$ and therefore $A \in T_1$.

Before considering the case of the Markov principle M let us recall the well-known fact (see [1]) that in the presence of the Church thesis the schema M is equivalent in HA to the schema M' mentioned in Fact 4. It means that the theories $HA+CT+M$ and $HA+CT+M'$ have the same deducible formulas. Hence it is enough to show that each instance of the schema M' is in T_2, but this follows from Facts 4 and 2.

We have proved that any sentence derivable in HA just as each instance of the schemata CT and M are in T_i ($i = 1, 2$). Since the deduction theorem holds for HA and HRO^-, it follows from Facts 1–5 that if F is deducible in $HA+CT+M$ from $F_1, F_2, \ldots$, then $(F^D)^\circ$ is deducible in HA from $(F_1^D)^\circ, (F_2^D)^\circ, \ldots$. Therefore if F is deducible in $HA+CT+M$ from $F_1, F_2, \ldots$ and $F_1, F_2, \ldots$ are in T_2, i.e.

$(F_1^D)^\circ, (F_2^D)^\circ, \ldots$ are deducible in HA, then $HA \vdash (A^D)^\circ$, that is $F \in \mathsf{T}_2$. Hence T_2 is closed under derivability in $HA+CT+M$. For the same reason, if $F_1, F_2, \ldots$ are in T_1, i.e. $(F_1^D)^\circ, (F_2^D)^\circ, \ldots$ are true in the standard model of arithmetic, then $(F^D)^\circ$ is also true because classical truth is stable under derivability in HA. Whence it follows that T_1 is closed under derivability in $HA+CT+M$.

An arithmetic theory T is said to have *the existential property* if for every closed formula of the form $\exists x F(x)$ in T there exists a natural number n such that $F(n) \in \mathsf{T}$.

Theorem 2. T_1 *and* T_2 *have the existential property.*

Proof. It is known (see [7],[1]) that HA has the existential property. Now the theorem is a trivial consequence of the definitions.

6 Internally Enumerable Theories

Following [8] we shall say that an arithmetic formula $A(x_1, \ldots, x_n)$ is *decidable* in an arithmetic theory T if for each natural numbers $k_1, \ldots, k_n$ either $A(k_1, \ldots, k_n)$ is in T, or $\neg A(k_1, \ldots, k_n)$ is in T. A set of natural numbers A is said to be *enumerable in* T if there exists decidable in T formula $B(x, y)$ such that $A = \{k|(\exists n)[B(k, n) \in \mathsf{T}]\}$. An arithmetic theory T is said to be *internally enumerable* if the set of Gödel numbers of the theorems of T is enumerable in T.

Theorem 3. 1) *The set of Gödel numbers of the theorems of* T_2 *is* Σ_1-*complete.*
2) *The theory* T_2 *is internally enumerable.*

Proof. 1) Σ_1-completeness of the set under consideration means that this set is a Σ_1-set and each Σ_1-set is 1-1-reducible to T_2. By definition, a set $A \subseteq N$ is a Σ_1-set if there exists arithmetic formula $\exists y F(x, y)$ with $F(x, y)$ quantifier-free such that for each natural n, $n \in A$ if and only if the formula $\exists y F(n, y)$ is true in the standard model of arithmetic. It is well known that Σ_1-sets are exactly the recursively enumerable sets. Obviously, T_2 is recursively enumerable. Let A be an arbitrary Σ_1-set, $F(x, y)$ be a quantifier-free formula such that for each natural n, $n \in A$ if and only if $\exists y F(n, y)$ is true in the standard model of arithmetic. It is well known that arithmetic formula of the form $\exists y F(n, y)$ is true in the standard model of arithmetic if and only if $HA \vdash \exists y F(n, y)$. By the definition of the Gödel translation, $(\exists y F(n, y))^D$ is $\exists y F(n, y)$. This means that

$$n \in A \iff HA \vdash \exists y F(n, y) \iff HA \vdash ((\exists y F(n, y))^D)^\circ \iff \exists y F(n, y) \in \mathsf{T}_2.$$

Thus A is 1-1-reducible to T_2 and T_2 is Σ_1-complete.

2) It was mentioned just now that the set A of Gödel numbers of the formulas in T_2 is a Σ_1-set. Arguing as above, we see that there exists arithmetic formula $\exists y B(x, y)$ with $B(x, y)$ quantifier-free such that $k \in A \iff HA \vdash \exists y B(k, y)$ for each natural k. Since HA has the existential property, it follows that

$$k \in A \iff (\exists n) HA \vdash B(k, n).$$

Clearly, every closed quantifier-free formula belongs to T_2 if and only if it is provable in HA. Therefore, we have $A = \{k|(\exists n)[B(k,n) \in T_2]\}$. Evidently, the formula $B(x,y)$ is decidable in T_2. It follows that A is enumerable in T_2 and T_2 is internally enumerable.

Theorem 4. *The theory T_1 is not recursively enumerable.*

Proof. Let A be an arbitrary Σ_2-set. This means that there exists arithmetic formula $\exists y \forall z F(x,y,z)$ with $F(x,y,z)$ quantifier-free such that for each natural n, $n \in A$ if and only if the formula $\exists y \forall z F(n,y,z)$ is true in the standard model of arithmetic. Remark that $(\exists y \forall z F(n,y,z))^D$ is $\exists y \forall z F(n,y,z)$, therefore $((\exists y \forall z F(n,y,z))^D)^\circ$ is $\exists y \forall z F(n,y,z)$. Now we have that $n \in A$ if and only if the formula $((\exists y \forall z F(n,y,z))^D)^\circ$ is true in the standard model of arithmetic, that is, by definition, $\exists y \forall z F(n,y,z) \in T_1$. Thus $n \in A \iff \exists y \forall z F(n,y,z) \in T_1$ and A is 1-1-reducible to T_1. It follows that T_1 is not recursively enumerable because only recursively enumerable sets are 1-1-reducible to the recursively enumerable set and some Σ_2-sets are not recursively enumerable.

7 Predicate Logics of the Gödel Interpretation

Let T be an arbitrary set of arithmetic sentences. A closed predicate formula $\mathcal{A}(P_1, \ldots, P_n)$ with predicate variables $P_1, \ldots, P_n$ only will be called T-*valid* if $\mathcal{A}(F_1, \ldots, F_n) \in T$ for each arithmetic formulas $F_1, \ldots, F_n$. The set of all T-valid predicate formulas will be denoted by $L(T)$.

The following theorem was proved in [5]:

Theorem 5. *If a constructive arithmetic theory T is internally enumerable and has the existential property, then the logic $L(T)$ is Π_1^T-complete.*

If we combine this with Theorems 1–3, we get the following result.

Theorem 6. *The logic $L(T_2)$ is Π_2-complete.*

Proof. It follows from Theorems 1–3 that T_2 is a constructive theory that is internally enumerable and has the existential property. By Theorem 5 the logic $L(T_2)$ is $\Pi_1^{T_2}$-complete. Theorem 3 states that T_2 is Σ_1-complete. By standard arguments used in the arithmetic hierarchy and reducibility theory [6, §14.5] it is easy to prove that if A is Σ_1-complete, then every Π_1^A-complete set is Π_2-complete. This completes the proof.

It was proved in [4] that $T \leq_1 L(T)$ for every constructive arithmetic theory T. Combining this with Theorems 1 and 4, we get the following theorem.

Theorem 7. *The logic $L(T_1)$ is not recursively enumerable.*

Proof. By Theorem 4, T_1 is not recursively enumerable. Theorem 1 states that T_1 is a constructive theory, therefore $T_1 \leq_1 L(T_1)$ by the just now mentioned result. It follows that $L(T_1)$ is not recursively enumerable because only recursively enumerable sets are 1-1-reducible to the recursively enumerable set.

It is of interest to get a more precise characterization of the logic $L(T_1)$. The main problem here is characterizing arithmetic complexity of the theory T_1 itself. It is plausible that T_1 is not arithmetical. In that case the logic $L(T_1)$ is not arithmetical too.

References

1. A. G. Dragalin: Mathematical Intuitionism. An Introduction to the Proof Theory. Nauka, Moscow, 1979
2. K. Gödel: Über eine bisher noch nicht benützte Erweiterung des finiten Standpunktes. Dialectica **12** (1958) 280–287
3. S. C. Kleene: On the interpretation of intuitionistic number theory. J. Symbol. Log. **10** (1945) 109–124
4. V. E. Plisko: Constructive formalization of the Tennenbaum theorem and its applications. Mat. Zametki **48** (1990) 108–118
5. V. E. Plisko: On arithmetic complexity of certain constructive logics. Mat. Zametki **52** (1992) 94–104
6. H. Rogers, Jr.: Theory of Recursive Functions and Effective Computability. MIT Press, Cambridge, 1987
7. A. S. Troelstra (ed.): Metamathematical Investigation of Intuitionistic Arithmetic and Analysis. Lect. Notes. Math. **344** (1973) 1–485
8. V. A. Vardanyan: Bounds for the arithmetic complexity of predicate provability logics. Vopr. Kibernet. **136** (1988) 46–72

On the Completeness and Decidability of a Restricted First Order Linear Temporal Logic

Regimantas Pliuškevičius

Institute of Mathematics and Informatics
Akademijos 4, Vilnius 2600, LITHUANIA
email: regis@ktl.mii.lt

Abstract. The paper describes a class of FC-sequents of a restricted first order linear temporal logic with "next" and "always". It is shown that the first order linear temporal logic is finitary complete for FC-sequents. If FC-sequents are logically decidable, i.e., all sequents, constructed from all possible subformulas of logical formulas of FC-sequents are decidable in first-order logic, then FC-sequents are decidable in the restricted first order linear temporal logic. To obtain these results a saturation calculus [6, 7] for FC-sequents is constructed.

1 Introduction

In various temporal logics the most important are induction-like postulates. These postulates make the deductive tools for the logics very complicated. An especially complicated situation is in the first order case. However, an interesting cases of temporal reasoning usually involve quantifiers over rigid variables (that is, variables whose meanings are time-independent), which, of course, is much more complicated than just the reasoning on "fixed" decidable propositional statements. The main problem of dealing with a first-order temporal logic is that it is not recursively axiomatizable (see, e.g. [1, 5, 8]). In [4, 9] it was proved to become complete after adding the following ω-type rule of inference:

$$\frac{\Gamma \to \Delta, A; \ldots; \Gamma \to \Delta, \bigcirc^k A, \ldots}{\Gamma \to \Delta, \Box A} \ (\to \Box_\omega),$$

where $\bigcirc$ is a temporal operator "next", $\Box$ is a temporal operator "always". In some cases the first order linear temporal logic is finitary complete (see e.g. [5, 6, 7]). The main rule of inference in this case is the following:

$$\frac{\Gamma \to \Delta, R; R \to \bigcirc R; R \to A}{\Gamma \to \Delta, \Box A} \ (\to \Box).$$

(This rule corresponds to the induction axiom: $A \wedge \Box (A \supset \bigcirc A) \supset \Box A$; the formula R is called an *invariant formula*). Therefore we have an uncertain situation: in which case we *can* apply $(\to \Box)$ and when we *have* to apply $(\to \Box_\omega)$.

In [6, 7] saturation calculi were described for a restricted first-order linear temporal logic. The saturation calculi were developed specially for temporal logics containing induction-like postulates. Instead of induction-type postulates the saturation calculi contain a deductive saturation procedure, indicating some form of regularity in the derivations of the logic. The saturation calculi involve the process of calculation of some sequents called *k-th resolvents* analogously as in the resolution-like calculus. Therefore, induction-like rules of inference are replaced by a constructively defined deductive process of finding k-th resolvents. The saturation calculi are very powerful because they allow us to get derivations of valid sequents both in a finitary and an infinitary case. In the finitary case, we obtained a set of saturated sequents, showing that "almost nothing new" can be obtained if we continue the derivation process. In this case using only the unified saturation (i.e., based only on unification) we get a derivation of the empty k-th resolvent (in symbols: $Re^k(S) = \varnothing$). In the infinitary case, the similarity saturation calculus generates an infinite set of saturated sequents, showing that only "similar" sequents can be obtained continuing the derivation process.

The purpose of this paper is to generalize results of [5, 6, 7]. We describe some class of finite complete sequents (in brief: FC-sequents) of a restricted first order linear temporal logic and prove that the first order linear temporal logic is finitary complete for FC-sequents. The main property of an FC-sequent S is the following: either derivability of S reduces to the derivability of sequents $S_1, \ldots, S_n$ in the calculus without positive occurrence of $\square$ (i.e. the positive occurrences of $\square$ in S are not essential for the derivability of S), or starting for some k it is possible to generate $Re^l(S)$, i.e. the l-th ($l \geqslant k$) resolvent of FC-sequent S such that each member of $Re^l(S)$ has the shape $\Gamma, \square\Omega \to \Delta, \square\nabla$, where Γ, Δ consist of finite (up to renaming of variables) subformulas of Ω, ∇. If FC-sequents are decidable in a first-order logic, then FC-sequents are decidable in the restricted first order linear temporal logic. To obtain these results the saturation calculus with "catch" (i.e., with some stopping device) is constructed.

The saturation calculus with the catch involves some constructive simple stopping device for obtaining the upper bound of a maximal number of the k-th resolvent. The saturation calculus with the catch contains a more sophisticated separation rule (called a structural separation rule) which not only splits off the logical part from the temporal one in quasi-primary sequent S but also it splits off the temporal part according to the structure of S.

2 Description of infinitary sequent calculus $G^*_{L\omega}$ with implicit quantifiers

Saturation calculus *Sat* is founded by means of the infinitary calculus $G^*_{L\omega}$ containing the ω-type rule of inference and not containing explicitly the rules of inference for quantifiers.

Definition 2.1 (elementary formulas, formulas). Elementary formulas are expressions of the type $P(t_1, \ldots, t_n)$, where P is an n-place predicate symbol, $t_1, \ldots, t_n$ are arbitrary terms. Formulas are defined from the elementary formulas by means of logical symbols and the temporal operator $\square$, as usual. It is assumed that all predicate symbols are flexible (i.e., symbols that may change their value with time) and all function symbols are rigid (with time-independent meanings).

In a linear temporal logic we have the following equivalences: $\vDash \bigcirc(A \odot B) \equiv \ \vDash \bigcirc A \odot \bigcirc B$ $(\odot \in \{\supset, \wedge, \vee\})$; $\vDash \bigcirc \sigma A \equiv \vDash \sigma \bigcirc A$, $\sigma \in \{\urcorner, \square\}$. Therefore we can consider a formula containing occurrences of $\bigcirc$ ("next") entering only the formula $\bigcirc^k E$ (k-time "next" E), where E is an elementary formula and instead of the formula $\bigcirc^k E$ we use an "index" formula E^k. We also use the notation A^k for an arbitrary formula A in the following meaning.

Definition 2.2 (index, atomic formula). 1) If E is an elementary formula, $i, k \in \omega$, $k \neq 0$, then $(E^i)^k := E^{i+k}$ $(E^0 := E)$; E^l $(l \geqslant 0)$ is called an *atomic formula*; 2) $(A \odot B)^k := A^k \odot B^k$, $\odot \in \{\supset, \wedge, \vee\}$; 3) $(\sigma A)^k := \sigma A^k$, $\sigma \in \{\urcorner, \square, \forall x, \exists x\}$.

Definition 2.3 (sequent, memory formulas, catch of a sequent). A *sequent* is an expression of the form $\Gamma, \nabla_1^* \to \Delta, \nabla_2^*$, where Γ, Δ are arbitrary finite sets, ∇_1^*, ∇_2^* are arbitrary multisets of the expression of the shape A^* (where A is a formula containing variables), called a *memory formula*. The *catch of a sequent* S (denoted by $Ct_\square(S)$) is the sum of indices and occurrences of $\square$ in S governed by the occurrences of $\square$ in S.

Remark 2.1. The memory formulas will be used in the rule of renaming variables (see Definition 2.9).

Definition 2.4 (normal form, N-sequents). A sequent S will be in the *normal form*, (in brief: *N-sequent*), if

1) S contains neither positive occurrences of $\forall$ nor negative occurrences of $\exists$ (skolemization condition);

2) there are no occurrences of $\square$ in the scope of quantifiers (weak miniscopization condition);

3) all negative (positive) occurrences of $\square A$ in the sequent S are of the form $\square \forall \bar{x} A(\bar{x})$ $(\square \exists \bar{x} A(\bar{x})$, respectively), where $\bar{x} = x_1, \ldots, x_n$ $(n \geqslant 0)$ and $A(\bar{x})$ does not contain quantifiers (temporal prenex-form condition);

4) all formulas not containing $\square$ are of the form $Q\bar{x} A(\bar{x})$ $(Q \in \{\forall, \exists\}$, $\bar{x} = x_1, \ldots, x_n$ $(n \geqslant 0))$; where $A(\bar{x})$ does not contain quantifiers.

Theorem 2.1 *Let S be a sequent, then there exists S^+ such that S^+ is an N-sequent and S is satisfiable iff S^+ is satisfiable.*
Proof: follows from [3].

Definition 2.5 (quantifier-free form of an N-sequent). Let S be an N-sequent (we can assume, without loss of generality, that in S different occurrences of quantifiers bind different variables), then a sequent S^* obtained from S by dropping all the occurrences of $Q\bar{x}$ $(Q \in \{\forall, \exists\})$ will be called a *quantifier-free form* of the N-sequent.

Remark 2.2. It is easy to verify that the unique (up to the renaming of variables) quantifier-free form S^* corresponds to an N-sequent S and vice versa.

In this paper we consider only N-sequents in quantifier-free form.

Definition 2.6 (substitution, simple substitution, formula and sequent with substitution, LHS and RHS of substitution). A *substitution* is an expression of the form $\sigma = (x_1 \leftarrow t_1; \ldots; x_n \leftarrow t_n)$, where x_i is a variable, t_i is a term

$(1 \leqslant i \leqslant n)$. If $\forall i \ (1 \leqslant i \leqslant n)$ t_i is a variable, then such a substitution is called a simple one. Explicitly indicated variables (terms) $x_1, \ldots, x_n$ $(t_1, \ldots, t_n)$ in σ is called the *left-hand side (right-hand side,* respectively) (in short: *LHS and RHS*) of substitution σ. If A is a formula, $\sigma = (x_1 \leftarrow t_1; \ldots; x_n \leftarrow t_n)$ is a substitution, then the expression $A\sigma$ will be called a *formula with a substitution*, what means that all the occurrences of x_i are replaced by t_i $(1 \leqslant i \leqslant n)$. A sequent, each member of which is a formula or a formula with a substitution σ, is called a *sequent with a substitution* and denoted by $S\sigma$.

Definition 2.7 (axiomatic substitutions). Let A be an atomic formula, $\bar{p} = p_1, \ldots, p_n$; $\bar{q} = q_1, \ldots, q_n$; p_i, q_i be some terms, let $S = \Gamma, A(\bar{p})\sigma_1 \to A(\bar{q})\sigma, \Delta = \Gamma, A(\bar{t}) \to A(\bar{t}), \Delta$ (where $t = t_1, \ldots, t_n$; and t_i be some terms $(1 \leqslant i \leqslant n)$, then the substitution $\sigma_1 \circ \sigma$ is called an *axiomatic one*.

Definition 2.8 (rule (*subs*): substitution rule) Let A, $\bar{p}$, $\bar{q}$, $\bar{t}$, σ_1, σ be the same as in Definition 2.7. Then let us introduce the following rule:

$$\frac{\Gamma, A(\bar{p})\sigma_1 \to \Delta, A(\bar{q})\sigma_2}{\Gamma, A(\bar{p}) \to \Delta, A(\bar{q})} \ (subs),$$

where $(\sigma_1 \circ \sigma_2)$ is the mgu of $A(\bar{p})$, $A(\bar{q})$, where A is an atomic formula.

Remark 2.3. (a) The rule (*subs*) corresponds to the finding of the contrary pair in the resolution rule in a traditional resolution-like calculus.

(b) Instead of unification algorithm in the rule (*subs*) a simpler matching procedure can be used (see, e.g. [2]).

Definition 2.9 (rule (*r.v.*): rule of renaming of variables). Let us introduce the following rule:

$$\frac{\Gamma, A(\bar{p}), B^0(\bar{y}), B^*(\bar{x}), \nabla_1^* \to A(\bar{q}), C^0(\bar{y}), C^*(\bar{x}), \nabla_2^*}{S = \Gamma, A(\bar{p}), B^*(\bar{x}), \nabla_1^* \to A(\bar{q}), C^*(\bar{x}), \nabla_2^*} \ (r.v),$$

where (1) one of $B^*(\bar{x}), C^*(\bar{x})$ means the empty word, (2) $\bar{x} = x_1, \ldots, x_n$; $\bar{y} = y_1, \ldots, y_n$; $\bar{y} \notin S$; variables $\bar{x}$ are called renaming variables; $\bar{y}$ are called renamed ones; (3) (*subs*) is not applicable to S; (4) $B^0(\bar{y}) = B(\bar{y})$ $(C^0(\bar{y}) = C(\bar{y}))$, if either $A(\bar{p})$ is a positive (negative) or $A(\bar{q})$ is a negative (positive) occurrence of $B(\bar{x})$ $(C(\bar{x})$, respectively), otherwise $B^0(\bar{y}) = \varnothing$ $(C^0(\bar{y}) = \varnothing$, respectively) (condition (*)).

Remark 2.4. The rule (*r.v*) corresponds to the factorization rule in a traditional resolution-like calculus, and a duplicate of the main formulas in the rules of inference $(\forall \to), (\to \exists)$ in the sequent calculus.

Definition 2.10 (calculus $G_{L\omega}^*$). The *calculus* $G_{L\omega}^*$ is defined by the following postulates.

Axiom: $\Gamma, A \to A, \Delta$ or the premise of (*subs*).

Rules of inference:

1) temporal rules:

$$\frac{\{\Gamma \to \Delta, A^k\}_{k \in \omega}}{\Gamma \to \Delta, \Box A} \ (\to \Box_\omega) \qquad \frac{A, \Box A^1, \Gamma \to \Delta}{\Box A, \Gamma \to \Delta} \ (\Box \to);$$

2) logical rules of inference consist of traditional invertible rules of inference (for $\supset, \wedge, \vee, \neg$), besides, if the main formula of a rule of inference contains variables, then it is introduced into the list of memory formulas, for example,

$$\frac{A(\overline{x}), B(\overline{x}), \Gamma, (A(\overline{x}) \wedge B(\overline{x}))^*, \nabla_1^* \rightarrow \Delta, \nabla_2^*}{A(\overline{x}) \wedge B(\overline{x}), \Gamma, \nabla_1^* \rightarrow \Delta, \nabla_2^*} \ (\wedge \rightarrow);$$

3) implicit rules of inference for quantifiers: $(r.v)$ and $(subs)$;

4) structural rules: it follows from the definition of a sequent that $G_{L\omega}^*$ implicitly contains the structural rules "exchange" and "contraction".

Definition 2.11 (quasi-different axiomatic substitution, correct derivation in $G_{L\omega}^*$). Let $\sigma_1 = \overline{x}_1 \leftarrow \overline{t}_1$ and $\sigma_2 = \overline{x}_2 \leftarrow \overline{t}_2$ be two axiomatic substitutions. Then σ_1, σ_2 are called *quasi-different*, if $\overline{t} \neq \overline{t}_2$ and $\overline{x}_1 = \overline{x}_2$. A derivation D in $G_{L\omega}^*$ is *correct*, if D does not contain any two quasi-different axiomatic substitutions.

Remark 2.5. Let D be a derivation in $G_{L\omega}^*$ containing an application of $(r.v.)$. Then the condition $(*)$ in $(r.v.)$ means that either $A(\overline{p})$ or $A(\overline{q})$ are ancestors of the formula $B(\overline{x})$ or $C(\overline{x})$, which serves as the main formula of some rule of inference.

Theorem 2.2 (soundness and completeness of $G_{L\omega}^*$, admisibility of (cut) in $G_{L\omega}^*$). (a) *Let S be a quantifier-free variant of an N-sequent, then the sequent S is universally valid iff $G_{L\omega}^* \vdash S$; (b) $G_{L\omega}^* + (cut) \vdash S \Rightarrow G_{L\omega}^* \vdash S$.*
Proof. Point (a) is carried out in two steps. In the first step, the soundness and completeness of $G_{L\omega}$ with explicit rules for quantifiers are proved (see, e.g., [4]). In the second step, a deductive equivalence between the calculi $G_{L\omega}$ and $G_{L\omega}^*$ is proved. Point (b) follows from (a).

Definition 2.12 (primary and quasi-primary sequents, simple primary sequents) A sequent S is *primary*, if $S = \Sigma_1, \Pi_1^n, \Box\Delta_1, \nabla_1^* \rightarrow \Sigma_2, \Pi_2^n, \Box\Delta_2, \nabla_2^*$, where $\Sigma_i = \varnothing$ $(i \in \{1, 2\})$ or consist of logical formulas; $\Pi_i^n = \varnothing$ $(i \in \{1, 2\})$ or consist of atomic formulas with indices; $\Box\Delta_i = \varnothing$ $(i \in \{1, 2\})$ or consist of formulas of the shape $\Box B^k$ $(k \geqslant 0)$. If $k = n \geqslant 1$ in $\Box\Delta_i$ $(i \in \{1, 2\})$, then S is called a *quasi-primary* sequent. S is *simple*, if it does not contain positive occurrences of $\Box$.

Definition 2.13 (reduction of a sequent S to sequents $S_1, \ldots, S_n$). Let $\{i\}$ denote the set of rules of inference of a calculus I^*. The $\{i\}$-*reduction* (or, briefly, *reduction*) of S to $S_1, \ldots, S_n$ denoted by $R(S)\{i\} \Rightarrow \{S_1, \ldots S_n\}$ or briefly by $R(S)$, is defined to be a tree of sequents with the root S and leaves $S_1, \ldots, S_n$, and possibly some logical axioms such that each sequent in $R(S)$, different from S, is the "upper sequent" of the rule of inference in $\{i\}$ whose "lower sequent" also belongs to $R(S)$.

Definition 2.14 (calculi (I_1^*, I_2^* and Log). The calculus I_1^* is obtained from $G_{L\omega}^*$ replacing $(\rightarrow \Box_\omega)$ by the following one:

$$\frac{\Gamma \rightarrow \Delta, A; \ \Gamma \rightarrow \Delta, \Box A^1}{\Gamma \rightarrow \Delta, \Box A} \ (\rightarrow \Box^1).$$

I_2^* is obtained from I_1^* by dropping $(\to \square^1)$; Log is obtained from $G_{L\omega}^*$ by dropping temporal rules.

Remark 2.6. All rules of $I \in \{I_1^*, I_2^*, Log\}$ are invertible.

Lemma 2.1. *Let S be a sequent, then there exists $R(S)\{i\} \Rightarrow \{S_1, \ldots, S_n\}$ such that (1) $\forall j$ $(1 \leqslant j \leqslant n)$ S_j is a quasi-primary (primary) sequent; $\{i\}$ is the set of rules of the calculus I_1^* (the set of rules of inference of Log, respectively); (2) $G_{L\omega}^* \vdash S \Rightarrow G_{L\omega}^* \vdash S_j$ $(j = 1, \ldots, n)$.*
Proof: using the given derivation and bottom-up applying the rules of inference from $\{i\}$.

Definition 2.15 (proper primary and quasi-primary reduction, primary and quasi-primary reduction-tree). Let $P(S)\,(QP(S))$ be the set of primary (quasi-primary, respectively) sequents from Lemma 3.2. Then a set $P^*(S)$ $(Q^*P(S))$, obtained from $P(S)\,(QP(S)$, respectively) by dropping the sequents derivable in I_2^* and by dropping the sequents, which are subsumed by a sequent from $P(S)$ $(QP(S)$, respectively) is called a *proper primary (quasi-primary*, respectively) reduction of S. The derivation-tree consisting of bottom-up applications of the rules of inference of Log (of I_1^*) and having the sequent S as the root and the set $P^*(S)$ $(Q^*P(S)$, respectively) as the leaves, is called a *proper primary (quasi-primary) reduction-tree* of S and denoted by $P^*T(S)$ $(Q^*PT(S)$, respectively).

3 Description and investigation of saturation calculus Sat

In this section a finitary saturation calculus Sat will be introduced and investigated.

The object of consideration of Sat is quantifier-free forms of N-sequents of special shape, called FC-sequents.

Definition 3.1 (FC-sequents). Let S be an N-sequent, then a quantifier-free form S^* of S is called a finite complete sequent (in brief: an FC-sequent), if the following conditions are satisfied:

(1) S^* contains at most one positive occurrence of $\square$, and any negative occurrence of $\square$ does not contain a positive occurence of $\square$ (Horn-like condition);

(2) S^* has one of the following shapes (without loss of generality we can assume that S^* is a primary sequent, i.e. $S^* = \Sigma_1, \Pi_1^n, \square\Omega^k, \nabla_1^* \to \Sigma_2, \Pi_2^n, \square\Delta^k, \nabla_2^*$ $(n \geqslant 1,\ k \geqslant 0), \square\Delta^k = \varnothing$ or $\square\Delta = \square A^k)$:

(2.1) $\square\Delta^k = \varnothing$, i.e. S^* is simple;

(2.2) Ω^k and Δ^k do not have common predicate symbols;

(2.3) all atomic formulas from Ω with the index $k \geqslant 1$ have no function symbols; such an FC-sequent S^* is called functionally simple;

(2.4) if $A \in \square\Omega$, then all atomic formulas from A contain the same multiplicity of a function symbol f; in such a case, we say that the FC-sequent S^* is functionally continuous;

(2.5) let $\square B$ be a formula from $\square\Omega$; then all the atomic formulas from $\square B$ with indices and without them have different variables; in such a case, we say that the FC-sequent S^* consists of variable-isolated temporal formulas;

(2.6) let $\square\Omega$ contain a formula B of the shape $\square(N(x) \supset N^m(f^l(x)))$, $l \geqslant 1$, (such a formula is called an explicit implicative source of incompleteness); then $\square\Omega = B, \square\Omega_1$, where $\square\Omega_1$ is either $\square(N^m(f^l(y)) \supset N^1(y))$ or $\square(N^m(f^l(y_1)) \supset N^m(f^{l-1}(y_1))), \ldots, \square(N^m(f(y_l)) \supset N^1(y_l))$; in such a case, we say that the FC-sequent S^* contains explicit implicative temporal loop;

(2.7) let $\square\Omega$ contain formulas C, D of the shape $\square(N(x) \supset N^m(x))$, $\square(N(z) \supset N(f^l(z)))$ $(m, l \geqslant 1)$ (such formulas are called an implicit implicative source of incompleteness); then $\square\Omega = C, D, \square\Omega_1$, where $\square\Omega_1$ is the same as in point (2.6); in such a case, we say that the FC-sequent S^* contains an implicit implicative temporal loop;

(2.8) let $\square\Omega$ contain formulas C, D of the shape $\square(M(x) \equiv \daleth M^1(f^l(x)))$, $\square(N(y) \equiv N^1(g^l(y)))$ $(l \geqslant 1)$ (such formulas are called an equivalentive source of incompleteness); then $\square\Omega = C, D, \square\Omega_1$, where $\square\Omega_1 = \square(N^1(g^l(u)) \supset M^1(u)), \square(N^1(v) \supset M^1(f^l(v)))$; in such a case, we say that the FC-sequent S^* contains an equivalentive temporal loop.

Definition 3.2 (incompleteness constraint). Formulas composing $\square\Omega_1$ in points (2.6), (2.7), (2.8) of Definition 3.1 is called *incompleteness constraints*.

Definition 3.3 (minimal set of predicate symbols of Ω, structural separation rule: rule (SA_i^*)). The *minimal set of predicate symbols* of Ω is defined (and denoted by $Q(\Omega)$) as the follwing set $\{P^k | P^k$ enters Ω and k is the minimal index of the predicate symbol P from $\Omega\}$.

Let us introduce the following *structural separation rule*:

$$\frac{S_i^*}{S^*} (SA_i^*) \quad (i = 1 \text{ or } i = 2 \text{ or } i = 3 \text{ or } i = 4),$$

where S^* is a quasi-primary sequent , i.e., $S^* = \Sigma_1, \Pi_1^n, \square\Delta_1^n(\overline{x}), \nabla_1^* \to \Sigma_2, \Pi_2^n$, $\square\Delta_2^n(\overline{y}), \nabla_2^*(n \geqslant 1)$; $S_1^* = \Sigma_1, \nabla_1^* \to \Sigma_2, \nabla_2^*$; $S_2^* = \Pi_{11}, \nabla_1^* \to \Pi_{21}, \nabla_2^*$, if $\Pi_{j1} \subseteq \Pi_j$ $(j \in \{1, 2\})$ and $Q(\Pi_{j1}) \cap Q(\Delta_1) = \varnothing$; $S_3^* = \to \square\Delta_2(\overline{y}_1)$ or $S_3^* = \Pi_{12}, \square\Delta_1(\overline{x}), \nabla_1^* \to \Pi_{22}, \nabla_2^*$ $(Q(\Pi_{j2}) \cap Q(\Delta_1) \neq \varnothing)$; $\Pi_{j2} \subseteq \Pi_j$, $\{1, 2\})$, if $Q(\Delta_1) \cap Q(\Delta_2) = \varnothing$; $S_4^* = \Pi_{12}, \square\Delta_1(\overline{x}_1), \nabla_1^* \to \Pi_{22}, \square\Delta_2(\overline{y}_1), \nabla_2^*$, if $\Pi_{j2} \subseteq \Pi_j$ $(j \in \{1, 2\})$ and $Q(\Delta_1) \cap Q(\Pi_{j2}) \neq \varnothing$, moreover, if $\Delta_1 = \varnothing$, then $\Pi_{j2} = \varnothing$ and $S_3^* = S_4^*$; $\overline{x} = x_1, \ldots, x_n$; $\overline{x}_1 = x_1^1, \ldots, x_n^1$; $\overline{y} = y_1, \ldots, y_m$; $\overline{y}_1 = y_1^1, \ldots y_m^1$.

Remark 3.1. (a) A simple separation rule (A_i^*) has the shape

$$\frac{S_i^*}{S^*} (A_i^*) \quad (i = 1 \text{ or } i = 2),$$

where S^* is a quasi-primary sequent, i.e., $S^* = \Sigma_1, \Pi_1^n, \square\Omega^n(\overline{x}), \nabla_1^* \to \Sigma_2, \Pi_2^n$, $\square\Delta^n(\overline{y}), \nabla_2^*$; $S_1^* = \Sigma_1, \nabla_1^* \to \Sigma_2, \nabla_2^*$; $S_2^* = \Pi_1, \square\Omega(\overline{x}_1), \nabla_1^* \to \Pi_2, \square\Delta(\overline{y}_1), \nabla_2^*$; (b) The rule $(r.v.)$ is incorporated in to the rules (SA_i^*), (A_i^*).

Lemma 3.1 (disjunctional invertibility of (SA_i^*)). *Let S^* be a conclusion of* (SA_i^*), S_i^* *be a premise of* (SA_i^*), *then (1)* $G_{L\omega}^* \vdash S^* \Rightarrow Log \vdash S_j^*$ $(j \in \{1, 2\})$ *or (2)* $G_{L\omega}^* \vdash S^* \Rightarrow G_{L\omega}^* \vdash S_3^*$ *or (3)* $G_{L\omega}^* \vdash S^* \Rightarrow G_{L\omega}^* \vdash S_4^*$.
Proof: by induction on the height of the given derivation of S^*.

Definition 3.4 (matching formulas and sequents). We say that a formula A *matches* (see, also [2]) a formula B (in symbols $A \approx_m B$), if there exists a substitution σ such that $A = B\sigma$. We say that a sequent $S_1 = A_1, \ldots, A_n \to A_{n+1}, \ldots, A_{n+m}$ *matches* a sequent $S_2 = B_1, \ldots, B_n \to B_{n+1}, \ldots, B_{n+m}$ (in symbols: $S_1 \approx_m S_2$), if $\forall i$ $(1 \leqslant i \leqslant m)$ $A_i \approx_m B_i$.

Definition 3.5 (structural rule (W^*), m-subsumed sequent). Let us introduce the following rule of inference:

$$\frac{\Gamma \to \Delta}{\Pi, \Gamma' \to \Delta', \Theta} \ (W^*),$$

where $\Gamma \to \Delta \approx_m \Gamma' \to \Delta'$.

Let S_1 and S_2 be non-simple primary sequents. We say that a sequent S_1 *m-subsumes* a sequent S_2, if $\frac{S_1}{S_2}$ (W^*) (in symbols $S_1 \succcurlyeq mS_2$), or S_2 is *m-subsumed* by S_1. If S_1, S_2 do not contain variables, then we say that S_1 subsumes S_2 and denote it by $S_1 \succcurlyeq S_2$.

Definition 3.6 (subsumption rule, separated subsumption rule). Let us introduce the following *subsumption rule*:

$$\frac{\{S_1, \ldots, S_{n-1}, S_n, S_{n+1}\}}{\{S_1, \ldots, S_n\}} \ (Su),$$

where $\{S_1, \ldots, S_n\} = \emptyset$, if $n = 1$; $S_n \succcurlyeq mS_{n+1}$ and $S_1, \ldots, S_{n+1}$ are non-simple primary sequents. The bottom-up application of (Su) is denoted by (Su^-).

Let us introduce a *separated subsumption rule*:

$$\frac{\{S_1, \ldots, S_n, S_{n+1}, \ldots, S_{n+m}\}; \ \{S_1^*, \ldots, S_m^*\}}{\{S_1, \ldots, S_n\}} \ (SSu),$$

where $S_i^* \succcurlyeq mS_{n+i}$ $(1 \leqslant i \leqslant m)$; the set $\{S_1, \ldots, S_n, S_{n+1}, \ldots, S_{n+m}\}$ is called *passive part* of (SSu) and the set $\{S_1^*, \ldots, S_m^*\}$ is called an *active part* of (SSu). An inverse application of (SSu) is denoted by (SSu^-).

Definition 3.7 (subsumption-tree (ST)). The *subsumption-tree* of a primary sequent is defined as a deduction-tree (and denoted by (ST)), consisting of all possible applications of (Su^-).

Definition 3.8 (resolvent-tree $ReT(S)$ and resolvent $Re(S)$ of a sequent S). The *resolvent-tree* of a *sequent* S is defined by the following bottom-up deduction (denoted as $ReT(S)$) :

$$\cfrac{\cfrac{S_1, \quad \ldots\ldots, \quad S_k}{\cfrac{S_1^+}{S_1^*,} (SA_i^*) \qquad \cfrac{S_m^+}{S_n^*} (SA_i^*)} (ST)}{S} \ Q^*PT(S)$$

where $S_1^*, \ldots, S_n^*$ are quasi-primary sequents; moreover $S_j^* = \varnothing$ $(1 \leqslant j \leqslant n)$, if $Q^* P(S) = \varnothing$; $S_1^+, \ldots, S_m^+$ are some primary sequents; moreover S_l^+ $(1 \leqslant l \leqslant m)$ is empty, if either $(SA_i^*) \in \{(SA_1^*), (SA_2^*)\}$ or $(SA_i^*) = (SA_3^*)$ and $I_2^* \vdash S_l^+$, where S_l^+ is a simple sequent; and non-empty, if either $(SA_i^*) = (SA_4^*)$ or $(SA_i^*) = (SA_3^*)$ and S_l^+ is a non-simple one. If $\forall j, h$ $(1 \leqslant j \leqslant n; \ 1 \leqslant h \leqslant k)$ $S_j^* = \varnothing$ or $S_h = \varnothing$, then we say that *resolvent* of the sequent S is *empty* (in symbols: $Re(S) = \varnothing$). The set of primary sequents $S_1, \ldots, S_k$ $(k \geqslant 0)$ is called the *resolvent of* S and denoted as $Re(S)$. If $\exists_j$ $(1 \leqslant j \leqslant n)$ such that S_j^* is simple and $I_2^* \nvdash S_j^*$, then we say that the resolvent of the sequent S is false (in symbols $Re(S) = F$).

Remark 3.2. In Lemma 3.3 we show that the derivability in the induction-free calculus I_2^* of a simple sequent S reduces to the derivability of some logical parts of S in *Log*.

Definition 3.9 (*k*-th resolvent-tree $Re^k T(S)$ and *k*-th resolvent $Re^k(S)$ of a sequent S, subsumption index, maximal number of $Re^k(S)$). Let $k = 0$, then $Re^0(S) = Re^0 T(S) = P^*(S)$, i.e., 0-th resolvent of S is the proper primary reduction of S. Let $k \leqslant Ct_\square(S)$ and $Re^k(S) = \{S_1, \ldots, S_n\}$, then the $k+1$-resolvent tree $Re^{k+1} T(S)$ is defined by the following bottom-up deduction:

$$
\frac{\dfrac{R^* e^{k+1}(S) \qquad \overset{k}{\underset{i=0}{\bigcup}} Re^i(S)}{\underbrace{\dfrac{Re(S_1)}{S_1} ReT(S_1) \quad \cdots \quad \dfrac{Re(S_n)}{S_n} ReT(S_n)}_{Re^k(S)}} (ST)}{Re^{k+1}(S)} (SSu^-)
$$

The application of (ST) reduces the set $\overset{n}{\underset{i=1}{\bigcup}} Re(S_i)$ to a set $R^* e^{k+1}(S)$. The application of (SSu^-) reduces the set $R^* e^{k+1}(S)$ to a set of primary sequents called a $k+1$ *resolvent* of S and denoted by $Re^{k+1}(S)$. Moreover, the set $\overset{k}{\underset{i=0}{\bigcup}} Re^i(S)$ is the active part of the application of (SSu^-), and the set $R^* e^{k+1}(S)$ is the passive one of application of (SSu^-). If $\exists i (1 \leqslant i \leqslant n)$ such that $Re(S_i) = F$, then $Re^{k+1}(S) = F$. Let $Re^f(S)$ be the least resolvent from $\overset{k}{\underset{i=0}{\bigcup}} Re^i(S)$ which is actually active, i.e., $\exists S^* \in Re^f(S)$ and $S^* \succcurlyeq S^+ \in R^* e^{k+1}(S)$. Then f is called a *subsumption index*. The set $Re^{k+1}(S)$ is empty in two cases: (1) $\forall i$ $(1 \leqslant i \leqslant n)$ $Re(S_i) = \varnothing$ or (2) the application of (SSu^-) in $Re^{k+1} T(S)$ yields an empty set. Each $Re^k(S)$ $(k \in \omega)$ consists of primary sequents. The *maximal number* of $Re^k(S)$ (denoted by $Max\, Re^k(S)$) is coincidental with $Ct_\square(S) + 1$.

Definition 3.10 (derivation of the *m*-th resolvent $DRe^m(S)$, catch for $Re^m(S)$). *A derivation of the m-th resolvent* is defined by the following bottom-up deduction, denoted by $DRe^m(S)$:

$$DRe^m(S) \begin{cases} Re^m T(S) \\ \vdots \\ Re^0 T(S) \end{cases}.$$

The *catch* for $Re^m(S)$ is identical with $Ct_\Box(S) + 1$.

Definition 3.11 (termination of $DRe^k(S)$). The construction of $DRe^k(S)$ is *terminated*, if $Re^k(S) = \varnothing$ or $Re^k(S) = F$, where $k \leqslant Ct_\Box(S) + 1$.

Definition 3.12 (calculus *Sat*). The postulates of the calculus *Sat* consist of two parts:

- an auxiliary part, consisting of postulates of the calculus I_1^* and the rule (SA_i^*);
- the main part, consisting of the two following postulates:
  1) the rules of inference subsumption (Su) and (SSu);
  2) the axiom $Re^k(S) = \varnothing$.

Definition 3.13 (derivation in *Sat*, catch for derivation). The derivation of a sequent S in *Sat* is the same as $DRe^m(S)$. The aim of derivation in *Sat* consists in getting either $Re^k(S) = \varnothing$ (in this case, $Sat \vdash S$) or $Re^k(s) = F$ (in this case, $Sat \nvdash S$). It should be stressed that, in general, we can an get uncertain case, i.e., $Re^k(S) = ?$.

Definition 3.14 (saturated sets $Sat\{S\}$, saturated sequents) Let S be a sequent and $\exists n(Re^n(S) = 0)$, and let f be a subsumption index. Then the set $\overset{n-1}{\underset{i=f}{\cup}} Re^i(S)$ will be called a *saturated set* of the sequent S and denoted by $Sat\{S\}$.

Example 3.1. (a) Let $S = P(c), \Box A(x) \to \Box P(z)$, where $A(x) = P(x) \supset P_1^1(x)$, then $MaxRe^k(S) = 2$. It is easy to verify that $Re^0(S) = \{S\}$, $Re^1(S) = \{S_1\}$, where $S_1 = \Box A(x_1) \to \Box P(z_1)$; $Q^*P(S_1) = \{S_{11}; S_{12}; S_{13}; S_{14}\}$, where $S_{11} = \Box A^1(x_1) \to P(x_1), P(z_1)$; $S_{12} = P_1^1(x_1), \Box A^1(x_1) \to P(z_1)$; $S_{13} = \Box A^1(x_1) \to P(x_1), \Box P^1(z_1)$; $S_{14} = P_1^1(x_1), \Box A^1(x_1) \to \Box P^1(z_1)$. Since S_{11}, S_{12} are simple and $I_2^* \nvdash S_{1i}$ ($i \in \{1, 2\}$), $Re(S_1) = F$ and $Re^2(S) = F$. Therefore $Sat \nvdash S$.

(b) Let $S = P(c), \Box A(x, y) \to P(z)$, where $A(x, y) = P(x) \supset P^1(f(y))$ (i.e., S consists of variable-isolated temporal formulas), then $Max Re^k(S) = 2$ and $Re^0(S) = S$. It is easy to verify that $Re^1(S) = P(f(y))$, $A^*(x, y), \Box A(x_1, y_1) \to P(z_1)$ and $Re^2(S) = \varnothing$; the subsumption index $k = 1$ and $Sat \vdash S$.

(c) Let $S = P(x), \Box A(y) \to \Box P(z)$, where $A(y) = P(f(y)) \supset P^1(f(y))$ (i.e., S is functionally continuous), then $Max Re^k(S) = 2$ and $Re^0(S) = S$. It is easy to verify that $Re^1(S) = P(f(y))$, $A^*(y), \Box A(y_1) \to \Box P(z_1)$ and $Re^2(S) = \varnothing$; the subsumption index $k = 1$ and $Sat \vdash S$.

(d) Let $S = P(c), \Box A(x_1), \Box B(x_2) \to \Box P(c)$, where $A(x_1) = (P(x_1) \supset P^1(f(x_1))), B(x_2) = (P(f(x_2)) \supset P(x_2))$ (i.e., S contains an implicative source of incompleteness as well as an incompleteness constraint), then $MaxRe^k(S) = 2$. It is easy to verify that $Re^0(S) = S$; $Re^1(S) = \{S' = P(f(c)), \Box A(x_1^1), \Box B(x_2^1), A^*(x_1), B^*(x_2) \to \Box P(c)\}$; $R^*e^2(S) = \{S'' = P(f(c)), \Box A(x_1^2),$

$\Box B(x_2^2), A^*(x_1), A^*(x_1^1), B^*(x_2), B^*(x_2^1) \to \Box P(c)\}$. Since $S' \succcurlyeq mS''$, $Re^2(S) = \varnothing$; the subsumption index $k = 1$ and $Sat \vdash S$.

(e) Let $S = P(c), \Box(P(x) \supset P^1(f(x))) \to \Box(P(y))$, i.e., S contains the implicative source of incompleteness and do not contain an incompleteness constraint. It is easy to verify that $G^*_{L\omega} \vdash S$, but all $Re^k T(S)$ $(k \geqslant 2)$ contain only non-simple applications of (subs). Therefore from proposition $(*)$ (see Remark 3.4, later) we can conclude that $Sat \nvdash S$.

Definition 3.15 (resolvent subformulas). A set of *resolvent subformulas* of the formula A is denoted by $RSub(A)$.

1. $RSub(P^k(\bar{t})) = \begin{cases} \{P^m \mid 0 \leqslant m \leqslant k-1\}, & \text{if } k > 0, \\ \varnothing & \text{if } k = 0, \\ \bar{t} = t_1, \ldots, t_n, \quad n \geqslant 0 \end{cases}$

2. If $P(\bar{t}) \in RSub(A)$ and $\bar{t}$ differs from $\bar{t}$ at least by renaming of variables, then $P(\bar{t}_1) \in RSub(A)$;
3. $RSub(A \odot B) = RSub(A) \cup RSub(B)$, where $\odot \in \{\supset, \wedge, \vee\}$;
4. $RSub(\daleth A) = RSub(A)$;
5. $RSub(\Box A) = \{\Box A\} \cup RSub(A)$.

Let $S = A_1, \ldots, A_n \to A_{n+1}, \ldots, A_{n+m}$, then $RSub(S) = \overset{n+m}{\underset{i=1}{\cup}} RSub(A_i)$.

Definition 3.16 Let $\Omega_1, \ldots, \Omega_n, \ldots$ be a set of formulas, then Ω will be a *v-variant* of Ω_i $(i = 1, \ldots, n, \ldots)$, if $\Omega_1, \ldots, \Omega_n, \Omega$ coincide up to renaming of variables.

Definition 3.17 (v-coincidental sequents and formulas). Let $S^* = S_1(x_1, \ldots, x_n)$ and $S' = S_2 (y_1, \ldots, y_n)$ be sequents, $x_1, \ldots, x_n$ be variables entering the sequent S^*, $y_1, \ldots, y_n$ entering the sequent S'. We say that the sequents S^* and S' are *v-coincidental* (in symbols: $S^* \approx S'$) if $S_1(z_1, \ldots, z_n) = S_2(z_1, \ldots, z_n)$. We say that formulas A, B are v-coincidental, if $\to A \approx \to B$.

Definition 3.18 (v-finite set). Let M be a set of formulas, M^* be the set, obtained from M by dropping all but one v-coincidental formulas. The set M will be *v-finite*, if M^* is finite.

Remark 3.3. It follows from Definitions 3.16 and 3.18, that the set $RSub(S)$ is v-finite for an FC-sequent S.

Lemma 3.2 (on the shape of $Sat\{S\}$). *Let* $Sat \vdash S$, *where* $S = \Gamma, \Box\Omega \to \Delta, \Box A$ *and* Ω, A *have common predicate symbols and* $S_i \in Sat\{S\}$. *Then* $S_i = \Gamma_i, \Box\Omega_i \to \Delta_i, \Box A_i$, *where* $\Gamma_i, \Delta_i \subseteq RSub(\Omega)$, $\Omega(A)$ *is the v-variant of* Ω_i $(A_i,$ *respectively).*

Proof. Let $QRSub(A)$ stand for the set of formulas obtained from the set $RSub(A)$ (see Definition 3.15) replacing point 2 by the following: if $P(t) \in QRSub(A)$, then $P(t)\sigma \in QRSub(A)$. It is easy to verify that starting from some k, each member of $Re^l(S)(k \leqslant l)$ has the shape $\Gamma_i, \Box\Omega_i \to \Delta_i, \Box A_i$ $(\Gamma_i, \Delta_i$ is called parametrical formulas of $Re^l(S)$), where $\Omega_i(A_i)$ is the v-variant of $\Omega(A,$ respectively) and $\Gamma_i, \Delta_i \subseteq QRSub(\Omega)$. We prove that $\Gamma_i, \Delta_i \subseteq RSub(\Omega)$.

1. Let S be functionally simple or functionally continuous, then starting from some l all applications of the rule (subs) (in calculating $Re^n(S)$) are simple

(i.e. RHS of the substitution are variables). It means that $\Gamma_i, \Delta_i \subseteq RSub(\Omega)$.

2. Let S contain variable-isolated temporal formulas. Then, starting from some l, all applications of (subs) in calculating $Re^n(S), n \geqslant l$ are such that LHS of the substitution do not enter parametrical formulas of $Re^n(S)$. It means that $\Gamma_i, \Delta_i \subseteq RSub(\Omega)$.

3. S contains an explicit implicative temporal loop. Relying on the shape $\square\Omega$ we get that, starting from some l, all applications of the rule $(subs)$ (in calculating $Re^n(S), n \geqslant l$) whose LHS is the same as the variables from $\square\Omega(\bar{x}^m, \bar{y}^m)$ are simple. It means that $\Gamma_i, \Delta_i \subseteq RSub(\Omega)$.

4. Let S contain an equivalentive or implicit implicative temporal loop. Then the proof is the same as in the previous case.

Remark 3.4. From the proof of Lemma 3.2 we can derive the following proposition $(*)$; let $S = \Gamma, \square\Omega \to \Delta, \square A$, where Ω, A have common predicate symbols and $\exists n \ (Re^n(S) = \varnothing)$, (i.e., $Sat \vdash S$), then $\exists k$ such that $\forall l \ (k \leqslant l \leqslant n) \ Re^l T(S)$ contain only simple applications of $(subs)$.

Now we shall consider a derivability problem in the "induction-free" calculus I_2^* for simple FC-sequents. Instead of directly bottom-up applying the rules of inference of I_2^* we define a "degenerate" saturation procedure reducing the derivability of an FC-sequent in I_2^* to the derivability of some logical parts of S in Log.

Definition 3.19 (degenerate resolvent $dRe(S)$ and the degenerate resolvent tree $dReT(S)$ of a simple FC-sequent, degenerate subsumption rule (dSu), degenerate separated subsumption rule $(dSSu)$ and degenerate subsumption-tree (dST)). The *degenerate subsumption rule* (dSu) and *degenerate separated subsumption rule* $(dSSu)$ are obtained from the rules (Su), (SSu) (see Definition 3.5) replacing non-simple primary sequents by simple ones. The *degenerate subsumption-tree* (dST) is obtained from (ST) (see Definition 3.7 replacing (Su^-) by (dSu^-). The *degenerate resolvent-tree* $dReT(S)$ of a simple FC-sequent S is obtained from $ReT(S)$ constructed in Definition 3.8 replacing (1) $Q^*P(S)$ by $QP(S)$ (see Definition 2.14) and (2) (ST) by (dST) and (SSu^-) by $(dSSu^-)$. If all the applications of (SA_i^*) are logical, then we say that $dRe(S)$ is true (in brief: $dRe(S) = T$). The set $\{S_1, \ldots, S_k\}$ (see Definition 3.8) is $dRe(S)$.

Definition 3.20 (degenerate k-th resolvent tree $dRe^k T(S)$ and the degenerate k-th resolvent $dRe^k(S)$ of a simple FC-sequent). Let $k = 0$, then $dRe^0(S) = dRe^0 T(S) = P(S)$ (i.e., the primary reduction $P(S)$). Let $dRe^k(S) = \{S_1, \ldots, S_n\}$, then the degenerate $k + 1$-*resolvent-tree* $dRe^{k+1}(S)$ and $dRe^{k+1}(S)$ is obtained by deduction, indicated in Definition 3.9 replacing $Re(S_i)$, $ReT(S_i)$, $R^* e^{k+1}(S)$, $Re^i(S)$, (ST) by $dRe(S_i)$, $dReT(S_i)$, $dR^* e^{k+1}(S)$, $dRe^i(S)$, (dST), respectively. If $\forall i (1 \leqslant i \leqslant n) \ dRe(S_i) = T$, then $dRe^{k+1}(S) = T$. The set $dRe^{k+1}(S)$ is empty, if the application of $(dSSu^-)$ yields an empty set.

Algorithm $(*)$ (semi-decidable algorithm for derivability of a simple FC-sequent S in I_2^*).

(1) $k = 0$;

(2) let us calculate $dRe^k(S)$. If $dRe^k(S) = T$ (i.e., starting from k all

applications of (SA_i^*) are logical: the premise of (SA_i^*) is derivable in Log), then $Stop$ and, in this case, $I_2^* \vdash S$, otherwise 3;

(3) if $dRe^k(S) = \varnothing$, then $Stop$ and, in this case, $I_2^* \nvdash S$, otherwise $k := k+1$ and go to 2.

Lemma 3.3 *The derivability of a simple FC-sequent S in I_2^* reduces to the derivability of some sequents $S_1, \ldots, S_n$ in Log.*
Proof: follows from the construction of Algorithm $(*)$.

Definition 3.21 (logically decidable FC-sequents). FC-sequents are logically decidable, if all sequents of the shape $\Sigma_1, \nabla_1^* \to \Sigma_2, \nabla_2^*$ (where $\Sigma_1, \Sigma_2, \nabla_1, \nabla_2$ consist of all the possible subformulas of logical formulas of S) are decidable in the first-order logic.

Lemma 3.4 *Let S be a logically decidable simple FC-sequent, then Algorithm $(*)$ is a decision procedure for the derivability of the sequent S in I_2^*.*
Proof: follows from the fact that Algorithm $(*)$ always terminates for a logically decidable sequent.

Lemma 3.5 *Let S be a logically decidable FC-sequent, then S is decidable in the first order linear temporal logic.*
Proof: follows from the following facts: (1) all applications of (SA_i^*) are deterministic for a logically decidable sequent S_i; (2) it follows from Lemma 3.4 and the construction of $Re^k(S)$ that the process of calculation of $Re^k(S)$ always terminates for a logically decidable sequent S.

Lemma 3.6 *Let S be a FC-sequent, then $Sat \vdash S \Rightarrow G_{L\omega}^* \vdash S$.*
Proof. The proof consists of two parts. In the first part, the invariant calculus IN containing the invariant rule $(\to \square)$ (see Section 1) and "weak induction" rule $(\to \square^1)$ (see Definition 2.14) is constructed. Using the facts that the invariant formula can be constructed from the set of saturated sequents it is proved that $Sat \vdash S \Rightarrow IN \vdash S$. In the second part, using the facts that $(\to \square)$, $(\to \square^1)$ are admissible in $G_{L\omega}^*$, it is proved that $IN \vdash S \Rightarrow G_{L\omega}^* \vdash S$.

Lemma 3.7 *Let S be an FC-sequent, then $G_{L\omega}^* \vdash S \Rightarrow Sat \vdash S$.*
Proof. We can assume that S is a primary sequent, i.e. $S = \Sigma_1, \Pi_1^n, \square\Omega^k, \nabla_1^* \to \Sigma_2, \Pi_2^n, \square\Delta^k, \nabla_2^*$ ($n \geqslant 1$, $k \geqslant 0$). Let us consider different cases, depending on the shape of S.

1. $\square\Delta^k = \varnothing$, i.e. S is a simple sequent. In this case $I_2^* \vdash S$, therefore $Re^1(S) = \varnothing$ and $Sat \vdash S$.

2. $\square\Omega^k$ and $\square\Delta^k$ do not have common predicate symbols. It is easy to verify (using that $G_{L\omega}^* \vdash S$) that, in this case, $Re^1(S) = \varnothing$ or $Re^2(S) = \varnothing$, therefore $Sat \vdash S$.

3. S is of the shape indicated in points (2.3), (2.4), (2.5), (2.6), (2.7), (2.8) of Definition 3.1. Since $G_{L\omega}^* \vdash S$, we have that all simple sequents S_i^+ that appear in calculating $Re^n(S)$ are such that $I_2^* \vdash S_i^+$. Therefore, by Lemma 3.2, starting from some l either $Re^n(S) = \varnothing$ (i.e. $Sat \vdash S$) or each member of $Re^n(S)$ ($n \geqslant l$) is of the shape $\Gamma_i, \square\Omega_i \to \Delta_i, \square A_i$, where $\Gamma_i, \Delta_i \subseteq RSub(\Omega)$ (Ω is the v-variant of Ω_i). Since the set $RSub(\Omega)$ is v-finite, we must obtain the empty resolvent, i.e. $Sat \vdash^D S$ and $Ct_{\square}(D) \leqslant Ct_{\square}(S) + 1$.

Theorem 3.1. *The calculus Sat is sound and complete.*
Proof: follows from the soundness and completeness of $G^*_{L\omega}$ and Lemmas 3.6, 3.7.

References

1. H. Andreka, J. Nemeti, J. Sain: On the strenghth of temporal proof. LNCS **379** (1989) 135–144
2. K.R. Apt: On the unification free Prolog programs. LNCS **711** (1993) 1–19
3. M. Fisher: A normal form for first order temporal formulae. LNCS **607** (1992) 370–384
4. H. Kawai: Sequential calculus for a first order infinitary temporal logic. Zeitchr. fur Math. Logic und Grundlagen der Math. **33** (1987) 423–452
5. S. Merz: Decidability and incompleteness results for first order temporal logics of linear time. Journal of Applied Noc-Classical Logics **2** (1992) 139–156
6. R. Pliuškevičius: On saturated calculi for a linear temporal logic. LNCS **711** (1993) 640–649
7. R. Pliuškevičius: The saturated tableaux for linear miniscoped Horn-like temporal logic. Journal of Automated Reasoning **13** (1994) 51–67
8. A. Szalas: Concerning the semantic consequence relation in first-order temporal logic. Theoret. Comput. Sci. **47** (1986) 329–334
9. A. Szalas: A complete axiomatic characterization of first-order temporal logic of linear time. Theoret. Comp. Sci. **54** (1987) 199–214

Propositional Quantification
in Intuitionistic Logic

Tomasz Połacik

Institute of Mathematics, University of Silesia
Bankowa 14, 40007 Katowice, Poland
<polacik@us.edu.pl>

We investigate the fragment of second order propositional intuitionistic logic in the language of propositional variables p, q, r, etc., the standard propositional connectives $\neg, \vee, \wedge, \to$ and the propositional quantifiers $\exists p$, $\forall p$, $\exists q$, $\forall q$, etc. The constants $\bot$, $\top$ and equivalence $\leftrightarrow$ can be defined by means of the other connectives in the usual way. The notions of a (second order) formula and a free variable are standard; to denote formulae we shall use the letters F, G, etc. As usual, we identify the fragment of intuitionistic logic in the language of the standard propositional connectives $\neg, \vee, \wedge, \to$ with Heyting calculus, IPC.

The system IPC^2 corresponding to the fragment of second order logic in question can be defined by adding to the standard axiomatization of Heyting calculus the following schemata of axioms and rules (p is not free in G):

$$F(q) \to \exists p\, F(p),$$
$$\forall p\, F(p) \to F(q),$$
$$\frac{F(p) \to G}{\exists p\, F(p) \to G},$$
$$\frac{G \to F(p)}{G \to \forall p\, F(p)}.$$

The system IPC^2 can be regarded as minimal in the sense that it describes just the basic properties of propositional quantification. From among known properties of IPC^2 let us mention the definability of the quantifier $\exists$ and the connectives $\neg, \vee, \wedge$ in terms of $\forall$ and $\to$ (which contrasts to mutual undefinability of the connectives in IPC) and undecidability (in contrast to decidability of IPC). Note also that although IPC^2 is complete with respect to a variant of Kripke semantics (cf. [1]), intuitively it is incomplete since, e.g. the formula $\neg\forall p\,(p \vee \neg p)$, which is intuitively true intuitionistically, is not derivable in IPC^2. Hence, it follows that IPC^2 can not be complete with respect to any natural semantics for intuitionistic logic. So, in general, different semantics for IPC^2 can potentially lead to different sets of valid formulae and hence different properties of propositional quantification.

In [3] A. Pitts showed, using proof-theoretical methods, that propositional quantification can be modelled within Heyting calculus; recently S. Ghilardi and M. Zawadowski in [2] and, independently, A. Visser in [7] gave semantical proofs of the this property. In fact, it is proved that for every propositional variable p and every formula F of the language of IPC, the set of interpolants of F not involving p is not merely non-empty but contains the weakest and strongest elements with respect to the provability ordering in Heyting calculus. More precisely, the usual Interpolation Theorem for IPC can be strengthened in the following way: given a propositional variable p and a formula F, one can effectively find formulae $Ap\,F$ and $Ep\,F$ in the language of Heyting calculus, containing only variables not equal to p which occur in F such that for all formulae of G of IPC not involving p,

$$\text{IPC} \vdash G \to Ap\,F \quad \text{iff} \quad \text{IPC} \vdash G \to F,$$
$$\text{IPC} \vdash Ep\,F \to G \quad \text{iff} \quad \text{IPC} \vdash F \to G.$$

Consequently, we can model propositional quantification within Heyting calculus: the modelling in question is the identity on quantifier-free formulae and assigns $Ep\,F$ to $\exists p\,F$ and $Ap\,F$ to $\forall p\,F$. It is easy to see that all axioms and rules of IPC^2 are valid in this interpretation.

However, Pitts' interpretation can not be intuitively accepted as a natural semantics for intuitionistic second order logic since, for example, every formula of the form $\forall p\,(F \vee G) \to (\forall p\,F \vee \forall p\,G)$ — although intuitively unacceptable — is valid under this interpretation. Pitts' interpretation validates also all formulae of the form $\neg\neg\exists p\,F \to \exists p\,\neg\neg F$, which does not agree with the constructive meaning of existential quantification.

Since Pitts' quantifiers are definable in Heyting calculus, we can compare them with the specific meaning of $\forall$ and $\exists$ in any other interpretation of IPC^2. Moreover, we have

$$\text{IPC}^2 \vdash Ap\,F \to \forall p F \quad \text{and} \quad \text{IPC}^2 \vdash \exists p F \to Ep\,F,$$

so the formulae $Ap\,F \to \forall p F$ and $\exists p F \to Ep\,F$ are true in every model of second order propositional intuitionistic logic. We can thus ask whether a given interpretation of propositional quantification coincides with Pitts' interpretation i.e. whether the formulae $Ap\,F \leftrightarrow \forall p F$ and $\exists p F \leftrightarrow Ep\,F$ are valid with respect to the considered semantics.

We focus on the topological interpretation of IPC^2. Let us recall that it is defined as the extension of the standard topological interpretation of IPC (cf. [6]) as follows: Given the topological space T, we set an assignment which to the propositional variables $p, q, r, \ldots$ assigns open subsets of $T - P, Q, R, \ldots$ respectively. Then, relative to the chosen assignment, for each second order formula $F(p, \overline{q})$, we define, by induction on the complexity of F, an open set $F[P, \overline{Q}]$ (the symbols $\overline{q}$ and $\overline{Q}$ denote finite sequences of propositional variables and the corresponding open subsets of T respectively):

$$(\neg F')[P, \overline{Q}] = \text{int}\,(-F'[P, \overline{Q}])$$
$$(F' \vee F'')[P, \overline{Q}] = F'[P, \overline{Q}] \cup F''[P, \overline{Q}]$$
$$(F' \wedge F'')[P, \overline{Q}] = F'[P, \overline{Q}] \cap F''[P, \overline{Q}]$$
$$(F' \to F'')[P, \overline{Q}] = \text{int}\,(-F'[P, \overline{Q}] \cup F''[P, \overline{Q}])$$
$$(\exists p\, F')[\overline{Q}] = \bigcup\{F'[P, \overline{Q}] \;:\; P - \text{open}\}$$
$$(\forall p\, F')[\overline{Q}] = \text{int}\,\bigcap\{F'[P, \overline{Q}] \;:\; P - \text{open}\}$$

We say that a (second order) formula $F(\bar{q})$ is valid in a topological space T and write $T \models F$, if for all assignments for the propositional variables, $F[\bar{Q}] = T$.

We show that — although completely different in nature — the interpretation over any dense-in-itself metric space coincides with Pitts' interpretation when restricted to the monadic language. On the other hand, we prove that in the language of at least two variables the topological interpretation and Pitts' interpretation are different.

First we consider the monadic fragment of IPC^2. Let us fix the following enumeration of the formulae of the monadic formulae of IPC:

$$
\begin{aligned}
F_0(p) &= \bot, \\
F_1(p) &= p, \\
F_2(p) &= \neg p, \\
F_{2n+1}(p) &= F_{2n}(p) \vee F_{2n-1}(p), \\
F_{2n+2}(p) &= F_{2n}(p) \rightarrow F_{2n-1}(p), \\
F_\omega(p) &= \top,
\end{aligned}
$$

where $n \geq 1$. The formulae F_n are all, up to equivalence in IPC, monadic formulae of Heyting calculus. Note that with the provability ordering in IPC they give rise to the so-called Rieger-Nishimura lattice which is the single-generated free Heyting algebra.

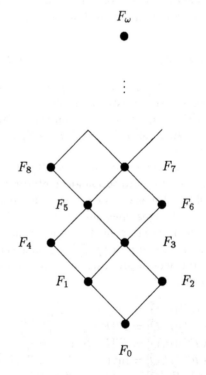

For the further reference, let us list a few sample formulae from our list:

$$F_3(p) = p \vee \neg p, \quad F_4(p) = \neg\neg p, \quad F_5(p) = \neg p \vee \neg\neg p, \quad F_6(p) = \neg\neg p \to p$$
$$F_7(p) = \neg\neg p \vee (\neg\neg p \to p), \quad F_9(p) = (\neg\neg p \to p) \vee ((\neg\neg p \to p) \to \neg p \vee \neg\neg p).$$

Note that for all monadic formulae $F(p)$ — not equivalent to $\top$, nor to $\bot$ — we have

$$\mathsf{IPC} \vdash \mathsf{A}p\, F \leftrightarrow \bot \qquad \text{and} \qquad \mathsf{IPC} \vdash \mathsf{E}p\, F \leftrightarrow \top.$$

So, in order to prove that in the monadic fragment Pitts' interpretation coincides with topological interpretation, we have to show that for a formula F (such that F is not equivalent to $\top$ nor to $\bot$) and a topological space T,

$$\mathrm{int} \bigcap \{F[P] : P - \mathrm{open}\} = \emptyset \qquad \text{and} \qquad \bigcup \{F[P] : P - \mathrm{open}\} = T.$$

The latter equality is trivially valid in every topological space. It is also easy to see that $\bigcap \{F[P] : P - \mathrm{open}\} = \emptyset$ for every topological space T and $F(p) = p, \neg p$ or $\neg\neg p$. So it is enough to prove this equality for all formulae $F(p)$ such that $\mathsf{IPC} \vdash p \vee \neg p \to F(p)$.

In order to show the required fact for universally quantified formulae, we introduce the notions of feasibility and realizability of subsets of topological spaces with respect to propositional formulae. These notions play a technical role and provide an efficient tool for proving our results. Although we shall refer to feasibility and realizability only in case of the monadic language, we give the general definitions.

Definition 1. Let $F(\overline{q})$ be a formula of the language of IPC and X a subset of a topological space T. We say that X *is feasible to* $F(\overline{q})$ if there is a sequence $\overline{Q}$ of open subsets of T such that $F[\overline{Q}] = X$.

Definition 2. Let $F(\overline{q})$ be a formula of the language of IPC and X a subset of a topological space T. We say that a sequence $\overline{Q}$ of open subsets of T *realizes* X *with respect to* $F(\overline{q})$ if $F[\overline{Q}] = X$ and $G[\overline{Q}] = T$ for every formula $G(\overline{q})$ of the language of IPC such that $\mathsf{IPC} \nvdash F \to G$ and $\mathsf{IPC} \nvdash G \to F$. We say that a subset X of T *is realizable with respect to* $F(\overline{q})$ if there is a sequence $\overline{Q}$ of open subsets of T which realizes X with respect to $F(\overline{q})$.

It is clear from the definition that if a set X is realizable with respect to a formula F, then it is also feasible to F. Of course, the converse need not hold. Note also that only open subsets of topological spaces can be feasible or realizable.

We restate Definition 2 to clarify the notion of realizability in the case of the monadic language.

Corollary 3. *An open set P realizes a subset X of a topological space T with respect to a formula $F_m(p)$ if $F_m[P] = X$ and one of the following holds:*

(i) $m = 0$ *and* $X = \emptyset$ *and* P *is an arbitrary open subset of* T,

(ii) $m = 1$ and $P = \emptyset$,
(iii) $m = 2$ and $P = T$,
(iv) $m = 2n + 1$, $n \geq 1$ and $F_{2n+2}[P] = T$,
 (v) $m = 2n + 2$, $n \geq 1$ and $F_{2n}[P] = T$,
(vi) $m = \omega$ and $X = T$ and P is an arbitrary open subset of T.

Corollary 3 becomes more readable if referred to the appropriate fragment of Rieger-Nishimura lattice:

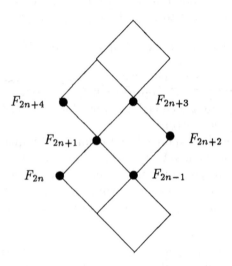

Note that, for $m \geq 3, m \neq 4$, an open subset P of a space T realizes X with respect to F_m, if and only if the Heyting algebra $\mathcal{A}$ generated by P (as a subalgebra of the algebra of the family of open subsets of the space T) is finite, strongly compact with $F_m[P] = X$ as its greatest non-unit element. Moreover, P realizes X with respect to F_m if and only if there is an embedding ϕ of $\mathcal{A}$ into (the algebra of open subsets of) the space T with $\phi(x) = X$ and $\phi(1) = T$, where x and 1 denote the greatest non-unit element and the unit of $\mathcal{A}$ respectively.

Now we aim at answering the question: Which subsets of topological spaces are feasible and realizable with respect to particular F_m's where $m \geq 3$ and $m \neq 4$? Note that when P and X are open subsets of an arbitrary topological space and $n \geq 1$, then P realizes X with respect to F_{2n+1} iff P realizes X with respect to F_{2n+4}. Moreover, if P realizes X with respect to F_{2n+1}, then $F_{2n-1}[P] = X$, which means that X is feasible to F_{2n-1}. So, it is sufficient to answer our question only in case of realizability with respect to F_m's where m is an odd number and $m \geq 3$.

Notice, in this case only open and dense sets can be feasible or realizable with respect to F_m's. Thus, we can ask whether all of them are realizable with respect to the formulae in question. It turns out that in the case of F_3 and F_6 the answer is positive.

Theorem 4. *Let T be an arbitrary topological space and $X \subseteq T$. Let $m \in \{3, 6\}$. Then the following are equivalent:*

(i) *X is realizable with respect to F_m,*
(ii) *X is feasible to F_m,*
(iii) *X is open and dense.*

In contrast to the previous cases, feasibility and realizability of a (dense and open) set X with respect to F_5 requires an additional property.

Theorem 5. *In arbitrary topological space the following are equivalent:*

(i) *X is realizable with respect to F_5,*
(ii) *X is feasible to F_5,*
(iii) *X is open, dense and X can be decomposed into a union of two disjoint regularly open sets.*

Let us note the following consequence of Theorem 5:

Corollary 6. *Let T be an arbitrary topological space. Then, if $X \neq T$ is realizable with respect to F_5, then the set $-X$ disconnects the space.*

Theorem 5 suggests the question, whether all open and dense subsets of a particular topological space possess the property (iii). In general, the answer is negative — examples are provided by Corollary 6:

Example 1. Consider the space $[0, 1]$ and its open and dense subset $(0, 1]$. Since $\{0\}$ — which is the complement of the set $(0, 1]$ in the space $[0, 1]$ — does not disconnect $[0, 1]$, by Corollary 6, the set $(0, 1]$ is not feasible (and realizable with respect) to F_5. The same can be asserted about the sets $(0, 1)$ and $[0, 1)$.

Example 2. Let $n \geq 2$ and let Z be a subset of the Euclidean space $\mathbb{R}^n$; assume that Z is compact and homeomorphic to a proper subset of the sphere S^{n-1}. Then, since Z does not disconnect the space $\mathbb{R}^n$, the set $\mathbb{R}^n \setminus Z$ is not feasible (and realizable with respect) to F_5.

On the other hand, one can prove

Theorem 7. *In Cantor space and in the reals every open and dense subset is a union of two disjoint regularly open sets.*

So, every open and dense subset of Cantor space and of the reals is feasible and realizable with respect to F_5.

The problem of realizability and feasibility with respect to F_7 depends, as in the case of F_5, on some topological property. This property can be described as follows: given an arbitrary nowhere dense subset Z we can find a set T of isolated points such that Z is exactly the set of all the accumulation points of T, i.e. $Z = T^d$. Let us note the following

Lemma 8. *For every nowhere dense subset Z of an arbitrary dense-in-itself metric space, there is a set T of isolated points disjoint from Z and such that $T^d = Z$.*

So, considering F_7 we have to restrict to the class of dense-in-themselves metric spaces.

Theorem 9. *A dense and open subset of a dense-in-itself metric space is realizable with respect to F_7 iff it is realizable with respect to F_5.*

It follows that a subset X of the space T is realizable with respect to F_7 if it can be decomposed into a union of two disjoint regularly open space and T is dense-in-itself and metric. This condition is stronger that in the case of realizability with respect to F_5. One could expect that realizability of open and dense subsets with respect to particular F_m's, for sufficiently large m, would require stronger and stronger assumptions on topological properties of the space. Surprisingly, it turns out that only the assumption of metrizability suffices to prove that the problem of realizability with respect to F_m for sufficiently large m's has a uniform solution. The proof relies on the property which can be described intuitively in the following way: given an arbitrary nowhere dense subset Z of the topological space, we can find a family $\mathcal{F}$ of pairwise disjoint sets which are disjoint from cl Z and such that Z is the set of all the accumulation points of $\bigcup \mathcal{F}$; moreover, we require that all the elements of the family $\mathcal{F}$ can be contained in a given open subset Y of the space, disjoint from Z and such that any neighborhood of an arbitrary point of Z intersects Y. It turns out that this property is possessed by any dense-in-itself metric space. More precisely, we can prove

Lemma 10. *Let T be an arbitrary dense-in-itself metric space and let Y, Z be disjoint subsets of T such that $Z \subseteq$ cl Y. Then there is a set $T \subseteq Y$ and a family $\{K_x : x \in T\}$ of open neighbourhoods of x, for every $x \in T$, and such that*

(i) T is scattered and $T^d = $ cl Z;
(ii) cl $K_x \cap$ cl $Z = \emptyset$ for every $x \in T$;
(iii) cl $K_x \cap$ cl $K_y = \emptyset$ for all $x, y \in T$, $x \neq y$;
(iv) cl $(K \cap \bigcup_{x \in T}$ cl $K_x) = \bigcup_{x \in T}$ cl $(K \cap$ cl $K_x) \cup$ cl Z,
for all K such that $K \cap$ cl $K_x \neq \emptyset$ for every $x \in T$.

A generalization of the above result to topological spaces which are not metric is still an open problem. However, in view of the following example, significant improvements do not seem to be possible.

Example 3. Consider the space 2^m for $m \geq \aleph_1$. It is well-known that in 2^m every family of open, pairwise disjoint sets is at most countable. It yields the bound for the power of the set T to which Lemma 10 refers: T also can be at most countable and hence the power of the closure of T can be at most $2^{\aleph_0}$. On the other hand, in the space 2^m there are nowhere dense sets of the power greater than $2^{\aleph_0}$ and, obviously, cannot be the sets of the accumulation points of T.

Notice that every 2^m (for $m \geq \aleph_1$) is a compact, Hausdorff (and hence normal), totally disconnected space — nevertheless it is not metrizable.

Now we can state our next theorem.

Theorem 11. *Every open and dense subset of an arbitrary dense-in-itself metric space is realizable with respect to F_m for all $m \geq 9$, $m \neq 10$.*

To complete our investigation in case of the monadic language and universal quantification, we need the following theorem which can be proved by theorems 4, 5, 9 and 11.

Theorem 12. *Let T be an arbitrary dense-in-itself metric space and let $F(p)$ be an arbitrary monadic formula of Heyting calculus which is not provable in IPC. Then for every $x \in T$ there is an open set P_x such that $x \notin F[P_x]$.*

Now, let T be an arbitrary dense-in-itself metric space. Then for any monadic quantifier-free formula $F(p)$ which is not provable in Heyting calculus and for every $x \in T$, there is an open set P_x such that $x \notin F[P_x]$. Thus, we have

$$\text{int} \bigcap \{F[P] \; : \; P - \text{open}\} \subseteq \text{int} \bigcap \{F[P_x] \; : \; x \in T\} \subseteq \text{int} \bigcap_{x \in T} -\{x\} = \emptyset,$$

which proves

Theorem 13. *Let T be an arbitrary dense-in-itself metric space. Then the topological interpretation of the monadic fragment of IPC^2 over T coincides with Pitts' interpretation.*

We turn to the language of IPC^2 with at least two variables. We shall prove that, in this case, the standard topological meaning of quantifiers is not — in general — the same as the meaning of Pitts' quantifiers. However, instead of proving this negative result for the existential quantifier only (which would imply the negative result for the universal quantifier), we divide the problem into two separate cases of the two quantifiers. The reason is that in the definition of the existential quantifier,

$$\exists p \, F \; \leftrightarrow \; \forall q (\forall p \, (F \rightarrow q) \rightarrow q),$$

a new variable occurs and proceeding this way we would not be able to cover the case of the universal quantifier and the language of two variables.

First, let us note

Theorem 14. *For every topological space, the topological interpretation of the universal quantifier does not coincide with Pitts' interpretation when restricted to the language of at least two variables.*

To prove this fact, consider $F(p,q) = p \vee (p \to q)$. One can show that for every topological space T, $T \not\models \forall p\, F \to \mathsf{A}p\, F$.

Note that using the argument presented above, we prove in fact that under the standard topological interpretation of IPC^2 over any topological space T, the schema $\forall p\, (F \vee G) \to \forall p\, F \vee \forall p\, G$ is not generally valid. Recall that every formula of this form is valid under Pitts' interpretation.

Finally we turn to the problem of existential quantifier.

Theorem 15. *For every dense-in-itself metric space, the topological interpretation of the existential quantifier does not coincide with Pitts' interpretation when restricted to the language of two variables.*

In this case the counter-example is more complicated. Let us consider

$$G(q) = F_7(q) = \neg\neg q \vee (\neg\neg q \to q),$$
$$H(q) = F_6(q) = \neg\neg q \to q$$

and, in the end,

$$F(p,q) = (G \to (\neg p \to H \vee p \to H)) \to G.$$

Then, exploiting Lemma 10, one can show that for every dense-in-itself metric space T, $T \not\models \mathsf{E}p\, F \to \exists p\, F$.

References

1. D. M. Gabbay, *On 2nd order intuitionistic propositional calculus with full comprehension*, Archiv für mathematische Logik 16 (1974), 177–186.
2. S. Ghilardi, M. Zawadowski, *A sheaf representation and duality for finitely presented Heyting algebras*, Journal of Symbolic Logic 60 (1995), 911–939.
3. A. Pitts, *On an interpretation of second order quantification in the first order intuitionistic propositional logic*, Journal of Symbolic Logic 57 (1992), 33–52.
4. T. Połacik, *Propositional quantification in the monadic fragment of intuitionistic logic*, to appear in Journal of Symbolic Logic.
5. T. Połacik, *Pitts' quantifiers are not topological quantification*, preprint.
6. A. Tarski, *Der Aussagenkalkül und die Topologie*, Fundamenta Mathemeticae 31 (1938), 103–134.
7. A. Visser, *Bisimulations, Model Descriptions and Propositional Quantifiers*, Logic Group Preprint Series 161, Department of Philosophy, Utrecht University, Utrecht, 1996.

Sketch-as-Proof [*]

Norbert Preining[**]

University of Technology, Vienna, Austria

Abstract. This paper presents an extension of Gentzen's **LK**, called **L$_{PG}$K**, which is suitable for expressing projective geometry and for deducing theorems of plane projective geometry. The properties of this calculus are investigated and the cut elimination theorem for **L$_{PG}$K** is proven. A formulization of sketches is presented and the equivalence between sketches and formal proofs is demonstrated.

1 Introduction

Sketches are very useful things to illustrate the facts of a proof and to make the idea of a proof transparent. But sketches need not only be just a hint, they can, in certain cases, be regarded as a proof by itself. Projective geometry[1] is best for analyzing this relation between sketches and proofs.

The purpose of this paper is to bring an idea of what a sketch can do and to explain the relations between sketches and proofs. Therefore we extend Gentzen's **LK** to **L$_{PG}$K** and use the properties of **L$_{PG}$K** to formalize the concept of a sketch. We will then present a result on the equivalence of sketches and proofs.

2 A Short Introduction to Projective Geometry

The root of projective geometry is the parallel postulate introduced by Euclid (*c.* 300 B.C.). The believe in the absolute truth of this postulate remains unshakable till the 19th century when the founders of non-Euclidean geometry—Carl Friedrich Gauss (1777–1855), Nicolai Ivanovitch Lobachevsky (1793–1856), and Johann Bolyai (1802–1860)—concluded independently that a consistent geometry denying Euclid's parallel postulate could be set up. Nevertheless projective geometry was developed as an extension of Euclidean geometry; i.e., the parallel postulate was still used and a line was added to the Euclidean plane to contain the "ideal points", which are the intersection of parallel lines. Not till the end of the 19th century and the beginning of the 20th century, through the work of Felix Klein (1849–1925), Oswald Veblen (1880–1960), David Hilbert, and others, projective geometry was seen to be independent of the theory of parallels.

---

[*] Supported by the Austrian Research Fund (FWF Projekt P11934-MAT)
[**] University of Technology, Vienna, Austria
[1] We will understand under "projective geometry" the *plane* projective geometry and will loose the "plane" for simplicity.

Projective geometry was then developed as an abstract science based on its own set of axioms. For a more extensive discussion see [3], [2].

The projective geometry deals, like the Euclidian geometry, with points and lines. These two elements are primitives, which aren't further defined. Only the axioms tell us about their properties. We will use the expression *Point* (note the capital *P*) for the objects of projective geometry and *points* as usual for e.g. a point in a plane. The same applies to *Line* and *line*. The only predicate beside the equality is called *Incidence* and puts up a relation between Points and Lines.

Furthermore we must give some axioms to express certain properties of Points and Lines and to specify the behavior of the incidence on Points and Lines:

(PG1) For every two distinct Points there is one and only one Line, so that these two Points incide with this Line.

(PG2) For every two distinct Lines there is one and only one Point, so that this Point incides with the two Lines[2].

(PG3) There are four Points, which never incide with a Line defined by any of the three other Points.

2.1 Examples for Projective Planes

The projective closed Euclidean plane Π_{EP} The easiest approach to projective geometry is via the Euclidean plane. If we add one Point "at infinity" to each line and one "ideal Line", consisting of all these "ideal Points", it follows that two Points determine exactly one Line and two distinct Lines determine exactly one Point[3]. So the axioms are satisfied.

This projective plane is called Π_{EP} and has a lot of other interesting properties, especially that it is a classical projective plane.

The minimal Projective Plane One of the basic properties of projective planes is the fact, that there are seven distinct Points. Four Points satisfying axiom (PG3) and the three diagonal Points $([A_0B_0][C_0D_0]) =: D_1$ etc. (see. fig. 1). If we can set up a relation of incidence on these Points such as that the axioms (PG1) and (PG2) are satisfied, then we have a minimal projective plane. Fig. 1 defines such an incidence-table. This table has to be read carefully: The straight lines and the circle symbolize the Lines and the labeled points the Points of the minimal projective plane. There are no more Points, Lines, especially no intersections as the holes in the figure should suggest.

3   The Calculus $\mathbf{L_{PG}K}$

The calculus $\mathbf{L_{PG}K}$ is based on Gentzen's **LK**, but extends it by certain means. The usual notations as found in [6] are used.

---

[2] "one and only one" can be replaced by "one", because the fact that there is not more than one Point can be proven from axiom (PG1).

[3] More precise: The "ideal Points" are the congruence classes of the lines with respect to the parallel relation and the "ideal Line" is the class of these congruence classes.

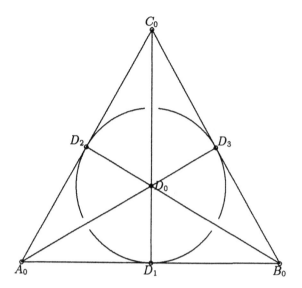

Fig. 1. Incidence Table for the minimal Projective Plane

3.1 The Language L_{PG} for $L_{PG}K$

The language for $L_{PG}K$ is a type language with two types, Points and Lines. These two types will be denoted with τ_P and τ_L, respectively. There are four individual constants of type τ_P: A_0, B_0, C_0, D_0, two function constants (the type is given in parenthesis): con:$[\tau_P, \tau_P \to \tau_L]$, intsec:$[\tau_L, \tau_L \to \tau_P]$, and two predicate constants (the type is given in parenthesis): $\mathcal{I}$:$[\tau_P, \tau_L]$, $=$.

There are free and bound variables of type τ_P and τ_L, which are P_i (free Points), X_i (bound Points), g_i (free Lines), x_i (bound Lines) and we will use the logical symbols $\neg$ (not), $\wedge$ (and), $\vee$ (or), $\supset$ (implies), $\forall_{\tau_P}$ (for all Points), $\forall_{\tau_L}$ (for all Lines), $\exists_{\tau_P}$ (there exists a Point), $\exists_{\tau_L}$ (there exists a Line).

The constants $A_0, \dots, D_0$ are used to denote the four Points obeying (PG3). We will use further capital letters, with or without sub- and superscripts, for Points[4] and lowercase letters, with or without sub- and superscripts, for Lines.

Furthermore we will use the notation $[PQ]$ for the connection con(P, Q) of two Points and the notation (gh) for the intersection intsec(g, h) of two Lines to agree with the classical notation in projective geometry. Finally $\mathcal{I}(P, g)$ will be written $P\mathcal{I}g$.

We also lose the subscript $_{\tau_P}$ and $_{\tau_L}$ in $\forall_{\tau_P}, \dots$, since the right quantifier is easy to deduce from the bound variable.

The formulization of terms, atomic formulas and formulas is a standard technique and can be found in [6].

---

[4] Capital letters are also used for formulas, but this shouldn't confuse the reader, since the context in each case is totally different.

3.2 The Rules and Initial Sequents of $\mathbf{L_{PG}K}$

Definition 1. A *logical initial sequent* is a sequent of the form $A \to A$, where A is atomic.

The *mathematical initial sequents* are formulas of one of the following forms:

1. $\to P\mathcal{I}[PQ]$ and $\to Q\mathcal{I}[PQ]$.
2. $\to (gh)\mathcal{I}g$ and $\to (gh)\mathcal{I}h$.
3. $\to x = x$ where x is a free variable.

The *initial sequents for* $\mathbf{L_{PG}K}$ are the logical initial sequents and the mathematical initial sequents.

The first two clauses are nothing else then (PG1) and (PG2). (PG3) is realized by a rule. The fact that $X = Y \to$ for $X, Y \in \{A_0, B_0, C_0, D_0\}$ and $X \neq Y$ can be deduced from this rule. The *rules for* $\mathbf{L_{PG}K}$ are structural rules, logical rules, cut rule (taken from **LK** for many-sorted languages), the following equality rules:

$$\frac{\Gamma \to \Delta, s = t \quad s = u, \Gamma \to \Delta}{t = u, \Gamma \to \Delta} \text{ (trans:left)}$$

$$\frac{\Gamma \to \Delta, s = t \quad \Gamma \to \Delta, s = u}{\Gamma \to \Delta, t = u} \text{ (trans:right)}$$

$$\frac{s = t, \Gamma \to \Delta}{t = s, \Gamma \to \Delta} \text{ (symm:left)} \qquad \frac{\Gamma \to \Delta, s = t}{\Gamma \to \Delta, t = s} \text{ (symm:right)}$$

$$\frac{\Gamma \to \Delta, s = t \quad s\mathcal{I}u, \Gamma \to \Delta}{t\mathcal{I}u, \Gamma \to \Delta} \text{ (id-}\mathcal{I}_{T_p}\text{:left)}$$

$$\frac{\Gamma \to \Delta, s = t \quad \Gamma \to \Delta, s\mathcal{I}u}{\Gamma \to \Delta, t\mathcal{I}u} \text{ (id-}\mathcal{I}_{T_p}\text{:right)}$$

$$\frac{\Gamma \to \Delta, u = v \quad s\mathcal{I}u, \Gamma \to \Delta}{s\mathcal{I}v, \Gamma \to \Delta} \text{ (id-}\mathcal{I}_{T_L}\text{:left)}$$

$$\frac{\Gamma \to \Delta, u = v \quad \Gamma \to \Delta, s\mathcal{I}u}{\Gamma \to \Delta, s\mathcal{I}v} \text{ (id-}\mathcal{I}_{T_L}\text{:left)}$$

$$\frac{\Gamma \to \Delta, s = t}{\Gamma \to \Delta, [su] = [tu]} \text{ (id-con:1)} \qquad \frac{\Gamma \to \Delta, u = v}{\Gamma \to \Delta, [su] = [sv]} \text{ (id-con:2)}$$

$$\frac{\Gamma \to \Delta, g = h}{\Gamma \to \Delta, (tg) = (th)} \text{ (id-int:1)} \qquad \frac{\Gamma \to \Delta, g = h}{\Gamma \to \Delta, (gt) = (ht)} \text{ (id-int:2)}$$

and the mathematical rules: (PG1-ID) and (Erase)

$$\frac{\Gamma \to \Delta, P\mathcal{I}g \quad \Gamma \to \Delta, Q\mathcal{I}g \quad P = Q, \Gamma \to \Delta}{\Gamma \to \Delta, [PQ] = g} \text{ (PG1-ID)}$$

$$\frac{\Gamma \to \Delta, X\mathcal{I}[YZ]}{\Gamma \to \Delta} \text{ (Erase)}$$

where $\neq (X, Y, Z)$ and $X, Y, Z \in \{A_0, B_0, C_0, D_0\}$

Finally proofs are defined as usual.

3.3 Sample Proofs in $L_{PG}K$

The Diagonal-points We will proof that the diagonal-point D_1 (cf. 2.1) is distinct from $A_0, \ldots, D_0$.

$$\frac{\rightarrow A_0 \neq D_1 \quad \rightarrow B_0 \neq D_1 \quad \rightarrow C_0 \neq D_1 \quad \rightarrow D_0 \neq D_1}{\rightarrow \neq (A_0, B_0, C_0, D_0, D_1)}$$

Each of the proof-parts is similar to the following for $\rightarrow A_0 \neq D_1$

$$\frac{\cfrac{\cfrac{A_0 = ([A_0 B_0][C_0 D_0]) \rightarrow A_0 = ([A_0 B_0][C_0 D_0]) \quad \rightarrow ([A_0 B_0][C_0 D_0])\mathcal{I}[C_0 D_0]}{A_0 = ([A_0 B_0][C_0 D_0]) \rightarrow A_0 \mathcal{I}[C_0 D_0]} \text{(atom)}}{\cfrac{A_0 = ([A_0 B_0][C_0 D_0]) \rightarrow}{\rightarrow A_0 \neq ([A_0 B_0][C_0 D_0])} \text{(}\neg\text{:right)}} \text{(Erase)}}$$

Identity of the Intersection-point We will proof the fact, that there is only one intersection-point of g and h, i.e, the dual fact of (PG1-ID).

$$\frac{\cfrac{\cfrac{\cfrac{\cfrac{\cfrac{P\mathcal{I}g \rightarrow P\mathcal{I}g \quad \rightarrow (gh)\mathcal{I}g \quad P = (gh) \rightarrow P = (gh)}{P\mathcal{I}g \rightarrow P = (gh), [P(gh)] = g} \text{(atom)} \quad \cfrac{P\mathcal{I}h \rightarrow P\mathcal{I}h \quad \rightarrow (gh)\mathcal{I}h \quad P = (gh) \rightarrow P = (gh)}{P\mathcal{I}h \rightarrow P = (gh), [P(gh)] = h} \text{(atom)}}{P\mathcal{I}g, P\mathcal{I}h \rightarrow P = (gh), g = h} \text{(atom)}}{g \neq h, P\mathcal{I}g, P\mathcal{I}h \rightarrow P = (gh)} \text{(}\neg\text{:left)}}{P\mathcal{I}g \wedge P\mathcal{I}h \wedge g \neq h \rightarrow P = (gh)} \text{(}\wedge\text{:left)}}{\rightarrow P\mathcal{I}g \wedge P\mathcal{I}h \wedge g \neq h \supset P = (gh)} \text{(}\supset\text{:right)}}{\rightarrow (\forall X)(\forall u)(\forall v)(X\mathcal{I}u \wedge X\mathcal{I}v \wedge u \neq v \supset X = (uv))} \text{(}\forall\text{:right)}}$$

4 On the Structure of Proofs in $L_{PG}K$

4.1 The Cut Elimination Theorem for $L_{PG}K$

We will refer to the equality rules, (PG1-ID) and (Erase) as (atom)-rules, because they only operate on atomic formulas and therefore they can be shifted above any logical rule (see Step 1 below). We will now transform any given proof in $L_{PG}K$ step by step into another satisfying some special conditions, especially that the new one contains no (Cut).

First we reduce the problem to a special class of proofs, the proofs in normal form[5]. A proof in this class is split into two parts $\mathcal{P}_1$ and $\mathcal{P}_2$

$$\vdots\ \mathcal{P}_1$$
$$\vdots\ \mathcal{P}_2$$
$$\Pi \rightarrow \Gamma$$

---

[5] This nomenclature is used only in this context and has no connection with any other "normalization".

where $\mathcal{P}_1$ is an (atom)-part with (atom)- and structural rules only and $\mathcal{P}_2$ is a logical part with logical and structural rules only. In the first part geometry is practiced in the sense that in this part the knowledge about projective planes is used. The second part is a logical part connecting the statements from the geometric part to more complex statements with logical connectives. It is easy to see, that for every proof in $\mathbf{L_{PG}K}$ there is a proof in normal form of the same endsequent.

Lemma 2. *For every proof in normal form with only one cut there is a proof in normal form of the same endsequent without a cut.*

PROOF (Sketch, detailed exposition in [5]):

STEP 1: We will start with the cut-elimination procedure as usual in **LK** as described e.g. in [6]. This procedure shifts a cut higher and higher till the cut is at an axiom where it can be eliminated trivially. Since in our case above all the logical rules there is the (atom)-part, the given procedure will only shift the cut in front of this part.

STEP 2: Now the cut is already in front of the (atom)-part:

$$
\cfrac{\vdots\ \mathcal{P}_1 \qquad\qquad \vdots\ \mathcal{P}_2}{\cfrac{\Pi_1 \to \Gamma_1, P(t,u) \quad P(t,u), \Pi_2 \to \Gamma_2}{\Pi \to \Gamma}}\ \text{(Cut)}
$$

First all the inferences not operating on the cut-formulas or one of its predecessors are shifted under the cut-rule. Then the rule from the right branch over the cut-rule are shifted on the left side by applying the dual rules[6] in inverse order. Finally we get on the right side either a logical axiom or a mathematical axiom. The case of a logical axiom is trivial, in case of the mathematical axiom the rules from the left side are applied in inverse order on the antecedent of the mathematical axiom which yields a cut-free proof. □

EXAMPLE: A trivial example should explain this method: The proof

$$
\cfrac{x_2 = x_3 \to x_2 = x_3 \quad \cfrac{x_1 = x_2 \to x_1 = x_2 \quad x_1 = u \to x_1 = u}{x_1 = x_2, x_1 = u \to x_2 = u}}{\cfrac{x_2 = x_3, x_1 = x_2, x_1 = u \to x_3 = u \qquad\qquad x_3 = u \to}{x_2 = x_3, x_1 = x_2, x_1 = u \to}}\ \text{(Cut)}
$$

will be transformed to

$$
\cfrac{x_1 = x_2 \to x_1 = x_2 \quad \cfrac{x_2 = x_3 \to x_2 = x_3 \quad x_3 = u \to}{x_2 = x_3, x_2 = u \to}}{x_1 = x_2, x_2 = x_3, x_1 = u \to}
$$

♡

---

[6] E.g. (trans:left) and (trans:right) are dual rules

Theorem 3 (Cut Elimination for $\mathbf{L_{PG}K}$). *If there is a proof of a sequent $\Pi \to \Gamma$ in $\mathbf{L_{PG}K}$, then there is also a proof without a cut.*

PROOF: By the fact that everything above a given sequent is a proof of this sequent and by using Lemma 2 and induction on the number of cuts in a proof we could eliminate one cut after another and end up with a cut-free proof. □

EXAMPLE: We will now present an example proof and the corresponding proof without a cut. We want to prove that for every line there is a point not on that line, in formula: $(\forall g)(\exists X)(X \not\!I g)$.

We will first give the proof in words and then in $\mathbf{L_{PG}K}$.

PROOF: (Words) When $A_0 \not\!I g$ then take A_0 for X. Otherwise $A_0 I g$. Next if $B_0 \not\!I g$ take B_0 for X. If also $B_0 I g$ then take C_0, since when A_0 and B_0 lie on g, then $g = [A_0 B_0]$ and $C_0 \not\!I [A_0 B_0] = g$ by (PG3). □

PROOF: ($\mathbf{L_{PG}K}$)

$$
\cfrac{
\cfrac{
A_0 I g \to A_0 I g
}{
\cfrac{
\to A_0 I g,\; A_0 \not\!I g
}{
\to A_0 I g \vee A_0 \not\!I g
}}
\qquad
\cfrac{
\cfrac{A_0 I g \to A_0 I g \qquad \cfrac{A_0 \not\!I g \to A_0 \not\!I g}{A_0 \not\!I g \to (\exists X)(X\not\!I g)}}{\qquad}
\qquad
\cfrac{
\cfrac{
\cfrac{B_0 I g \to B_0 I g}{\cfrac{\to B_0 I g,\; B_0\not\!I g}{\to B_0 I g \vee B_0\not\!I g}}
\qquad
\cfrac{\cfrac{B_0 I g \to B_0 I g}{\cfrac{B_0\not\!I g \to B_0\not\!I g}{B_0\not\!I g \to (\exists X)(X\not\!I g)}}\qquad \vdots\ \Pi_1}{\cfrac{B_0\not\!I g \to (\exists X)(X\not\!I g) \qquad A_0 I g, B_0 I g \to (\exists X)(X\not\!I g)}{B_0 I g \vee B_0\not\!I g,\; A_0 I g \to (\exists X)(X\not\!I g)}}
}{A_0 I g \to (\exists X)(X\not\!I g)}\ \text{(Cut)}
}{A_0 I g \vee A_0 \not\!I g \to (\exists X)(X\not\!I g)}
}{
\to (\exists X)(X\not\!I g)
}\ \text{(Cut)}
}{
\to (\forall g)(\exists X)(X\not\!I g)
}
$$

Π_1 :

$$
\cfrac{
\cfrac{A_0 I g \to A_0 I g \quad B_0 I g \to B_0 I g \quad A_0 = B_0 \to}{A_0 I g,\, B_0 I g \to g = [A_0 B_0]}
\qquad
\cfrac{C_0 \mathcal{I}[A_0 B_0] \to C_0 \mathcal{I}[A_0 B_0]}{C_0 \mathcal{I}[A_0 B_0] \to}\ \text{(Erase)}
}{
\cfrac{A_0 I g,\, B_0 I g,\, C_0 I g \to}{A_0 I g,\, B_0 I g \to (\exists X)(X\not\!I g)}
}
$$

□

The cut-elimination procedure[7] yields a cut-free proof of the same end-sequent:

$$
\cfrac{
\cfrac{A_0 I g \to B_0 I g \quad B_0 I g \to B_0 I g \quad A_0 = B_0 \to}{A_0 I g,\, B_0 I g \to g = [A_0 B_0]}
\qquad
\cfrac{C_0 \mathcal{I}[A_0 B_0] \to C_0 \mathcal{I}[A_0 B_0]}{C_0 \mathcal{I}[A_0 B_0] \to}\ \text{(Erase)}
}{
\cfrac{A_0 I g,\, B_0 I g,\, C_0 I g \to}{
\cfrac{\to A_0 \not\!I g,\, B_0 \not\!I g,\, C_0 \not\!I g}{
\cfrac{\to (\exists X)(X\not\!I g),\, (\exists X)(X\not\!I g),\, (\exists X)(X\not\!I g)}{
\cfrac{\to (\exists X)(X\not\!I g)}{
\to (\forall g)(\exists X)(X\not\!I g)
}}}}
}
$$

♡

---
[7] or a close look

With the cut-elimination theorem there are some consequences following, e.g. the mid-sequent-theorem for $\mathbf{L_{PG}K}$ and the term-depth-minimization of minimal proofs as found in [4].

5   The Sketch in Projective Geometry

Most of the proofs in projective geometry are illustrated by a sketch. But this method of a graphical representation of the maybe abstract facts is not only used in areas like projective geometry, but also in other fields like algebra, analysis and I have even seen sketches to support understanding in a lecture about large ordinals, which is highly abstract!

The difference between these sketches and the sketches used in projective geometry (and similar fields) is the fact, that proofs in projective geometry deal with geometric objects like Points and Lines, which are indeed objects we can imagine and draw on a piece of paper (which is not necessary true for large ordinals).

So the sketch in projective geometry has a more concrete task than only illustrating the facts, since it exhibits the incidences, which is the only predicate constant besides equality really needed in the formulization of projective geometry. It is a sort of proof by itself and so potentially interesting for a proof-theoretic analysis.

As a first example I want to demonstrate a proof of projective geometry, which is supported by a sketch. It deals with a special sort of mappings, the so called "collineation". This is a bijective mapping from the set of Points to the set of Points, which preserves collinearity. In a formula:

$$\mathrm{coll}(R, S, T) \supset \mathrm{coll}(R\kappa, S\kappa, T\kappa)$$

(Functions in projective geometry are written behind the variables!) The fact we want to proof is

$$\neg\mathrm{coll}(R, S, T) \supset \neg\mathrm{coll}(R\kappa, S\kappa, T\kappa)$$

That means, that not only collinearity but non-collinearity is preserved under a collineation.

The proof is relatively easy and is depicted in fig. 2: If $R\kappa$, $S\kappa$ and $T\kappa$ are collinear, then there exists a Point X' not incident with the Line defined by $R\kappa$, $S\kappa$, $T\kappa$. There exists a Point X, such that $X\kappa = X'$. This Point X doesn't lie on any of the Lines defined by R, S, T. Let $Q = ([RT][XS])$ then $Q\kappa\mathcal{I}[R\kappa S\kappa]$ and $Q\kappa\mathcal{I}[S\kappa X\kappa]$, that is $Q\kappa\mathcal{I}[S\kappa X']$ (since collinearity is preserved). So $Q\kappa = S\kappa$ (since $Q\kappa = ([R\kappa S\kappa][S\kappa X']) = S\kappa$), which is together with $Q \neq S$ a contradiction to the injectivity of κ.

This sketch helps you to understand the relation of the geometric objects and you can follow the single steps of the verbal proof.

If we are interested in the concept of the sketch in mathematics in general and in projective geometry in special then we must set up a formal description

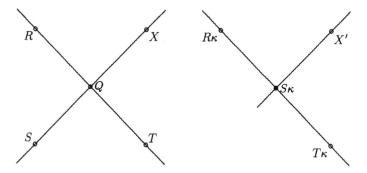

Fig. 2. Sketch of the proof $\neg\text{coll}(R, S, T) \supset \neg\text{coll}(R\kappa, S\kappa, T\kappa)$

of what we mean by a sketch. This is necessary if we want to be more concrete on facts on sketches.

We will give only a short description of what a sketch is and refer the interested reader to [5] for a detailed exposition of the formulization.

A sketch is coded as a quadruple $(\mathcal{M}, \mathcal{E}_+, \mathcal{E}_-, Q)$, where $\mathcal{M}$ is a set of terms with certain limitations, $\mathcal{E}_+$ is a set of positive atomic formulas with the predicate $\mathcal{I}$ over the set $\mathcal{M}$, $\mathcal{E}_-$ is a set of negated atomic formulas with the predicate $\mathcal{I}$ over the set $\mathcal{M}$ and Q is a set of equalities. All these sets have to obey certain requirements ensuring the consistency.

But a sketch is only a static concept, nothing could happen, we cannot "construct". So we want to give some actions on a sketch, which construct a new sketch with more information. These actions on sketches should reflect the actions done by geometrician when drawing, i.e. developing, a sketch. After these actions are defined we can explain what we mean by a construction in this calculus for construction.

The actions primarily operate on the set $\mathcal{E}_+$, since the positive facts are those which are really constructed in a sketch. But on the other hand there are some actions to add negative facts to a sketch. This is necessary for formalizing the elementary way of proving a theorem by an indirect approach.

The actions are:

- Connection of two Points X, Y
- Intersection of two Lines g, h
- Assuming a new Object C in general position
- Giving the Line $[XY]$ a name $g := [XY]$
- Giving the Point (gh) a name $P := (gh)$
- Identifying two Points u and t
- Identifying two Lines l and m
- Using a "Lemma": Adding a positive literal $t\mathcal{I}u$
- Using a "Lemma": Adding a negative literal $t\mathcal{\not I}u$
- Using a "Lemma": Adding a negative literal $t \neq u$

To deduce a fact with sketches we connect the concept of the sketch and the concept of the actions into a new concept called construction. This construction will deduce the facts.

A construction is coded as a rooted and directed tree with a sketch attached to each node[8] and certain demands on its structure.

Finally it is possible to define what a construction deduces by observing the formulas in the leafs of the tree.

We want to mention that great parts of the actions can be automatized so that the constructor can concentrate on the construction. We want to develop a program incorporating these ideas which produces a proof from a sketch.

6 An Example of a Construction

In this section we want to give an example proved on the one hand within $\mathbf{L_{PG}K}$ and on the other hand within the calculus of construction given in the last section. Although a lot of the concepts mentioned in this part weren't introduced, we give the full listing to give the reader a hint on what is really happening.

We want to prove the fact that the diagonal-point $D_1 := ([A_0 B_0][C_0 D_0])$ and the diagonal-point $D_2 := ([A_0 C_0][B_0 D_0])$ are distinct. See fig. 3 for the final sketch, i.e. we have already constructed all the necessary objects from the given Points A_0, B_0, C_0, D_0. This step is relatively easy and there are no problems with any of the controls.

The construction tree is depicted in fig. 3 and the respective labels can be found in the table on p. 11. Note the bold formulas in $\mathcal{E}_+^5, \mathcal{E}_-^5$ and in $\mathcal{E}_+^6, \mathcal{E}_-^6$, which yield the contradiction.

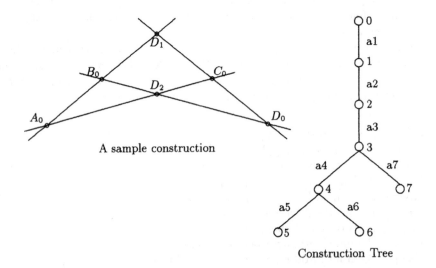

A sample construction

Construction Tree

Fig. 3. Sketch and Construction Tree

---

[8] Actually a semisketch, but don't bother about ...

In the following lists and in the figure not all formulas are mentioned, especially such formulas unnecessary for the construction are not listed. For the construction tree compare with fig. 3. We can see, that the case-distinction after $D_1 = D_2$ yields a contradiction in any branch, therefore we could deduce with the construction that $D_1 \neq D_2$, since this formula is in all leafs, which are not contradictious.

$$a1 = \quad ([A_0 B_0][C_0 D_0])$$

$\mathcal{M}^0 = \{A_0, B_0, C_0, D_0, [A_0 B_0], \dots\}$

$\mathcal{M}^1 = \{A_0, B_0, C_0, D_0, [A_0 B_0], \dots, ([A_0 B_0][C_0 D_0])\}$

$\mathcal{E}^0_+ = \{A_0 \mathcal{I}[A_0 B_0], \dots\}$

$\mathcal{E}^1_+ = \{A_0 \mathcal{I}[A_0 B_0], \dots, ([A_0 B_0][C_0 D_0]) \mathcal{I}[A_0 B_0], \dots\}$

$\mathcal{E}^0_- = \{A_0 \neq B_0, \dots, A_0 \not\!\mathcal{I}[C_0 D_0]\}$

$\mathcal{E}^1_- = \{A_0 \neq B_0, \dots, A_0 \not\!\mathcal{I}[C_0 D_0]\}$

$Q^0 = \{A_0 = A_0, \dots, [A_0 B_0] = [A_0 B_0], \dots\}$

$Q^1 = Q^0 \cup \{([A_0 B_0][C_0 D_0])\}$

$a2 = \quad ([A_0 C_0][B_0 D_0])$

$\mathcal{M}^2 = \{A_0, B_0, C_0, D_0, [A_0 B_0], \dots, ([A_0 B_0][C_0 D_0]), ([A_0 C_0][B_0 D_0])\}$

$\mathcal{E}^2_+ = \{A_0 \mathcal{I}[A_0 B_0], \dots, ([A_0 B_0][C_0 D_0]) \mathcal{I}[A_0 B_0], ([A_0 C_0][B_0 D_0]) \mathcal{I}[A_0 C_0], \dots\}$

$\mathcal{E}^2_- = \{A_0 \neq B_0, \dots, A_0 \not\!\mathcal{I}[C_0 D_0]\}$

$Q^2 = Q^1 \cup \{([A_0 C_0][B_0 D_0])\}$

$a3 = \quad g := [A_0 B_0], h := [C_0 D_0], l := [A_0 C_0], m := [B_0 D_0], D_1 := (gh), D_2 := (lm)$

$\mathcal{M}^3 = \{A_0, B_0, C_0, D_0, g, h, l, m, D_1, D_2\}$

$\mathcal{E}^3_+ = \{A_0 \mathcal{I}g, B_0 \mathcal{I}g, C_0 \mathcal{I}h, D_0 \mathcal{I}h, A_0 \mathcal{I}l, C_0 \mathcal{I}l, B_0 \mathcal{I}m, D_0 \mathcal{I}m, D_1 \mathcal{I}g, D_1 \mathcal{I}h, D_2 \mathcal{I}l, D_2 \mathcal{I}m\}$

$\mathcal{E}^3_- = \{C_0 \not\!\mathcal{I}g, D_0 \not\!\mathcal{I}g, A_0 \not\!\mathcal{I}h, B_0 \not\!\mathcal{I}h, B_0 \not\!\mathcal{I}l, D_0 \not\!\mathcal{I}l, A_0 \not\!\mathcal{I}m, C_0 \not\!\mathcal{I}m, A_0 \neq B_0, \dots\}$

$Q^3 = \{A_0 = A_0, \dots, g = g, h = h, l = l, m = m, D_1 = D_1, D_2 = D_2\}$

$a4 = \quad D_1 = D_2$

$\mathcal{E}^4_+ = \{A_0 \mathcal{I}g, B_0 \mathcal{I}g, C_0 \mathcal{I}h, D_0 \mathcal{I}h, A_0 \mathcal{I}l, C_0 \mathcal{I}l, B_0 \mathcal{I}m, D_0 \mathcal{I}m, D_1 \mathcal{I}g, D_1 \mathcal{I}h, D_1 \mathcal{I}l, D_1 \mathcal{I}m\}$

$\mathcal{E}^4_- = \{C_0 \not\!\mathcal{I}g, D_0 \not\!\mathcal{I}g, A_0 \not\!\mathcal{I}h, B_0 \not\!\mathcal{I}h, B_0 \not\!\mathcal{I}l, D_0 \not\!\mathcal{I}l, A_0 \not\!\mathcal{I}m, C_0 \not\!\mathcal{I}m, A_0 \neq B_0, \dots\}$

$Q^4 = \{A_0 = A_0, \dots, g = g, h = h, l = l, m = m, D_1 = D_1, D_2 = D_2, D_1 = D_2\}$

$a5 = \quad g = l$

$\mathcal{E}^5_+ = \{A_0 \mathcal{I}g, B_0 \mathcal{I}g, C_0 \mathcal{I}h, D_0 \mathcal{I}h, \mathbf{C_0 \mathcal{I}g}, B_0 \mathcal{I}m, D_0 \mathcal{I}m, D_1 \mathcal{I}g, D_1 \mathcal{I}h, D_1 \mathcal{I}m\}$

$\mathcal{E}^5_- = \{\mathbf{C_0 \not\!\mathcal{I}g}, D_0 \not\!\mathcal{I}g, A_0 \not\!\mathcal{I}h, B_0 \not\!\mathcal{I}h, B_0 \not\!\mathcal{I}g, D_0 \not\!\mathcal{I}g, A_0 \not\!\mathcal{I}m, C_0 \not\!\mathcal{I}m, A_0 \neq B_0, \dots\}$

$Q^5 = \{A_0 = A_0, \dots, g = g, h = h, l = l, m = m, D_1 = D_1, D_2 = D_2, D_1 = D_2, g = l\}$

$a6 = \quad A_0 = D_1$

$\mathcal{E}^6_+ = \{A_0 \mathcal{I}g, B_0 \mathcal{I}g, C_0 \mathcal{I}h, D_0 \mathcal{I}h, A_0 \mathcal{I}l, C_0 \mathcal{I}l, B_0 \mathcal{I}m, D_0 \mathcal{I}m, A_0 \mathcal{I}h, \mathbf{A_0 \mathcal{I}m}\}$

$\mathcal{E}^6_- = \{C_0 \not\!\mathcal{I}g, D_0 \not\!\mathcal{I}g, A_0 \not\!\mathcal{I}h, B_0 \not\!\mathcal{I}h, B_0 \not\!\mathcal{I}l, D_0 \not\!\mathcal{I}l, \mathbf{A_0 \not\!\mathcal{I}m}, C_0 \not\!\mathcal{I}m, A_0 \neq B_0, \dots\}$

$Q^6 = \{A_0 = A_0, \dots, g = g, h = h, l = l, m = m, D_1 = D_1, D_2 = D_2, D_1 = D_2, A_0 = D_1, A_0 = D_2\}$

$a7 = \quad D_1 \neq D_2$

$\mathcal{E}^7_+ = \{A_0 \mathcal{I}g, B_0 \mathcal{I}g, C_0 \mathcal{I}h, D_0 \mathcal{I}h, A_0 \mathcal{I}l, C_0 \mathcal{I}l, B_0 \mathcal{I}m, D_0 \mathcal{I}m, D_1 \mathcal{I}g, D_1 \mathcal{I}h, D_2 \mathcal{I}l, D_2 \mathcal{I}m\}$

$\mathcal{E}^7_- = \{C_0 \not\!\mathcal{I}g, D_0 \not\!\mathcal{I}g, A_0 \not\!\mathcal{I}h, B_0 \not\!\mathcal{I}h, B_0 \not\!\mathcal{I}l, D_0 \not\!\mathcal{I}l, A_0 \not\!\mathcal{I}m, C_0 \not\!\mathcal{I}m, A_0 \neq B_0, \dots, D_1 \neq D_2\}$

$Q^7 = \{A_0 = A_0, \dots, g = g, h = h, l = l, m = m, D_1 = D_1, D_2 = D_2\}$

Table 1. Table of the construction tree nodes

We will now give also a short description of what is happening in this tree: The initial sketch is

$$\mathcal{M} = \{A_0, B_0, C_0, D_0, [A_0B_0], \dots\}$$
$$\mathcal{E}_+ = \{A_0\mathcal{I}[A_0B_0], \dots, D_0\mathcal{I}[C_0D_0]\}$$
$$\mathcal{E}_- = \{A_0\not\mathcal{I}[B_0C_0], \dots, D_0\not\mathcal{I}[A_0B_0]\}$$

After constructing the points D_1 and D_2 and with the shortcuts $[A_0B_0] = g$, $[C_0D_0] = h$, $[A_0C_0] = l$, $[B_0D_0] = m$ we obtain

$$\mathcal{M} = \{A_0, B_0, C_0, D_0, g, h, l, m, D_1, D_2, \dots\}$$
$$\mathcal{E}_+ = \{A_0\mathcal{I}g, B_0\mathcal{I}g, C_0\mathcal{I}h, D_0\mathcal{I}h,$$
$$A_0\mathcal{I}l, C_0\mathcal{I}l, B_0\mathcal{I}m, D_0\mathcal{I}m,$$
$$D_1\mathcal{I}g, D_1\mathcal{I}h, D_2\mathcal{I}l, D_2\mathcal{I}m\}$$
$$\mathcal{E}_- = \{C_0\not\mathcal{I}g, D_0\not\mathcal{I}g, A_0\not\mathcal{I}h, B_0\not\mathcal{I}h,$$
$$B_0\not\mathcal{I}l, D_0\not\mathcal{I}l, A_0\not\mathcal{I}m, C_0\not\mathcal{I}m,$$
$$A_0 \neq B_0, \dots\}$$

We now want to add $D_1 \neq D_2$. For this purpose we identify D_1 and D_2 and put the new sets through the contradiction procedure. We will now follow the single steps:

$$D_2 \leftarrow D_1 \tag{1}$$

and as a consequence

$$D_2\mathcal{I}l \Rightarrow D_1\mathcal{I}l \tag{1a}$$
$$D_2\mathcal{I}m \Rightarrow D_1\mathcal{I}m \tag{1b}$$

and so we get the critical constellation $(A_0, D_1; g, l)$

$$A_0\mathcal{I}g, A_0\mathcal{I}l, D_1\mathcal{I}g, D_1\mathcal{I}l \tag{C}$$

Inquiring the first solution $g = l$ yields

$$l \leftarrow g \tag{1.1}$$

and as a consequence

$$C_0\mathcal{I}l \Rightarrow C_0\mathcal{I}g \tag{1.1a}$$

which is a contradiction to

$$C_0\not\mathcal{I}g \in \mathcal{E}_- \tag{1.1b}$$

Inquiring the second solution $D_1 = A_0$ yields

$$D_1 \leftarrow A_0 \tag{1.2}$$

and as a consequence

$$D_1 \mathcal{I}m \Rightarrow A_0 \mathcal{I}m \qquad (1.2a)$$

which is a contradiction to

$$A_0 \bar{\mathcal{I}}m \in \mathcal{E}_- \qquad (1.2b)$$

Since these are all the critical constellations and a contradiction is derived for each branch, the assumption that $D_1 = D_2$ is wrong and $D_1 \neq D_2$ can be added to $\mathcal{E}_-$.

We will now give a proof in $\mathbf{L_{PG}K}$ which corresponds to the above construction. The labels in this proof will not be the rules of $\mathbf{L_{PG}K}$, but references to the above lines.

Π_1 :

$$\frac{\dfrac{(1.1)}{g = l \to g = l \quad \to C_0 \mathcal{I}l}}{\dfrac{g = l \to C_0 \mathcal{I}g}{g = l \to}\,(1.1b)}\,(1.1a)$$

Π_2 :

$$\frac{\dfrac{(1.2)}{A_0 = D_1 \to A_0 = D_1} \quad \dfrac{\dfrac{(1)}{D_1 = D_2 \to D_1 = D_2 \quad \to D_2 \mathcal{I}m}}{D_1 = D_2 \to D_1 \mathcal{I}m}\,(1b)}{\dfrac{A_0 = D_1, D_1 = D_2 \to A_0 \mathcal{I}m}{A_0 = D_1, D_1 = D_2 \to}\,(1.2b)}\,(1.2a)$$

Π_3 :

$$\frac{\dfrac{(1.1)}{\dfrac{\to A_0 \mathcal{I}g \quad \to A_0 \mathcal{I}l \quad g = l \to g = l \quad \to D_1 \mathcal{I}g}{\to g = l, A_0 = (gl)}} \quad \dfrac{\dfrac{(1)}{D_1 = D_2 \to D_1 = D_2 \quad \to D_2 \mathcal{I}l}}{\dfrac{D_1 = D_2 \to D_1 \mathcal{I}l}{D_1 = D_2 \to g = l, D_1 = (gl)}}\,(1a) \quad \dfrac{(1.1)}{g = l \to g = l}}{\dfrac{D_1 = D_2 \to g = l, A_0 = D_1}{D_1 = D_2 \to g = l \vee A_0 = D_1}}$$

Π_1 examines the branch when $g = l$, Π_2 examines the branch when $A_0 = D_1$, and Π_3 deduces that either $g = l$ or $D_1 = A_0$ under the assumption that $D_1 = D_2$ has to be true. The final proof is

$$\frac{\dfrac{\vdots \Pi_3}{D_1 = D_2 \to g = l \vee A_0 = D_1} \quad \dfrac{\dfrac{\vdots \Pi_1}{g = l \to} \quad \dfrac{\vdots \Pi_2}{A_0 = D_1, D_1 = D_2 \to}}{g = l \vee A_0 = D_1, D_1 = D_2 \to}}{D_1 = D_2 \to}\,(\text{Cut})$$

From this example we can see that construction and proof are very similar in this case. In the next section we want to prove the general result that any construction can be transformed into a proof and vice versa.

7   The Relation between Sketches and Proofs

It is possible to translate a "proof" by construction into a proof in $\mathbf{L_{PG}K}$ and it is also possible to show the equivalence of these to concepts.

Theorem 4. *For any sequent proven in $\mathbf{L_{PG}K}$ there is a set of constructions deducing the same sequent and vice versa.*

PROOF: The proof depends on Herbrand's theorem in the original form and can be found in [5].

According to Tarski (cf. [7]) the abstract theory of projective geometry is undecidable. This result should yield interesting consequences on the relation between sketches and proofs.

8   Closing Comments

We hope that this first analysis of projective geometry from a proof-theoretic point of view opens up a new interesting way to discuss features of projective geometry, which is widely used in a lot of applied techniques. Especially the fact, that the sketches drawn by geometers have actually the same strength as the proofs given in a formal calculus, puts these constructions in a new light. Till now they were considered as nothing more then hints to understand the formal proof by exhibiting you the incidences. But they can be used as proves by themselves.

Therefore we want to develop an automatic sketching tool producing proofs in $\mathbf{L_{PG}K}$ and further analyze $\mathbf{L_{PG}K}$, since e.g. the various interpolation theorems and the discussion of Beth's definability theorem should yield interesting consequences on projective geometry and the way new concepts are defined in projective geometry.

Furthermore we want to discuss some other systems, especially for a treatise on generalized sketches, which should compare to generalizations of proofs (see [1]).

References

1. Matthias Baaz. Generalization of proofs and term complexity. In *Logic Colloquium '94, Main Lecture*, Clermont, France, 1994. Unpublished manuscript.
2. H. S. M. Coxeter. *Projective Geometry*. Springer-Verlag, 1994.
3. David Hilbert. *Grundlagen der Geometrie*. B. G. Teubner, 1899.
4. J. Krajíček and P. Pudlák. The number of proof lines and the size of proofs in first order logic. *Arch. Math. Logic*, pages 69–84, 1988.
5. Norbert Preining. Sketch-as-proof, a proof-theoretic analysis of axiomatic projective geometry. Master's thesis, University of Technology, Vienna, Austria, 1996. ftp://ftp.logic.tuwien.ac.at/pub/preining/sketch-as-proof.ps.
6. Gaisi Takeuti. *Proof Theory*. North Holland, 1987.
7. Alfred Tarski. On essential undecidability. *Journal of Symbolic Logic*, 14:75–78, 1949.

Translating Set Theoretical Proofs into Type Theoretical Programs

Anton Setzer

Department of Mathematics, Uppsala University,
P.O. Box 480, S-751 06 Uppsala, Sweden
email: setzer@math.uu.se

Abstract. We show how to translate proofs of Π_2^0-sentences in the fragment of set theory $\mathrm{KPI}_\mathrm{U}^+$, which is an extension of Kripke-Platek set theory, into proofs of Martin-Löf's Type Theory using a cut elimination argument. This procedure is effective. The method used is an implementation of techniques from traditional proof theoretic analysis in type theory. It follows that $\mathrm{KPI}_\mathrm{U}^+$ and $\mathrm{ML}_1\mathrm{W}$ show the same Π_2^0-sentences and have therefore the same programs. As a side result we get that Π_2^0-sentences provable in intensional and extensional version of $\mathrm{ML}_1\mathrm{W}$ are the same, solving partly a problem posed by M. Hofmann.

1 Introduction

Mathematics is usually developed on the basis of set theory. When trying to use type theory as a new basis for mathematics, most of mathematics has to be reformulated. This is of great use, because the step to programs is direct and one can expect to obtain good programs, provided we have an implementation of Martin-Löf's as a real programming language. However, it seems that many mathematicians will continue to work in set theory. Even when changing to type theory for the formalization, often the proofs will be developed first having classical set theory in the background.

The advantage of set theory is that it is highly flexible – e.g. there are no problems with subtyping – and that it allows to write down expressions without having to care about the type of the objects. Therefore methods for transferring directly set theoretical arguments into type theory could not only make the step from traditional mathematics to type theory easier, but as well help to program in type theory.

Π_2^0-sentences can be considered as specifications of programs and proofs of such sentences in Martin-Löf's Type Theory are implementations of such programs. In this article we will prove that all Π_2^0-sentences provable in the Kripke-Platek style set theory $\mathrm{KPI}_\mathrm{U}^+$ are theorems of $\mathrm{ML}_1\mathrm{W}$, Martin-Löf's Type Theory with W-type and one universe as well. $\mathrm{KPI}_\mathrm{U}^+$ is Kripke Platek set theory with urelemente (the natural numbers), one admissible (admissibles are the recursive analogues of cardinals) closed under the step to the next admissible (a recursive inaccessible) and finitely many admissibles above it. The translation works for the usual variations of type theory (intensional vs. extensional, different formulations of the identity type). Since in [Se93] we have shown that all arithmetical sentences provable in ML_1W are theorems of KPI_U^+ it follows that KPI_U^+ and ML_1W (and the variations of ML_1W) have the same Π_2-theorems. Therefore we solve at the same time at least partly a question posed by M. Hofmann in a talk given in Munich, whether the Π_2^0-sentences of extensional and intensional type theory are the same.

It follows that transferring programs to ML_1W from proof theoretically stronger set theories is no longer possible, therefore our result is the best possible.

Our method is heavily built on transfinite induction. In [Se97] the author has shown that ML_1W proves transfinite induction up to the ordinals $\psi_{\Omega_1}(\Omega_{1+n})$, therefore as well up to $\alpha_n := \psi_{\Omega_1}(\epsilon_{\Omega_{1+n}+1}) + 1$. Transfinite induction up to α_n is exactly what we need in order to analyze $KPI_{U,n}^+$ ($KPI_U^+ = \bigcup_{n \in \omega} KPI_{U,n}^+$). Now it is just necessary to formalize this analysis in ML_1W using that we have transfinite induction up to α_n, and to extract the validity of Π_2^0-sentences from the cut free proofs.

We use here techniques from proof theory. These are based on the ordinal analysis of KPI as formalized by Buchholz in [Bu92]. Although it is very difficult to understand these cut elimination arguments, we think that it is not crucial to have full insight in what is going on there in order to be able to work with the techniques presented in this article.

The main problem we had to overcome for writing this article was to formalize modern proof theoretical methods (Buchholz' $\mathcal{H}$-controlled derivations), which are carried out in full set theory, in type theory. We solved this problem using proof trees with a correctness predicate.

Extensions of the result. The methods used here can be extended to all recent proof theoretical studies using infinitary derivations and ordinal analysis. Only, the type theory is not available yet in all cases, except for Mahlo universes. (For Mahlo, the author has given a formulation in [Se96]). Further, one can see easily that the well-foundedness of the W-type is not needed really here, since we have always a descent in ordinals assigned to the nodes of the tree. (For the RS*-derivations, $||\Gamma||$ is descending). Therefore, by replacing the W-type by a recursive object obtained using the recursion theorem, which can be defined in PRA, one shows with nearly the same proof that $PRA + TI(OT_n)$ and $KPI_{U,n}^+$ have the same Π_2-theorems.

The restriction to *arithmetical* Π_2^0-sentences is not crucial. The quantifiers in such sentences could range as well over bigger sets like lists, free algebras or sets built using Σ and sums. We have in this article added to KPI_U^+ natural numbers as urelemente. By adding the types mentioned before as urelemente as well the proofs should go through without essential modifications.

Practicability It should not be too complicated to program the method used here directly in type theory, the only exception is the well ordering proof, which will be used here and seems to be too long for practical applications. However, one can think of conservative extensions of ML_1W by adding types the elements of which represent ordinal denotations and rules for transfinite induction. Then everything shown here can be easily implemented in Martin-Löf's Type Theory.

Other approaches. Independently, W. Buchholz has taken a different approach for obtaining the same result, by using denotation systems for derivations (extending [Bu91]). He uses transfinite induction as well. His approach has the advantage of giving directly executable programs, whereas our method has the advantage of being very perspicuous and explicit.

The other approach for extracting programs from classical proofs are based on the A-translation. This can even be carried out for full set theory, as shown by Friedman (a good presentation can be found in [Be85] Sect. 1VIII.3). A lot of research is carried out for extracting practical programs using the A-translation, see for instance [BS95] or [Sch92]. However, since Martin-Löf's Type Theory is already a programming language, we believe that our approach allows to switch more easily between classical proofs and direct programming. Further, in KPI_U^+ one has constructions corresponding precisely to the different type constructors in type theory, so with our method we have good control over the strength of the methods used.

2 General Assumptions

Assumption 1. (a) We assume some coding of sequences of natural numbers. $<k_0, \ldots, k_l>$ denotes the sequence $k_0, \ldots, k_l$ and $(k)_i$ the i-th element (beginning with $i = 0$) of the sequence k.

(b) In the following we will omit the use of Gödel-brackets.

(c) In type theory a class is an expression $\{x \mid \phi\}$ such that we can prove $x \in N \Rightarrow \phi$ set, and $s \eta \{x \mid \phi\} := \phi[x := s]$. We identify primitive recursive sets with the class corresponding to it.

(d) In type theory, let $\mathcal{P}(N) := N \to U$. For $A \in \mathcal{P}(N)$, $a \in_U A := T(A(u))$ (where $T(s)$ is the set represented by an element $s \in U$).

We introduce OT, the set of ordinal notations as in [Se97]:

Definition 2. (a) Let OT be defined as in [Se97], Definition 3.9. We define OT_n by:

$0, I \eta \text{ OT}_n$.

If $\alpha, \beta \eta \text{ OT}_n$, $\gamma ='_{NF} \alpha + \beta \lor \gamma =_{NF} \phi_\alpha\beta \lor \gamma =_{NF} \Omega_\gamma \lor \gamma =_{NF} \psi_\beta\gamma$, $\gamma \eta \text{ OT} \cap \epsilon_{\Omega_{I+n}+1}$, then $\gamma \eta \text{ OT}_n$.

In the following α, β, γ denote elements of OT_n.

(b) We restrict the ordering $\prec$ on OT to OT_n (replace $\prec$ by $\prec \cap \text{OT}_n \times \text{OT}_n$).

(c) Let $\text{ML}_1 W$ be Martin-Löf's Type Theory with W-type and one universe, as for instance formulated in [Se97] (we can use other formulations of the identity type as well).

(d) For arithmetical sentences ϕ, let $\widehat{\phi}$ the canonical interpretation of ϕ in $\text{ML}_1 W$.

Theorem 3. If $\text{ML}_1 W \vdash n \in N \Rightarrow \phi(n)$ type, then $\text{ML}_1 W \vdash \forall k \eta \text{ OT}_n.((\forall l \prec k.\phi(k)) \to \phi(l)) \to \forall k \eta \text{ OT}_n.\phi(k)$.

Proof: Let $\mathcal{W}' := \mathcal{W}_{n+1}$ as in [Se97], Definition 5.37. Then $\mathrm{OT}_{n_0} \subset \mathcal{W}'$. Let $\psi(x) := x \; \eta \; \mathrm{OT}_n \to \phi(x)$. Then $\mathrm{Prog}(\mathcal{W}_0, (x)\psi(x))$, $\forall k \in \mathcal{W}_0.\psi(k)$ and we get the assertion.

3 The Set Theory $\mathrm{KPI}^+_{\mathrm{U},n}$

We introduce an extension of Kripke-Platek set theory $\mathrm{KPI}^+_{\mathrm{U},n}$. The best references for this theory are [Ba75] which is an excellent introduction to Kripke-Platek set theory and [Jä86], in which Kripke-Platek set theory is extended by adding a predicate for admissibles – the recursive analogue of regular cardinals – and many variations of such theories are analyzed proof theoretically. Our definition is in the spirit of [Jä86] and adds axioms, which assert the existence of one inaccessible I (which is an admissible closed under the step to the next admissible) and finitely many admissibles above it.

Definition 4. of the theory $\mathrm{KPI}^+_{\mathrm{U},n}$

(a) The language of $\mathrm{KPI}^+_{\mathrm{U},n}$ consists of infinitely many number variables, infinitely many set variables, symbols for finitely many primitive recursive relations (on the natural numbers) P of arbitrary arity (the corresponding primitive recursive relation is denoted by $\widetilde{\mathrm{P}}$), the relations Ad, $\overline{\mathrm{Ad}}$, $\in$ and $\notin$ (the latter are written infix) and the logical symbols $\wedge, \vee, \forall, \exists$.

In the following $x^{\mathrm{nat}}, y^{\mathrm{nat}}, z^{\mathrm{nat}}, u^{\mathrm{nat}}, v^{\mathrm{nat}}$ denote number variables and x^{set}, y^{set}, ... denote set variables, to which we might add (this will apply to all future such conventions) indices or accents. $x^?, y^?, \ldots$ denote either a set or a number variable. In a formula, we will omit the superscripts nat, set and ? after the first occurrence of the variable.

We assume that $\mathrm{P}_=$ (the equality of natural numbers), $\mathrm{P}_<$ (the $<$-relation on $\mathbb{N}$) $\perp$ (the 0-ary false relation) are represented in the languge and that for every primitive recursive relation P represented in the language, the negation of this relation $\overline{\mathrm{P}}$ is represented as well.

$\top := \overline{\perp}$.

(b) Number terms are $\mathrm{S}^k(0)$ and $\mathrm{S}^k(x^{\mathrm{nat}})$, where $k \in \mathbb{N}$, $\mathrm{S}^0(r) := r$, $\mathrm{S}^{k+1}(r) := \mathrm{S}(\mathrm{S}^k(r))$. The set terms are the set variables. r^{nat}, s^{nat} and t^{nat} denote number terms. The superscript will be omitted again after the first occurrence. We define $\mathrm{val}(\mathrm{S}^k(0)) := k$.

(c) Prime formulas are $\mathrm{P}(t_1^{\mathrm{nat}}, \ldots, t_k^{\mathrm{nat}})$, where P is a symbol for a k-ary primitive recursive relation (the correspongind relation will be denoted by $\widetilde{\mathrm{P}}$), $\mathrm{Ad}(x^{\mathrm{set}})$, $\overline{\mathrm{Ad}}(x^{\mathrm{set}})$, $r^{\mathrm{set}} \in y^{\mathrm{set}}$, $r^{\mathrm{nat}} \in y^{\mathrm{set}}$, $r^{\mathrm{set}} \notin y^{\mathrm{set}}$, $r^{\mathrm{nat}} \notin y^{\mathrm{set}}$.

(d) Formulas are prime-formulas, and, if ϕ and ψ are formulas, then $\phi \wedge \psi$, $\phi \vee \psi$, $\forall x^?.\phi$, $\exists x^?.\phi$ are formulas, too.

(e) We define the negation of a formula by the de Morgan rules:
$\neg \mathrm{P}(t_1^{\mathrm{nat}}, \ldots, t_k^{\mathrm{nat}}) := \overline{\mathrm{P}}(t_1^{\mathrm{nat}}, \ldots, t_k^{\mathrm{nat}})$, $\neg(r \in y^{\mathrm{set}}) := r \notin y$, $\neg \mathrm{Ad}(x^{\mathrm{set}}) := \overline{\mathrm{Ad}}(x)$, $\neg(\psi \wedge \phi) := \neg(\psi) \vee \neg(\phi)$, $\neg(\forall x^?.\phi) := \exists x^?.\neg \phi$, $\neg(\neg(\phi)) := \phi$ otherwise. The set of free variables $\mathrm{FV}(A)$ of a formula A is defined as usual.

(f) $\phi \to \psi := \neg\phi \lor \psi$,

$\phi \leftrightarrow \psi := (\phi \to \psi) \land (\psi \to \phi)$.

$\forall x^? \in y^{\mathrm{set}}.\phi := \forall x^?.x \in y^{\mathrm{set}} \to \phi$,

$\exists x^? \in y^{\mathrm{set}}.\phi := \exists x^?.x \in y^{\mathrm{set}} \land \phi$.

$x^{\mathrm{set}} \subset y^{\mathrm{set}} := (\forall u^{\mathrm{nat}} \in x.u \in y) \land (\forall z^{\mathrm{set}} \in x.z \in y)$,

$x^{\mathrm{set}} = y^{\mathrm{set}} := (x \subset y \land y \subset x \land (\mathrm{Ad}(x) \leftrightarrow \mathrm{Ad}(y)))$.

$r^{\mathrm{nat}} = s^{\mathrm{nat}} := \mathrm{P}_=(r,s)$.

$r < s := \mathrm{P}_<(r,s)$.

$\mathrm{trans}(x^{\mathrm{set}}) := \forall y^{\mathrm{set}} \in x.((\forall u^{\mathrm{nat}} \in y.u \in x) \land (\forall z^{\mathrm{set}} \in y.z \in x))$.

(g) The quantifier $\forall x^{\mathrm{set}}$ in front of a formula ϕ is bounded, if $\forall x.\phi$ can be written as $\forall x^{\mathrm{set}} \in y^{\mathrm{set}}.\psi$, similarly for $\exists x^{\mathrm{set}}$. Quantifieres over number variables are always bounded.

(h) A formula is arithmetical, if it contains neither set terms nor set variables; it is a $\Sigma_1^{\mathrm{arith}}$-formula, if it is arithmetical and all sub-formulas starting with universal number quantifiers are of the form $\forall x^{\mathrm{nat}}.x^{\mathrm{nat}} < s^{\mathrm{nat}} \to \phi$; it is an arithmetical Π_2^0-formula, if it is of the form $\forall x^{\mathrm{nat}}.\phi$ where ϕ is a $\Sigma_1^{\mathrm{arith}}$-formula.

A formula is Δ_0, if it contains only bounded set-quantifiers. It is Σ_1, if it contains no unbounded universal set-quantifiers.

If ϕ is a formula, let $\phi^{y^{\mathrm{set}}}$ be the result of replacing in ϕ every unbounded set-quantifier $\genfrac{}{}{0pt}{}{\forall}{\exists} x^{\mathrm{set}}.\psi$ by $\genfrac{}{}{0pt}{}{\forall}{\exists} x^{\mathrm{set}} \in y^{\mathrm{set}}.\psi$ (note that number quantifiers remain unchanged).

(i) Γ, Δ denote multi-sets of formulas. $\Gamma, \Delta := \Gamma \cup \Delta$, $\Gamma, \phi := \Gamma \cup \{\phi\}$.

(j) The logical rules of $\mathrm{KPI}_{\mathrm{U},n}^+$ are

$$\Gamma, \phi, \neg\phi \qquad \frac{\Gamma, \phi \quad \Gamma, \psi}{\Gamma, \phi \land \psi} \qquad \frac{\Gamma, \phi}{\Gamma, \phi \lor \psi} \qquad \frac{\Gamma, \psi}{\Gamma, \phi \lor \psi}$$

$$\frac{\Gamma, \phi}{\Gamma, \forall x^?.\phi} \ (\text{if } x \notin \mathrm{FV}(\Gamma)) \qquad \frac{\Gamma, \phi[x^{\mathrm{nat}} := t^{\mathrm{nat}}]}{\Gamma, \exists x^{\mathrm{nat}}.\phi} \qquad \frac{\Gamma, \phi[x^{\mathrm{set}} := y^{\mathrm{set}}]}{\Gamma, \exists x^{\mathrm{set}}.\phi}$$

$$\frac{\Gamma, \phi \quad \Gamma, \neg\phi}{\Gamma}.$$

(k) Axioms of $\mathrm{KPI}_{\mathrm{U},n}^+$

$\forall \vec{z}^?$ stands in the following for $\forall z_1^?, \dots, z_l^?$ for some l.

The set axioms are

(Ext$_1$) $\quad \forall x^{\mathrm{set}}, y^{\mathrm{set}}, z^{\mathrm{set}}.(x = y \to x \in z \to y \in z)$,

$\qquad\qquad \forall x^{\mathrm{nat}}, y^{\mathrm{nat}}, z^{\mathrm{set}}.(x = y \to x \in z \to y \in z)$.

(Ext$_2$) $\quad \forall x^{\mathrm{set}}.\forall y^{\mathrm{set}}.(x = y \to \mathrm{Ad}(x) \to \mathrm{Ad}(y))$.

(Found) $\quad \forall \vec{z}^?.[(\forall x^{\mathrm{set}}.(\forall y^{\mathrm{set}} \in x^{\mathrm{set}}.\phi(y,\vec{z})) \to \phi(x,\vec{z})) \to$

$\qquad\qquad \forall x^{\mathrm{set}}.\phi(x,\vec{z})]$.

(Pair) $\quad \forall x^?, y^?.\exists z^{\mathrm{set}}.x \in z \land y \in z$.

(Union) $\quad \forall x^{\mathrm{set}}.\exists y^{\mathrm{set}}.\forall z^{\mathrm{set}} \in x.((\forall u^{\mathrm{nat}} \in z.(u \in y)) \land$

$\qquad\qquad (\forall v^{\mathrm{set}} \in z.(v \in y)))$.

($\Delta_0 -$ Sep) $\quad \forall \vec{z}^?.\forall x^{\mathrm{set}}.\exists y^{\mathrm{set}}.[[\forall u^{\mathrm{nat}} \in y.\phi_{\mathrm{nat}}(u,\vec{z})] \land$

$\qquad\qquad [\forall v^{\mathrm{set}} \in y.v \in x \land \phi_{\mathrm{set}}(v,\vec{z})] \land$

$$[\forall u^{\mathrm{nat}}.(\phi_{\mathrm{nat}}(u,\vec{z}) \to u \in y)] \wedge$$
$$[\forall v^{\mathrm{set}} \in x.(\phi_{\mathrm{set}}(v,\vec{z}) \to v \in y)]]$$

where ϕ_{nat} and ϕ_{set} are $\varDelta_0$-formulas .

$$(\varDelta_0 - \mathrm{Coll}) \quad \forall \vec{z}^?.\forall x^{\mathrm{set}}.[(\forall y^{\mathrm{nat}} \in x.\exists v^{\mathrm{set}}.\phi_{\mathrm{nat}}(y,v,\vec{z})) \wedge$$
$$(\forall u^{\mathrm{set}} \in x.\exists v^{\mathrm{set}}.\phi_{\mathrm{set}}(u,v,\vec{z}))] \to \exists w^{\mathrm{set}}.$$
$$[(\forall y^{\mathrm{nat}} \in x.\exists v^{\mathrm{set}} \in w.\phi(y,v,\vec{z})) \wedge$$
$$(\forall u^{\mathrm{set}} \in x.\exists v^{\mathrm{set}} \in w.\phi(u,v,\vec{z}))]$$

where ϕ_{nat} and ϕ_{set} are $\varDelta_0$-formulas .

(Ad.1) $\forall x^{\mathrm{set}}.\mathrm{Ad}(x) \to (\mathrm{trans}(x) \wedge x \neq \emptyset)$.

(Ad.2) $\forall x^{\mathrm{set}}, y^{\mathrm{set}}.((\mathrm{Ad}(x) \wedge \mathrm{Ad}(y)) \to$
$$(x \in y \vee x = y \vee y \in x)) .$$

(Ad.3) $\forall x^{\mathrm{set}}.(\mathrm{Ad}(x) \to \phi^x)$, where ϕ is an instance of an axiom
(Pair), (Union), $(\varDelta_0 - \mathrm{Sep})$ or $(\varDelta_0 - \mathrm{Coll})$.

$(+)_n$ $\exists x^{\mathrm{set}}, x_1^{\mathrm{set}}, \ldots, x_n^{\mathrm{set}}.\mathrm{Ad}(x) \wedge$
$$(\forall y \in x.\exists z^{\mathrm{set}} \in x.(\mathrm{Ad}(z) \wedge y \in z)) \wedge$$
$$\mathrm{Ad}(x_1) \wedge \cdots \wedge \mathrm{Ad}(x_n) \wedge$$
$$x \in x_1 \wedge x_1 \in x_2 \wedge \cdots \wedge x_{n-1} \in x_n .$$

The arithmetical axioms are:
Some formulas $\forall \vec{x}^{\mathrm{nat}}.\exists \vec{y}^{\mathrm{nat}}.\phi(\vec{x},\vec{y})$, where ϕ is quantifier free and for some primitive recursive functions $f_1, \ldots, f_l$ $\mathrm{ML}_1\mathrm{W}$ proves $\forall \vec{k} \in \mathbb{N}.\hat{\phi}(\vec{n}, f_1(\vec{k}), \ldots, f_l(\vec{k}))$. Additionally induction: $\phi(0) \wedge \forall x^{\mathrm{nat}}.(\phi(x) \to \phi(S(x))) \to \forall x^{\mathrm{nat}}.\phi(x)$.

(1) $\mathrm{KPI}_{\mathrm{U}}^+$ is the union of all theories $\mathrm{KPI}_{\mathrm{U},n}^+$ ($n \in \omega$).

4   Formalization of the Infinitary System RS

Let from now on be $n_0 \in \omega$ fixed.
As in [Bu92], we introduce set terms indexed by variables, where
$[x^{\mathrm{set}} \in \mathrm{L}_\alpha : \phi(x)]$ stands for the the set $\{x \in \mathrm{L}_\alpha \mid \phi(x)\}$, and $[x^{\mathrm{nat}} : \phi(x)]$ for $\{x \in \mathbb{N} \mid \phi(x)\}$ (RS stands for ramified set theory):

Definition 5. We define the RS-terms and RS-formulas as follows:

(a) $\mathcal{T}_{\mathrm{nat}} := \{S^k(0) \mid k \in \mathbb{N}\}$.

(b) We define inductively simultaneously for all $\alpha \in \mathrm{OT}_{n_0}$ the sets of terms $\mathcal{T}'_\alpha$ and of formulas FOR'_α by:

$\mathrm{L}_\alpha \in \mathcal{T}'_\alpha$.
If $\phi, \psi \in \mathrm{FOR}'_\alpha$ and $x^{\mathrm{set}} \in \mathrm{FV}(\phi) \vee y^{\mathrm{nat}} \in \mathrm{FV}(\psi)$, then $[x \in \mathrm{L}_\alpha : \phi^{\mathrm{L}_\alpha}] \cup [y : \psi^{\mathrm{L}_\alpha}] \in \mathcal{T}'_\alpha$.
If $r, s \in \mathcal{T}'_\alpha$, $t \in \mathcal{T}'_{\mathrm{nat}}$, then $r \in s$, $r \notin s$, $t \in s$, $t \notin s$, $\mathrm{Ad}(r)$, $\overline{\mathrm{Ad}}(r) \in \mathrm{FOR}'_\alpha$.
If $\phi, \psi \in \mathrm{FOR}'_\alpha$, then $\phi \wedge \psi$, $\phi \vee \psi$, $\forall x^?.\phi$, $\exists x^?.\psi \in \mathrm{FOR}'_\alpha$.

Here the free variables $FV(A)$ of $A \in \mathcal{T}'_\alpha$ or $A \in FOR'_\alpha$ and ϕ^r are defined as usual or before.

(c) $\mathcal{T}_\alpha := \{t \in \mathcal{T}'_\alpha \mid FV(t) = \emptyset\}$, $FOR_\alpha := \{A \in FOR'_\alpha \mid FV(A) = \emptyset\}$.
$\mathcal{T}_{set} := \bigcup_{\alpha \in OT_{n_0}} \mathcal{T}_\alpha$, $\mathcal{T} := \mathcal{T}_{set} \cup \mathcal{T}_{nat}$, $FOR := \bigcup_{\alpha \in OT_{n_0}} FOR_\alpha$.
The elements of FOR are called RS-formulas.
An RS-formula is Δ_0, if it contains no unbounded set-quantifiers, Σ_1, if it contains no unbounded universal set-quantifiers, and $\Sigma(\alpha)$, if it is of the form $\psi^{L_\alpha}[x_1^{set} := t_1, \ldots, x_n^{set} := t_n]$ for some Σ_1-formula ψ of KPI_{U,n_0}^+ and some $t_i \in \mathcal{T}_{\prec\alpha}$. In the last situation $\phi^{t,\alpha} := \psi^t$. Further $\phi^{\beta,\alpha} := \phi^{L_\beta,\alpha}$, $\phi^\beta := \phi^{L_\beta}$.

(d) $FOR^{\Delta_0} := \{\phi \in FOR \mid \phi \ \Delta_0\text{-formula}\}$, $FOR_\alpha^{\Delta_0} := FOR^{\Delta_0} \cap FOR_\alpha$.

(e) For $t \in \mathcal{T}$, $A \in FOR$ we define $K(t), K(A) \subset OT_{n_0}$:
$K(t^{nat}) := \emptyset$.
$K(L_\alpha) := \{\alpha\}$, $K([x^{set} \in L_\alpha : \phi] \cup [y^{nat} : \psi]) := \{\alpha\} \cup K(\phi) \cup K(\psi)$.
$K(A)$ is the union of $K(t)$ for all terms $t \in \mathcal{T}$ which occur A.
$|A| := \max K(A)$ for $A \in FOR \cup \mathcal{T}$.

(f) In the following $r^{set}, s^{set}, t^{set}$ denote elements of $\mathcal{T}_{set}$, and $r^?, s^?, t^?$ elements of $\mathcal{T}_{set} \cup \mathcal{T}_{nat}$.

Note that elements of $\mathcal{T}$ and FOR are finite objects, which can be coded easily in Martin-Löf Type Theory in such a way that all the above mentioned sets, relations and functions are primitive recursive.

In order to assign infinitary formulas to $s^? \in t^{set}$ we define first the auxiliary formula $s^? \overset{\circ}{\in} t^{set}$:

Definition 6. (a) For $s, t \in \mathcal{T}$ such that $|s| \prec |t|$ we define $s \overset{\circ}{\in} t$:
$s \overset{\circ}{\in} L_\alpha := \top$,
$s^{set} \overset{\circ}{\in} [x^{set} \in L_\alpha : \phi(x)] \cup [y^{nat} : \psi(y)] := \phi[x := s]$,
$s^{nat} \overset{\circ}{\in} [x^{set} \in L_\alpha : \phi(x)] \cup [y^{nat} : \psi(y)] := \psi[y := s]$.

(b) We assign to formulas ϕ in FOR^{Δ_0} expressions $\phi \simeq \bigwedge_{\iota \in J} \phi_\iota$ or $\phi \simeq \bigvee_{\iota \in J} \phi_\iota$, where $J \subset \mathcal{T}$, as follows:
If $\widetilde{P}(val(s_1^{nat}), \ldots, val(s_k^{nat}))$ is false then $P(s_1^{nat}, \ldots, s_k^{nat}) :\simeq \bigvee_{\iota \in \emptyset} \phi_\iota$.
$(\phi_0 \vee \phi_1) := \bigvee_{\iota \in \{0,1\}} \phi_\iota$.
$r^{set} \in s^{set} :\simeq \bigvee_{t^{set} \in \mathcal{T}_{\prec|s|}} (t \overset{\circ}{\in} s \wedge t = r)$.
$r^{nat} \in s^{set} :\simeq \bigvee_{t^{nat} \in \mathcal{T}_{nat}} (t \overset{\circ}{\in} s \wedge t = r)$.
$\exists x^{nat}.\phi(x) :\simeq \bigvee_{s^{nat} \in \mathcal{T}_{nat}} \phi(s)$.
$\exists x^{set} \in t^{set}.\phi :\simeq \bigvee_{s^{set} \in \mathcal{T}_{\prec|t|}} (s \overset{\circ}{\in} t \wedge \phi[x := s])$.
$Ad(s^{set}) :\simeq \bigvee_{t^{set} \in J} (t = s)$ with $J := \{L_\kappa \mid \kappa \in R \wedge \kappa \preceq |s|\}$.
In all other cases, we have for some J, ψ_ι, $\neg\phi \simeq \bigvee_{\iota \in J} \psi_\iota$ and define $\phi :\simeq \bigwedge_{\iota \in J} (\neg\psi_\iota)$.
If $\phi \simeq \bigvee_{\iota \in J} \phi_\iota$, we call ϕ an $\vee$-formula, and if $\phi \simeq \bigwedge_{\iota \in J} \phi_\iota$, ϕ an $\wedge$-formula. In both situations let $Index(\phi) := J$, $\phi[\iota] := \phi_\iota$. Note that we can primitive recursively decide, whether ϕ is an $\vee$ or $\wedge$-formula, and for the J as above whether $\iota \in J$. Further $\phi[\iota]$ is primitive recursive in ϕ and ι.

We write $\bigwedge_{\iota \in J} \phi_\iota$ for any formula ϕ such that $\phi \simeq \bigwedge_{\iota \in J} \phi_\iota$, similar for $\bigvee_{\iota \in J} \phi_\iota$.

(c) We define $\mathrm{rk}(\theta)$ for $\theta \in \mathrm{FOR} \cup \mathcal{T}$ by

$\mathrm{rk}(L_\alpha) := \omega \cdot (\alpha + 1)$,

$\mathrm{rk}([x^{\mathrm{set}} \in L_\alpha : \phi] \cup [y^{\mathrm{nat}} : \psi]) := \max\{\omega \cdot \alpha + 1, \mathrm{rk}(\phi[x := L_0]) + 2, \mathrm{rk}(\psi[y := 0]) + 2\}$,

$\mathrm{rk}(\mathrm{Ad}(t^{\mathrm{set}})) := \mathrm{rk}(t) + 5$,

$\mathrm{rk}(s^? \in t^{\mathrm{set}}) := \max\{\mathrm{rk}(s) + 6, \mathrm{rk}(t) + 1\}$,

$\mathrm{rk}(\exists x^{\mathrm{set}} \in t.\phi) := \max\{\mathrm{rk}(t), \mathrm{rk}(\phi[x := L_0]) + 2\}$,

$\mathrm{rk}(\exists x^{\mathrm{nat}}.\phi) := \mathrm{rk}(\phi[x := 0]) + 2$,

$\mathrm{rk}(\phi_0 \vee \phi_1) := \max\{\mathrm{rk}(\phi_0), \mathrm{rk}(\phi_1)\} + 1$,

$\mathrm{rk}(\neg\phi) := \mathrm{rk}(\phi)$ otherwise.

[Bu92], Lemma 1.9 and Definitions 1.11, 1.12, 2.1. can be shown and defined accordingly. The functions $\alpha \mapsto \alpha^{\mathrm{R}}$ and $\Gamma \mapsto \|\Gamma\|$ (considered as functions on the codes) are primitive recursive.

Lemma 7. *Assume* $\mathrm{ML}_1\mathrm{W} \vdash B \in \mathrm{N} \to \mathcal{P}(\mathrm{N})$, $\mathrm{ML}_1\mathrm{W} \vdash \Phi \in \mathrm{N}^3 \to \mathrm{U}$, $\mathrm{ML}_1\mathrm{W} \vdash \Psi \in \mathrm{N} \to \mathrm{U}$.
Let $\Gamma \in \mathcal{P}(\mathrm{N}) \to \mathcal{P}(\mathrm{N})$,
$\Gamma(M) := \{k \in \mathrm{N} \mid \Psi(k) \wedge \forall l \in_{\mathrm{U}} B(k).\exists l' \in_{\mathrm{U}} M.\Phi(k, l, l')\}$.
(the interpretation is that $B(k)$ is the set of predecessors of a node k, $\Psi(k)$ is the condition at every nodes, and $\Phi(k, l, l')$ is the condition relating a node with its predecessors). Then we can define $\mathbb{I}$ *such that* $\mathrm{ML}_1\mathrm{W} \vdash \mathbb{I} \in \mathcal{P}(\mathrm{N})$, *and we can prove in* $\mathrm{ML}_1\mathrm{W}$:
$\Gamma(\mathbb{I}) \subset \mathbb{I}$, *and for every subclass A of N we have* $\Gamma(A) \subset A \to A \subset \mathbb{I}$.

Proof: Define $\mathrm{W}_\Gamma := \mathrm{W}k \in \mathrm{N}.\tau(k)$ with $\tau(k) := \Sigma l \in \mathrm{N}.(l \in_{\mathrm{U}} B(k))$.
Let for $\sup(k, s) \in \mathrm{W}_\Gamma$,
$\mathrm{index}(\sup(r, s)) := r$, $\mathrm{pred}(\sup(r, s), a) := s(a)$,
$\mathrm{LocCor}(\sup(k, s)) := \Psi(k) \wedge \forall l \in \mathrm{N}.\forall p \in (l \in_{\mathrm{U}} B(k)).\Phi(k, l, \mathrm{index}(s(<l, p>)))$.
Define $w \prec^1_{\mathrm{W}} \sup(k', s) :\Leftrightarrow \exists r \in \tau(k).s(r) = w$.
Let $w \preceq_{\mathrm{W}} w' :\Leftrightarrow \exists l \in \mathrm{N}.\exists f \in \mathrm{N} \to \mathrm{W}_\Gamma.f(0) = w' \wedge f(l) = w \wedge \forall i < l.f(i+1) \prec^1_{\mathrm{W}} f(i)$.
Let for $w \in \mathrm{W}_\Gamma$, $\mathrm{Cor}(w) :\Leftrightarrow \forall w' \preceq_{\mathrm{W}} w.\mathrm{LocCor}(w')$.
Let $\mathbb{I} := \{k \mid \exists w \in \mathrm{W}_\Gamma.\mathrm{Cor}(w) \wedge \mathrm{index}(w) = k\}$.
Then one easily sees that $\mathbb{I}$ fulfills the conditions of the theorem.

Definition 8. (a) As in [Bu92] we define the infinitary system RS^* as the collection of all derivations generated by five inference rules ($\phi, \phi_\iota \in \mathrm{FOR}^{\Delta_0}$, $\Gamma \subset \mathrm{FOR}^{\Delta_0}$):

$$(\wedge)^* \quad \frac{\cdots \vdash^\rho \Gamma, \phi_\iota \cdots (\iota \in J)}{\vdash_\rho \Gamma, \bigwedge_{\iota \in J} \phi_\iota}$$

$$(\vee)^* \quad \frac{\vdash_\rho \Gamma, \phi_{\iota_0}, \ldots, \phi_{\iota_k}}{\vdash_\rho \Gamma, \bigvee_{\iota \in J} \phi_\iota} \quad \text{(if } \iota_0, \ldots, \iota_k \in J \text{ and}$$

$$K(\iota_0, \ldots, \iota_k) \subset K(\Gamma, \bigvee_{\iota \in J} \phi_\iota))^*$$

$(\mathrm{Ad})^*$ $\dfrac{\cdots \vdash_\rho \Gamma, \phi[x^{\mathrm{set}} := L_\kappa] \cdots (\kappa \preceq |t^{\mathrm{set}}|)}{\vdash_\rho \Gamma, \mathrm{Ad}(t) \to \phi[x := t]}$, $\quad$ if $\mathrm{rk}(\phi[a := L_0]) \prec \rho$

$(\mathrm{Ref})^*$ $\quad \Gamma, \phi \to \exists x^{\mathrm{set}} \in L_\kappa.\phi^{x,\kappa}$, $\quad$ if $\phi \in \Sigma(\kappa) \wedge \kappa \in R \wedge \rho \neq 0$

$(\mathrm{Found})^*$ $\quad \Gamma, \exists x^{\mathrm{set}} \in L_\alpha((\forall y^{\mathrm{set}} \in x.\phi[z := y]) \wedge \neg\phi[z := x])$,
$\qquad \forall x^{\mathrm{set}} \in L_\alpha.\phi[z := x]$ $\quad$ if $\rho \neq 0$.

(More precisely $(\bigwedge)^*$ should be read as: if $\phi \simeq \bigwedge_{\iota \in J} \phi_\iota$, then
$\dfrac{\cdots \vdash^\rho \Gamma, \phi_\iota \cdots (\iota \in J)}{\vdash_\rho \Gamma, \phi}$ is a rule etc.)

(b) We formalize (a) in Martin-Löf Type Theory as follows:
In order to get unique predecessors, we replace the information on the nodes by sequences $<rule, \rho, \Gamma>$, where $rule$ is of the form $< \bigwedge, \phi>$, $< \bigvee, \phi, \iota_0, \ldots, \iota_l>$, $<\mathrm{Ad}, \phi, x, t^{\mathrm{set}}, \kappa>$, $<\mathrm{Ref}, \phi, x, \kappa>$ or $<\mathrm{Found}, \phi, x, y, z, \alpha>$.
We want to apply Lemma 7 to B, Ψ_ρ, Φ. We define
$B(<< \bigwedge, \phi>, \Gamma>) := \mathrm{Index}(\phi)$,
$\Psi_\rho(<< \bigwedge, \phi>, \Gamma>) := \phi \in \Gamma \wedge \phi \wedge -\mathrm{Formula}$,
$\Phi(< \bigwedge, \phi, \Gamma>, \iota, p) := (((p)_1 = (\Gamma \setminus \{\phi\}) \cup \{\phi[\iota]\}) \vee ((p)_1 = \Gamma \cup \{\phi[\iota]\}))$.
$B(<< \bigvee, \phi, \iota_0, \ldots, \iota_l>, \Gamma>) := \{0\}$,
$\Psi_\rho(<< \bigvee, \phi, \iota_0, \ldots, \iota_l>, \Gamma>) := \phi \in \Gamma \wedge \phi \vee -\mathrm{Formula} \wedge$
$\iota_0, \ldots, \iota_l \in J \wedge K(\iota_0, \ldots, \iota_k) \subset K(\Gamma, \bigvee_{\iota \in J} \phi)^*$,
$\Phi(<< \bigvee, \phi, \iota_0, \ldots, \iota_l>, \Gamma>, \iota, p) := (((p)_1 = (\Gamma \setminus \{\phi\}) \cup \{\phi[\iota_0], \ldots, \phi[\iota_l]\}) \vee$
$((p)_1 = \Gamma \cup \{\phi[\iota_0], \ldots, \phi[\iota_l]\}$.
The other rules are treated in a similar way.
Then with the set $\mathbb{I}_\rho$ as in Lemma 7 defined for B, Ψ_ρ, Φ, $\{(p)_1 | p \in \mathbb{I}\}$ is the set of sequences derivable in RS, and we define $\vdash_\rho^* \Gamma :\Leftrightarrow \exists p \in \mathbb{I}_\rho.(p)_1 = \Gamma$.

(c) $q \vdash^* \Gamma :\Leftrightarrow q \vdash_0^* \Gamma$.

Lemmata and Theorems 2.4–2.9 of [Bu92] follow now with nearly the same proofs. The only modifications to be made are to define $[s^{\mathrm{nat}} \neq t^{\mathrm{nat}}]$ and to add instances for the case $A = P(t_1^{\mathrm{nat}}, \ldots, t_m^{\mathrm{nat}})$ in Lemma 2.7. Further we can easily prove that for all arithmetical axioms ϕ, except the induction theorem we have $\vdash^* \phi^\lambda$. The only case, where some work is necessary is to give a cut-free proof of the induction axiom, and the reader can easily find such a proof, so for every instance ϕ of the induction axiom we have $\vdash^* \phi^\lambda$.

5 $\mathcal{H}$-controlled Derivations

The next step is to formalize $\mathcal{H}$-controlled derivations as in [Bu92]. However, this is only necessary for operators $\mathcal{H}_\gamma[\theta]$, where $\mathcal{H}_\gamma$ is defined in [Bu92], Definition 4.3. Further note that $\mathcal{H}_\gamma(X)$ is needed only for finite sets X. We formalize $\mathcal{H}_\gamma$ first:

Definition 9. (a) $\gamma \in C(\alpha, \beta) :\Leftrightarrow \gamma \prec \beta \vee \gamma \ \eta \ \{0, I\} \vee \exists \delta, \rho \in C(\alpha, \beta).\gamma ='_{NF}$
$\delta + \rho \vee \gamma =_{NF} \Omega_\delta \vee (\gamma =_{NF} \psi_\delta \rho \wedge \rho \prec \alpha)$,
where $=_{NF}$ is defined as in Definition 3.11 of [Se97].
$C(\alpha, \beta)$ can be defined easily as a primitive recursive set.

(b) For X being a finite subset of N we define $\mathcal{H}_\gamma(X) := \{\gamma \in OT_{n_0} \mid \forall \alpha, \beta \in OT_{n_0}.((X \cap OT_{n_0} \subset C(\alpha, \beta) \wedge \gamma \prec \alpha) \rightarrow \gamma \in C(\alpha, \beta))\}$.
Note that the condition $X \cap OT_{n_0} \subset C(\alpha, \beta)$ is primitive recursive, since X is finite.

(c) $\mathcal{H}_\gamma[\theta](X) := \mathcal{H}_\gamma(k(\theta) \cup X)$.
$\alpha \in \mathcal{H}_\gamma[\theta] :\Leftrightarrow \alpha \in \mathcal{H}_\gamma[\theta](\emptyset)$.

We check easily that for $C_\kappa(\alpha)$, as defined in [Se97], Definition 3.9., we have $C_\kappa(\alpha) = C(\alpha, \psi_\kappa \alpha)$. The properties in [Bu92], Lemma 4.4 b–d, 4.5–4.7 follow now directly from the properties of the ordinal denotation system in [Se97].

Definition 10. (see Theorem 3.8 of [Bu92]).
Inductive definition of $\mathcal{H}_\gamma[\theta] \vdash^\alpha_\rho \Gamma$:
Assume $\{\alpha\} \subset k(\Gamma) \subset \mathcal{H}_\gamma[\theta]$. Then we can conclude $\mathcal{H}_\gamma[\theta] \vdash \Gamma$, iff one of the following cases holds:

($\bigwedge$) $\bigwedge_{\iota \in J} \phi_\iota \in \Gamma \wedge \forall \iota \in J.\exists \alpha_\iota \prec \alpha.(\mathcal{H}_\gamma[\theta, \iota] \vdash^{\alpha_\iota}_\rho \Gamma, \phi_\iota)$.

($\bigvee$) $\bigvee_{\iota \in J} \phi_\iota \in \Gamma \wedge \exists \iota_0 \in J.\exists \alpha_0 \prec \alpha.((\mathcal{H}_\gamma[\theta] \vdash^{\alpha_0}_\rho \Gamma, \phi_{\iota_0}) \wedge \iota_0 \preceq \alpha)$.

(Cut) $rk(\psi) \prec \rho \wedge \exists \alpha_0 \prec \alpha.((\mathcal{H}_\gamma[\theta] \vdash^{\alpha_0}_\rho \Gamma, \psi) \wedge (\mathcal{H}_\gamma[\theta] \vdash^{\alpha_0}_\rho \Gamma, \neg\psi))$.

(Ref) $(\exists z^{set} \in L_\kappa.\phi^{(z,\kappa)}) \in \Gamma \wedge (\mathcal{H}_\gamma[\theta] \vdash^{\alpha_0}_\rho \Gamma, \phi) \wedge \alpha_0 + 1 \prec \alpha \wedge \phi \in \Sigma(\kappa) \wedge \kappa \in R$.

One sees easily that we can formalize $\mathcal{H}$-controlled derivations in a similar way as in Definition 8.
Now in [Bu92] Lemma 3.9, 3.13–3.17 with $\mathcal{H}$ replaced by $\mathcal{H}_\gamma[\theta]$ and by omitting all conditions on $\mathcal{H}$ (which are fulfilled), and Lemma 3.10, 3.11 with $\mathcal{H}$ replaced by $\mathcal{H}_\gamma$ and again by omitting conditions on $\mathcal{H}$, further Lemma 4.7, Theorem 4.8 and the Corollary follow with the same proofs and can be formalized in $ML_1 W$. Theorem 3.12 reads now as follows:

Theorem 11. *For every theorem ϕ of KPI^+_{U,n_0} there exists an $m < \omega$ such that with $\lambda := \Omega_{I+m}$ for all $\gamma \ \mathcal{H}_\gamma \vdash^{\omega^{\lambda+m}}_{\lambda+m} \phi^\lambda$.*

Theorem 12. *For every arithmetical formula ϕ, if $KPI^+_{U,n_0} \vdash \phi$, then $\mathcal{H}_\beta \vdash^\gamma \phi$ for some $\gamma \prec \epsilon_{\Omega_{I+n_0}+1}$.*

Proof: Let $\lambda := \Omega_{I+n_0}$.
From $KPI^+_{U,n_0} \vdash \phi$ follows by Theorem 11 $\mathcal{H}_0 \vdash^{\omega^{\lambda+m}}_{\lambda+m} \phi$, by [Bu92] 3.16 (adapted to our setting) $\mathcal{H}_0 \vdash^\alpha_{\lambda+1} \phi$ for some $\alpha \prec \epsilon_{\lambda+1}$, by [Bu92] 4.8 $\mathcal{H}_{\hat\alpha} 0 \vdash^{\psi_{\Omega_1}\hat\alpha}_{\psi_{\Omega_1}\hat\alpha} \phi$ with $\hat\alpha := \omega^{\lambda+1+\alpha_0} \prec \epsilon_{\lambda+1}$, by [Bu92] 3.16 with $\gamma := \varphi_{\psi_{\Omega_1}\hat\alpha}(\psi_{\Omega_1}\hat\alpha) \ \aleph_{\hat\alpha} \vdash^\gamma_0 \phi$, let $\beta := \hat\alpha$.

Lemma 13. *If $\mathcal{H}_\rho[\theta] \vdash^\alpha_\rho \Gamma, \bigwedge_{\iota \in J} \phi_\iota$, then $\mathcal{H}_\rho[\theta, \iota] \vdash^\alpha_\rho \Gamma, \phi_\iota$.*

Proof: If $\phi := \bigwedge_{\iota \in J} \phi_\iota$ is not the main formula of the last premise, the assertion follows by IH and the same rule.

Otherwise we have the case of last rule $(\bigwedge)$, $\mathcal{H}_\rho[\theta, \iota] \vdash^{\alpha_\iota}_\rho \Gamma, \phi_\iota$, or $\mathcal{H}_\rho[\theta, \iota] \vdash^{\alpha_\iota}_\rho \Gamma, \phi, \phi_\iota$, in which case by IH we conclude the first case. By [Bu92] Lemma 3.9 (a) follows the assertion.

6 Result

Definition 14. We define a primitive recursive relation e rel l, where l is considered as the code of a formula of RS-formula (and Gödel-brackets are omitted):
e rel l is false, if l is not the code of a Σ_1^{arith}-formula.
e rel $P(S^{k_1}(0), \ldots, S^{k_l}(0)) \leftrightarrow e = 0 \wedge \tilde{P}(k_1, \ldots, k_l)$.
e rel $\phi \wedge \psi \leftrightarrow \exists l, k.e = <l, k> \wedge l$ rel $\phi \wedge k$ rel ψ.
e rel $\phi \vee \psi \leftrightarrow \exists l, k.(e = <l, k> \wedge ((l = 0 \wedge k \text{ rel } \phi) \vee (l = 1 \wedge k \text{ rel } \psi)))$.
e rel $\exists x^{\text{nat}}.\phi \leftrightarrow \exists l, k.e = <l, k> \wedge k$ rel $\phi[x := S^l(0)]$.
  e rel $\phi_1, \ldots, \phi_n :\Leftrightarrow e$ rel $(\phi_1 \vee \cdots \vee \phi_n)$.

Lemma 15. (a) For every formula $\phi \in \Sigma_1^{\text{arith}}$ such that $\text{FV}(\phi) \subset \{x_1^{\text{nat}}, \ldots, x_l^{\text{nat}}\}$ we have
$$\text{ML}_1\text{W} \vdash \forall k_1, \ldots, k_l \in \text{N}.\forall n \in \text{N}.((n \text{ rel } \phi[x_1 := S^{k_1}(0), \ldots, x_l := S^{k_l}(0)])$$
$$\rightarrow \hat{\phi}[x_1 := k_1, \ldots, x_l := k_l]).$$
(b) $\forall \Gamma \in \Sigma_1^{\text{arith}}.\forall \alpha, \rho, \theta.\mathcal{H}_\rho[\theta] \vdash^\alpha_0 \Gamma \rightarrow \exists n.n \text{ rel } \Gamma$.

  Proof: b: by an easy induction on the rules. Note that only the rules $(\bigvee)$ and $(\bigwedge)$ occur.

Theorem 16. Let $\phi = \forall x^{\text{nat}}.\psi$, $\psi \in \Sigma_1^{\text{arith}}$. Assume $\text{KPI}^+_{\text{U},n_0} \vdash \phi$. Then $\text{ML}_1\text{W} \vdash \hat{\phi}$.

Proof: By Theorem 11 follows $\mathcal{H}_\rho \vdash^\alpha_0 \phi$. Assume $k \in \text{N}$. By Lemma 13 follows $\mathcal{H}_\rho \vdash^\alpha_0 \psi[x := S^k(0)]$. Then by Lemma 15 follows $\hat{\psi}[x := k]$, therefore $\forall x.\psi$.

Corollary 17. ML_1W proves the consistency of $\text{KPI}^+_{\text{U},n_0}$.

References

[Ba75] Barwise, J.: *Admissible Sets and Structures. An Approach to Definability Theory*, Springer, 1975.

[Be85] Beeson, M.: *Foundations of Constructive Mathematics* Springer, Berlin, 1985

[BS95] Berger, U. and Schwichtenberg, H.: *Program Extraction from Classical Proofs.* In: D. Leivant (Ed.): *Logic and Computational Complexity, LCC '94, Indianapolis, October 1994*, Springer Lecture Notes in Computer Science 960, pp. 77–97.

[Bu91] Buchholz, W.: *Notation systems for infinitary derivations.* Arch. Math. Logic (1991) 30:277–296.

[Bu92] Buchholz, W.: *A simplified version of local predicativity.* In: Aczel, P. et al. (Eds.): *Proof Theory.* Cambridge University Press, Cambridge, 1992, pp. 115–147.

[Jä86] Jäger, G.: *Theories for Admissible Sets: A Unifying Approach to Proof Theory.* Bibliopolis, Naples, 1986.

[Sch92] Schwichtenberg, H.: *Proofs as programs.* In:Aczel, P. et al. (Eds.): *Proof Theory.* Cambridge University Press, Cambridge, 1992, pp. 81–113.

[Se93] Setzer, A.: *Proof theoretical strength of Martin-Löf Type Theory with W-type and one universe.* PhD-thesis, Munich, 1993.

[Se96] Setzer, A.: *Extending Martin-Löf Type Theory by one Mahlo-Universe.* Submitted.

[Se97] Setzer, A.: *Well-ordering proofs for Martin-Löf Type Theory with W-type and one universe.* Submitted.

Denotational Semantics for Polarized (But-non-constrained) *LK* by Means of the Additives

Lorenzo Tortora de Falco*

Équipe de Logique Mathématique, Université Paris VII

Abstract. Following the idea that a classical proof is a superimposition of 'constructive' proofs, we use the linear connective "&" to define an embedding P^a of polarized but non-constrained *LK* in linear logic, which induces a denotational semantics. A classical cut-elimination step is now a cut-elimination step in one of the 'constructive slices' superimposed by the classical proof.

1 Introduction

Much work has been done in the past 6-7 years to extract the computational content from classical proofs. On the proof-theoretical side, let's quote for example [Girard 91] (*LC*), [Parigot 91] (*FD*), [DJS 95] (LK^{tq} and its restrictions) and [BarBer 95] (the symmetric λ-calculus).

With the notable exception of the symmetric λ-calculus of [BarBer 95] which enjoys only strong normalization, all the systems previously mentioned enjoy the usual good computational properties (strong normalization and confluence) and have a denotational semantics. To obtain such good properties, it is necessary to cope with the "non-determinism" of classical cut-elimination: this is done by equipping classical formulas with an index, a colour or a polarity (depending on the personal taste of the authors), which will tell us how to perform cut-elimination in case of conflict.

But it turned out (perhaps a bit surprisingly) that adding this information is not enough to fix classical logic: there are rules that are "difficult" to handle, in the sense that if we don't restrict their application to the premises of a particular form, then the elimination of certain 'logical cuts' (Gentzen 's key-cases) yields *two* proofs. What is catastrophic is the fact that the identification of these two proofs leads to the collapse of the whole system: we have to identify all the proofs of a given sequent (see [Girard 91] and [DJS 95]).

The solution proposed by the works previously mentioned is either to avoid such difficult rules (this is possible keeping completeness w.r.t. classical provability, see [Parigot 91] and [BarBer 95]) or to restrict the application of such difficult rules to the premises which lead to key-cases the elimination of which is unproblematic (see [Girard 91] and [DJS 95]).

A trace of this restriction is the fact that the translations from *LK* (with polarized or coloured formulas) to *LC* or LK^{tq} (and its subsystems) are far from

---

* tortora@logique.jussieu.fr.

being deterministic: there are two possible ways to translate the difficult rules with certain premises. The same holds for the translations in linear logic (LL), which are the actual starting point of both the works [Girard 91] and [DJS 95]. However, these translations suggest that the computational object which lies 'behind' a classical proof *is not* a linear proof, but rather a superimposition of linear proofs. This work is a first attempt to turn this intuition into a mathematical statement: "there exists a denotational semantics of polarized (but non-constrained) LK in coherent spaces".

Trying to take seriously the idea of superimposition, we define inductively an embedding P^a of polarized LK-proofs in LL, which is an extension of the P-embedding defined by Girard in [Girard 91]: when we meet a "difficult" rule, then, instead of choosing one of the two possible linear translations, we superimpose them.

Technically speaking, the obvious candidate to superimpose proofs is the linear connective "&". Following one of the main points of [Tortora 96] (namely the "independence lemma"), to make sure that the components of the additive boxes introduced by P^a do not interact during cut-elimination, we keep these additive boxes at exponential depth zero. This is done by choosing as main doors for these additive boxes a "purely negative" formula (in the sense of [Girard 91]), namely $\perp \& \perp$: this makes the additive boxes "commute" (semantically speaking) with all the exponential ones.

Thus, we translate the classical proof α by the linear proof $P^a(\alpha)$, and then, by "delaying" as much as possible all the additive boxes introduced by P^a, we obtain a superimposition of linear proofs with the same denotational semantics of $P^a(\alpha)$.

This suggests to define a classical cut-elimination step as a cut-elimination step in *one* of the "slices" that the classical proof superimposes. We obtain in such a way a denotational semantics (induced by P^a) for polarized LK, which still satisfies all the computational isomorphisms of LC (see [Girard 91]).

2 Preliminaries: linear logic and classical logic, additives

2.1 Linear logic and classical logic

The use of LL as a tool to analize classical cut-elimination is at the core of [Girard 91] and [DJS 95].

Girard observes that if A and B are linear formulas of type "!" (resp. of type "?"), that is if $A \vdash !A$ (resp. $?A \vdash A$) is provable in LL, then so are the formulas $A \otimes B$ and $A \oplus B$ (resp. $A \wp B$ and $A \& B$), and that A is of type "!" (resp. "?") iff $A^\perp$ is of type "?" (resp. "!"). This suggests a translation P from classical to linear logic, which translates each classical formula by a linear formula of type "!" or of type "?". To apply structural operations directly to formulas of type "?" (and not only to formulas starting with "?"), Girard slightly modifies the coherent denotational semantics of LL and introduces the semantics of correlation spaces for the linear fragment necessary to interpret classical logic. He also proves that the main properties of classical connectives can be recovered at the computational level (i.e. are isomorphisms in the semantics).

Danos, Joinet and Schellinx work directly with special translations in LL (called "decorations") which suggest a cut-elimination procedure (the tq-protocol) for different versions of coloured second order classical sequent calculus (LK^{tq}, LK^{η}, LK^{η}_{pol}). They show that Girard 's LC is a subsystem of LK^{η}_{pol} (up to the stoup/no stoup formulation, see subsection 2.3), that the P-embedding defined by Girard is suitable for the whole (first order) LK^{η}_{pol}, and that the cut-elimination procedure defined by Girard for LC is the tq-protocol restricted to LC.

In [QT 96a], we prove that the P-embedding induces a denotational semantics (to which we will also refer to as the P-semantics) in coherent spaces for first order LK^{η}_{pol} (then for LC): if π and π' are LK^{η}_{pol}-proofs, from $\pi \to_{tq} \pi'$ we deduce $[P(\pi)] = [P(\pi')]$ (where $[d]$ denotes the coherent semantics of the linear proof d). In [QT 96b], we clarify the links between first order LK^{η}_{pol} and LC: LC is obtained from the multiplicative fragment of LK^{η}_{pol} by decomposing the multiplicative conjunction between two negative formulas into an additive one and a multiplicative conjunction with "V" (the classical unit for the multiplicative conjunction).

2.2 About the additives

In this subsection, we briefly recall the independence lemma, proven in [Tortora 96]. Although we will not use this result (so that this subsection is not, strictly speaking, necessary to understand the content of the paper), it has been one of the main sources of inspiration for the present work. We assume here that the reader has some acquaintance with the proof-nets of LL defined in [Girard 87].

If T and T' are two proof-nets s.t. $T \twoheadrightarrow T'$ (i.e. T' is obtained from T by applying some steps of cut-elimination), then any additive box B of T' comes from exactly one additive box $\overleftarrow{B}$ of T: we say that B is a residue of $\overleftarrow{B}$ in T'.

Lemma 1 *(Independence lemma).* *Let T, T' be two proof-nets, s.t. $T \twoheadrightarrow T'$. Let $\mathbb{B}$ be an additive box with exponential depth 0 in T, and $\mathbb{B}'$ a residue of $\mathbb{B}$ in T'. If $\mathbb{B}'$ has additive depth p and $B_1, ..., B_p$ are the additive boxes containing $\mathbb{B}'$ in T', then for all residue $\mathbb{B}''$ of $\mathbb{B}$ in T' (different from $\mathbb{B}'$), there exists $i \in \{1, ..., p\}$ s.t. $\mathbb{B}''$ is a subproof-net of the component of B_i which does not contain $\mathbb{B}'$.*

A possible way of reading this lemma (connected to the present paper) is the following: "normalization preserves superimposition *at exponential depth* 0".

2.3 Notations and conventions

We shall use the conventions of [Girard 91], thus distinguish between *positive* and *negative* formulas and take a one-sided version of sequent calculus, rather than distinguish between formulas coloured q and t and take a two-sided version of sequent calculus like in [DJS 95].

For classical formulas we take the following conventions:

$P, Q, R,...$ will denote positive formulas

$L, M, N,...$ will denote negative formulas

$A, B, C,...$will denote formulas that may be either positive or negative.

We say that a formula (or, better said, an occurrence of formula) A in a proof π of any subsystem of LK is *main*, if A is either one of the (two) conclusions of an axiom, or the main conclusion of a logical rule of π.

In the sequel we will call LC the extension of Girard 's original system to first order LK^{η}_{pol}. (We don't call it LK^{η}_{pol} because we will use a version which makes use of "the stoup", and this is more in the spirit of LC, and also because we don't want to create confusions with the system LK_{pol} defined in the next section).

To be precise one should note that there are LC-rules that are not (strictly speaking) rules of LK^{η}_{pol}. This is because the constraint on proofs induced by Girard's stoup is (slitghly) less strong than [DJS 95]'s η-constraint: an attractive formula (i.e. necessary positive in the one-sided version) which is active in an irreversible rule has to be main following [DJS 95], and only main up to reversible, structural and (some kind of) cut rules on the context, following [Girard 91]. From the semantical point of view the two constraints are the same: to each LC-proof one can associate several LK^{η}_{pol}-proofs, but the denotational semantics induced by the P-embedding in LL equates all of them. But syntactically speaking, in LC there are for example two ways of performing certain logical steps: if $\gamma = cut(\alpha, \beta)$ is a proof of $\vdash \Gamma, \Delta; \Pi$ obtained from the proof α of $\vdash \Gamma; R \wedge_m Q$ and the proof β of $\vdash \Delta, \neg R \vee_m \neg Q; \Pi$ by applying a logical cut rule, and if we call γ_1 and γ_2 the two proofs obtained from γ by performing the two cuts with active formulas $R/\neg R$, $Q/\neg Q$ in the two possible ways, then γ_1 and γ_2 might have two syntactically different normal forms w.r.t. the tq-protocol of cut-elimination (of course the normal forms will anyway be equal in the semantics).

So, what we call LC in this paper, is the system LK^{η}_{pol} (with the stronger η-constraint on proofs), the only *notational* difference being that we will write sequents making use of the stoup (this is particularly helpful for our purposes, as it will appear clearly in the next section).

In the whole paper, we will apply directly the structural operations to any linear formula A of type "?", implicitly meaning (if A does not actually start with "?") that we first apply a dereliction to A, then the structural operation to $?A$, and finally we cut with the proof of $?A \vdash A$ (with $\vdash A, !A^{\perp}$ in the one-sided version). One can also think directly in terms of correlation spaces (which, by the way, is proven to be the same in [QT 96a]).

If B and C are the immediate subformulas of A, an η-proof of A is the proof of $\vdash \neg A, A$ consisting in the axioms $\vdash \neg B, B$ and $\vdash \neg C, C$, and precisely one instance of each of the logical rules introducing A (or $\neg A$)'s main connective. We will also call "piece(s) of η-proof" the subproof(s) of η, obtained from η by removing the last (reversible) rule.

3 LK_{pol} and the embedding P^a in linear logic

We consider here a version of the sequent calculus LK where each formula comes equipped with a polarity, in the style of [Girard 91].

The real difference between the approach we present here and the ones previously mentioned lies in the fact that we *do not* restrict the rules of the calculus LK. Both in [Girard 91] and in [DJS 95] there is a restriction in the application of certain rules ("the irreversible ones in which one of the active formulas of one of the premises is positive"): as stated, this is the reason for the introduction of "the stoup" in [Girard 91] and of the "η-constraint" in [DJS 95].

3.1   The sequent calculus LK_{pol}

LK_{pol} is nothing but the usual calculus LK (where we distinguish between additive and multiplicative rules) for first order classical logic. However, in LK_{pol}, each formula comes equipped with a polarity (it is positive or negative), which is completely determined by the polarity of the atoms: the polarity of compound formulas is obtained by looking at the main connective of the formula, if it is $\vee_m$ or $\wedge_a$ (resp. $\wedge_m$ or $\vee_a$) then the formula is negative (resp. positive). Of course, for any formula A, A and $\neg A$ have different polarities (see [Girard 91] and [DJS 95] for more details).

Moreover, we "decorate" (not in the technical sense of [DJS 95]!) each sequent of this polarized LK by means of the symbol ";", where on the right of ";" there is at most one *positive* formula. This is of course in the spirit of [Girard 91]. However, our calculus does not put any constraint on proofs. The following system (which will be called LK_{pol}) shows how it is possible to "decorate" by means of ";", in a unique way, *any* polarized proof α of LK, in such a way that all the sequents of α have the form $\vdash \Gamma; \Pi$, where Γ is a set of (polarized) formulas and Π is a set containing at most one formula, which has to be positive.

Axiom:

$$\vdash \neg P; P$$

Cut-rules:

$$
\begin{array}{cccc}
\gamma & \delta & \gamma & \delta \\
\vdots & \vdots & \vdots & \vdots \\
\vdash \Gamma; P \quad \vdash \Delta, \neg P; \Pi & & \vdash \Gamma, P; \Pi \quad \vdash \Delta, \neg P; \Pi' & \\
\hline
\vdash \Gamma, \Delta; \Pi & & \vdash \Gamma, \Delta, \Pi'; \Pi &
\end{array}
$$

Structural rules:

$$
\begin{array}{cc}
\beta & \beta \\
\vdots & \vdots \\
\vdash \Gamma; \Pi & \vdash \Gamma, A, A; \Pi \\
\hline
\vdash \Gamma, A; \Pi & \vdash \Gamma, A; \Pi
\end{array}
$$

Logical rules:

1) constants:

$$\vdash; V \qquad \vdash \neg F; \Pi$$

2) unary logical rules:

$$
\begin{array}{cc}
\beta & \beta \\
\vdots & \vdots \\
\dfrac{\vdash \Gamma, A, B; \Pi}{\vdash \Gamma, A \vee_m B; \Pi} & \dfrac{\vdash \Gamma, A; P}{\vdash \Gamma, A \vee_m P;}
\end{array}
$$

$$
\begin{array}{ccc}
\beta & \beta & \beta \\
\vdots & \vdots & \vdots \\
\dfrac{\vdash \Gamma; P}{\vdash \Gamma; P \vee_a B} & \dfrac{\vdash \Gamma, P; \Pi}{\vdash \Gamma, P \vee_a B; \Pi} & \dfrac{\vdash \Gamma, N; \Pi}{\vdash \Gamma, \Pi; N \vee_a B}
\end{array}
$$

3) binary logical rules:
For the connective $\wedge_a$ we have

$$
\begin{array}{cccc}
\gamma & \delta & \gamma & \delta \\
\vdots & \vdots & \vdots & \vdots \\
\vdash \Gamma, A, \Pi'; \Pi & \vdash \Gamma, B, \Pi; \Pi' & \vdash \Gamma, A; \Pi & \vdash \Gamma, B; \Pi \\
\end{array}
$$
$$
\dfrac{\vdash \Gamma, A, \Pi'; \Pi \qquad \vdash \Gamma, B, \Pi; \Pi'}{\vdash \Gamma, A \wedge_a B, \Pi, \Pi';} \qquad \dfrac{\vdash \Gamma, A; \Pi \qquad \vdash \Gamma, B; \Pi}{\vdash \Gamma, A \wedge_a B; \Pi}
$$

where we obviously suppose $\Pi \neq \Pi'$, and

$$
\dfrac{\vdash \Gamma, A; \Pi \qquad \vdash \Gamma, \Pi; P}{\vdash \Gamma, A \wedge_a P, \Pi;} \qquad \dfrac{\vdash \Gamma; P \qquad \vdash \Gamma; Q}{\vdash \Gamma, P \wedge_a Q;}
$$

The case of the connective $\wedge_m$ is the more complicated, and we have to distinguish all the possible cases for the polarity of the active formulas:

$$
\dfrac{\vdash \Gamma; P \qquad \vdash \Delta; Q}{\vdash \Gamma, \Delta; P \wedge_m Q} \qquad \dfrac{\vdash \Gamma, P; \Pi \qquad \vdash \Delta, Q; \Pi'}{\vdash \Gamma, \Delta, P \wedge_m Q, \Pi, \Pi';} \qquad \dfrac{\vdash \Gamma; P \qquad \vdash \Delta, Q; \Pi}{\vdash \Gamma, \Delta, P \wedge_m Q; \Pi}
$$

$$
\dfrac{\vdash \Gamma, M; \Pi \qquad \vdash \Delta, N; \Pi'}{\vdash \Gamma, \Delta, \Pi, \Pi'; M \wedge_m N} \qquad \dfrac{\vdash \Gamma, P; \Pi \qquad \vdash \Delta, N; \Pi'}{\vdash \Gamma, \Delta, P \wedge_m N, \Pi'; \Pi} \qquad \dfrac{\vdash \Gamma; P \qquad \vdash \Delta, N; \Pi'}{\vdash \Gamma, \Delta, \Pi'; P \wedge_m N}
$$

First order quantifiers:

$$
\begin{array}{cc}
\beta & \beta \\
\vdots & \vdots \\
\cfrac{\vdash \Gamma, A; \Pi}{\vdash \Gamma, \forall x A; \Pi} & \cfrac{\vdash \Gamma; P}{\vdash \Gamma, \forall x P;}
\end{array}
$$

with the usual restriction "x is not free in Γ, Π", and

$$
\begin{array}{ccc}
\beta & \beta & \beta \\
\vdots & \vdots & \vdots \\
\cfrac{\vdash \Gamma, N[t/x]; \Pi}{\vdash \Gamma, \Pi; \exists x N} & \cfrac{\vdash \Gamma; P[t/x]}{\vdash \Gamma; \exists x P} & \cfrac{\vdash \Gamma, P[t/x]; \Pi}{\vdash \Gamma, \exists x P; \Pi}
\end{array}
$$

We can add to these rules the "classical dereliction", which allows to deduce $\vdash \Gamma, \Pi$; from a proof of $\vdash \Gamma; \Pi$.

3.2 The embedding P^a of LK_{pol} in linear logic

We are going to define P^a, an extension of the P-embedding defined by Girard in [Girard 91], which allows to translate in LL also all the rules of LK_{pol} which are not rules of LC: the irreversible rules ($\vee_a$, $\wedge_m$ and $\exists$) with a positive active formula which is not main.

We translate formulas in the usual P-way (see [Girard 91] and [DJS 95]), and the sequent $\vdash \Gamma_{neg}, \Gamma_{pos}; \Pi$ by the sequent $\vdash P(\Gamma_{neg}), ?P(\Gamma_{pos}), P(\Pi), \overrightarrow{\perp}$, where $\overrightarrow{\perp}$ indicates a certain number of occurrences of the formula $\perp\&\perp$, and Γ_{neg} (resp. Γ_{pos}) is the subset of Γ containing all the negative (resp. positive) formulas of Γ.

Remark: Observe that $\perp$ and $\perp\&\perp$ are linear formulas of type "?", so that (following our conventions) they can be active in a structural rule, or context in a promotion rule.

In the sequel we will write directly A instead of $P(A)$ and $?\Gamma$ instead of $P(\Gamma_{neg}), ?P(\Gamma_{pos})$. The embedding P^a from the set of LK_{pol}-proofs to the set of linear proofs is defined inductively in the following way:
- for the rules of LK_{pol} which are also rules of LC we translate in the obvious way, and the previous remark says precisely that the set of formulas $\overrightarrow{\perp}$ does not introduce any problem.

However, even among these rules, there is a rule which has to be carefully translated: it is the $\wedge_a$. We proceed in the following way (if the main conclusion of the rule $\wedge_a$ is $A \wedge_a B$, suppose for example that A is positive and not in the stoup, B is negative, and the two formulas in the two stoups are different):

$$P^a(\gamma) \qquad\qquad\qquad P^a(\delta)$$

$$\vdots \qquad\qquad\qquad\qquad \vdots$$

$$\cfrac{\cfrac{\vdash ?\Gamma, ?P, ?\Pi', \Pi, \vec{\bot}_1}{\vdash ?\Gamma, ?P, ?\Pi', ?\Pi, \vec{\bot}_1}}{\vdash ?\Gamma, ?P, ?\Pi', ?\Pi, \vec{\bot}_1, \vec{\bot}_2}\ W \qquad \cfrac{\cfrac{\vdash ?\Gamma, N, ?\Pi, \Pi', \vec{\bot}_2}{\vdash ?\Gamma, N, ?\Pi, ?\Pi', \vec{\bot}_2}}{\vdash ?\Gamma, N, ?\Pi', ?\Pi, \vec{\bot}_1, \vec{\bot}_2}\ W$$

$$\vdash ?\Gamma, ?P\&N, ?\Pi', ?\Pi, \vec{\bot}_1, \vec{\bot}_2$$

The important point here is that we do not "mix" formulas coming from $\vec{\bot}_1$ and formulas coming from $\vec{\bot}_2$, even if they are the same (see also the remark following definition 5).

- for the rules of LK_{pol} that are not rules of LC, let's consider some of the rules introducing the connective $\wedge_m$ which contain all the difficult cases:

• suppose that in the premisses the active formula P is out of the stoup and the main conclusion of the rule is $P \wedge_m N$, then we translate by

$$\cfrac{\vdash N^\bot, N}{\cfrac{\vdash ?N^\bot, N}{\cfrac{\vdash ?N^\bot, !N}{}}}$$

$$P^a(\delta) \qquad\qquad P^a(\gamma) \qquad \cfrac{\vdash P, P^\bot \quad \vdash ?N^\bot, !N}{\vdash P\otimes !N, P^\bot, ?N^\bot}$$

$$\vdots \qquad\qquad \vdots \qquad \cfrac{}{\vdash ?(P\otimes !N), P^\bot, ?N^\bot}$$

$$\cfrac{\vdash ?\Delta, N, \Pi', \vec{\bot}_2}{\cfrac{\vdash ?\Delta, N, ?\Pi', \vec{\bot}_2}{\vdash ?\Delta, !N, ?\Pi', \vec{\bot}_2}} \qquad \cfrac{\vdash ?\Gamma, ?P, \Pi, \vec{\bot}_1 \qquad \cfrac{\vdash ?(P\otimes !N), !P^\bot, ?N^\bot}{\vdash ?\Gamma, ?(P\otimes !N), ?N^\bot, \Pi, \vec{\bot}_1}}{}$$

$$\vdash ?\Gamma, ?\Delta, ?(P\otimes !N), ?\Pi', \Pi, \vec{\bot}_1, \vec{\bot}_2$$

By inspection of cases, one can check that there is a unique (semantically speaking) canonical way to translate all the rules introducing $\wedge_m$, except in case both the active formulas are positive and out of the stoup.

• suppose now that in the premisses both the active formulas P and Q are out of the stoup (this is *the* crucial case), then we translate by:

Remark: All the (proof-theoretical) solutions that we know to the problem of extracting the computational content from the proofs of classical logic *choose* as the translation of the rule $\wedge_m$ with two positive active formulas that are in the stoup *one* of the two linear subproofs premisses of the last rule introducing $\perp \& \perp$. The upshot of this work is precisely *not to choose* but rather to superimpose. This corresponds to the basic idea that a proof in LK is a superimposition of proofs of linear logic (or of LC).

4   Canonical form

Definition 2 *We say that the rule R of LK_{pol} is difficult iff $R = \wedge_m$ and both the active formulas in the premises of R are positive and are not in the stoup.*

Remark: It is proven in [Girard 91] and in [DJS 95] that each rule R of LK_{pol} can be translated by "some" rules of LC:
(i) if R is already an LC-rule then the translation is straightforward
(ii) if R is not a rule of LC (and if it is not difficult), then one cuts the premise(s) of the rule with a piece of η-proof of the conclusion of R. This procedure (which is also called a way of "constraining" R) might involve a choice in the order of performance of two cut rules; but this choice has no computational content: the two proofs corresponding to the two possible choices are interpreted by the same clique in the P-semantics
(iii) if R is a difficult rule, then one translates it like in case (ii), but now the order of performance of the two cuts *does* matter: not only the two proofs corresponding to the two possible choices are interpreted by different cliques in the P-semantics, but it might very well be the case that they cannot be identified by *any* (non degenerated) denotational semantics.
One has two different ways of constraining each difficult rule, which correspond exactly to the two different ways of translating R in LL through the P-embedding.
This means (as expected) that there exists a classical version of P^a (which will be called C^a) that translates proofs of LK_{pol} into proofs of LC. It is the identity on formulas, and acts on sequents and proofs like P^a: just substitute "$\neg V$" for "$\perp$", "$\wedge_a$" for "$\&$", and erase the exponential rules (through C^a we are constraining the proofs of LK_{pol} in "all the possible different ways"). For any LK_{pol}-proof α, $P^a(\alpha)$ and $P(C^a(\alpha))$ have the same denotational semantics.
We decided to focus on P^a (rather than on C^a) simply to stress the relation with LL.

Remark: Let α be a proof of LK_{pol} with k difficult rules. Let's call α^c the set of LC-proofs obtained from α by constraining all its rules in all the possible *different* ways. From the previous remark we deduce that there are 2^k such LC-proofs (because there are two different ways of constraining each difficult rule of α).
The reader should note that α^c is uniquely determined only from the *semantical* point of view: to each LK_{pol}-proof α is associated a unique set of 2^k cliques of

the coherent space interpreting the conclusion of α. However, as stated in (ii) of the previous remark, there might be different syntactical representations of "the same" proof of α^c. From now on, α^c will denote any such set of LC-proofs.

Definition 3 *Let α be an LK_{pol}-proof with k difficult rules and $\alpha^c = \{\alpha_1, \ldots, \alpha_{2^k}\}$. We'll call slice of α any LC-proof belonging to α^c.*

Definition 4 *Let α be an LK_{pol}-proof with k difficult rules and $\alpha^c = \{\alpha_1, \ldots, \alpha_{2^k}\}$. For all $i \in \{1, \ldots, 2^k\}$, we define $P^+(\alpha_i)$ as the proof $P(\alpha_i)$, to which we added k occurrences of the rule introducing $\perp$.*

Definition 5 *Let α be an LK_{pol}-proof with conclusion $\vdash \Gamma; \Pi$ and k difficult rules $R_1, \ldots, R_k$, and let $\alpha^c = \{\alpha_1, \ldots, \alpha_{2^k}\}$.*
To each $i \in \{1, \ldots, 2^k\}$ we can associate a sequence $\perp_{i_1}^1, \ldots, \perp_{i_k}^k$, where for all $j \in \{1, \ldots, k\}$ $i_j \in \{1, 2\}$, in such a way that R_j is constrained in the same way in α_l and α_m iff $l_j = m_j$.
This gives a bijection between $\{P^+(\alpha_i) : i \in \{1, \ldots, 2^k\}\}$ and $\{\perp_{i_1}^1, \ldots, \perp_{i_k}^k : i_j \in \{1, 2\}\}$.
We define the proof $\&(P^+(\alpha_i))$ of LL by building successively:
- the 2^{k-1} proofs with conclusion $\vdash ?\Gamma, \Pi, \perp_1^1 \& \perp_2^1, \perp_{i_2}^2, \ldots, \perp_{i_k}^k$, where $1 \le i_j \le 2$
- the 2^{k-2} proofs with conclusion $\vdash ?\Gamma, \Pi, \perp_1^1 \& \perp_2^1, \perp_1^2 \& \perp_2^2, \perp_{i_3}^3, \ldots, \perp_{i_k}^k$, where $1 \le i_j \le 2$

$$\vdots$$

- the proof $\&(P^+(\alpha_i))$ with conclusion $\vdash ?\Gamma, \Pi, \perp_1^1 \& \perp_2^1, \perp_1^2 \& \perp_2^2, \ldots, \perp_1^{k-1} \& \perp_2^{k-1}, \perp_1^k \& \perp_2^k$.
(The following example, in case $k = 2$, should clarify the construction).

Example: Suppose that in the previous definition $k = 2$. Then we have $\alpha^c = \{\alpha_1, \alpha_2, \alpha_3, \alpha_4\}$, and for $i \in \{1, 2, 3, 4\}$ we have that $P^+(\alpha_i)$ is an LL-proof with conclusion $\vdash ?\Gamma, \Pi, \perp_{i_1}^1, \perp_{i_2}^2$ where $i_1, i_2 \in \{1, 2\}$. Suppose for example that the conclusion of $P^+(\alpha_1)$ is $\vdash ?\Gamma, \Pi, \perp_1^1, \perp_1^2$, the conclusion of $P^+(\alpha_2)$ is $\vdash ?\Gamma, \Pi, \perp_1^1, \perp_2^2$, the conclusion of $P^+(\alpha_3)$ is $\vdash ?\Gamma, \Pi, \perp_2^1, \perp_1^2$, the conclusion of $P^+(\alpha_4)$ is $\vdash ?\Gamma, \Pi, \perp_2^1, \perp_2^2$.
Then, to obtain the proof $\&(P^+(\alpha_i))$, we first build the 2 ($= 2^{2-1}$) proofs $P^+(\alpha_1) \& P^+(\alpha_3)$ with conclusion $\vdash ?\Gamma, \Pi, \perp_1^1 \& \perp_2^1, \perp_1^2$ and $P^+(\alpha_2) \& P^+(\alpha_4)$ with conclusion $\vdash ?\Gamma, \Pi, \perp_1^1 \& \perp_2^1, \perp_2^2$, and then the proof $\&(P^+(\alpha_i))$ with conclusion $\vdash ?\Gamma, \Pi, \perp_1^1 \& \perp_2^1, \perp_1^2 \& \perp_2^2$.

Remark: In the previous definition, it is crucial to observe that, in the construction of the proof $\&(P^+(\alpha_i))$, the formulas active in the additive contractions of the rules $\&$ (that is the context formulas) which are different from $?\Gamma, \Pi$ are $\perp_1^i \& \perp_2^i$, or $\perp_1^i$, or $\perp_2^i$.
The fact that there are no additive contractions between $\perp_l^i$ and $\perp_m^i$ with $l \ne m$ means that there is no "mixing" between two "$\perp$" which stem for two different ways of constraining the rule R_i.
Observe also that all the linear proofs $\&(P^+(\alpha_i))$ associated to the LK_{pol}-proof α are semantically equal.

Proposition 6 *(canonical form)*
Let α be an LK_{pol}-proof, let $R_1 \ldots R_k$ be its difficult rules, and let $\alpha^c = \{\alpha_1, \ldots, \alpha_{2^k}\}$.
Then $[P^a(\alpha)] = [\&(P^+(\alpha_i))]$, where if d is a linear proof then $[d]$ is its denotational semantics.

proof: The idea is that, because the formulas $\perp\&\perp$ introduced by the embedding P^a are all formulas of the sequent conclusion of $P^a(\alpha)$ and because they are reversible and purely negative, it is possible to apply the rules introducing them "at the end" of the proof. The main point is of course the commutation of the rules introducing $\perp\&\perp$ with the promotion rule. But remember that the rule $\perp$ commutes to the promotion, and that $\&$ is reversible in a very strong sense: the denotational semantics of the proof obtained by applying a structural operation to the formula $A\&B$ (where A and B are linear formulas of type "?") *is* the denotational semantics of the proof obtained by applying first the same structural operation to A and to B and then by applying the rule introducing $\&$ (see [Girard 91] and [QT 96a] for more details).
A detailed proof can be given by induction on the proof α, i.e. by considering all the possible cases for the last rule R of α. $\qquad\qquad\square$

Definition 7 *Let α be a proof of LK_{pol} with k difficult rules and let $\alpha^c = \{\alpha_1, \ldots, \alpha_{2^k}\}$.*
We shall call canonical form of α, and we shall denote by $\bigwedge_a(\alpha_1, \ldots, \alpha_{2^k})$ the proof of LC obtained by applying to $\{\alpha_1, \ldots, \alpha_{2^k}\}$ exactly the same construction applied in definitions 4 and 5 to $\{P(\alpha_1), \ldots, P(\alpha_{2^k})\}$: just substitute $\neg V$ for $\perp$ and $\bigwedge_a$ for $\&$.

Corollary 8 *Let α be a proof of LK_{pol} with k difficult rules and let $\alpha^c = \{\alpha_1, \ldots, \alpha_{2^k}\}$. Then $[P^a(\alpha)] = [P(\bigwedge_a(\alpha_1, \ldots, \alpha_{2^k}))]$*
(or, otherwise stated, $[P(C^a(\alpha))] = [P(\bigwedge_a(\alpha_1, \ldots, \alpha_{2^k}))]$).

proof: It is the LC-counterpart of the canonical form theorem (proposition 6).
$\square$

Remark: The corollary says (in particular) that all the canonical forms of an LK_{pol}-proof α have the same denotational semantics.

5 Cut-elimination in LK_{pol}

Definition 9 *Let α and β be two proofs in LK_{pol}. We will say that β is obtained from α by applying one cut-elimination step, and we will write $\alpha \to \beta$, iff there exist a canonical form $\bigwedge_a(\alpha_1, \ldots, \alpha_{2^k})$ of α and a canonical form $\bigwedge_a(\beta_1, \ldots, \beta_{2^h})$ of β s.t.:*
- $k = h$
- there exists $i \in \{1, \ldots, 2^k\}$ such that in LC $\alpha_i \to \beta_i$ (that is β_i is obtained from α_i by applying one cut-elimination step of LC)
- $\forall j \in \{1, \ldots, 2^k\}$, $j \neq i$, we have $\alpha_j = \beta_j$.
$\twoheadrightarrow$ will denote, as usual, the reflexive and transitive closure of $\to$.
An LK_{pol}-proof is said to be normal when it is an LC-proof without cuts.

Remark: (*i*) All the reducts of a proof α of LK_{pol} can be "materialized" only through their canonical forms.

(*ii*) If α is an LK_{pol}-proof without cuts, then its canonical forms are not necessary cut-free (take any proof without cuts containing a rule which is not an LC-rule). Intuitively, this means that such a proof is not (yet) completely explicit.

Theorem 10 *If in LK_{pol} $\alpha \twoheadrightarrow \beta$, then $[P^a(\alpha)] = [P^a(\beta)]$, i.e. P^a is a denotational semantics for LK_{pol}.*

proof: To prove that P^a induces a denotational semantics for LK_{pol}, it is clearly enough to prove that if $\alpha \to \beta$, then $[P^a(\alpha)] = [P^a(\beta)]$. But remember that it is proven in [QT 96a] that P is a denotational semantics for LC, so that from corollary 8 we deduce: $[P^a(\alpha)] = [P(\bigwedge_a(\alpha_1, \ldots, \alpha_{2^k}))] = [P(\bigwedge_a(\beta_1, \ldots, \beta_{2^k}))] = [P^a(\beta)]$. $\qquad\square$

Remark: Observe that, because to an LK_{pol}-proof α we can associate (syntactically) different sets α^c, the cut-elimination procedure previously defined does not enjoy the Church-Rosser property. For the same reason, although cut-elimination in LK_{pol} seems to be strongly normalizing, this does not follow immediately from the same result for LC proven in [DJS 95] and [JST 96].

This seems to indicate that (following a suggestion of [Girard 91]), one should perhaps try to find a better syntax (which would identify two different ways of representing "the same" proof of α^c) for classical logic.

5.1   Gentzen 's cut-elimination is not explicitation

This is an informal subsection, where we give two examples, the aim of which is simply to suggest a possible retrospective analysis of Gentzen 's procedure of cut-elimination in the framework of LK_{pol}. What turns out is that this procedure is not (in general) compatible with the P^a-denotational semantics. Intuitively, if we assume that $P^a(\alpha)$ is the implicit content of the LK_{pol}-proof α, then Gentzen's cut-elimination does something different from simply making explicit what is already implicit in the proof. Sometimes it interacts with explicitation by making choices (like in example 1: it "does too much"), and sometimes it adds some "new" implicit content (like in example 2: "it doesn't do enough"): in both cases this procedure deeply modify the computational content of the proof.

example 1: The proof α (where R is supposed to be the unique difficult rule)

$$
\begin{array}{ccc}
\beta & \gamma & \delta \\
\vdots & \vdots & \vdots \\
\vdash P; \Pi \quad\quad \vdash Q; \Pi' & & \vdash \neg P, \neg Q; \Pi'' \\
\hline
\vdash P \wedge_m Q, \Pi, \Pi'; & R & \vdash \neg P \vee_m \neg Q; \Pi'' \\
\hline
& \vdash \Pi, \Pi', \Pi''; &
\end{array}
$$

is substituted by one of the two following proofs α_1 and α_2:

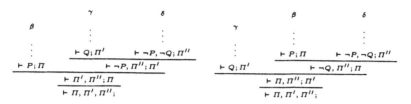

It is possible to check that each of these two proofs is a reduct (w.r.t. the cut-elimination of LC) of one of the two slices of α; choosing one of them means interacting with the process of cut-elimination of LK_{pol}. In LL this interaction is a cut between $\&(P^+(\alpha_1), P^+(\alpha_2))$ and one of the (two) proofs of $1 \oplus 1$ (which "selects" one of the two slices of the reduct of α).

example 2: The proof α

$$
\frac{\dfrac{\vdash \neg R; R \qquad \vdash \neg R; R}{\dfrac{\vdash R, R; \neg R \wedge_m \neg R}{\vdash R; \neg R \wedge_m \neg R}} \qquad \overset{\beta}{\underset{\vdots}{\vdash \neg R, P \wedge_m Q;}}}{\vdash P \wedge_m Q; \neg R \wedge_m \neg R}
$$

is substituted by the proof α'

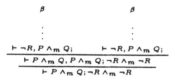

Suppose that $P \wedge_m Q$ in the subproof β is the conclusion of the unique difficult rule of α; then α has two slices while α' has four (because it has *two* difficult rules). Observe also that the two slices of α reduce to two (of the four) slices of α', and that the slices of α' which are not reducts of the slices of α cannot (in general) be identified with the reducts of the two slices of α without producing (algorithmic) inconsistency: to identify them means to identify all the proofs of a given sequent.

The point is that for β we have two possible choices for the translation of the difficult rule (let 's call them $P(\beta_{PQ})$ and $P(\beta_{QP})$) and in $P^a(\alpha)$ when we reduce the cut following the contraction on $?P(R)$, we have to contract all the context formulas ... including $\bot\&\bot$ introduced by the translation!

This means that the reduct of $P^a(\alpha)$ choose either $P(\beta_{PQ})$ or $P(\beta_{QP})$ for *both* the copies of β, while the P-image of one of the slices of α' will contain $P(\beta_{PQ})$ and $P(\beta_{QP})$ as subproofs.

Actually, for α' to have the same implicit content of α, one has to introduce the notion of "synchronization" of two difficult rules, and to synchronize the two difficult rules of α'. Technically speaking, to synchronize two difficult rules (which of course don't need to be equal, i.e. to have the same active formulas and conclusions) means precisely to contract the two "$\bot\&\bot$" introduced by the P^a-translation of the two rules.

We see that in this case Gentzen 's procedure "doesn't do enough" to keep the same implicit content.

6   Computational isomorphisms in LK_{pol}

Definition 11 *We say that R and Q (resp. M and N) are (ϕ, ψ)-isomorphic in LK_{pol}, and we write $R \stackrel{\phi,\psi}{\simeq} Q$, (resp. $M \stackrel{\phi,\psi}{\simeq} N$) if there exist a proof ϕ without difficult rules of $\vdash \neg R; Q$ (resp. a proof ϕ without difficult rules of $\vdash M; \neg N$) and a proof ψ without difficult rules of $\vdash \neg Q; R$ (resp. a proof ψ without difficult rules of $\vdash N; \neg M$) s. t.*
- the coherent semantics of the proof $cut(\phi, \psi)$ of $\vdash \neg R; R$ (resp. $\vdash M; \neg M$) is the identity of $P(R)$ (resp. of $P(M)$)
- the coherent semantics of the proof $cut(\psi, \phi)$ of $\vdash \neg Q; Q$ (resp. $\vdash N; \neg N$) is the identity of $P(Q)$ (resp. of $P(N)$).

Remark: We ask positive formulas in the conclusions of ϕ and ψ to be in the stoup, so that the cuts between ϕ and ψ are interpreted as composition of functions in the coherent semantics. This ensures us that when $A \stackrel{\phi,\psi}{\simeq} B$, then $P(A)$ and $P(B)$ are isomorphic as coherent spaces.

Theorem 12 *In LK_{pol} we have the same computational isomorphisms than in LC.*

proof: Girard proves (in [Girard 91]) that the correlation semantics of LC satisfies a certain number of isomorphisms. In [QT 96b], we prove that these isomorphisms are also satisfied if one takes the (slightly more syntactical) previous definition of isomorphism, and simply by considering the usual coherent semantics induced by the P-embedding (see [QT 96a]). The fact that this still holds for LK_{pol} (which contains LC as a subsystem) is straightforward: the definition of isomorphism is the same! □

Remark: (i) Observe however that, in spite of the previous theorem, the situation in LK_{pol} is different from the one we have in LC. Consider for example the associativity of conjuction.
If we take three proofs α_1 of $\vdash P;$, α_2 of $\vdash Q;$ and α_3 of $\vdash R;$, then the two proofs β of $\vdash (P \wedge_m Q) \wedge_m R;$ and γ of $\vdash P \wedge_m (Q \wedge_m R);$ that one can get simply by performing the two conjuctions in a different order, have both four slices. Let 's distinguish them by the order of application of the cut-rules of LK_{pol} between the pieces of η-proofs and the proofs α_1, α_2 and α_3.
The slice QPR (for example) of β is not a slice of γ. But this is only apparently contradictory with the isomorphism $(P \wedge_m Q) \wedge_m R \stackrel{\phi,\psi}{\simeq} P \wedge_m (Q \wedge_m R)$. Actually, roughly speaking, there exists a proof of $\vdash P \wedge_m (Q \wedge_m R);$ containing "the same" slices of β. To convince himself, the reader may observe that what one eventually does to get the proof β of $\vdash (P \wedge_m Q) \wedge_m R;$ from α_1, α_2 and α_3 is to cut α_1, α_2 and α_3 with the canonical proof of $\vdash \neg P, \neg Q, \neg R; (P \wedge_m Q) \wedge_m R$ in four of the six possible ways. Then, one obtains a proof β' of $\vdash P \wedge_m (Q \wedge_m R);$ with "the same" slices of β by cuting α_1, α_2 and α_3 with the canonical proof of $\vdash \neg P, \neg Q, \neg R; P \wedge_m (Q \wedge_m R)$ in *the same* four ways (which are *not* the four ways used to obtain the four slices of γ).

This last proof β' is precisely a reduct of the proof obtained by cuting β with the (canonical) proof ϕ of $\vdash \neg(P \wedge_m Q) \vee_m \neg R; P \wedge_m (Q \wedge_m R)$.

(*ii*) If one really wants to have simoultaneously, for the proofs of $\vdash P \wedge_m (Q \wedge_m R)$; and of $\vdash (P \wedge_m Q) \wedge_m R$; obtained from α_1, α_2 and α_3 of (*i*), all the six possible slices, then one has to take a version of LK_{pol} with difficult rules of arbitrary arity. But observe that this is not necessary to have a computational isomorphism between $A \wedge_m (B \wedge_m C)$ and $(A \wedge_m B) \wedge_m C$, whatever the polarities of A, B and C are.

7 Further developments

From the technical point of view, the next step (which looks difficult) would be to find a better syntax for LK_{pol}, identifying all the canonical forms of a given proof.

What seems interesting (in the framework of LK_{pol}) is to define some kind of relationship between the slices superimposed by an LK_{pol}-proof.

The notion of synchronization of two (or more) difficult rules might also play an important role. It is a very natural notion, which is technically translated by one (or more) contraction(s) on the $\perp \& \perp$ introduced by P^a (as stated in example 2 of 5.1). To take difficult rules of arbitrary arity might also increase the flexibility of the calculus.

Finally, following an old idea of [Girard 91], one might also try to "understand" Gentzen 's cross-cut procedure in the framework of LK_{pol}.

References

[BarBer 95]   Barbanera F., Berardi S., A Symmetric Lambda Calculus for "classical" Program extraction, to appear in *I&C*, 1995

[DJS 95]   Danos V., Joinet J.-B., Schellinx H., A new deconstructive logic: linear logic, to appear in the Journal of Symbolic Logic, 1995

[Girard 87]   Girard J.Y., Linear logic, Theoretical Computer Science, 50:1-102, 1987.

[Girard 91]   Girard J.Y., A new constructive logic: classical logic. Mathematical Strucures in Computer Science, 1(3):255-296, 1991.

[JST 96]   Joinet J.-B., Schellinx H., Tortora de Falco L., Strong Normalization for "all-style" LK^{tq}, extended abstract, Lecture Notes in Artificial Intelligence 1071, p. 226-243, Springer-Verlag 1996

[Parigot 91]   Parigot M., Free Deduction: an analysis of computation in classical logic. In: Voronkov, A, editor, Russian Conference on Logic Programming, pp. 361-380. Springer Verlag. LNAI 592. 1991.

[QT 96a]   Quatrini M., Tortora de Falco L., Polarisation des preuves classiques et renversement, Compte-Rendu de l'Académie des Sciences Paris, t. 323, Série I, p. 113-116, 1996

[QT 96b]   Quatrini M., Tortora de Falco L., Reversion and isomorphisms in classical logic: from LK_{pol}^{η} to LC, manuscript, 1996

[Tortora 96]   Tortora de Falco L., Generalized Standardization lemma for the additives, vol. 3 Electronic Notes in Theoretical Computer Science, 1996

The Undecidability of Simultaneous Rigid
E-Unification with Two Variables

Margus Veanes

Uppsala University Computing Science Department
P.O. Box 311, S-751 05 Uppsala, Sweden

Abstract. Recently it was proved that the problem of simultaneous
rigid E-unification, or SREU, is undecidable. Here we show that 4 rigid
equations with ground left-hand sides and 2 variables already imply undecidability. As a corollary we improve the undecidability result of the
$\exists^*$-fragment of intuitionistic logic with equality. Our proof shows undecidability of a very restricted subset of the $\exists\exists$-fragment. Together with
other results, it contributes to a complete characterization of decidability
of the prenex fragment of intuitionistic logic with equality, in terms of
the quantifier prefix.

1 Introduction

Recently it was proved that the problem of simultaneous rigid E-unification
(SREU) is undecidable [11]. This (quite unexpected) undecidability result has
lead to other new undecidability results, in particular that the $\exists^*$-fragment of
intuitionistic logic with equality is undecidable [13,15]. Here we show that 4 rigid
equations[1] with ground left-hand sides and 2 variables already imply undecidability. As a corollary we improve the undecidability result of the $\exists^*$-fragment
of intuitionistic logic with equality. Namely that the $\exists\exists$-fragment is undecidable. In fact, our proof shows undecidability of a very restricted subset of the
$\exists\exists$-fragment. Together with the result that the $\exists$-fragment is decidable [6], it contributes to a complete characterization of decidability of the prenex fragment of
intuitionistic logic with equality, in terms of the quantifier prefix.

1.1 Background of SREU

Simultaneous rigid E-unification was proposed by Gallier, Raatz and Snyder [21]
as a method for automated theorem proving in classical logics with equality. It
can be used in automatic proof methods, like semantic tableaux [18], the connection method [3] or the mating method [1], model elimination [32], and others
that are based on the Herbrand theorem, and use the property that a formula
φ is valid (i.e., $\neg\varphi$ is unsatisfiable) iff all paths through a matrix of φ are inconsistent. This property was first recognized by Prawitz [38] (for first-order logic
without equality) and later by Kanger [28] (for first-order logic with equality).

---

[1] It has been noted by Gurevich and Veanes that 3 rigid equations suffices [25].

In first-order logic with equality, the problem of checking the inconsistency of the paths results in SREU. Before SREU was proved to be undecidable, there were several faulty statements of its decidability, e.g. [19,24].

1.2 Outline of the Paper

In Section 2 we introduce the notations used in this paper and briefly explain the background material. In Section 3 we prove the main result of this paper (Theorem 8), that implies immediately the undecidability result of a very restricted case of SREU. In Section 4 we use this result to obtain undecidability of a restricted subset of the $\exists\exists$-fragment of intuitionistic logic with equality. Finally, the current status about SREU is summarized and some open problems are listed in Section 5.

2 Preliminaries

We introduce here the main notions and definitions used in this paper. Given a signature Σ, i.e., a set of function symbols with fixed arities, the set of all ground (or closed) terms over Σ is denoted by $\mathcal{T}_\Sigma$. Unless otherwise stated it is always assumed that Σ is nonempty, finite and includes at least one constant (function symbol of arity 0). We also assume certain familiarity with basic notions from term rewriting [16], regarding ground rewriting systems. By a *substitution* we understand a function from variables to *ground* terms and a substitution is always denoted by θ. An application of θ on a variable x is written as $x\theta$ instead of $\theta(x)$.

2.1 Finite Tree Automata

Finite tree automata [17,39] is a natural generalization of classical finite automata to automata that accept or recognize trees of symbols, not just strings. Here we adopt a definition of tree automata based on rewrite rules. This definition is used for example by Dauchet [4].

▶ A *tree automaton* or *TA* is a quadruple $A = (Q, \Sigma, R, F)$ where
  - Q is a finite set of constants called *states*,
  - Σ is a *signature* or an *input alphabet* disjoint from Q,
  - R is a set of *rules* of the form $\sigma(q_1, \ldots, q_n) \to q$, where $\sigma \in \Sigma$ has arity $n \geq 0$ and $q, q_1, \ldots, q_n \in Q$,
  - $F \subseteq Q$ is the set of *final states*.
  A is called a *deterministic* TA or DTA if there are no two different rules in R with the same left-hand side.

Note that if A is deterministic then R is a reduced set of ground rewrite rules, i.e., for any rule $s \to t$ in R t is irreducible and s is irreducible with respect to $R \setminus \{s \to t\}$. So R is a ground canonical rewrite system. In this context terms are also called trees. A set of terms (or trees) is called a *forest*.

▶ The forest *recognized* by a TA $A = (Q, \Sigma, R, F)$ is the set

$$T(A) = \{\, t \in \mathcal{T}_\Sigma \mid (\exists q \in F)\, t \xrightarrow{*}_R q \,\}.$$

A forest is called *recognizable* if it is recognized by some TA.

We assume that the reader is familiar with classical automata theory and we follow Hopcroft and Ullman [27] in that respect.

2.2   Simultaneous Rigid E-Unification

A *rigid equation* is an expression of the form $E \mathrel{\vdash\!\!\!\!-} s \approx t$ where E is a finite set of equations, called the *left-hand side* of the rigid equation, and s and t are arbitrary terms. A *system* of rigid equations is a finite set of rigid equations. A substitution θ is a *solution of* or *solves* a rigid equation $E \mathrel{\vdash\!\!\!\!-} s \approx t$ if

$$\vdash\ (\bigwedge_{e \in E} e\theta) \Rightarrow s\theta \approx t\theta,$$

and θ is a *solution of* or *solves* a system of rigid equations if it solves each member of that system. Here $\vdash$ is classical or intuitionistic provability (for this class of formulas both provabilities coincide). The problem of solvability of systems of rigid equations is called *simultaneous rigid E-unification* or SREU for short. Solvability of a single rigid equation is called *rigid E-unification*. Rigid E-unification is known to be decidable, in fact NP-complete [20]. The following simple lemma is useful.

Lemma 1. *Let $A = (Q, \Sigma, R, F)$ be a DTA, f a binary function symbol, and c_1 and c_2 constants not in Q or Σ. There is a set of ground equations E such that for all θ such that $x\theta \in \mathcal{T}_\Sigma$, θ solves $E \mathrel{\vdash\!\!\!\!-} f(c_1, x) \approx c_2$ iff $x\theta \in T(A)$.*

Proof. Let $E = R \cup \{\, f(c_1, q) \to c_2 \mid q \in F \,\}$. It follows easily that E is a canonical rewrite system, and since c_2 is irreducible with respect to E we have in particular for all $t \in \mathcal{T}_\Sigma$, that (cf [16, Section 2.4])

$$E \vdash f(c_1, t) \approx c_2 \quad \Leftrightarrow \quad f(c_1, t) \xrightarrow{*}_E c_2.$$

But

$$f(c_1, t) \xrightarrow{*}_E c_2 \quad \Leftrightarrow \quad (\exists q \in F)\, t \xrightarrow{*}_R q.$$

The rest is obvious. □

3   Minimal Undecidable Case of SREU

We present yet another proof of the undecidability of SREU. At the end of this section we give a brief summary of the other proofs. The main idea behind this proof is based on a technique that we call *shifted pairing* after Plaisted [37].

The idea is to express repetition explicitly by a sequence of strings (like IDs of a TM). The first string of the sequence fulfills some initial conditions, the last string some final conditions and another sequence is used to check that the consequtive strings of the first sequence satisfy some relationship (like validity of a computation step).

A similar technique was used already by Goldfarb in the proof of the unde-cidability of second-order unification [23] (which was by reduction of Hilbert's tenth problem) and later, adopted from that proof, also in the third proof of the undecidability of SREU by Degtyarev and Voronkov [13] (which was also by reduction of Hilbert's tenth problem). There the key point was to explicitly represent the history of a multiplication process.

Shifted pairing bears also certain similarities to the technique that is used to prove that any recursively enumerable set of strings is given by the intersection of two (deterministic) context free languages [27, Lemma 8.6].

3.1 Overview of the Construction

We consider a fixed Turing machine M,

$$M = (Q_M, \Sigma_{\text{in}}, \Sigma_{\text{tape}}, \delta, q_0, \text{b}, \{q_{\text{acc}}\}).$$

We can assume, without loss of generality, that the final ID of M is simply q_{acc} (and that $q_0 \neq q_{\text{acc}}$), i.e., the tape is always empty when M enters the final state. We construct a system $S_M(x, y)$ of four rigid equations:

$$S_M(x, y) = \{ \ E_{\text{id}} \mathrel{\vdash} c'_{\text{id}} \cdot x \approx c_{\text{id}}, \tag{1}$$

$$E_{\text{mv}} \mathrel{\vdash} c'_{\text{mv}} \cdot y \approx c_{\text{mv}}, \tag{2}$$

$$\Pi_1 \mathrel{\vdash} x \approx y, \tag{3}$$

$$\Pi_2 \mathrel{\vdash} x \approx (q_0 \cdot e_0) \cdot y \ \} \tag{4}$$

where all the left-hand sides are ground, c'_{id}, c_{id}, c'_{mv} and c_{mv} are constants, and $q_0 \cdot e_0$ is a word that represents the initial ID of M with empty input string (ϵ). We prove that M accepts ϵ iff S_M is solvable. This establishes the undecidability result because all the steps in the construction are effective.

The main idea behind the rigid equations is roughly as follows. Assume that there is a substitution θ that solves the system.

– From θ being a solution of (1), it follows that $x\theta$ represents a sequence

$$(v_0, v_1, \ldots, v_m)$$

of IDs of M, where v_m is the final ID of M.

– From θ being a solution of (2), it follows that $y\theta$ represents a sequence

$$((w_0, w_0^+), (w_1, w_1^+), \ldots, (w_n, w_n^+))$$

of *moves* of M, i.e., $w_i \vdash_M w_i^+$ for $0 \leq i \leq n$.

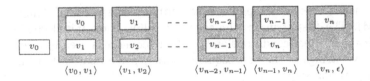

Fig. 1. $(\langle v_0, v_1 \rangle, \langle v_1, v_2 \rangle, \ldots, \langle v_n, \epsilon \rangle)$ is a shifted pairing of $(v_0, v_1, \ldots, v_n)$.

- From θ being a solution of (3) it follows that $n = m$ and $v_i = w_i$ for $0 \leq i \leq m$.
- And finally, from θ being a solution of (4) it follows that v_0 is the initial ID and $v_i = w_{i-1}^+$ for $1 \leq i \leq m$.

The combination of the last two points is the so-called "shifted pairing" technique. This is illustrated by Figure 1. The outcome of this shifted pairing is that $x\theta$ is a valid computation of M with input ϵ, and thus M accepts ϵ. Conversely, if M accepts ϵ then it is easy to construct a solution of the system. We now give a formal construction of the above idea.

3.2 Words and Trains

Words are certain terms that we choose to represent strings with, and trains are certain terms that we choose to represent sequences of strings with. We use the letters v and w to stand for strings of constants. Let . be a binary function symbol. We write it in infix notation and assume that it associates to the right. For example $t_1 . t_2 . t_3$ stands for the term $.(t_1, .(t_2, t_3))$.

▶ We say that a (ground) term t is a *c-word* if it has the form

$$a_1 . a_2 . \cdots . a_n . c$$

for some $n \geq 0$ where each a_i and c is a constant. A *word* is a c-word for some constant c.

We use the following convenient shorthand notation for words. Let t be the word $a_1 . a_2 . \cdots . a_n . c$ and v the string $a_1 a_2 \cdots a_n$. We write $v . c$ for t and say that t *represents* v.

▶ A term t is called a *c-train* if it has the form

$$t_1 . t_2 . \cdots . t_n . c$$

for some $n \geq 0$ where each t_i is a word and c is a constant. If $n = 0$ then t is said to be *empty*. The t_i's are called the *words of* t. A *train* is a c-train for some constant c.

By the *pattern* of a train

$$(v_1 . c_1) . (v_2 . c_2) . \cdots . (v_n . c_n) . c$$

we mean the string $c_1 c_2 \cdots c_n$. Let $\mathcal{V} = \{V_i\}_{i \in I}$ be a finite family of regular sets of strings over a finite set Σ of constants, where I is a set of constants disjoint from Σ. Let U be a regular set of strings over I and let c be a constant not in Σ or I.

▶ We let $\mathrm{Tn}(\mathcal{V}, U, c)$ denote the set of all c-trains t such that the pattern of t is in U and, for $i \in I$, each i-word of t represents a string in V_i.

Example 2. Consider the set $\mathrm{Tn}(\{V_a, V_b, V_c\}, ab^*c, \Lambda)$. This is the set of all Λ-trains t such that the first word of t is an a-word representing a string in V_a, the last word of t is a c-word representing a string in V_c and the middle ones (if any) are b-words representing strings in V_b.

We say that a set of trains *has a regular pattern* if it is equal to some set $\mathrm{Tn}(\mathcal{V}, U, c)$ with $\mathcal{V}$, U and c as above. The following theorem is proved in Veanes [40].

Theorem 3 (Train Theorem). *Any set of trains with a regular pattern is recognizable and a DTA that recognizes this set can be obtained effectively.*

3.3 Representing IDs and Moves

Recall that an *ID* of M is any string in $\Sigma_{\mathrm{tape}}^* Q_M \Sigma_{\mathrm{tape}}^*$ that doesn't end with a blank ($\mathfrak{b}$). Let us assign arity 0 to all the tape symbols (Σ_{tape}) and all the states (Q_M) of M, and let Σ denote the signature consisting of all those constants, the binary function symbol . and four new constants e_0, e_1, e_{acc} and Λ.

ID-trains IDs are represented by e-words, where e is one of e_0, e_1 or e_{acc}. In particular, the initial ID is represented by the word $q_0 \cdot e_0$. The final ID is represented by the word $q_{\mathrm{acc}} \cdot e_{\mathrm{acc}}$ and all the other IDs are represented by corresponding e_1-words. The term

$$(q_0 \cdot e_0) \cdot (v_1 \cdot e_1) \cdot (v_2 \cdot e_1) \cdot \cdots \cdot (v_n \cdot e_1) \cdot (q_{\mathrm{acc}} \cdot e_{\mathrm{acc}}) \cdot \Lambda$$

is called an *ID-train*. By using Theorem 3 let

$$A_{\mathrm{id}} = (Q_{\mathrm{id}}, \Sigma, R_{\mathrm{id}}, F_{\mathrm{id}})$$

be a DTA that recognizes the set of all ID-trains. Let c_{id}' and c_{id} be new constants and (1) the rigid equation provided by Lemma 1, i.e., for all θ such that $x\theta \in \mathcal{T}_\Sigma$,

$$\theta \text{ solves } (1) \quad \Leftrightarrow \quad x\theta \in T(A_{\mathrm{id}}).$$

Move-trains Let c_{ab} be a new constant for each pair of constants a and b in the set $\Sigma_{\text{tape}} \cup Q_M$. Let also e_2 and Λ' be new constants. Let now Γ be a signature that consists of $\cdot$, all those c_{ab}'s, e_2 and Λ'.

For and ID w of M we let w^+ denote the successor of w with respect to the transition function of M. For technical reasons it is convenient to let $q_{\text{acc}}^+ = \epsilon$, i.e., the successor of the final ID is the empty string. The pair (w, w^+) is called a *move*. Let $w = a_1 a_2 \cdots a_m$ and $w^+ = b_1 b_2 \cdots b_n$ for some $m \geq 1$ and $n \geq 0$. Note that $n \in \{m - 1, m, m + 1\}$. Let $k = \max(m, n)$. If $m < n$ let $a_k = \flat$ and if $n < m$ let $b_k = \flat$, i.e., pad the shorter of the two strings with a blank at the end.

▶ We write $\langle w, w^+ \rangle$ for the string $c_{a_1 b_1} c_{a_2 b_2} \cdots c_{a_k b_k}$ and say that the e_2-word $\langle w, w^+ \rangle . e_2$ *represents* the move (w, w^+). By a *move-train* we mean any term

$$t = t_1 . t_2 . \cdots . t_n . \Lambda'$$

such that each t_i represents a move.

Example 4. Take $\Sigma_{\text{in}} = \{0, 1\}$, and let $\mathsf{q}, \mathsf{p} \in Q_M$. Assume that the transition function δ is such that, when the tape head points to a blank and the state is q then a 1 is written to the tape, the tape head moves left and M enters state p, i.e., $\delta(\mathsf{q}, \flat) = (\mathsf{p}, 1, \mathrm{L})$. Let the current ID be $00\mathsf{q}$, i.e., the tape contains the string 00 and the tape head points to the bank following the last 0. So $(00\mathsf{q}, 0\mathsf{p}01)$ is a move. This move is represented by the term $c_{00} . c_{0\mathsf{p}} . c_{\mathsf{q}0} . c_{\flat 1} . e_2$, or $\langle 00\mathsf{q}, 0\mathsf{p}01 \rangle . e_2$ if we use the above notation.

It is easy to see that the set of all strings $\langle w, w^+ \rangle$ where w is an ID, is a regular set. By using Theorem 3 let

$$A_{\text{mv}} = (Q_{\text{mv}}, \Gamma, R_{\text{mv}}, F_{\text{mv}})$$

be a DTA that recognizes the set of all move-trains. Let c'_{mv} and c_{mv} be new constants and (2) the rigid equation provided by Lemma 1, i.e., for all θ such that $y\theta \in \mathcal{T}_\Gamma$,

$$\theta \text{ solves (2)} \quad \Leftrightarrow \quad y\theta \in T(A_{\text{mv}}).$$

3.4 Shifted Pairing

We continue with the contruction of S_M. What has remained to define is Π_1 and Π_2. These are defined as sets of equations corresponding to the following canonical rewrite systems.

$$\Pi_1 = \{ c_{ab} \to a \mid a, b \in \Sigma_{\text{tape}} \cup Q_M \} \cup$$
$$\{ e_1 \to e_0, \ e_2 \to e_0, \ e_{\text{acc}} \to e_0, \ \Lambda' \to \Lambda, \ \flat . e_0 \to e_0 \}$$
$$\Pi_2 = \{ c_{ab} \to b \mid a, b \in \Sigma_{\text{tape}} \cup Q_M \} \cup$$
$$\{ e_1 \to e_0, \ e_2 \to e_0, \ e_{\text{acc}} \to e_0, \ \Lambda' \to \Lambda, \ \boldsymbol{\flat . e_0 \to e_0, \ e_0 . \Lambda \to \Lambda} \}$$

The differences between Π_1 and Π_2 are indicated in boldface.

Lemma 5. *If θ solves (3) and (4) then $x\theta, y\theta \in \mathcal{T}_{\Sigma \cup \Gamma}$.*

Proof. By induction on the size of $x\theta$ [40]. □

Lemma 6. *If θ solves $S_M(x,y)$ then $x\theta \in \mathcal{T}_\Sigma$ and $y\theta \in \mathcal{T}_\Gamma$.*

Proof. Assume that θ solves $S_M(x,y)$. Obviously $x\theta \in \mathcal{T}_{\Sigma \cup Q_{id}}$ since θ solves (1). By Lemma 5 we know also that $x\theta \in \mathcal{T}_{\Sigma \cup \Gamma}$. But Σ, Γ and Q_{id} don't share any constants. So $x\theta \in \mathcal{T}_\Sigma$. A similar argument shows that $y\theta \in \mathcal{T}_\Gamma$. □

Lemma 7. *If θ solves $S_M(x,y)$ then $x\theta$ is an ID-train and $y\theta$ is a move-train.*

Proof. By Lemma 6, the definition of A_{id} and A_{mv} and Lemma 1. □

We have now reached the main theorem of this paper.

Theorem 8. *$S_M(x,y)$ is solvable iff M accepts ϵ.*

Proof. ($\Rightarrow$) Let θ be a substitution that solves $S_M(x,y)$. By using Lemma 7 we get that $x\theta$ and $y\theta$ have the following form:

$$x\theta = (v_0 \cdot e_0) \cdot (v_1 \cdot e_1) \cdot \cdots \cdot (v_{m-1} \cdot e_1) \cdot (v_m \cdot e_{acc}) \cdot \Lambda$$
$$y\theta = (\langle w_0, w_0^+\rangle \cdot e_2) \cdot (\langle w_1, w_1^+\rangle \cdot e_2) \cdot \cdots \cdot (\langle w_n, w_n^+\rangle \cdot e_2) \cdot \Lambda'$$

where all the v_i's and w_i's are IDs of M, $v_0 = q_0$ and $v_m = q_{acc}$.

Since θ solves (3) it follows that the normal forms of $x\theta$ and $y\theta$ under Π_1 must coincide. The normal form of $x\theta$ under Π_1 is

$$(v_0 \cdot e_0) \cdot (v_1 \cdot e_0) \cdot \cdots \cdot (v_{m-1} \cdot e_0) \cdot (v_m \cdot e_0) \cdot \Lambda.$$

The normal form of $y\theta$ under Π_1 is

$$(w_0 \cdot e_0) \cdot (w_1 \cdot e_0) \cdot \cdots \cdot (w_{n-1} \cdot e_0) \cdot (w_n \cdot e_0) \cdot \Lambda.$$

Note that each term $\langle w_i, w_i^+\rangle \cdot e_2$ reduces first to $w_i' \cdot e_0$ where $w_i' = w_i$ or $w_i' = w_i\flat$. The extra blank at the end is removed with the rule $\flat \cdot e_0 \to e_0$. So

$$v_0 = q_0, \quad v_n = q_{acc}, \quad v_i = w_i \ (0 \le i \le n = m). \tag{5}$$

Since θ solves (4) it follows that the normal forms of $x\theta$ and $(q_0 \cdot e_0) \cdot y\theta$ under Π_2 must coincide. The normal form of $x\theta$ under Π_2 is the same as under Π_1 because $x\theta$ doesn't contain any constants from Γ and the rule $e_0 \cdot \Lambda \to \Lambda$ is not applicable. From $w_n = q_{acc}$ follows that $w_n^+ = \epsilon$ and thus $\langle w_n, w_n^+\rangle \cdot e_0 = c_{q_{acc}\flat} \cdot e_0$. But

$$(c_{q_{acc}\flat} \cdot e_0) \cdot \Lambda \longrightarrow_{\Pi_2} (\flat \cdot e_0) \cdot \Lambda \longrightarrow_{\Pi_2} e_0 \cdot \Lambda \longrightarrow_{\Pi_2} \Lambda.$$

The normal form of $(q_0 \cdot e_0) \cdot y\theta$ under Π_2 is thus

$$(q_0 \cdot e_0) \cdot (w_0^+ \cdot e_0) \cdot (w_1^+ \cdot e_0) \cdot \cdots \cdot (w_{n-1}^+ \cdot e_0) \cdot \Lambda.$$

It follows that

$$w_i^+ = v_{i+1} \ (0 \le i < n). \tag{6}$$

From (5) and (6) follows that $(v_0, v_1, \ldots, v_n)$ is a valid computation of M, and thus M accepts ϵ.

($\Leftarrow$) Assume that M accepts ϵ. So there exists a valid computation

$$(v_0, v_1, \ldots, v_n)$$

of M where $v_0 = q_0$, $v_n = q_{acc}$ and $v_i^+ = v_{i+1}$ for $0 \le i < n$. Let θ be such that $x\theta$ is the corresponding ID-train and $y\theta$ the corresponding move-train. It follows easily that θ solves $S_M(x, y)$. $\square$

The "shifted pairing" technique that is used in Theorem 8 is illustrated in Figure 1.

The following result is an immediate consequence of Theorem 8 because all the constructions involved with it are effective.

Corollary 9. *SREU is undecidable if the left-hand sides are ground, there are two variables and four rigid equations.*

It was observed by Gurevich and Veanes that the two DTAs A_{id} and A_{mv} can be combined into one DTA (by using elementary techniques of finite tree automata theory [22]), and by this way reducing the number of rigid equtions in S_M into *three* [25]. It is still an open question if SREU with *two* rigid equations is decidable.

3.5 Previous Undecidability Proofs of SREU

The first proof of the udecidability of SREU [11] was by reduction of the monadic semi-unification [2] to SREU. This proof was followed by two alternative (more transparent) proofs by the same authors, first by reducing second-order unification to SREU [10,15], and then by reducing Hilbert's tenth problem to SREU [14]. The undecidability of second-order unification was proved by Goldfarb [23]. Reduction of second-order unification to SREU is very simple, showing how close these problem are to each other. Plaisted took the Post's Correspondence Problem and reduced it to SREU [37]. From his proof follows that SREU is undecidable already with ground left-hand sides and three variables. He uses several function symbols of arity 1 and 2.

3.6 Herbrand Skeleton Problem

The *Herbrand skeleton problem of multiplicity n* is a fundamental problem in automated theorem proving [7], e.g., by the method of matings [1], the tableaux method [18], and others. It can can be formulated as follows:

Given a quantifier free formula $\varphi(x)$, does there exist a sequence of ground terms $t_1, \ldots, t_n$ such that the disjunction $\varphi(t_1) \vee \cdots \vee \varphi(t_n)$ is valid?

The undecidability of this problem was established recently by Voda and Komara [41] by a technique similar to the one used in the reduction of Hilbert's tenth problem to SREU [14]. Their proof is very complicated and (contrary to their claim) it is shown in Gurevich and Veanes [25] by using a novel logical lemma that the Herbrand skeleton problem of any fixed multiplicity reduces easily to SREU. As a corollary (by using Theorem 8) improving the result in [41], by proving the undecidability of this problem for a restricted Horn fragment of classical logic (where variables occur only positively).

4 Undecidability of the ∃∃-fragment of Intuitionistic Logic with Equality

Undecidability of the ∃*-fragment of intuitionistic logic with equality was established recently by Degtyarev and Voronkov [13,15]. We obtain the following improvement of this result. Let $\vdash_i$ stand for provability in intuitionistic predicate calculus with equality and let $\vdash_c$ stand for provability in classical predicate calculus (with equality).

Theorem 10. *The class of formulas in intuitionistic logic with equality of the form* $\exists x \exists y \, \varphi(x, y)$ *where* φ *is quantifier free, and*

- *the language contains (besides constants) a function symbol of arity* ≥ 2,
- *the only connectives in* φ *are '$\wedge$' and '$\Rightarrow$' and*
- *the antecedents of all implications in* φ *are closed,*

is undecidable.

Proof. Let $S_M(x, y)$ be the system of rigid equations given by Theorem 8. So

$$S_M(x, y) \quad = \quad \{\, E_i \vdash s_i \approx t_i \mid 1 \leq i \leq 4 \,\},$$

where each E_i is a set of (ground) equations. Let $\psi_i = \bigwedge_{e \in E_i} e$ for $1 \leq i \leq 4$. Note that each ψ_i is closed. Let

$$\varphi(x, y) \quad = \quad \bigwedge_{1 \leq i \leq 4} (\psi_i \Rightarrow s_i \approx t_i).$$

The construction of φ from S_M and thus from M is clearly effective. To prove the theorem it is enough to prove the following statement:

$$\epsilon \in L(M) \quad \Leftrightarrow \quad \vdash_i \exists x \exists y \varphi(x, y).$$

($\Rightarrow$) Assume $\epsilon \in L(M)$. By Theorem 8 there is a substitution θ that solves $S_M(x, y)$. By definition, this means that $\vdash_c \varphi(x\theta, y\theta)$. But

$$\vdash_c \varphi(x\theta, y\theta) \Rightarrow \vdash_i \varphi(x\theta, y\theta)$$

for this particular class of formulas. The rest is obvious.

($\Leftarrow$) Assume that $\vdash_i \exists x \exists y \varphi(x, y)$. By the explicit definabilty property of intuitionistic logic there are ground terms t and s such that $\vdash_i \varphi(t, s)$ and thus $\vdash_c \varphi(t, s)$. Let θ be such that $x\theta = t$ and $y\theta = s$. It follows that θ solves the system $S_M(x, y)$, and thus $\epsilon \in L(M)$ by Theorem 8. □

A closely related problem is the *skeleton instantiation problem* (the problem of existence of a derivation with a given skeleton in a given proof system). Voronkov shows that SREU is polynomially reducible to this problem [42, Theorem 3.12] (where the actual proof system under consideration is a sequent calculus $LJ^{\approx}$ for intuitionistic logic with equality). Moreover, the basic structure of the skeleton is determined by the number of variables in the SREU problem and the number of rigid equations in it. The above corollary implies that this problem is undecidable already for a very restricted class of skeletons.

In Degtyarev, Gurevich, Narendran, Veanes and Voronkov [6] it is proved that the ∃-fragment of intuitionistic logic with equality is decidable. For further results about the prenex fragment see Degtyarev, Matiyasevich and Voronkov [9], Degtyarev and Voronkov [12] and Voronkov [43,42]. Decidabilty problems for some other fragments of intuitionistic logic with and without equality were studied by Orevkov [35,36], Mints [34] and Lifschitz [31].

5 Current Status and Open problems

The first decidability proof of rigid E-unification is given in Gallier, Narendran, Plaisted and Snyder [20]. Recently a simpler proof, without computational complexity considerations, has been given by de Kogel [5]. We start with the *solved cases*:

- Rigid E-unification with ground left-hand side is NP-complete [30]. Rigid E-unification in general is NP-complete and there exist finite complete sets of unifiers [19,20].
- SREU with one variable and a fixed number of rigid equations is P-complete [6].
- If all function symbols have arity ≤ 1 (the *monadic* case) then SREU is PSPACE-hard [24]. If only one unary function symbol is allowed then the problem is decidable [8,9]. If only constants are allowed then the problem is NP-complete [9] if there are at least two constants.
- About the monadic case it is known that if there are more than 1 unary function symbols then SREU is decidable iff it is decidable with just 2 unary function symbols [9].
- If the left-hand sides are ground then the monadic case is decidable [26]. Monadic SREU with one variable is PSPACE-complete [26].
- The word equation solving [33] (i.e., unification under associativity), which is an extremely hard problem with no interesting known computational complexity bounds, can be reduced to monadic SREU [8].
- Monadic SREU is equivalent to a non-trivial extension of word equations [26].
- Monadic SREU is equivalent to the decidability problem of the prenex fragment of intuitionistic logic with equality with function symbols of arity ≤ 1 [12].
- In general SREU is undecidable [11]. Moreover, SREU is undecidable under the following restrictions:
  - The left-hand sides of the rigid equations are ground [37].

- Furthermore, there are only two variables and three rigid equations with fixed ground left-hand sides [25].
- SREU with one variable is decidable, in fact EXPTIME-complete [6].

Note also that SREU is decidable when there are no variables, since each rigid equation can be decided for example by using any congruence closure algorithm or ground term rewriting technique. Actually, the problem is then P-complete because the uniform word problem for ground equations is P-complete [29]. The *unsolved cases* are:

? Decidability of monadic SREU [26].
? Decidability of SREU with *two* rigid equations.

Both problems are highly non-trivial.

References

1. P.B. Andrews. Theorem proving via general matings. *Journal of the Association for Computing Machinery*, 28(2):193–214, 1981.
2. M. Baaz. Note on the existence of most general semi-unifiers. In *Arithmetic, Proof Theory and Computation Complexity*, volume 23 of *Oxford Logic Guides*, pages 20–29. Oxford University Press, 1993.
3. W. Bibel. *Deduction. Automated Logic.* Academic Press, 1993.
4. M. Dauchet. Rewriting and tree automata. In H. Comon and J.P. Jouannaud, editors, *Term Rewriting (French Spring School of Theoretical Computer Science)*, volume 909 of *Lecture Notes in Computer Science*, pages 95–113. Springer Verlag, Font Romeux, France, 1993.
5. E. De Kogel. Rigid *E*-unification simplified. In P. Baumgartner, R. Hähnle, and J. Posegga, editors, *Theorem Proving with Analytic Tableaux and Related Methods*, number 918 in Lecture Notes in Artificial Intelligence, pages 17–30, Schloß Rheinfels, St. Goar, Germany, May 1995.
6. A. Degtyarev, Yu. Gurevich, P. Narendran, M. Veanes, and A. Voronkov. The decidability of simultaneous rigid *E*-unification with one variable. UPMAIL Technical Report 139, Uppsala University, Computing Science Department, March 1997.
7. A. Degtyarev, Yu. Gurevich, and A. Voronkov. Herbrand's theorem and equational reasoning: Problems and solutions. In *Bulletin of the European Association for Theoretical Computer Science*, volume 60. October 1996. The "Logic in Computer Science" column.
8. A. Degtyarev, Yu. Matiyasevich, and A. Voronkov. Simultaneous rigid *E*-unification is not so simple. UPMAIL Technical Report 104, Uppsala University, Computing Science Department, April 1995.
9. A. Degtyarev, Yu. Matiyasevich, and A. Voronkov. Simultaneous rigid *E*-unification and related algorithmic problems. In *Eleventh Annual IEEE Symposium on Logic in Computer Science (LICS'96)*, pages 494–502, New Brunswick, NJ, July 1996. IEEE Computer Society Press.
10. A. Degtyarev and A. Voronkov. Reduction of second-order unification to simultaneous rigid *E*-unification. UPMAIL Technical Report 109, Uppsala University, Computing Science Department, June 1995.

11. A. Degtyarev and A. Voronkov. Simultaneous rigid E-unification is undecidable. UPMAIL Technical Report 105, Uppsala University, Computing Science Department, May 1995.

12. A. Degtyarev and A. Voronkov. Decidability problems for the prenex fragment of intuitionistic logic. In *Eleventh Annual IEEE Symposium on Logic in Computer Science (LICS'96)*, pages 503–512, New Brunswick, NJ, July 1996. IEEE Computer Society Press.

13. A. Degtyarev and A. Voronkov. Simultaneous rigid E-unification is undecidable. In H. Kleine Büning, editor, *Computer Science Logic. 9th International Workshop, CSL'95*, volume 1092 of *Lecture Notes in Computer Science*, pages 178–190, Paderborn, Germany, September 1995, 1996.

14. A. Degtyarev and A. Voronkov. Simultaneous rigid E-unification is undecidable. In H. Kleine Büning, editor, *Computer Science Logic. 9th International Workshop, CSL'95*, volume 1092 of *Lecture Notes in Computer Science*, pages 178–190, Paderborn, Germany, September 1995, 1996.

15. A. Degtyarev and A. Voronkov. The undecidability of simultaneous rigid E-unification. *Theoretical Computer Science*, 166(1–2):291–300, 1996.

16. N. Dershowitz and J.-P. Jouannaud. Rewrite systems. In J. Van Leeuwen, editor, *Handbook of Theoretical Computer Science*, volume B: Formal Methods and Semantics, chapter 6, pages 243–309. North Holland, Amsterdam, 1990.

17. J. Doner. Tree acceptors and some of their applications. *Journal of Computer and System Sciences*, 4:406–451, 1970.

18. M. Fitting. First-order modal tableaux. *Journal of Automated Reasoning*, 4:191–213, 1988.

19. J. Gallier, P. Narendran, D. Plaisted, and W. Snyder. Rigid E-unification: NP-completeness and applications to equational matings. *Information and Computation*, 87(1/2):129–195, 1990.

20. J.H. Gallier, P. Narendran, D. Plaisted, and W. Snyder. Rigid E-unification is NP-complete. In *Proc. IEEE Conference on Logic in Computer Science (LICS)*, pages 338–346. IEEE Computer Society Press, July 1988.

21. J.H. Gallier, S. Raatz, and W. Snyder. Theorem proving using rigid E-unification: Equational matings. In *Proc. IEEE Conference on Logic in Computer Science (LICS)*, pages 338–346. IEEE Computer Society Press, 1987.

22. F. Gécseg and M. Steinby. *Tree Automata*. Akadémiai Kiodó, Budapest, 1984.

23. W.D. Goldfarb. The undecidability of the second-order unification problem. *Theoretical Computer Science*, 13:225–230, 1981.

24. J. Goubault. Rigid E-unifiability is DEXPTIME-complete. In *Proc. IEEE Conference on Logic in Computer Science (LICS)*. IEEE Computer Society Press, 1994.

25. Y. Gurevich and M. Veanes. Some undecidable problems related to the Herbrand theorem. UPMAIL Technical Report 138, Uppsala University, Computing Science Department, March 1997.

26. Y. Gurevich and A. Voronkov. The monadic case of simultaneous rigid E-unification. UPMAIL Technical Report 137, Uppsala University, Computing Science Department, 1997. To appear in Proc. of *ICALP'97*.

27. J. E. Hopcroft and J. D. Ullman. *Introduction to Automata Theory, Languages and Computation*. Addison-Wesley Publishing Co., 1979.

28. S. Kanger. A simplified proof method for elementary logic. In J. Siekmann and G. Wrightson, editors, *Automation of Reasoning. Classical Papers on Computational Logic*, volume 1, pages 364–371. Springer Verlag, 1983. Originally appeared in 1963.

29. D. Kozen. Complexity of finitely presented algebras. In *Proc. of the 9th Annual Symposium on Theory of Computing*, pages 164–177, New York, 1977. ACM.

30. D. Kozen. Positive first-order logic is NP-complete. *IBM J. of Research and Development*, 25(4):327–332, 1981.

31. V. Lifschitz. Problem of decidability for some constructive theories of equalities (in Russian). *Zapiski Nauchnyh Seminarov LOMI*, 4:78–85, 1967. English Translation in: Seminars in Mathematics: Steklov Math. Inst. 4, Consultants Bureau, NY-London, 1969, p.29–31.

32. D.W. Loveland. Mechanical theorem proving by model elimination. *Journal of the Association for Computing Machinery*, 15:236–251, 1968.

33. G.S. Makanin. The problem of solvability of equations in free semigroups. *Mat. Sbornik (in Russian)*, 103(2):147–236, 1977. English Translation in American Mathematical Soc. Translations (2), vol. 117, 1981.

34. G.E. Mints. Choice of terms in quantifier rules of constructive predicate calculus (in Russian). *Zapiski Nauchnyh Seminarov LOMI*, 4:78–85, 1967. English Translation in: Seminars in Mathematics: Steklov Math. Inst. 4, Consultants Bureau, NY-London, 1969, p.43–46.

35. V.P. Orevkov. Unsolvability in the constructive predicate calculus of the class of the formulas of the type $\neg\neg\forall\exists$ (in Russian). *Soviet Mathematical Doklady*, 163(3):581–583, 1965.

36. V.P. Orevkov. Solvable classes of pseudo-prenex formulas (in Russian). *Zapiski Nauchnyh Seminarov LOMI*, 60:109–170, 1976. English translation in: Journal of Soviet Mathematics.

37. D.A. Plaisted. Special cases and substitutes for rigid E-unification. Technical Report MPI-I-95-2-010, Max-Planck-Institut für Informatik, November 1995.

38. D. Prawitz. An improved proof procedure. In J. Siekmann and G. Wrightson, editors, *Automation of Reasoning. Classical Papers on Computational Logic*, volume 1, pages 162–201. Springer Verlag, 1983. Originally appeared in 1960.

39. J.W. Thatcher and J.B. Wright. Generalized finite automata theory with an application to a decision problem of second-order logic. *Mathematical Systems Theory*, 2(1):57–81, 1968.

40. Margus Veanes. *On Simultaneous Rigid E-Unification*. PhD thesis, Computing Science Department, Uppsala University, 1997.

41. P.J. Voda and J. Komara. On Herbrand skeletons. Technical report, Institute of Informatics, Comenius University Bratislava, July 1995. Revised January 1996.

42. A. Voronkov. On proof-search in intuitionistic logic with equality, or back to simultaneous rigid E-Unification. UPMAIL Technical Report 121, Uppsala University, Computing Science Department, January 1996.

43. A. Voronkov. Proof-search in intuitionistic logic based on the constraint satisfaction. UPMAIL Technical Report 120, Uppsala University, Computing Science Department, January 1996. Updated March 11, 1996.

The Tangibility Reflection Principle
for Self-Verifying Axiom Systems

Dan E. Willard *

Abstract: At the 1993 meeting of the Kurt Gödel Society, we described some axiom systems that could verify their own consistency and prove all Peano Arithmetic's Π_1 theorems. We will extend these results by introducing some new axiom systems which use stronger definitions of self-consistency.

Introduction: Define an axiom system α to be **Self-Justifying** iff

i) one of α's theorems will assert α's consistency (using some reasonable definition of consistency),

ii) and the axiom system α is in fact consistent.

Rogers [22] and Jeroslow [8] have noted that Kleene's Fixed Point Theorem implies every r.e. axiom system α can be easily extended into a broader system α^* which satisfies condition (i). Kleene's proposal [9] was essentially for the system α^* to contain all α's axioms plus the one additional axiom sentence:

There is no proof of 0=1 from the union of α with *"THIS SENTENCE"*.

The catch is that α^* can be inconsistent even while its added axiom justifies α^*'s consistency. Thus α^* will typically violate Part-ii of the definition of Self-Justification. This problem arises even for systems α *weaker than* Peano Arithmetic, [4, 5, 7, 20, 23, 24]. For instance, Pudlák [20] has shown that no extension of Robinson's System Q can verify its own Hilbert consistency, and Robert Solovay (private communications [24]) has used some methods of Nelson and Wilkie-Paris [16, 28] to generalize Pudlák's theorem for most systems which merely recognize Successor as a total function. (The footnote[1] gives the detailed statement of Solovay's theorem.) Also, Löb's Theorem [15] can be generalized (see [30, 31]) to state that *even after* the Subtraction and Division primitives replace the Addition and Multiplication, it is impossible to develop

---

*SUNY-Albany Email=dew@cs.albany.edu, Support by NSF Grant CCR 9302920

[1]If a system α is consistent, recognizes Successor as a total function and can prove all Peano Arithmetic's Π_1 theorems about Subtraction and Division, then Solovay's theorem asserts α cannot prove the non-existence of a Hilbert-proof of 0=1 from itself.

axiom systems α that can verify (1)'s reflection principle for all sentences Ψ and also prove Peano Arithmetic's Π_1 theorems about Subtraction and Division.

$$\{ \ \exists \, y \quad \mathrm{Prf}_\alpha \, (\, \lceil \Psi \rceil \, , \, y \,) \ \} \quad \supset \quad \Psi \tag{1}$$

Let $x - y = 0$ when $x < y$, and $\lfloor \frac{x}{y} \rfloor = x$ when $y = 0$. Subtraction and Division (with Integer Rounding), thus defined, are total functions. All the Π_1 sentences of Arithmetic can be trivially rewritten with Subtraction and Division so replacing Addition and Multiplication. (This revised notation will allow an axiom system to evade many implications of the Incompleteness Theorem.)

More precisely, let $\mathrm{Tangible}(x)$ be a formula which asserts that the number x is not of unusually large size, i.e. its size is "tangible". For fixed constant $k \geq 2$, three possible definitions of tangibility are given below.

a. $\mathrm{TangPred}\,(x) \quad = \quad \{ \ \exists v \ \ x = v - 1 \ \}$

b. $\mathrm{TangDiv}_k\,(x) \quad = \quad \{ \ \exists v \ \ x < \frac{v}{k} \ \}$

c. $\mathrm{TangRoot}_k\,(x) \quad = \quad \{ \ \exists v \ \ x < v^{1/k} \ \}$

Let $\lceil \Psi \rceil$ denote Ψ's Gödel number, and $\mathrm{Prf}_\alpha \, (\, x \, , \, y \,)$ be a Δ_0 formula indicating y is the Gödel number of a proof from axiom system α of the theorem with Gödel number x. For a prenex normal sentence Ψ, let Ψ^i denote a sentence identical to Ψ except that each universally quantified variable in Ψ^i is bounded by i. Define Ψ's **Tangibility Reflection Principle** to be the assertion:

$$\forall x \ \{ \ [\ \exists \, y \quad \mathrm{Prf}_\alpha \, (\, \lceil \Psi \rceil \, , \, y \,) \wedge \mathrm{Tangible}(x) \ ] \quad \supset \quad \Psi^x \ \} \tag{2}$$

We will illustrate several examples of axiom systems α and formulae $\mathrm{Tangible}(x)$ such that α can prove the validity of (2) for every prenex normal sentence Ψ.

Part of the reason that the Reflection Principle (2) is interesting is that we can prove that the stronger principle (1) is never feasible for all sentences Ψ simultaneously (see [30, 31]). Equation (2)'s reflection principle is thus a necessary compromise: Its pleasing aspect is that it indicates that an axiom system α can know that when it proves an arbitrary sentence Ψ then Ψ is at least valid over the range of small numbers of so-called "tangible" size.

The Reflection Principle (2) is especially interesting when Ψ is a Σ_1 sentence. In this case, $\Psi \equiv \Psi^0$, and α can trivially deduce (3) from (2):

$$\{ \ \exists \, y \quad \mathrm{Prf}_\alpha \, (\, \lceil \Psi \rceil \, , \, y \,) \ \} \quad \supset \quad \Psi \tag{3}$$

Moreover, if Ψ is the sentence $0=1$ then α can trivially deduce (4) from (3).

$$\forall y \quad \neg \, \mathrm{Prf}_\alpha \, (\, \lceil 0 = 1 \rceil \, , \, y \,) \tag{4}$$

The above two types of self-justification sentences were the focus of our IS(A) and IS*(A) axiom systems in our prior paper [29]. The tangibility reflection principle thus subsumes our prior results.

Other Comments About the Prior Literature: Since the time of Gödel, logicians have recognized that some price must be always paid when one desires an axiom system to be able to verify some form of its own consistency. One possible sacrifice consists of using some type of weakened definition of *self-consistency*. (For instance, Feferman refers to such weak definitions of self-consistency as formulations which are "numerically correct" but "intensionally excessively weak" [5].) The earliest example of such a definition appears in the Kriesel-Taekeuti analysis [14, 16, 20, 26, 28] of Second-Order generalizations of Gentzen's Sequent Calculus, where systems are displayed that can verify their own weak cut-free definition of their self-consistency. More recent examples of weakened definitions of self-consistency appear in Nelson's analysis of Robinson's System Q [16] and in Pudlák's more general results [7, 20, 21] about finitely axiomatized sequential theories. These results show how First-Order Systems can verify their self-consistency when consistency is weakly defined to be a cut-free form of consistency local to "Definable Cuts". (In Pudlak's analysis, the Wilkie-Paris form of "Restricted Hilbert Proof" [28] is also permissible.) Another interesting result is the analysis of $I\Sigma_0$ by Wilkie-Paris [28], which showed that $I\Sigma_0 + SuperExp$ could verify $I\Sigma_0$'s Hilbert consistency, but that $I\Sigma_0 + \Omega_n$ could not even verify $I\Sigma_0$'s Herbrand consistency.

Our research differs from the preceding articles by *not weakening* the definition of self-consistency. Instead, we weaken the base axiom system α, by removing α's axiom that Multiplication (and in some instances Addition) are total functions. The advantage of replacing Multiplication and Addition with Division and Subtraction is our axiom systems can then verify definitions of self-consistency as natural as Eq. (4) (with "Prf" designating Hilbert provability).

The resulting loss of the axioms that Multiplication and Addition are total functions is obviously a non-trivial sacrifice. However, a pleasing aspect of our formalisms is that they will support (2)'s *"Tangibility Reflection Principle"*. Another pleasing aspect is that our systems will have a capacity to prove all Peano Arithmetic's Π_1 theorems, when these Π_1 theorems are rewritten in a form with Subtraction and Division replacing Addition and Multiplication.

2. MAIN FORMALISM: Define $F(a_1, a_2, ...a_j)$ to be a **Non-Growth Function** iff $F(a_1, a_2, ...a_j) \leq Maximum(2, a_1, a_2, ...a_j)$ holds always. Our axiom systems will employ a set of seven non-growth functions, called the **Grounding Functions**. They will include Integer Subtraction and Division (defined in Section 1), $Predecessor(x) = Max(x - 1, 0)$, $Maximum(x, y)$, $Logarithm(x) = \lceil Log_2(x + 1) \rceil$, $Root(x, y) = \lfloor x^{1/y} \rfloor$ and $Count(x, j)$ designating the number of "1" bits among x's rightmost j bits.

We will use mostly conventional notation. A *term* is defined to be a constant, variable or function symbol (whose input arguments are other terms). If t is a term then the quantifiers in the wffs $\forall v \leq t \ \Psi(v)$ and $\exists v \leq t \ \Psi(v)$ will be called *bounded quantifiers*. They are semantically equivalent to the formulae $\forall v (v \leq t \supset \Psi(v))$ and $\exists v (v \leq t \land \Psi(v))$. A formula Φ will be called Δ_0^-

iff all its quantifiers are bounded. A formula Υ will be called Π_1^- iff it is written in the form $\forall v_1 \, \forall v_2 \, ... \, \forall v_n \, \Phi(v_1, v_2, ...v_n)$, where $\Phi(v_1, v_2, ...v_n)$ is Δ_0^-.

Our definitions of Δ_0^- and Π_1^- formulae are similar to the conventional definitions of Δ_0 and Π_1 Arithmetic Formulae, except that the terms in Δ_0^- and Π_1^- formulae are built out of Grounding functions (rather from Addition and Multiplication). Every conventional Π_1 formula can be trivially translated into an equivalent Π_1^- formula (using its Subtraction and Division primitives).

The acronym "ISTR" stands for "Introspective Semantics with Tangible Reflection". The **"ISTR MAPPING"** is defined as a transformation that maps an initial axiom system A onto an axiom system $\text{ISTR}(A)$, such that the latter can simultaneously recognize its own consistency and prove all of A's Π_1^- theorems. $\text{ISTR}(A)$ consists of the following four groups of axioms:

Group-Zero: $\text{ISTR}(A)$ contains one constant symbol $\bar{n}$ for each natural number $n \geq 0$. The Group-Zero axioms define these constants formally. They will include $\bar{1} \neq \bar{0}$, and for each $n > 0$, the axioms: $\text{Predecessor}(\, \bar{n} \,) = \overline{n-1}$, $\overline{2n} - \bar{n} = \bar{n}$ and $\overline{2n+1} - \bar{n} - \bar{1} = \bar{n}$.

Group-1: The Π_2^- axiom (below) indicates that Bitwise-Or is a total function:

$$\forall x \, \forall y \, \exists z \, \forall i \leq \text{Log}[\text{Max}(x,y,z)] \; \{\text{Bit}(z,i) = \text{Max}[\text{Bit}(x,i), \text{Bit}(y,i)]\} \qquad (5)$$

All the other Group-1 axioms will be Π_1^- sentences. They will assign the "=" and "<" predicates their usual logical properties. Also, Group-1 will include a finite set of axioms defining the Grounding functions, so that for each function F and set of constants $\bar{k}$, $\bar{c}_1$, $\bar{c}_2$, ...$\bar{c}_m$, the combination of the Group-Zero and Group-1 axioms will imply $F(\bar{c}_1, \bar{c}_2, ...\bar{c}_m) = \bar{k}$ when this sentence is true. Any finite set of Π_1^- sentences that meet the preceding conditions is adequate. (See [31] for an example.)

Group-2: Let $\lceil \Phi \rceil$ denote Φ's Gödel number, and $\text{Prf}_A(x,y)$ denote a Δ_0 formula indicating y is a proof from axiom system A of x. For each Π_1^- sentence Φ, the Group-2 schema will contain an axiom:

$$\forall y \; \{ \text{Prf}_A (\, \lceil \Phi \rceil , y \,) \supset \Phi \} \qquad (6)$$

Group-3: If Ψ is a prenex sentence, let Ψ_z^x denote a sentence identical to Ψ except that each universally quantified variable in Ψ_z^x is bounded by x and each existentially quantified variable in Ψ_z^x is bounded by z. Let $\text{HilbPrf}_\alpha(x,y)$ indicate that y is a Hilbert-proof from the axiom system α of x. Define **SIZE**$(y) = c$ when c is the largest stored constant in the proof y. For each prenex sentence Ψ, $\text{ISTR}(A)$ will contain a corresponding " Ψ−axiom", which is equivalent to the sentence:

$$\forall x \forall y \forall z \, \{ \, [\, \text{HilbPrf}_{\text{ISTR}(A)} (\, \lceil \Psi \rceil, y) \wedge \text{Size}(y) < x \leq \tfrac{z}{2}] \supset \Psi_z^x \, \} \qquad (7)$$

Section 4 gives the formal encoding of (7). It uses a Fixed Point Principle because the axiom (7) refers to an axiom system which includes itself.

3. MAIN THEOREMS: Say an axiom system A is **Regularly Consistent** iff $\text{Prf}_A(x, y)$ has a Δ_0^- encoding and all A's Π_1^- theorems are valid. This section will prove ISTR(A) is consistent whenever A is Regularly Consistent, and that ISTR(A) supports the Tangibility Reflection Principle.

Lemma 3.1 Let Φ denote a prenex normal sentence, and Φ_j^i denote a sentence identical to Φ except that each universally quantified variable in Φ_j^i is bounded by i, and its existentially quantified variables are similarly bounded by j. Also let Φ^i be an abbreviation for Φ_∞^i. This notation implies:

A) $a \leq b \wedge \Phi_a^i \supset \Phi_b^i$

B) $a \geq b \wedge \Phi_j^a \supset \Phi_j^b$

C) $\Phi_j^i \supset \Phi^i$

D) $a \geq b \wedge \Phi^a \supset \Phi^b$

Proof. Immediate from Φ_j^i's definition. $\square$

Notation. $L(x)$ denotes the number (below) whose length is the same as x's length and which consists of a sequence of repetitions of the bit "1".

$$L(x) = 2^{\lfloor \text{Log}_2(x) \rfloor + 1} - 1 \qquad (8)$$

Say a Gödel encoding method satisfies **Conventional Sizing** constraints iff it is impossible for its encoded proofs p to contain a constant symbol representing a number $c \geq p/2$. This simply means that:

$$\forall p \quad \text{Size}(p) \leq \frac{p}{2} - 1 \qquad (9)$$

Theorem 3.2. Suppose ISTR(A)'s Group-3 axioms use an encoding method satisfying (9). The regular consistency of A then implies ISTR(A) is consistent.

Proof by Contradiction. Suppose for the sake of contradiction that ISTR(A) is inconsistent but A is regularly consistent. The latter implies all ISTR(A)'s Group-zero, Group-1 and Group-2 axioms are valid in the Standard Model of the Natural Numbers. Hence, at least one of ISTR(A)'s Group-3 axioms is invalid in the Standard Model (as otherwise ISTR(A) would be consistent). The generic from of ISTR(A)'s Group-3 axioms appears below:

$$\forall x \, \forall y \, \forall z \, \{ \, [\, \text{HilbPrf}_{\text{ISTR}(A)} (\, \ulcorner \Psi \urcorner, y) \wedge \text{Size}(y) < x \leq \frac{z}{2}] \supset \Psi_z^x \, \} \quad (10)$$

Since some Group-3 axiom is invalid, there exists a tuple (Ψ, i, p, t) satisfying:

$$\text{HilbPrf}_{\text{ISTR}(A)} (\, \ulcorner \Psi \urcorner, p) \wedge \text{Size}(p) < i \leq \frac{t}{2} \wedge \neg \Psi_t^i \qquad (11)$$

Since (11) indicates $t \geq 2i > L(i)$, Lemma 3.1A yields $\neg\Psi_t^i \supset \neg\Psi_{L(i)}^i$, and thus

$$\text{HilbPrf}_{\text{ISTR}(A)} (\, \ulcorner \Psi \urcorner, p) \wedge \text{Size}(p) < i \wedge \neg \Psi_{L(i)}^i \qquad (12)$$

The paragraph (above) has thus established that *at least one triple* (Ψ, p, i) *exists* which satisfies the condition (12). Let (Ψ, p, i) therefore denote the particular triple satisfying (12) with *minimal value* in its third component. Our proof by contradiction will be completed by constructing from (Ψ, p, i) another such triple (Φ, q, j) with $j < i$ that also satisfies this condition.

We first observe that since $\Psi^i_{L(i)}$ is invalid in the Standard Model of the Natural Numbers, so must also $\Psi^{L(i)}_{L(i)}$ be invalid (by Lemma 3.1.B combined with the $i \leq L(i)$ inequality). Let $M_{L(i)}$ denote the model of the finite set of integers $0, 1, 2, \ldots L(i)$. Since $\Psi^{L(i)}_{L(i)}$ is invalid in the standard (infinite) model of the Natural Numbers and all Ψ's constant symbols represent numbers bounded by $\text{SIZE}(p) < i \leq L(i)$, it follows that:

** The sentence Ψ is invalid in the *finite* model $M_{L(i)}$.

Fact 3.3 The proof p, associated with the triple (Ψ, p, i), must employ some Group-3 axiom of $\text{ISTR}(A)$ which is invalid in the model $M_{L(i)}$.

Proof of Fact 3.3. Our first claim is that all $\text{ISTR}(A)$'s Group-zero, Group-1 and Group-2 axioms *appearing in the proof* p are valid in the model $M_{L(i)}$. Since Bitwise-Or is a total function locally within the model $M_{L(i)}$, the Group-1 axiom in Eq (5) is certainly valid in the model $M_{L(i)}$. All the other Group-zero, 1 and 2 axioms appearing in the proof p are Π^-_1 sentences, having the special property that all their constant symbols represent integers no greater than $L(i)$ (see footnote[2]). From the regular consistency of A, we may assume that all the Group-2 axioms are valid in the standard model of the natural numbers. Clearly, all the Group-zero and Group-1 axioms are also valid in the standard model of the natural numbers. Since each of these axioms are valid Π^-_1 sentences whose constant symbols are no greater than $L(i)$, they must (see footnote[3]) also be valid locally in the (finite) model $M_{L(i)}$.

However since ** indicated that p produces a proof of a theorem which is invalid in the model $M_{L(i)}$, it must follow that *some axiom from the proof* p is invalid in $M_{L(i)}$. This invalid axiom must be a Group-3 axiom because all other possibilities have been precluded. □

We will now use Fact 3.3 to finish Theorem 3.2's proof. Let Eq. (13) denote the Group-3 axiom which (by Fact 3.3) is invalid in $M_{L(i)}$:

$$\forall x \, \forall y \, \forall z \, \{ \, [\text{HilbPrf}_{\text{ISTR}(A)}(\lceil \Phi \rceil, y) \wedge \text{Size}(y) < x \leq \frac{z}{2}] \, \supset \, \Phi^x_z \, \} \quad (13)$$

---

[2]The constant symbols are so bounded because $\text{Size}(p) < i \leq L(i)$.

[3]The reason that "they must be also valid in the model $M_{L(i)}$" is that all the terms in a Π^-_1 sentence are built out of the seven Grounding Functions (all of which are *NON-GROWTH* functions). Since the relevant Π^-_1 sentences are valid in the Standard (infinite) Model, employ constant symbols that are no greater than $L(i)$ and all their function symbols represent non-growth functions, these Π^-_1 sentences are automatically valid also in the (finite) model $M_{L(i)}$ by a completely trivial argument.

Section 4 proves that "HilbPrf$_{\text{ISTR}(A)}$($\lceil\Phi\rceil$, y)" and "Size(y) < x" have Δ_0^- encodings. These encodings, combined with the assumption that (13) is invalid *locally in the model* $M_{L(i)}$, imply (see footnote [4]) there exists (j, q, r) satisfying:

$$\text{HilbPrf}_{\text{ISTR}(A)}(\lceil\Phi\rceil, q) \wedge \text{Size}(q) < j \leq \frac{r}{2} \wedge \neg\Phi_r^j \wedge \text{MAX}(j, q, r) \leq L(i) \quad (14)$$

Since (14) indicates $j \leq \frac{r}{2}$, we get $L(j) < r$ and then from Lemma 3.1.A $\neg[\Phi_r^j] \supset \neg[\Phi_{L(j)}^j]$. Hence from (14), there exists q and j satisfying both the inequality $j \leq \frac{r}{2} \leq \lfloor\frac{L(i)}{2}\rfloor \leq i - 1$ and

$$\text{HilbPrf}_{\text{ISTR}(A)}(\lceil\Phi\rceil, q) \wedge \text{Size}(q) < j \wedge \neg\Phi_{L(j)}^j \quad (15)$$

The triple (Φ, q, j) thus contradicts the previously assumed minimality of (Ψ, p, i) because $j \leq i-1$ and the triple satisfies Equation (15). This contradiction shows that ISTR(A) must be consistent because the opposite hypothesis contradicts (Ψ, p, i)'s required minimality. $\square$

Theorem 3.4 The ISTR(A) axiom system can prove the validity of its Tangibility Reflection Principle, using the TangDiv$_2$(x) definition of Tangibility.

Proof. Lemma 3.1.C indicates $\Psi_z^x \supset \Psi^x$. Since ISTR(A) can verify Lemma 3.1.C, ISTR(A) can use this formula to simplify the right side of Equation (7)'s Group-3 axiom, thereby deriving:

$$\forall x \, \forall y \, \forall z \, \{ \, [\text{HilbPrf}_{\text{ISTR}(A)}(\lceil\Psi\rceil, y) \wedge \text{Size}(y) < x \leq \frac{z}{2}] \supset \Psi^x \, \} \quad (16)$$

Using the fact that Equation (9) implies Size(y) $\leq \frac{y}{2} - 1$ and substituting $z = y$ and $x = \frac{y}{2}$ into (16), ISTR(A) can further deduce that

$$\forall y \, \{ \, \text{HilbPrf}_{\text{ISTR}(A)}(\lceil\Psi\rceil, y) \supset \Psi^{y/2} \, \} \quad (17)$$

Using Lemma 3.1.D to simplify (17) produces the yet further derivation that:

$$\forall y \, \forall x \leq \frac{y}{2} \, \{ \, \text{HilbPrf}_{\text{ISTR}(A)}(\lceil\Psi\rceil, y) \supset \Psi^x \, \} \quad (18)$$

The combination of (16), (18) and Size(y) $\leq \frac{y}{2} - 1$ imply the validity of (19). (This is because (18) implies (19) is valid when $x \leq y/2$, and similarly Line (16) combined with Size(y) $\leq \frac{y}{2} - 1$ implies (19) is valid when $x > y/2$.)

$$\forall x \, \forall y \, \forall z \, \{ \, [\text{HilbPrf}_{\text{ISTR}(A)}(\lceil\Psi\rceil, y) \wedge x < \frac{z}{2}] \supset \Psi^x \, \} \quad (19)$$

---

[4] The Δ_0^- formulae encoding "HilbPrf$_{\text{ISTR}(A)}$($\lceil\Phi\rceil$, y)" and "Size(y) < x" can be structured so that the only physical constants lying in them are say 0, 1, or $\lceil\Phi\rceil$. Since $\lceil\Phi\rceil < y$ when y is a proof of Φ, all the bounded quantified variables in these Δ_0^- formulae are less than Max($x, y, 1$). Because the Grounding functions are non-growth functions, any term built out of such tightly bounded variables and grounding functions is also bounded by Max($x, y, 1$). These facts, combined with the assumption that (13) is invalid *locally in the model* $M_{L(i)}$, imply that there exists a three integers (j, q, r) satisfying Equation (14).

Lastly, it is easy for ISTR(A) to recognize that (19) is equivalent to its TangDiv$_2$ Reflection Principle, since (19) is just simply another notation for rewriting (2)'s Reflection Principle. As each step in this proof can be verified internally by ISTR(A), it follows that ISTR(A) has the capacity to recognize the validity of its TangDiv$_2$ Reflection Principle for each prenex sentence Ψ. $\square$

4. THE GÖDEL ENCODING OF THE GROUP-3 AXIOMS:

This section will show ISTR(A)'s Group-3 axioms have ideally terse Π_1^- encodings. (The techniques described herein are also applicable to the IS(A) and ISREF(A) systems in our prior papers [29, 30], which claimed but did not prove such encodings were possible.) We will use the following notation:

i. Subst (g, h) will denote Gödel's classic substitution formula, which yields TRUE when g is an encoding of a formula, and h is an encoding of a sentence which replaces all occurrence of free variables in g with simply a constant, whose value equals the integer g.

ii. UNION(A) will denote the union of the Group-Zero, 1 and 2 axioms.

iii. SubstHilbPrf$_\alpha(s, y, g)$ is a formula stating y is a proof of s from the union of axiom system α with the one further axiom whose Gödel number is the integer h satisfying Subst(g, h). SubstHilbPrf$_{\text{UNION}(A)}(s, y, g,)$ denotes the version of SubstHilbPrf where α is UNION(A).

Let IS$_H(A)$ denote an axiom system identical to ISTR(A) except that IS$_H(A)$'s Group-3 axiom scheme instead consists of one single sentence, which can be "informally" viewed as asserting:

$$\forall y \quad \neg \text{HilbPrf}_{\text{IS}_H(A)}(\ulcorner 0 = 1 \urcorner , y) \tag{20}$$

The Group-3 axiom of IS$_H(A)$ is trivial to "formalize". Let $\Theta(g)$ denote:

$$\forall y \quad \{ \ \neg \text{SubstHilbPrf}_{\text{UNION}(A)} (\ulcorner 0 = 1 \urcorner , y , g) \ \} \tag{21}$$

Let N denote the Gödel number of $\Theta(g)$ Then the "formal" encoding of IS$_H(A)$'s Group-3 axiom (20) can be simply defined as $\Theta(N)$.

The goal in this section is to describe the analog of $\Theta(N)$ needed by the more complex Group-3 axioms of ISTR(A). Its Group-3 schema will be more complicated than IS$_H(A)$'s Group-3 axiom (20) primarily because ISTR(A)'s Group-3 schema will consist of an infinite number of distinct axiom-sentences.

Define a **Pseudo-Formula** to be a logical construct which differs from conventional formula by allowing the three pseudo-symbols, denoted as " $\clubsuit$ ", " $\ulcorner \clubsuit \urcorner$ ", and " $\clubsuit_x^z$ ", to appear anywhere in the pseudo-formula. Usually, $\Xi(\clubsuit)$ will denote a pseudo-formula. The symbol " $\clubsuit$ " will be a place-holder where an arbitrary prenex normal sentence Φ may be later inserted into the "pseudo-formula" Ξ . Similarly, the symbols " $\ulcorner \clubsuit \urcorner$ ", and " $\clubsuit_x^z$ ", will be place-holders for Φ's Gödel number $\ulcorner \Phi \urcorner$, and Section 2's symbol " Φ_x^z ". In particular, our discussion will include a formal encoding of the primitives:

iv. PseudoTransform$_1$(h, r) indicating that h is a pseudo-formula and r is a conventional formula, where for some Π_1^- sentence Φ each appearance of " $\lceil \clubsuit \rceil$ " in h is replaced by an appearance of " $\lceil \Phi \rceil$ " in r and each appearance of " $\clubsuit$ " is replaced by " Φ " .

v. PseudoTransform$_2$(h, r) indicating that h is a pseudo-formula and r is a conventional formula, where for some prenex normal sentence Φ each appearance of " $\lceil \clubsuit \rceil$ " in h is replaced by an appearance of " $\lceil \Phi \rceil$ " in r and each appearance of " $\clubsuit_z^x$ " is replaced by " Φ_z^x " .

vi. ExHilbPrf$_\alpha$ (s, y, g) is a generalization of item (iii)'s SubstHilbPrf$_\alpha$ formula, designed to take into account both pseudo-formulae and item (i)'s Subst relation. In particular, let h denote the unique Gödel number of a *pseudo-sentence* satisfying Subst(g, h), where g is the Gödel number of a *pseudo-formula*. Then ExHilbPrf$_\alpha$ (s, y, g) will indicate that y is a Hilbert-style proof of the sentence s, where each axiom a used in the proof y satisfies either $a \in \alpha$ or PseudoTransform$_2$(h, a).

vii. ConSize(x, y) will indicate that y encodes a list of formulae and that each physical constant in the list y represents a quantity $< x$.

Section 2 had indicated that ISTR(A) would contain one Group-3 axiom for each prenex sentence Ψ. These axioms were "informally" defined as:

$$\forall x \forall y \forall z \; \{ \text{HilbPrf}_{\text{ISTR}(A)}(\lceil \Psi \rceil, y) \wedge \text{Size}(y) < x \wedge x \le \frac{z}{2} \supset \Psi_z^x\} \qquad (22)$$

In order to formally encode (22), we will need the pseudo-formula Ξ ($\clubsuit$) below. Note (23) differs from (22) by replacing " Ψ " with " $\clubsuit$ ", replacing " HilbPrf$_{\text{ISTR}(A)}$ ($\bullet$, y) " with " ExHilbPrf$_{\text{UNION}(A)}$ ($\bullet$, y , g) ", and replacing " Size(y) $< x$ " with " ConSize(x, y) ".

$$\forall x \forall y \forall z \{\text{ExHilbPrf}_{\text{UNION}(A)}(\lceil \clubsuit \rceil, y, g) \wedge \text{ConSize}(x, y) \wedge x \le \frac{z}{2} \supset \clubsuit_z^x\} \qquad (23)$$

Let N denote the Gödel number of (23). Then the *informal expression* " Prf$_{\text{ISTR}(A)}$ (s , y) " can be *formally encoded* as being the particular formula: " ExHilbPrf$_{\text{UNION}(A)}$ (s , y , N) ". This implies that the *Formal Gödel Encoding* of ISTR(A)'s Group-3 axioms (from (22)) is:

$$\forall x \forall y \forall z \; \{ \text{ExHilbPrf}_{\text{UNION}(A)}(\lceil \Psi \rceil, y, N) \wedge \text{ConSize}(x, y) \wedge x \le \frac{z}{2} \supset \Psi_z^x\} \qquad (24)$$

It should be evident to a reader familiar with prior Gödel encodings that (24)'s two formulae ExHilbPrf$_{\text{UNION}(A)}$ ($\lceil \Psi \rceil$, y , N) and ConSize(x, y) can be formally encoded in the language of ISTR(A). In itself, this is an interesting fact because it immediately establishes that Equation (24)'s Group-3 axiom can be formally encoded as a fully well defined sentence. The remainder of this section will prove the tighter theorem that Equation (24)'s Group-3 axiom schema has a Π_1^- encoding.

Our discussion will use some well-known properties of Linear Computational Hierarchy (LinH), described by Hájek-Pudlák, Krajíček, and Wrathall [7, 11, 33]. LinH is the linear-time analog of the Polynomial-Time Hierarchy. It designates those languages that can be recognized in $O(N)$ time by a multi-tape machine that has access to an arbitrary but finite pre-specified number of levels of non-determinism. LinH's formal definition can be found in any of [7, 11, 33].

Let us define a byte as a sequence of six bits, and assume that all our Gödel encodings of proofs are integers that represent a finite length string of non-zero bytes. We will assume that the constant symbol representing the natural number i is encoded as a base-32 number consisting of a sequence of $\lceil 1+\text{Log}_{32}(i+1) \rceil$ such bytes. The symbol representing the i-th variable will also be represented by a sequence of $\lceil 1+\text{Log}_{32}(i+1) \rceil$ bytes. All other symbols (such as " $\forall$ ", " $\exists$ ", " $\wedge$ ", " $\neg$ ",....) will be encoded by a single non-zero byte.

LinH is discussed in great detail by Hájek-Pudlák, Krajíček, and Wrathall [7, 11, 33]. The last part of this section will explain how every LinH decision procedure has a Δ_0^- encoding. In this context, Theorem 4.1 will imply Eq (24)'s two predicates $\text{ConSize}(x, y)$ and $\text{ExHilbPrf}_{\text{UNION}(A)}(s, y, h)$ are Δ_0^- encodable (since they correspond to the 12-th and 16-th formulae below).

Theorem 4.1 There exists LinH algorithms to determine whether a byte-string (or in some instances whether a collection of several byte-strings) satisfies each of the sixteen itemized conditions below.

1. *Term(s)* indicating that the byte-string s is a term built out of Section 2's seven Grounding functions.

2. *Formula(s)* indicating that the byte-string s is a formula in the first-order logical language employed by the axiom system $\text{ISTR}(A)$.

3. *List(s)* indicating that the byte-string s is a list of formulae.

4. *HilbertDeduction(s,i)* indicating s satisfies $\text{List}(s)$, and that the i-th formula listed in s is a deduction from earlier formulae from this list, using the conventional Hilbert-deductive rules of Modus Ponens and Generalization.

5. *PrenexSentence(s)* indicating that the string s is a prenex sentence.

6. Π_1-*Sentence(s)* indicating that the byte-string s is a Π_1^- sentence.

7. *Replace(s,t,x,y)* indicating that the byte-string s is identical to the byte-string t, *except that* every appearance of the substring y in t is replaced by an appearance of x in s.

8. *Replace$_k$($s, t, x_1, x_2, ...x_k, y_1, y_2, ...y_k$)* designating the natural generalization of $\text{Replace}(s, t, x, y)$, indicating that the string s is identical to the byte-string t except that now k *distinct different substrings* $y_1, y_2, ...y_k$ from t have their every appearance in t replaced by the substrings $x_1, x_2, ...x_k$ in s.

9. *LogicalAxiom(s)* indicating s is a "logical axiom" in a Hilbert proof.

10. *Transform(a,b)* indicating that the byte-string a satisfies $\text{Formula}(a)$ and that if Ψ designates this formula then b is a byte-string that represents the physical constant in $\text{ISTR}(A)$'s language that encodes $\lceil \Psi \rceil$.

11. $Subst(g, h)$ designating Gödel's substitution relation (i.e. see Item (i)).

12. $ConSize(x, y)$ whose definition was given in Item (vii).

13. $PseudoTransform_1(h, r)$ and $PseudoTransform_2(h, r)$ whose definitions are given in Items (iv) and (v).

14. $UnionAxiom(s)$ indicating s is a Group-zero, Group-1 or Group-2 axiom.

15. $Group3Test(s, g)$ indicating there exists some h satisfying both $Subst(g, h)$ and PseudoTransform$_2(h, s)$. (We will employ Group3Test in a context where g corresponds to a special fixed constant $\bar{m}$ that serves as a template for generating Group-3 axioms. Group3Test$(s, \bar{m})$ will then indicate s is a Group-3 axiom.)

16. $ExHilbPrf_{UNION(A)}(s, y, g)$ as defined via *Items (ii)* and *(vi)*.

Description of LinH Decision Procedures for Representing Formulae (1) through (6): It is well known that context-free grammars can be recognized by LinH decision procedures (i.e. the needed LinH procedures require merely a non-deterministic Turing Machine with one tape and one extra stack [33]). Since formulae (1) through (3) represent context-free grammars, these formulae have LinH representations.

The decision algorithm for Formula (4) uses a 2-step procedure. It will first employ Formula (3) for checking that s satisfies List(s). If the preceding answer is affirmative then s will be accepted if a non-deterministic search confirms that its i-th sentence is a valid deduction from prior sentences in this list. (The latter involves searching for two integers, j and k, which are less than i and have the property that the i-th sentence either follows from the j-th sentence by Generalization, or from the j-th and k-th sentence by Modus Ponens). Since (4)'s procedure involves only performing one non-deterministic search after executing (3)'s procedure, it follows that (4)'s decision procedure lies at merely one level higher than (3)'s decision procedure in the LinH Hierarchy.

A similar argument will show (5) and (6) have LinH procedures. This is because a sentence s is prenex (or Π_1^-) if it satisfies Formula(s) and additionally satisfies a few extra conditions which can be trivially confirmed by moving a few levels higher up in the LinH Hierarchy. (For example, s will be a prenex sentence if all its quantifiers appear on its left side, and if a non-deterministic search procedure is unable to identify any variable in s which is free.) □

Description of LinH Decision Procedures for Representing Formulae (7) through (9): It is trivial to construct LinH decision procedures for Formulae (7) and (8). The decision procedure for (9) is easy to build, using the procedures (1), (2) and (8). An example will illustrate this point.

All the common formulations of Hilbert-style proofs (cf. [7]) involve only a finite number of skeleton axiom schema used to generate the logical axioms. (Depending on one's notation, there are usually between four and six such schema.) For example, one of the logical schema is typically:

$$* * * \quad ((A \supset (B \supset C)) \supset ((A \supset B) \supset (A \supset C))$$

We will first illustrate how to construct a LinH algorithm, called Check(s), for testing for whether a byte-string s is a formula of the Hilbert-type above.

Let t denote the string *** . The first step of Check(s) will call the procedure Formula(s) to determine whether or not s is a byte-string encoding a formula. If the answer is affirmative, the second step of Check(s) will seek to non-deterministically construct three strings a , b , and c satisfying the condition Replace$_3(s, t, a, b, c, \mathcal{A}, \mathcal{B}, \mathcal{C})$. The string s will be accepted by the procedure Check(s) when these conditions are met. (This nondeterministic search for (a, b, c) runs in O(Log(s)) time because $a \leq s$, $b \leq s$, and $c \leq s$.)

The formal LinH decision algorithm for implementing LogicalAxiom(s) will employ the procedure Check(s) for determining whether or not s is a logical axiom of the type *** above. Since typical encodings of Hilbert-style proofs employ four-to-six logical axiom schemes, we will also need to employ a few straightforward analogs of Check(s), called perhaps Check$_1(s)$ Check$_2(s)$ Check$_3(s)$ and Check$_4(s)$, to verify whether or not s is one of these other types of logical axioms. The combination of these test-searches will provide a LinH decision algorithm for implementing LogicalAxiom(s). □

Description of LinH Decision Procedures for Representing Formulae (10) through (12): It is trivial to construct a LinH decision procedure for Transform(a, b). Moreover, the LinH procedure for implementing Subst(g, h) is trivial to construct by making subroutine calls to the procedures Transform, Formula and Replace (given in items (2), (7) and (10)). Similarly, an implementation of the ConSize(x, y) decision procedure will rest essentially on making subroutine calls to List(y) and Transform(a, b). (In the interests of brevity, we will omit giving a more detailed description of the Decision Procedures (10) through (12): Their overall structure is quite similar to the earlier Decision Procedures (1) through (9).) □

Description of LinH Procedures for PseudoTransform$_1(h, r)$ and PseudoTransform$_2(h, r)$: Both procedures are similar. Therefore we will describe only PseudoTransform$_1(h, r)$'s decision procedure. It will begin by calling the procedure Formula(r) to verify r is a formula. If the preceding answer is affirmative, then the ordered pair (h, r) will be accepted if a non-deterministic search (running in O(Log($h + r$)) time) can find an ordered pair of strings (a, b) with $a < r$ and $b < r$ satisfying each of Transform(a, b), Π_1−Sentence(a) and Replace$_2$(h , r , "♣" , "⌈♣⌉" , a , b). □

Description of UnionAxiom(s)'s LinH Decision Procedure: For i equal to 0, 1 or 2, let Group$_i(s)$ be a formula indicating that s is a Group-i type axiom. It is trivial to devise a LinH algorithm that tests for whether Group$_0(s)$ is satisfied. Since there exists only a finite number of allowed Group-1 axioms, there certainly exists a LinH algorithm that can test for whether Group$_1(s)$ is satisfied. Employing PseudoTransform$_1$, it is easy to devise a LinH algorithm that also tests for whether Group$_2(s)$ is satisfied. (This is because for some fixed constant $\bar{m}$, which serves as a "template" for "generating" the full set of Group-2 axioms, Group$_2(s)$ can be defined as being the operation

PseudoTransform$_1(\bar{m}, s)$.) The decision procedure UnionAxiom(s) is then defined as the operation Group$_0(s)\vee$Group$_1(s)\vee$Group$_2(s)$. $\square$

Description of Group3Test(s, g)'s LinH Decision Procedure: This procedure will accept the input (s, g) when a nondeterministic search, running in time $O(\text{Log}(s))$, can construct a sting h such that both Subst(g, h) and PseudoTransform$_2(h, s)$ are satisfied. (Since for some fixed constant c, the inequality Log(h) $< c+$Log(s) holds when PseudoTransform$_2(h, s)$ is satisfied, the $O(\text{Log}(s))$ bound on the search's non-deterministic running time is sufficient to test for whether or not h exists.) $\square$

Description of ExHilbPrf$_{\text{UNION}(A)}$(s, y, g)'s LinH Decision Procedure: Let t_i denote the i-th sentence in the list y. The LinH Decision Procedure for ExHilbPrf$_{\text{UNION}(A)}$(s, y, g) will begin by testing for whether List(y) is satisfied and the last sentence in y is s. If these conditions are met, it will accept the string (s, y, g) if for each $i \leq$Length(y) some one of the four conditions of: 1) HilbertDeduction(y, i), 2) LogicalAxiom(t_i), 3)UnionAxiom(t_i), or 4) Group3Test(t_i, g) are satisfied.

The reason that this 4-part test will have a LinH implementation is that each of items (1) through (4) are LinH primitives. Thus, we can test for whether or not every $i \leq$Length(y) satisfies one of these four conditions by having a nondeterministic search attempt to find some $i \leq$Length(y) which fails all four of these conditions simultaneously $\square$

Theorem 4.2 The Group-3 axioms in Equation (24) have Π_1^- encodings.

Proof. It is implicit from the prior literature that each LinH Decision Procedure can be represented by a Δ_0^- formula. In particular, Wrathall [33] showed how to translate LinH Decision Procedures into Rudimentary Formulae, and Benett [1] showed how Exponentiation's graph has a Δ_0 encoding. Both the Hájek-Pudlák and Krajíček textbooks [7, 11] have noted that these results imply every LinH Decision Procedure has a Δ_0 encoding. Although the particular topic of Δ_0^- formulae was not explicitly addressed by the prior literature, it easily implies that all LinH Decision Procedure are also Δ_0^- encodable. (This is because two of the seven Grounding functions are Log(x) and Count(x, i), implying Δ_0^- formulae have *built-in* primitives for representing the two essential constructs of Exponentiation and "NUON" [7].)

Because Parts 12 and 16 of Theorem 4.1 formalized LinH decision procedures for representing the formulae ExHilbPrf $_{\text{UNION}(A)}$($\lceil\Psi\rceil, y, N$) and ConSize(x, y), these predicates must thus have Δ_0^- encodings (by the preceding paragraph). This implies that Equation (24) has a Π_1^- encoding. $\square$.

5. GENERALIZATIONS: Let SemPrf$_\alpha(x, y)$ be a Δ_0 formula indicating y is the Gödel number of a Semantic Tableaux proof from axiom system α of the theorem x. Let $\lambda > 0.1$ denote some fixed constant, and IS$^\lambda(A)$ be an axiom system which differs from ISTR(A) by having its Group-1 schema recognize Addition as a total function and having its Group-3 schema weakened by containing axioms of the form:

$$\forall x \, \forall y \, \forall z \quad \{ \ \text{SemPrf}_{\text{IS}^\lambda(A)} \, (\ \lceil \Psi \rceil \, , \, y \,) \ \wedge \ y^\lambda < \frac{z}{x} \ \supset \ \Psi_z^x \ \} \quad (25)$$

Also, define ISREF(A) to be a modification of ISTR(A) whose Group-1 schema does not recognize Bitwise-Or as a total function, but which employs the stronger form of Group-3 schema below:

$$\forall x \, \forall y \quad \{ \ \text{HilbPrf}_{\text{ISREF}(A)} \, (\ \lceil \Psi \rceil \, , \, y \,) \ \wedge \ x > \text{Size}(y) \ \supset \ \Psi_{x-1}^{x-1} \ \} \quad (26)$$

IS$^\lambda(A)$ and ISREF(A) are proven in [30, 31] to satisfy analogs of Theorem 3.2's consistency property. Also, Theorem 3.4 is generalized in [30, 31] to establish that ISREF(A) supports the TangPred version of the Tangibility Reflection Principles, and IS$^\lambda(A)$ supports the TangRoot variant of Equation (2) (with "Prf" then designating a Semantic Tableaux proof). Thus, there is a trade-off where ISREF(A) has access to a stronger version of Tangibility Reflection Principle at the cost of dropping the axiom that Bitwise-Or is a total function, while IS$^\lambda(A)$ has a weaker reflection principle but can recognize Addition as a total function. One other reason ISREF(A) is interesting is that some of its properties are connected to the open questions about NP (see [30]).

The axiom that Bitwise-Or is a total function may provide ISTR(A) with epistemological advantages over ISREF(A) (where this axiom is absent). This is because Bitwise-Or allows one to represent Finite-Set-Union as a total function, when one views an integer as a Bit-Vector. Moreover, the Group-2 axioms allow ISTR(A) to trivially recognize Set-Intersection (represented by Bitwise-And) and Set Subtraction (represented by Bitwise-Subtraction) as also total functions. Thus, ISTR(A) appears to be philosophically curious when it recognizes *simultaneously* its Hilbert consistency, the validity of the preceding basic finite-set operations, and also the validity of all Peano Arithmetic's Π_1^- theorems.

There are two other topics that should be briefly mentioned. Let $Pred^N(x)$ denote a function that consists of N iterations of the Predecessor Operation. Say an axiom system α is **Infinitely Far-Reaching** iff it there exists a finite subset of axioms $S \subset \alpha$ such that for *arbitrary* N the *fixed (!) finite set* of axioms S is sufficient to prove $\exists x \, Pred^N(x) = 1$.

For an arbitrary regularly consistent axiom system A , it is possible to devise a self-justifying axiom system $I(A)$ that is infinitely far reaching, able to prove all A's Π_1^- theorems and *also* able to verify that *"There is no Hilbert-proof of 0=1 from I(A)"* (see [32]). This is somewhat surprising because Solovay's Extension of Pudlák's theorem [20] (described in the opening paragraph) states that the italicized phrase is not provable when $I(A)$ recognizes either Successor or any other monotonically growing operation as a total function. Thus, [32]'s axiom system must, by necessity, use a formalism that is Infinitely-Far-Reaching but still too weak to recognize Successor as a total function.

One other notion in [32] is the concept of a *"Self-Justifying Axiom Galaxy"*. For an arbitrary regularly consistent axiom system A, such a galaxy $G(A)$ is defined as uncountably large set of distinct axiom systems α such that each $\alpha \in G(A)$ is consistent, capable of proving all A's Π_1^- theorems, and capable of verifying the TangDiv$_2$ consistency of every axiom system in $G(A)$ (including α *itself.*) The advantage of the Galaxy concept is that one can assign a Lebesgue Probability Measure over its set of members. In some carefully encoded sense [32], one can then demonstrate that each $\alpha \in G(A)$ can verify that the set of systems in $G(A)$ that support the assumption that Exponentiation is a total function has a Lebesque Probability Measure of 1. Yet no individual $\alpha \in G(A)$ can prove that Successor is a total function because it can not determine *with 100% certainty (!)* whether or not itself has such a property. (This positive result is again a very tight fit with the complementary negative results established by Pudlák and Solovay. From the latter, we know that it is impossible for any such α to prove Successor is a total function.)

6. OVERALL PERSPECTIVE: We focussed mostly on the ISTR(A) axiom system in Sections 2-4 (due to the obvious limitations on page space). Section 5 indicated that this formalism fits into a broader theory (described in [29, 30, 31, 32]).

Both ISTR(A) and Section 5's other self-justifying systems are closely linked to the research of Pudlák and Solovay. The first paragraph of this article had indicated Pudlák [20] proved no finite sequential extension of Robinson's System Q could verify its own Hilbert consistency and Solovay had generalized Pudlák's theorem for systems merely recognizing Successor as a total function. Our positive results, concerning the potential of Self-Justifying Systems, closely complements the negative results of Pudlák and Solovay, which categorizes very broadly the extent of the Second Incompleteness Theorem. In conjunction, the tight fit between these positive and negative results provide a fairly detailed explanation of when Self-Justification is feasible and when it is not.

References

[1] J. Benett, Ph. D. Dissertation, Princeton University.

[2] A. Bezboruah and J. Shepherdson, "Gödel's Second Incompleteness Theorem for Q", J of Symbolic Logic 41 (1976), 503-512.

[3] S. Buss, "Polynomial Hierarchy and Fragments of Bounded Arithmetic", 17th ACM Symposium on Theory of Comp (1985) pp. 285-290

[4] S. Buss, Bounded Arithmetic, Princeton Ph. D. dissertation published in Proof Theory Lecture Notes, Vol. 3, published by Bibliopolic (1986).

[5] S. Feferman, "Arithmetization of Metamathematics in a General Setting", Fund Math 49 (1960) pp. 35-92.

[6] P. Hájek, "On the Interpretatability of Theories Containing Arithmetic (II)", Comm. Math. Univ. Carol. 22 (1981) pp.595-594.

[7] P. Hájek and P. Pudlák, Metamathematics of First Order Arith, SpringerVerlag 1991

[8] R. Jeroslow, "Consistency Statements in Formal Mathematics", Fundamentae Mathematicae 72 (1971) pp. 17-40.

[9] S. Kleene, "On the Notation of Ordinal Numbers", JSL 3 (1938), pp. 150-156.

[10] J. Krajíček, "A Note on the Proofs of Falsehoods", Arch Math Logik 26 (1987) pp. 169-176.

[11] J. Krajíček, Bounded Propositional Logic and Compl. Theory, Cambridge U Press 1995

[12] J. Krajíček and P. Pudlák, "Propositional Proof Systems, The Consistency of First-Order Theories and The Complexity of Computation", JSL 54 (1989) pp 1063-1079.

[13] G. Kreisel, "A Survey of Proof Theory, Part I" in Journal of Symbolic Logic 33 (1968) pp. 321-388 and " Part II" in Proceedings of Second Scandinavian Logic Symposium (1971) North Holland Press (with Fenstad ed.), Amsterdam

[14] G. Kreisel and G. Takeuti, "Formally self-referential propositions for cut-free classical analysis and related systems", Dissertations Mathematica 118, 1974 pp. 1 -55.

[15] M. Löb, A Solution to a Problem by Leon Henkin, JSL 20 (1955) pp. 115-118.

[16] E. Nelson, Predicative Arithmetic, Mathematical Notes, Princeton Press, 1986.

[17] R.Parikh, "Existence and Feasibility in Arithmetic", JSL 36 (1971), pp.494-508.

[18] J. Paris and C. Dimitracopoulos, "A Note on the Undefinability of Cuts", JSL 48 (1983) pp. 564-569.

[19] J. Paris and A. Wilkie, "Δ_0 Sets and Induction", Proceedings of the Jadswin Logic Conference (Poland), Leeds University Press (1981) pp. 237-248.

[20] P. Pudlák, "Cuts Consistency Statements and Interpretations", JSL 50 (1985) 423-442.

[21] P. Pudlák, "On the Lengths of Proofs of Consistency", Kurt Gödel Soceity Proc. (1996).

[22] H. Rogers, Theory of Rec. Funct. & Effective Comp., McGraw Hill 1967, see pp 186-188.

[23] C. Smorynski, "The Incompleteness Theorem", Handbook Math Logic, 821-866, 1983.

[24] R. Solovay, Private Communications (April 1994) proving Footnote 1's statement. (See Appendix A of [31] for more details about Solovay's Extension of Pudlák's theorem [20].)

[25] R. Statman, "Herbrand's theorem and Gentzen's Notion of a Direct Proof", Handbook on Math Logic, (1983) pp. 897-913.

[26] G. Takeuti, "On a Generalized Logical Calculus", Japan J. Math 23 (1953) 39-96.

[27] G. Takeuti, Proof Theory, Studies in Logic Volume 81, North Holland, 1987.

[28] A. Wilkie and J. Paris, "On the Scheme of Induction for Bounded Arithmetic", Annals Pure Applied Logic (35) 1987, pp. 261-302.

[29] D. Willard, "Self-Verifying Axiom Systems", in Proceedings of the Third Kurt Gödel Symposium (1993) pp. 325-336, published in Springer-Verlag LNCS (Vol. 713).

[30] D. Willard, "Self-Reflection Principles and NP-Hardness", in Proceedings of 1996 DIMACS Workshop on Feasible Arithmetic and Length of Proof (AMS Press).

[31] D. Willard, "Self-Verifying Axiom Systems, the Incompleteness Theorem and Related Reflection Principles" manuscript (1997).

[32] D. Willard, "Some Curious Exceptions to the Second Incompleteness Theorem", longer manuscript which is the unabridged version of the present conference paper.

[33] C. Wrathall, Rudimentary Predicates and Relative Computation, Siam J. on Computing 7 (1978), pp. 194-209.

Upper Bounds for Standardizations and an Application

Hongwei Xi

Department of Mathematical Sciences
Carnegie Mellon University
Pittsburgh, PA 15213, USA

email: hwxi+@cs.cmu.edu

Abstract. We first present a new proof for the standardization theorem, a fundamental theorem in λ-calculus. Since our proof is largely built upon structural induction on lambda terms, we can extract some bounds for the number of β-reduction steps in the standard β-reduction sequences obtained from transforming a given β-reduction sequences. This result sharpens the standardization theorem. As an application, we establish a superexponential bound for the lengths of β-reduction sequences from any given simply typed λ-terms.

1 Introduction

The standardization theorem of Curry and Feys [CF58] is a fundamental theorem in λ-calculus, stating that if u reduces to v for λ-terms u and v, then there is a *standard* β-reduction from u to v. Using this theorem, we can readily prove the normalization theorem, i.e., a λ-term has a normal form if and only if the leftmost β-reduction sequence from the term is finite. The importance of lazy evaluation in functional programming languages largely comes from the normalization theorem. Moreover, the standardization theorem can be viewed as a syntactic version of sequentiality theorem in [Ber78]. For instance, it can be readily argued that *parallel or* is inexpressible in λ-calculus by using the standardization theorem. In fact, a syntactic proof of the sequentiality theorem can be given with the help of the standardization theorem.

There have been many proofs of the standardization theorem in the literature such as the ones in [Mit79], [Klo80], [Bar84] and [Tak95]. In the presented proof we intend to find a bound for standardizations, namely, to measure the length of standard β-reduction sequence obtained from a given β-reduction sequence. This method presents a concise and more accurate formulation of the standardization theorem. As an application, we establish a super exponential bound on the number of β-reduction steps in β-reduction sequences from any given simply typed λ-terms. This not only strengthens the strong normalization theorem in the simply typed λ-calculus $\lambda^{\rightarrow}$, but also yields more understanding on $\mu(t)$, the number of steps in a longest β-reduction sequence from a given simply typed λ-term t. Since $\mu(t)$ can often be used as an induction order, its structure plays a key role in understanding related inductive proofs.

The structure of the paper is as follows. Some preliminaries are explained in Section 2. Our proof of the standardization theorem is presented in Section 3, and some upper bounds for standardizations are extracted in Section 4. In Section 5, we establish a bound for the lengths of *beta*-reduction sequences from any given simply typed λ-terms. Finally, some related work is mentioned and a few remarks are drawn in Section 6.

We point out that many proof details in this paper are omitted for the sake of limited space. We shall soon submit a full version of the paper to a journal for its being further refereed.

2 Preliminaries

We give a brief explanation on the notations and terminology used in this paper. Most details not included here appear in [Bar84].

Definition 1. *(λ-terms)* The set Λ of λ-terms is defined inductively as follows.

- (variable) There are infinitely many variables $x, y, z, \ldots$ in Λ; variables are the only subterms of themselves.
- (abstraction) If $t \in \Lambda$ then $(\lambda x.t) \in \Lambda$; u is a subterm of $(\lambda x.t)$ if u is $(\lambda x.t)$ or a subterm of t.
- (application) If $t_0, t_1 \in \Lambda$ then $t_0(t_1) \in \Lambda$; u is a subterm of $t_0(t_1)$ if u is $t_0(t_1)$ or a subterm of t_i for some $i \in \{0, 1\}$.

The set $FV(t)$ of free variables in t is defined as follows.

$$FV(t) = \begin{cases} \{x\} & \text{if } t = x \text{ for some variable } x; \\ FV(t_0) - \{x\} & \text{if } t = (\lambda x.t_0); \\ FV(t_0) \cup FV(t_1) & \text{if } t = t_0(t_1). \end{cases}$$

The set Λ_I of λI-terms is the maximal subset of Λ such that, for every term $t \in \Lambda_I$, if $(\lambda x.t_0)$ is a subterm of t then $x \in FV(t_0)$.

$u\{x := v\}$ stands for substituting v for all free occurrences of x in u. α-conversion or renaming bounded variables may have to be performed in order to avoid name collisions. Rigorous definitions are omitted here. We assume some basic properties on substitution such as the substitution lemma (Lemma 2.1.16 [Bar84]).

Definition 2. *(β-redex, β-reduction and β-normal form)* a β-redex is a term of form $(\lambda x.u)(v)$; $u\{x := v\}$ is the contractum of the β-redex; $t \longrightarrow_\beta t'$ stands for a β-reduction step where t' is obtained from replacing some β-redex in t with its contractum; a β-normal form is a term in which there is no β-redex.

Let $\longrightarrow_\beta{}^n$ stand for n steps of β-reduction, and $\longrightarrow\!\!\!\!\rightarrow_\beta$ stand for some steps of β-reduction, which could be 0. Usually there are many different β-redexes in a term t; a β-redex r_1 in t is to the left of another β-redex r_2 in t if the first symbol of r_1 is to the left of that of r_2.

Definition 3. *(Multiplicity)* Given a β-redex $r = (\lambda x.u)(v)$; the multiplicity $m(r)$ of r is the number of free occurrences of the variable x in u.

Definition 4. *(β-Reduction Sequence)* Given a β-redex r in t; $t \xrightarrow{r}_\beta u$ stands for the β-reduction step in which β-redex r gets contracted; $[r_1] + \cdots + [r_n]$ stands for a β-reduction sequence of the following form.

$$t_0 \xrightarrow{r_1}_\beta t_1 \xrightarrow{r_2}_\beta \cdots \xrightarrow{r_n}_\beta t_n$$

Conventions We use $\sigma, \tau, \ldots$ for β-reduction sequences;

$$\sigma : t \longrightarrow\!\!\!\!\twoheadrightarrow_\beta t' \quad \text{and} \quad t \xrightarrow{\sigma}\!\!\!\!\twoheadrightarrow_\beta t'$$

for a β-reduction sequence from t to t'; $|\sigma|$ for the length of σ, namely, the number of β-reduction steps in σ, which might be 0.

Definition 5. *(Concatenation)* Given $\sigma : t_0 \longrightarrow\!\!\!\!\twoheadrightarrow_\beta t_1$ and $\tau : t_1 \longrightarrow\!\!\!\!\twoheadrightarrow_\beta t_2$; $\sigma + \tau$ stands for the concatenation of σ and τ, namely, $\sigma + \tau : t_0 \xrightarrow{\sigma}\!\!\!\!\twoheadrightarrow_\beta t_1 \xrightarrow{\tau}\!\!\!\!\twoheadrightarrow_\beta t_2$.

Conventions Let $\sigma : u \longrightarrow\!\!\!\!\twoheadrightarrow_\beta v$ and $C[\ ]$ be a context, then σ can also be regarded as the β-reduction sequence which reduces $C[u]$ to $C[v]$ in the obvious way. In other words, we may use σ to stand for $C[\sigma]$. An immediate consequence of this is that $\sigma + \tau$ can be regarded as $C_1[\sigma] + C_2[\tau]$ for some proper contexts C_1 and C_2.

We now introduce the concept of *residuals* of β-redexes. The rigorous definition of this notion can be found in [Hue94]. Let $\mathcal{R}$ be a set of β-redexes in a term t, $r = (\lambda x.u)(v)$ in $\mathcal{R}$, and $t \xrightarrow{r}_\beta t'$. This β-reduction step affects β-redexes r' in $\mathcal{R}$ in the following way. We assume that all bound variables in u are distinct from the free variables in v.

- r' is r. Then r' has no residual in t'.
- r' is in v. All copies of r' in $u\{x := v\}$ are called residuals of r' in t';
- r' is in u. Then $r'\{x := v\}$ in $u\{x := v\}$ is the only residual of r' in t';
- r' contains r. Then the residual of r' is the term obtained by replacing r in r' with $u\{x := v\}$;
- Otherwise, r' is not affected and is its own residual in t'.

The residual relation is transitive.

Definition 6. *(Developments)* Given a λ-term t and a set $\mathcal{R}$ of redexes in t; if $\sigma : t \longrightarrow\!\!\!\!\twoheadrightarrow_\beta u$ contracts only β-redexes in $\mathcal{R}$ or their residuals, then σ is a development.

Definition 7. *(Involvement)* Given $t \longrightarrow\!\!\!\!\twoheadrightarrow_\beta u$ and $\sigma : u \longrightarrow\!\!\!\!\twoheadrightarrow_\beta v$; a β-redex in t is involved in σ if some of its residuals is contracted in σ.

Definition 8. *(Head β-redex)* Given t of form $\lambda x_1 \ldots \lambda x_m.r(t_1) \ldots (t_n)$, where r is a β-redex and $m, n \geq 0$; r is called the head β-redex in t; a β-reduction is a head β-reduction if the contracted β-redex is a head β-redex.

Proposition 9. *We have the following.*

- *Let r_h be the head β-redex in t; if $t \xrightarrow{r}_\beta u$ for some $r \neq r_h$ then r_h has exactly one residual in u, which is the head β-redex of u.*
- *If $\sigma : t \twoheadrightarrow_\beta u$ contains a head β-reduction, then t contains a head β-redex r_h and r_h is involved in σ.*

Proof. Please see Section 8.3 [Bar84] for proofs. □

3   The Proof of Standardization Theorem

Standardization theorem of Curry and Feys [CF58] states that any β-reduction sequence can be standardized in the following sense.

Definition 10. *(Standard β-Reduction Sequence)* Given a β-reduction sequence

$$\sigma : t = t_0 \xrightarrow{r_0}_\beta t_1 \xrightarrow{r_1}_\beta t_2 \xrightarrow{r_2}_\beta \cdots;$$

σ is standard if for all $0 \leq i < j$, r_j is not a residual of some β-redex to the left of r_i.

Lemma 11. *Given $t \xrightarrow{\sigma_s}_\beta u \xrightarrow{r}_\beta v$, where σ_s is standard and r is the residual of the head β-redex in t; then we can construct a standard β-reduction sequence $t \xrightarrow{\tau}_\beta v$ with $|\tau| \leq 1 + \max\{m(r), 1\} \cdot |\sigma_s|$.*

Proof. Let r_h be the head β-redex in t, and we proceed by structural induction on t.

- $t = (\lambda x.t_0)$. By the induction hypothesis on t_0, this case is trivial.
- $t = t_0(t_1)$, and r_h is in t_0. Then we may assume $\sigma_s = \sigma_0 + \sigma_1$, where $t_i \xrightarrow{\sigma_i}_\beta u_i$ are standard for $i = 0, 1$, and $u = u_0(u_1)$. Note that r must be in u_0 by Proposition 9 (1). Hence $v = v_0(u_1)$, where $u_0 \xrightarrow{r}_\beta v_0$. By induction hypothesis, we can construct a standard β-reduction sequence $t_0 \xrightarrow{\tau_0}_\beta v_0$ with $|\tau_0| \leq 1 + \max\{m(r), 1\} \cdot |\sigma_0|$. Let $\tau = \tau_0 + \sigma_1$, then $t \xrightarrow{\tau}_\beta v$ is standard. It can be readily verified that $|\tau| \leq 1 + \max\{m(r), 1\} \cdot |\sigma_s|$.
- $t = (\lambda x.t_0)(t_1)$, and r_h is t. Then we can assume $\sigma_s = \sigma_0 + \sigma_1$, where $t_i \xrightarrow{\sigma_i}_\beta u_i$ are standard for $i = 0, 1$, and $r = u = (\lambda x.u_0)u_1$. Hence, $v = u_0\{x := u_1\}$. Let $\sigma_0 = [r_1] + \cdots + [r_n]$, and $\sigma_0^* = [r_1^*] + \cdots + [r_n^*]$, where $r_j^* = r_j\{x := t_1\}$ for $j = 1, \ldots, n$. Then we know $\sigma_0^* : t_0\{x := t_1\} \twoheadrightarrow_\beta u_0\{x := t_1\}$ is also standard. Notice that $\sigma_0^* + \sigma_1 + \cdots + \sigma_1$ is a β-reduction sequence which reduces $t_0\{x := t_1\}$ to $v = u_0\{x := u_1\}$, where σ_1 occurs $m(r)$ times and each σ_1 reduces one occurrence of t_1 in $u_0\{x := t_1\}$ to u_1. If a β-redex contracted in some σ_1 is to the left of some r_j^*, then all β-redexes contracted in that σ_1 are to the left of that r_j^*. Hence, we can move that σ_1 to the front of r_j^*. In this way, we can construct a standard β-reduction sequence from $t_0\{x := t_1\}$ to $v = u_0\{x := u_1\}$ in the following form.

$$\sigma_s^* = \cdots + [r_1^*] + \cdots + \ldots + \cdots + [r_n^*] + \cdots,$$

where $\cdots$ stands for a β-reduction sequence of form $\sigma_1 + \cdots + \sigma_1$, which may be empty, and r_j^* may also denote their corresponding residuals. Hence $\tau = [r_h] + \sigma_s^*$ is a standard β-reduction sequence from t to v. Notice

$$|\tau| = 1 + |\sigma_s^*| = 1 + |\sigma_0^*| + m(r) * |\sigma_1| \le 1 + \max\{m(r), 1\} \cdot |\sigma_s|.$$

$\square$

Lemma 12. *Given* $t \xrightarrow{\sigma_s}_\beta u \xrightarrow{r}_\beta v$, *where* σ_s *is standard; then we can construct a standard β-reduction sequence* $t \xrightarrow{\tau}_\beta v$ *with* $|\tau| \le 1 + \max\{m(r), 1\} \cdot |\sigma_s|$.

Proof. The proof proceeds by induction on $|\sigma_s|$ and the structure of t, lexicographically ordered. By Corollary 8.3.8 [Bar84], t is of form

$$\lambda x_1 \ldots \lambda x_m . t_0(t_1) \ldots (t_n),$$

where $m, n \ge 0$, and t_0 is either a variable or an λ-abstraction. We have two cases.

- $\sigma_s + [r]$ contains no head β-reduction. Then we may assume σ_s to be of form

$$\sigma_{0,s} + \sigma_{1,s} + \cdots + \sigma_{n,s},$$

  where $\sigma_{i,s}$ are standard β-reduction sequences from t_i to u_i for $i = 0, 1, \ldots, n$ and $u = \lambda x_1 \ldots \lambda x_m . u_0(u_1) \ldots (u_n)$. Note that r must be in some u_k. Let $u_k \xrightarrow{r}_\beta v_k$. By induction hypothesis, there exists a standard β-reduction sequence τ_k from t_k to v_k with $|\tau_k| \le 1 + \max\{m(r), 1\} \cdot |\sigma_{k,s}|$. Let $\tau = \sigma_{0,s} + \cdots + \tau_k + \cdots + \sigma_{n,s}$, then τ is standard.
- $\sigma_s + [r]$ contains some head β-reduction. By Proposition 9 (2), let r_h be the head β-redex in t, which is involved in $\sigma_s + [r]$. We have two subcases.
  - r is the residual of r_h. Then by Lemma 11 we are done.
  - r_h is involved in σ_s. Since σ_s is standard,

  $$\sigma_s = [r_h] + \sigma_{h,s} : t \xrightarrow{r_h}_\beta t_h \xrightarrow{\sigma_h}_\beta u$$

  for some standard β-reduction sequence $\sigma_{h,s}$. Note $|\sigma_{h,s}| < |\sigma_s|$. By induction hypothesis, we can construct a standard β-reduction sequence $\tau_h : t_h \xrightarrow{}_\beta v$. Clearly, $\tau = [r_h] + \tau_h : t \xrightarrow{}_\beta v$ is a standard β-reduction sequence.

In either case, it can be immediately verified that $|\tau| \le 1 + \max\{m(r), 1\} \cdot |\sigma_s|$.

$\square$

Theorem 13. (Standardization) *Every finite β-reduction sequence can be standardized.*

Proof. Given $t \xrightarrow{\sigma}_\beta v$, we proceed by induction on $|\sigma|$. If σ is empty then σ is standard. Now assume $\sigma = \sigma' + [r]$, where $t \xrightarrow{\sigma'}_\beta u \xrightarrow{r}_\beta v$. By induction hypothesis, we can construct a standard β-reduction sequence $t \xrightarrow{\sigma'_s}_\beta u$. Hence, Lemma 12 yields the result. $\square$

4 The Upper Bounds

It is clear from the previous proofs that we actually have an algorithm to transform any β-reduction sequences into standard ones. Let $std(\sigma)$ denote the standard β-reduction sequence obtained from transforming a given β-reduction sequence σ. We are ready to give some upper bounds on the number of β-reduction steps in $std(\sigma)$.

Theorem 14. (Standardization with bound) *Given a β-reduction sequence*

$$\sigma : t \longrightarrow\!\!\!\!\!\twoheadrightarrow_\beta u,$$

where $\sigma = [r_0] + [r_1] + \cdots + [r_n]$ *for some* $n \geq 1$, *then there exists a standard β-reduction sequence* $t \xrightarrow{\sigma_s}\!\!\!\!\!\twoheadrightarrow_\beta u$ *with* $|\sigma_s| \leq (1 + \max\{m(r_1), 1\}) \cdots (1 + \max\{m(r_n), 1\})$.

Proof. Let $\sigma_0 = [r_0]$, $\sigma_i = [r_0] + [r_1] + \cdots + [r_i]$ and $l_i = |std(\sigma_i)|$ for $i = 0, 1, \ldots, n$. By Lemma 12, we have $l_{i+1} \leq 1 + \max\{m(r_{i+1}), 1\} \cdot l_i$ for $i = 0, 1, \ldots, n-1$ according to the proof of Theorem 13. Note $1 \leq l_i$, and thus, for $i = 0, 1, \ldots, n-1$,

$$l_{i+1} \leq 1 + \max\{m(r_{i+1}), 1\} \cdot l_i \leq (1 + \max\{m(r_{i+1}), 1\}) \cdot l_i.$$

Since $l_0 = 1$, this yields $l_n \leq (1 + \max\{m(r_1), 1\}) \cdots (1 + \max\{m(r_n), 1\})$. Note $\sigma = \sigma_n$. Let $\sigma_s = std(\sigma_n)$, then we are done. $\square$

Clearly, this simple bound is not very tight. With a closer study, a tighter but more complex bound can be given in the same fashion. Unlike many earlier proofs in the literature, our proof of the standardization theorem does not use the *finiteness of developments theorem* (FD). In this respect, our proof is similar to the one in [Tak95]. We remark that the uses of FD in other proofs are inessential in general and can be avoided in proper ways. As a matter of fact, Theorem 14 can be modified to show that all developments are finite, following the application in the next section. We will not pursue in this direction since the work in [dV85] has produced an exact bound for finiteness of developments.

Given $\sigma : t \longrightarrow\!\!\!\!\!\twoheadrightarrow_\beta u$, we can also give a bound on $|std(\sigma)|$ in terms of $|\sigma|$ and the size of t defined below.

Definition 15. The size $|t|$ of a term t is defined inductively as follows.

$$|t| = \begin{cases} 1 & \text{if } t \text{ is a variable;} \\ 1 + |t_0| & \text{if } t = (\lambda x.t_0); \\ |t_0| + |t_1| & \text{if } t = t_0(t_1). \end{cases}$$

Proposition 16. *If* $t \longrightarrow_\beta u$ *then* $|u| < |t|^2$.

Proof. A structural induction on t yields the result. $\square$

Corollary 17. *Given* $\sigma : t \longrightarrow\!\!\!\!\!\twoheadrightarrow_\beta u$; *then there is a standard β-reduction sequence* $\sigma_s : t \longrightarrow\!\!\!\!\!\twoheadrightarrow_\beta u$ *with* $|\sigma_s| < |t|^{2^{|\sigma|}}$.

Proof. This clearly holds if $|\sigma| = 1$. Now assume $\sigma = [r_0] + [r_1] + \cdots + [r_{n-1}]$: $t \twoheadrightarrow_\beta u$. By Proposition 16, we have $|r_i| < |t|^{2^i}$ for $i = 1, \ldots, n-1$, which yields $1 + \max\{m(r_i), 1\} \le |t|^{2^i}$ for $i = 1, \ldots, n-1$. By Theorem 14, we can construct a standard β-reduction sequence $\sigma_s : t \twoheadrightarrow_\beta u$ with $|\sigma_s| \le |t|^{2^1} \cdots |t|^{2^{n-1}} < |t|^{2^n}$. $\qquad\square$

Now we introduce a lemma which will be used in the next section.

Lemma 18. *If $\sigma : t \twoheadrightarrow_\beta u$ is a development, then $|u| < 2^{|t|}$.*

Proof. This can be verified by a structural induction on t. $\qquad\square$

5 An Application

It is a well-known fact that the simply typed λ-calculus $\lambda^\to$ enjoys strong normalization. Given a simply typed λ-term t, let $\mu(t)/\nu(t)$ be the length of a longest/shortest β-reduction sequence which reduces t to a β-normal form. In this section, as an application of our previous result, we will give an upper bound on $\mu(t)$. We first show for simply typed λ-terms t that $\nu(t)$ is not an elementary. This can bring us some feeling on how tight our upper bound on $\mu(t)$ is. Among various proofs showing the strong normalization property of $\lambda^\to$, some proofs such as the ones in [Gan80a] and [Sch91] can yield superexponential upper bounds on $\mu(t)$. Gandy invented a semantic approach in [Gan80a], which is called *functional interpretations* and has its traces in many following papers such as [Sch82], [Pol94] and [Kah95]. In [Sch91], Schwichtenberg adopted a syntactic approach from [How80], which bases on cut elimination in intuitionistic logic.

Compared with other related methods in the literature, our following syntactic method is not only innovative but also yields a quite intelligible and tight bound. It also exhibits a nice way in $\lambda^\to$ to transform strong normalization into weak normalization, simplifying a much involved transformation in [Sch91]. Therefore, the new transformation has its own value in this respect.

Definition 19. *(Simple Types and $\lambda^\to$-terms)* Types are formulated in the following way.

- Atomic types are types.
- If U and V are types then $U \to V$ is a type.

$\lambda^\to$-terms are defined inductively as follows.

- (variable) For each type U, there are infinitely many variables $x^U, y^U, \ldots$ of that type.
- (abstraction) If v is of type V and x does occur free in v then $(\lambda x^U.v)$ is of type $U \to V$.
- (application) If u is of type $U \to V$ and v is of type U, then $u(v)$ is of type V.

We often omit the type superscript of a variable if this causes no confusion. On the other hand, superscripts may be used to indicate the types of $\lambda^\to$-terms.

5.1 $\nu(t)$ is not elementary for $\lambda^{\rightarrow}$-terms t

It is obvious that we can code $\lambda^{\rightarrow}$-terms with some elementary functions; $\nu(t)$ can then be regarded as a functions defined on the codings of $\lambda^{\rightarrow}$-terms; we omit the detailed treatment.

Definition 20. Given an atomic type A; let $U_0 = A$ and $U_{n+1} = U_n \rightarrow U_n$ for $n \geq 0$; let $\bar{i}_n = \lambda f^{U_{n+1}} \lambda x^{U_n}.f(\cdots(f(x))\cdots)$, where $i, n \geq 0$ and f occurs i times; let $s_n = \bar{2}_n(\bar{2}_{n-1})\cdots(\bar{2}_0)$ for $n \geq 0$; let function $\text{tower}(n, m)$ be defined as follows.

$$\text{tower}(n, m) = \begin{cases} m & \text{if } n = 0; \\ 2^{\text{tower}(n-1,m)} & \text{if } n > 0. \end{cases}$$

The following properties follow immediate.

Proposition 21. For every $n \geq 0$,

- $|s_n| = 5(n+1)$;
- $\bar{i}_{n+1}\bar{j}_n \longrightarrow_\beta \bar{k}_n$ for $k = j^i$;
- $s_n \longrightarrow_\beta \bar{k}_0$ for $k = \text{tower}(n+1, 1)$.

Theorem 22. $\nu(t)$ is not elementary.

Proof. Let σ be a shortest β-reduction sequence which reduces s_n to its β-normal form, which is $\bar{k}_0$ for $k = \text{tower}(n+1, 1)$ by the Church-Rosser theorem. Hence, it follows from Proposition 16 that

$$\text{tower}(n+1, 1) < |\bar{k}_0| < |s_n|^{2^{\nu(s_n)}} = (5n+1)^{2^{\nu(s_n)}}$$

for $n \geq 0$. Since $\text{tower}(n+1, 1)$ is not elementary, $\nu(\cdot)$ cannot be elementary, either. $\qquad\square$

5.2 A bound on $\mu(t)$ for $\lambda^{\rightarrow}I$-terms t

Since the leftmost β-reduction sequence from any λI-term t is a longest one among all β-reduction sequences from t, it goes straightforward to establish a bound on $\mu(t)$ for $\lambda^{\rightarrow}I$-terms t if we can find any normalization sequences for them. In order to get a tighter bound, the key is to find shorter normalization sequences. We start with a weak normalization proof due to Turing according to [Gan80], which can also be found in many other literatures such as [GLT89].

Definition 23. The rank $\rho(T)$ of a simple type T is defined as follows.

$$\rho(T) = \begin{cases} 0 & \text{if } T \text{ is atomic}; \\ 1 + max\{\rho(T_0), \rho(T_1)\} & \text{if } T = T_0 \rightarrow T_1. \end{cases}$$

The rank $\rho(r)$ of a β-redex $r = (\lambda x^U.v^V)u^U$ is $\rho(U \rightarrow V)$, and the rank of a term t is

$$\hat{\rho}(t) = \begin{cases} \langle 0, 0 \rangle \text{ if } t \text{ is in } \beta\text{-normal form; or} \\ \langle k = \max\{\rho(r) : r \text{ is a } \beta\text{-redex in } t\}, \\ \quad \text{the number of } \beta\text{-redexes } r \text{ in } t \text{ with } \rho(r) = k \rangle, \text{ otherwise.} \end{cases}$$

The ranks of terms are lexically ordered.

Notice that a β-redex has a redex rank, which is a number, and also has a term rank, which is a pair of numbers.

Observations Now let us observe the following.

- If $t \longrightarrow_\beta t'$ and β-redex r' in t' is a residual of some β-redex r in t, then $\rho(r') = \rho(r)$.
- Given $t = t[r]$ with $\hat\rho(t) = \langle k, n \rangle$, where $r = (\lambda x^V.u^U)v^V$ is a β-redex with $\rho(r) = k$ and no β-redexes in r have rank k. Then $\hat\rho(t') < \hat\rho(t)$ for $t = C[r] \longrightarrow_\beta t' = C[u\{x := v\}]$. This can be verified by counting the number of β-redexes in t' with rank k. It is easy to see that any β-redex in t' which is not a residual must have rank $\rho(U)$ or $\rho(V)$, which is less than k. Hence, a β-redex in r' with rank k must be a residual of some β-redex r_1 in t with rank k. Note r_1 has only one residual in t' since r_1 is not in r. This yields $\hat\rho(t') < \hat\rho(t)$ since $\hat\rho(t')$ is either $\langle k, n-1 \rangle$ or $\langle k', n' \rangle$ for some $k' < k$.

Lemma 24. *Given t with $\hat\rho(t) = \langle k, n \rangle$ for some $k > 0$; then we can construct a development $\sigma : t \longrightarrow\!\!\!\!\rightarrow_\beta u$ such that $|\sigma| = n$ and $\hat\rho(u) = (k', n')$ for some $k' < k$.*

Proof. Following the observations, we can always reduce innermost β-redexes with rank k until there exist no β-redexes with rank k. This takes n steps and reaches a term with a less rank. $\qquad\square$

Definition 25. We define

$$m(\sigma) = \begin{cases} 1 & \text{if } |\sigma| = 0; \\ (1 + \max\{m(r_1), 1\}) \cdots (1 + \max\{m(r_n), 1\}) & \text{if } \sigma = [r_1] + \cdots + [r_n]. \end{cases}$$

Clearly, $m(\sigma_1 + \sigma_2) = m(\sigma_1)m(\sigma_2)$.

Theorem 26. *If t is a $\lambda^\rightarrow$-term with $\hat\rho(t) = \langle k, n \rangle$ for some $k > 0$, then there exists $\sigma : t \longrightarrow\!\!\!\!\rightarrow_\beta u$ such that u is in β-normal form and*

$$m(\sigma) < \mathrm{tower}(1, \sum_{i=1}^{k}(\mathrm{tower}(i-1, |t|))^2).$$

Proof. By Lemma 24, there exists a development $\sigma' : t \longrightarrow\!\!\!\!\rightarrow_\beta t'$ with $|\sigma'| = n$ and $\hat\rho(t') = \langle k', n' \rangle$ for some $k' < k$. Let $\sigma' = [r_1] + \cdots + [r_n]$, then $1 + m(r_n) < 2^{|t|}$ by Lemma 18. Hence, $m(\sigma') < 2^{n|t|} < 2^{|t|^2}$ since $n < |t|$ clearly holds. Now let us proceed by induction on k.

- $k = 1$. Since t' is in β-normal form, let $\sigma = \sigma'$ and we are done.
- $k > 1$. By induction hypothesis, there exists $\sigma'' : t' \longrightarrow\!\!\!\!\rightarrow_\beta u$ such that u is in β-normal form and $m(\sigma'') < \mathrm{tower}(1, \sum_{i=1}^{k-1}(\mathrm{tower}(i-1, |t'|))^2)$. Let $\sigma = \sigma' + \sigma''$, then

$$\begin{aligned} m(\sigma) &= m(\sigma')m(\sigma'') \\ &< \mathrm{tower}(1, |t|^2)\mathrm{tower}(1, \sum_{i=1}^{k-1}(\mathrm{tower}(i-1, |t'|))^2) \\ &< \mathrm{tower}(1, \sum_{i=1}^{k}(\mathrm{tower}(i-1, |t|))^2) \end{aligned}$$

since $|t'| < 2^{|t|}$ by Lemma 18. $\qquad\square$

It is a well-known fact that the leftmost β-reduction sequence from a λI-term t is a longest one if t has a β-normal form.

Corollary 27. *Given any simply typed $\lambda^{\to}I$-term t with $\hat{\rho}(t) = \langle k, n \rangle$; every β-reduction sequence from t is of length less than $\mathrm{tower}(k + 1, |t|)$.*

Proof. It can be verified that the result holds if $|t| \leq 3$. For $|t| > 3$, we have

$$\mathrm{tower}(1, \sum_{i=1}^{k}(\mathrm{tower}(i - 1, |t|))^2) \leq \mathrm{tower}(k + 1, |t|).$$

By Theorem 26, there exists $t \xrightarrow{\sigma}_\beta u$ such that $m(\sigma) < \mathrm{tower}(k + 1, |t|)$ and u is in β-normal form. This yields that $\mathrm{std}(\sigma) < \mathrm{tower}(k + 1, |t|)$ by Theorem 14. Since t is a $\lambda^{\to}I$-term, the leftmost β-reduction sequence from t is a longest one. This concludes the proof. □

Notice that the leftmost β-reduction sequence from t may not yield a longest one if t is not a $\lambda^{\to}I$-term. Therefore, the proof of Corollary 27 cannot go through directly for all $\lambda^{\to}$-terms.

5.3 An upper bound on $\mu(t)$ for $\lambda^{\to}$-terms t

Our following method is to transform a $\lambda^{\to}$-term t into a $\lambda^{\to}I$-term $[\![t]\!]$ such that $\mu(t) \leq \mu([\![t]\!])$. Since we have already established a bound for $\mu([\![t]\!])$, this bound certainly works for $\mu(t)$.

Lemma 28. *Given $t = r(u_1)\ldots(u_n)$ and $t_0 = u\{x := v\}(u_1)\ldots(u_n)$, where $r = (\lambda x.u)(v)$; if t_0 and v are strongly normalizable, then t is strongly normalizable and $\mu(t) \leq 1 + \mu(t_0) + \mu(v)$.*

Proof. Let $\sigma : t \longrightarrow\!\!\!\!\to_\beta t^*$ be a β-reduction sequence, and we verify that $|\sigma| \leq 1 + \mu(t_0) + \mu(v)$. Clearly, we can assume that β-redex r is involved in σ. Then $\sigma = \sigma_1 + [r'] + \sigma_2$ is of the following form.

$$t \xrightarrow{\sigma_1}_\beta (\lambda x.u')(v')(u_1')\ldots(u_n') \xrightarrow{r'}_\beta u'\{x := v'\}(u_1')\ldots(u_n') \xrightarrow{\sigma_2}_\beta t^*,$$

where $\sigma_1 = \sigma_u + \sigma_v + \sigma_{u_1} + \cdots + \sigma_{u_n}$ for $u \xrightarrow{\sigma_u}_\beta u'$, $v \xrightarrow{\sigma_v}_\beta v'$, $u_1 \xrightarrow{\sigma_{u_1}}_\beta u_1'$, ..., and $u_n \xrightarrow{\sigma_{u_n}}_\beta u_n'$. Let $\tau_1 : u\{x := v\} \longrightarrow\!\!\!\!\to_\beta u\{x := v'\}$ be the β-reduction sequence which reduces each occurrence of v in $u\{x := v\}$ to v' by following σ_v, and let $\tau_2 : u\{x := v'\} \longrightarrow\!\!\!\!\to_\beta u'\{x := v'\}$ be the β-reduction sequence which reduces $u\{x := v'\}$ to $u'\{x := v'\}$ by following σ_u. Clearly, $|\tau_1| = m(r)|\sigma_v|$ and $|\tau_2| = |\sigma_u|$. Also let $\tau : u\{x := v\}(u_1)\ldots(u_n) \longrightarrow\!\!\!\!\to_\beta t^*$ be $\tau_1 + \tau_2 + \sigma_{u_1} + \cdots + \sigma_{u_n} + \sigma_2$, then $|\tau| \leq \mu(t_0)$ by definition. Note

$$|\sigma| = |\sigma_1 + [r'] + \sigma_2| = |\sigma_u| + |\sigma_v| + |\sigma_{u_1}| + \cdots + |\sigma_{u_n}| + 1 + |\sigma_2| \leq 1 + |\tau| + |\sigma_v|.$$

By definition, $|\sigma_v| \leq \mu(v)$. Hence, $|\sigma| \leq 1 + \mu(t_0) + \mu(v)$. □

Definition 29. *(Transformation)* To facilitate the presentation, we assume that there exist constants $\langle , \rangle$ of type $U \to (V \to U)$ for all types U and V. Let $\langle u, v \rangle$ denote $\langle , \rangle(u)(v)$.

$$\llbracket t \rrbracket = \begin{cases} t, \text{ if } t \text{ is a variable;} \\ \lambda x \lambda y_1^{U_1} \ldots \lambda y_m^{U_m}.\langle \llbracket t_0 \rrbracket(y_1) \cdots (y_m), x \rangle, \text{ if } t = (\lambda x.t_0), \\ \quad \text{where } t_0 \text{ has type } U_1 \to \cdots \to U_m \to V \text{ and } V \text{ is atomic;} \\ \llbracket t_0 \rrbracket(\llbracket t_1 \rrbracket), \text{ if } t = t_0(t_1). \end{cases}$$

We will see clearly that $\langle , \rangle$ can always be replaced by a free variable of the same type without altering our following argument..

Proposition 30. *For every $\lambda^\to$-term t of type T, we have the following.*

1. $\llbracket t \rrbracket$ *is a $\lambda^\to I$-term of type T;*
2. $\llbracket t\{x^U := u\} \rrbracket = \llbracket t \rrbracket\{x^U := \llbracket u \rrbracket\}$ *for any $\lambda^\to$-term u of type U;*
3. $\mu(t) \leq \mu(\llbracket t \rrbracket)$.

Proof. (1) and (2) can be readily proven by structural induction on t. By (1) and Corollary 27, we know $\mu(\llbracket t \rrbracket)$ exists for every $\lambda^\to$-term t. We now proceed to show (3) by induction on $\mu(\llbracket t \rrbracket)$ and the structure of $\llbracket t \rrbracket$, lexicographically ordered.

- $t = \lambda x.u$. By induction, $\mu(t) = \mu(u) \leq \mu(\llbracket u \rrbracket) \leq \mu(\llbracket t \rrbracket)$.
- $t = x(u_1)\ldots(u_n)$, where x is some variable. Note $\mu(\llbracket t \rrbracket) = x(\llbracket u_1 \rrbracket)\ldots(\llbracket u_n \rrbracket)$. By induction hypothesis,

$$\mu(t) = \mu(u_1) + \ldots + \mu(u_n) \leq \mu(\llbracket u_1 \rrbracket) + \ldots \mu(\llbracket u_n \rrbracket) = \mu(\llbracket t \rrbracket).$$

- $t = r(u_1)\ldots(u_n)$, where $r = (\lambda x.u)(v)$. By definition,

$$\llbracket t \rrbracket = \llbracket r \rrbracket(\llbracket u_1 \rrbracket)\ldots(\llbracket u_n \rrbracket) \text{ and } \llbracket r \rrbracket = (\lambda x \lambda y_1 \ldots \lambda y_m.\langle \llbracket u \rrbracket(y_1)\ldots(y_m), x \rangle)(\llbracket v \rrbracket).$$

Hence,

$$\llbracket r \rrbracket \longrightarrow_\beta \lambda y_1 \ldots \lambda y_m.\langle \llbracket u \rrbracket\{x := \llbracket v \rrbracket\}(y_1)\ldots(y_m), \llbracket v \rrbracket \rangle.$$

Since $\llbracket u \rrbracket(y_1)\ldots(y_m)$ is of atomic type, $m \geq n$. This yields

$$\llbracket t \rrbracket \twoheadrightarrow_\beta \lambda y_{n+1} \ldots \lambda y_m.\langle \llbracket u \rrbracket\{x := \llbracket v \rrbracket\}(\llbracket u_1 \rrbracket)\ldots(\llbracket u_n \rrbracket)(y_{n+1})\ldots(y_m), \llbracket v \rrbracket \rangle.$$

By (2), $\llbracket u \rrbracket\{x := \llbracket v \rrbracket\} = \llbracket u\{x := v\} \rrbracket$, and thus,

$$\llbracket u \rrbracket\{x := \llbracket v \rrbracket\}(\llbracket u_1 \rrbracket)\ldots(\llbracket u_n \rrbracket) = \llbracket u\{x := v\} \rrbracket(\llbracket u_1 \rrbracket)\ldots(\llbracket u_n \rrbracket)$$
$$= \llbracket u\{x := v\}(u_1)\ldots(u_n) \rrbracket.$$

By induction hypothesis,

$$\mu(u\{x := v\}(u_1)\ldots(u_n)) \leq \mu(\llbracket u\{x := v\}(u_1)\ldots(u_n) \rrbracket).$$

Therefore, by Lemma 28,

$$\mu(t) \leq 1 + \mu(u\{x := v\}(u_1)\ldots(u_n)) + \mu(v)$$
$$\leq 1 + \mu(\llbracket u\{x := v\}(u_1)\ldots(u_n) \rrbracket) + \mu(\llbracket v \rrbracket) \leq \mu(\llbracket t \rrbracket).$$

$\square$

Corollary 31. *Given any simply typed $\lambda^\rightarrow$-term t with $\hat{\rho}(t) = \langle k, n \rangle$; every reduction sequence from t is of length less than* $\mathrm{tower}(k + 1, (2k + 3)|t|)$.

Proof. Given a subterm $\lambda x.u$ of type $U = U_1 \rightarrow \cdots \rightarrow U_m \rightarrow V$ in t, where V is atomic, we can simply transform $\lambda x.u$ into $\lambda x.\llbracket u \rrbracket$ if $k < \rho(U)$ since no β-redexes with rank greater than k can occur in any β-reduction sequence of t; if $\rho(U) \leq k$, we have

$$\|\llbracket \lambda x.u \rrbracket\| = |\lambda x \lambda y_1 \ldots \lambda y_m.\langle \llbracket u \rrbracket (y_1) \ldots (y_m), x \rangle| = \|\llbracket u \rrbracket\| + 2m + 3 \leq \|\llbracket u \rrbracket\| + 2k + 3.$$

Thus, it can be readily shown that $\|\llbracket t \rrbracket\| \leq (2k + 3)|t|$. Also it can be immediately verified by the definition that if $\rho(t) = \langle k, n \rangle$ for some k and n then $\rho(\llbracket t \rrbracket) = \langle k, n \rangle$. By Corollary 27, we have

$$\mu(\llbracket t \rrbracket) < \mathrm{tower}(k + 1, \|\llbracket t \rrbracket\|) \leq \mathrm{tower}(k + 1, (2k + 3)|t|).$$

This yields $\mu(t) < \mathrm{tower}(k + 1, (2k + 3)|t|)$ by Proposition 30 (3). □

6 Related Work and Conclusion

For those who know the strong equivalence relation $\cong$ on β-reductions in [Bar84], originally due to Berry and Lévy, it can be verified that $\sigma \cong \mathrm{std}(\sigma)$ for all β-reduction sequences σ.

There is a short proof of the standardization theorem due to Mitschke [Mit79], which analyses the relation between head and internal β-reductions. It shows any β-reduction sequence can be transformed into one which starts with head β-reductions followed by internal β-reductions. In this formulation, it is not easy to extract a bound directly from the proof. Our proof is a variant of Mitschke's proof. Lemma 11 simplifies the process which commutes head β-reduction with internal β-reductions, illuminating on why this process halts eventually. In this respect, our proof resembles a proof in [Tak95], where Takahashi exploited the notion of parallel β-reduction to show the termination of the commutation process.

There are also two proofs due to Klop [Klo80], to which the present proof bears some connection. Though all these proofs aim at commuting the contracted leftmost β-redexes to the front, our proof uses a different strategy to show the termination of such commutations. While Klop focuses on the strong equivalence relation $\cong$, we establish Lemma 11 by a structural induction without using the finiteness developments theorem. This naturally yields an upper bound for standardizations.

In our application, an upper bound is given for the lengths of β-reduction sequences in $\lambda^\rightarrow$. This is a desirable result since $\mu(t)$, the length of a longest β-reduction sequence from t, can often be used as an induction order in many proofs. Gandy mentions a similar bound in [Gan80a] but details were left out. His semantic method, which aims at giving strong normalization proofs, is quite different from ours. Schwichtenberg presents a similar bound in [Sch91] using an approach adopted from [How80]. His method of transforming $\lambda^\rightarrow$-terms into $\lambda^\rightarrow I$-terms closely relates to our presented method but is very much involved. It seems – in the author's opinion – that such involvedness is not only unnecessary but also obscures the merits in Schwichtenberg's proof. In addition, the proof of *finiteness of developments* theorem by de Vrijer [dV85] yields an exact bound for the lengths of developments, and thus, is casually related to our proof of the standardization theorem with bound.

In Gentzen's sequent calculus, there exists a similar bound for the sizes of cut-free proofs obtained from cut elimination. Mints [Min79] (of which I have only learned the abstract) showed a way of computing the maximum length of a β-reduction from the length of a standard β-reduction sequence. In this respect, our work can be combined with his to show the maximum length of a β-reduction sequence from the length of an *arbitrary* one. This also motivates our planning on establishing a similar bound for the first-order λ-calculus with dependent types. On the other hand, Theorem 22 suggests that a lower bound for $\mu(t)$ have a similar superexponential form, and this makes it a challenging task to sharpen our presented bound for $\mu(t)$, although it seems to be somewhat exaggerated. Also Statman proved that $\lambda^\rightarrow$ is not elementary [Sta79].

7 Acknowledgement

I gratefully acknowledge Richard Statman's efforts on reading the paper and providing his comments. I also thank Frank Pfenning and Peter Andrews for their support and for providing me a nice work environment.

References

[Bar76] H.P. Barendregt et al. (1976), Some notes on lambda reduction, *Preprint No. 22, University of Utrecht, Department of mathematics*, pp. 13-53.

[Bar84] H.P. Barendregt (1984), The Lambda Calculus: Its Syntax And Semantics, *North-Holland publishing company, Amsterdam.*

[Ber78] G. Berry (1978), Séquentialité de l'évaluation formelle des λ-expressions, *Proc. 3-e Colloque International sur la Programmation, Paris.*

[Chu41] A. Church, (1941), The calculi of lambda conversion, *Princeton University Press, Princeton.*

[CF58] H.B. Curry and R. Feys (1958), Combinatory Logic, *North-Holland Publishing Company, Amsterdam.*

[dV85] R. de Vrijer (1985), A direct proof of the finite developments theorem, *Journal of Symbolic Logic*, 50:339-343.

[Gan80] R.O. Gandy (1980), An early proof of normalization by A.M. Turing, *To: H.B. Curry: Essays on combinatory logic, lambda calculus and formalism*, edited by J.P. Seldin and J.R. Hindley, Academic press, pp. 453-456.

[Gan80a] R.O. Gandy (1980), Proofs of Strong Normalization, *To: H.B. Curry: Essays on Combinatory logic, lambda calculus and formalism*, edited by J.P. Seldin and J.R. Hindley, Academic press, pp. 457-478.

[GLM92] G. Gonthier, J.J. Lévy and P.-A. Melliès (1992), An abstract standardization theorem, in *Proceedings of Logic in Computer Science*, pp. 72–81.

[GLT89] J.-Y. Girard et al. (1989), Proofs and types, *Cambridge Press*, 176 pp.

[Hue94] Gérard Huet (1994), Residual Theory in λ-Calculus: A Formal Development, *Journal of Functional Programming Vol. 4*, pp. 371–394.

[Hin78] J.R. Hindley (1978), Reductions of residuals are finite, *Trans. Amer. Math. Soc. 240*, pp. 345-361.

[How80] W. Howard (1980), Ordinal analysis of terms of finite type, *Journal of Symbolic Logic*, 45(3):493-504.

[Hyl73] J.M.E. Hyland (1973), A simple proof of the Church-Rosser theorem, *Typescript, Oxford University*, 7 pp.

[Kah95] Stefan Kahrs (1995), Towards a Domain Theory for Termination Proofs, *Laboratory for Foundation of Computer Science*, 95-314, Department of Computer Science, The University of Edinburgh.

[Klo80] J.W. Klop (1980), Combinatory reduction systems, *Ph.D. thesis, CWI, Amsterdam, Mathematical center tracts, No. 127.*

[Lév78] J.-J. Lévy (1978), Réductions correctes et optimales dans le lambda calcul, *Thèse de doctorat d'état, Université Paris VII.*

[Min79] G.E. Mints (1979), A primitive recursive bound of strong normalization for predicate calculus (in Russian with English abstract), *Zapiski Naucnyh Seminarov Leningradskogo Otdelenija Matematiceskogo Instituta im V.A. Steklova Akademii Nauk SSSR (LOMI) 88*, pp. 131-135.

[Mit79] G. Mitschke (1979), The standardization theorem for the λ-calculus, *Z. Math. Logik Grundlag. Math. 25*, pp. 29-31.

[Pol94] J. van de Pol (1994), Strict functionals for termination proofs, *Lecture Notes in Computer Science 902*, edited by J. Heering, pp. 350-364.

[Sta79] Richard Statman (1979), The typed λ-calculus is not elementary, *Theoretical Computer Science 9*, pp. 73-81.

[Sch82] H. Schwichtenberg (1982), Complexity of normalization in the pure typed lambda-calculus, *The L.E.J. Brouwer Centenary Symposium*, edited by A.S. Troelstra and D. van Dalen, North-Holland publishing company, pp. 453-457.

[Sch91] H. Schwichtenberg (1991), An upper bound for reduction sequences in the typed lambda-calculus, *Archive for Mathematical Logic*, 30:405-408.

[Tak95] Masako Takahashi (1995), Parallel Reductions in λ-Calculus, *Information and Computation 118*, pp. 120–127.

[Wad76] C.P. Wadsworth (1976), The relation between computational and denotational properties for Scott's D_∞-models of λ-calculus, *SIAM Journal of Computing*, 5(3):488-521.

[Xi96a] H. Xi (1996), An induction measure on λ-terms and its applications, *Technical report 96-192, Department of Mathematical Sciences, Carnegie Mellon University*, Pittsburgh.

[Xi96b] H. Xi (1996), Separating developments in λ-calculus, *Manuscripts.*

Springer
and the
environment

At Springer we firmly believe that an international science publisher has a special obligation to the environment, and our corporate policies consistently reflect this conviction.
We also expect our business partners – paper mills, printers, packaging manufacturers, etc. – to commit themselves to using materials and production processes that do not harm the environment. The paper in this book is made from low- or no-chlorine pulp and is acid free, in conformance with international standards for paper permanency.

Lecture Notes in Computer Science

For information about Vols. 1–1211

please contact your bookseller or Springer-Verlag